AF779992

Advances in Virus Detection with Sensors

Detection Science Series

Editor-in-chief:
Michael Thompson, *University of Toronto, Canada*

Series editors:
Damien Arrigan, *Curtin University, Australia*
Subrayal Reddy, *University of Central Lancashire, UK*
Mengsu (Michael) Yang, *City University of Hong Kong, Hong Kong*

For a list of titles in this series see:
rsc.li/detection-science

How to obtain future titles on publication:
A standing order plan is available for this series. A standing order will bring delivery of each new volume immediately on publication.

For further information please contact:
Book Sales Department, Royal Society of Chemistry, Thomas Graham House, Science Park, Milton Road, Cambridge, CB4 0WF, UK
Telephone: +44 (0)1223 420066, Fax: +44 (0)1223 420247
Email: booksales@rsc.org
Visit our website at books.rsc.org

Advances in Virus Detection with Sensors

Edited by

Adil Denizli
Hacettepe Üniversitesi, Turkey
Email: denizli@hacettepe.edu.tr

ROYAL SOCIETY OF **CHEMISTRY**

Detection Science Series No. 32

Hardback ISBN: 978-1-83767-776-4
EPUB ISBN: 978-1-83767-778-8
PDF ISBN: 978-1-83767-777-1
Print ISSN: 2052-3068
Electronic ISSN: 2052-3076

A catalogue record for this book is available from the British Library

© The Royal Society of Chemistry 2026

All rights reserved

Apart from fair dealing for the purposes of research for non-commercial purposes or for private study, criticism or review, as permitted under the Copyright, Designs and Patents Act 1988 and the Copyright and Related Rights Regulations 2003, this publication may not be reproduced, stored or transmitted, in any form or by any means, including but not limited to training or developing any form of artificial intelligence model or tool, without the prior permission in writing of the Royal Society of Chemistry or the copyright owner, or in the case of reproduction in accordance with the terms of licences issued by the Copyright Licensing Agency in the UK, or in accordance with the terms of the licences issued by the appropriate Reproduction Rights Organization outside the UK. Enquiries concerning reproduction outside the terms stated here should be sent to the Royal Society of Chemistry at the address printed on this page.

Whilst this material has been produced with all due care, The Royal Society of Chemistry cannot be held responsible or liable for its accuracy and completeness, nor for any consequences arising from any errors or the use of the information contained in this publication. The publication of advertisements does not constitute any endorsement by The Royal Society of Chemistry or Authors of any products advertised. The views and opinions advanced by contributors do not necessarily reflect those of The Royal Society of Chemistry which shall not be liable for any resulting loss or damage arising as a result of reliance upon this material.

The Royal Society of Chemistry is a charity, registered in England and Wales, Number 207890, and a company incorporated in England by Royal Charter (Registered No. RC000524), registered office: Burlington House, Piccadilly, London W1J 0BA, UK, Telephone: +44 (0) 20 7437 8656.

For further information see our website at www.rsc.org

For general enquiries, please contact books@rsc.org

For EU product safety enquiries, please email books@rsc.org or contact Royal Society of Chemistry Worldwide (Germany) GmbH, Römischer Hof, Unter den Linden 10, 10117 Berlin.

Preface

The 21st century has witnessed the emergence and re-emergence of viral threats that have profoundly shaped global health, economies, and societal norms. From recurring influenza strains to novel pandemics, like coronavirus disease (COVID)-19, the need for rapid, accurate, and scalable virus detection has become more urgent than ever. This growing demand has placed sensor technologies at the forefront of scientific innovation, offering powerful tools for early diagnosis, environmental surveillance, and real-time health monitoring.

Advances in Virus Detection with Sensors aims to present a comprehensive and multidisciplinary overview of the current state and future directions of sensor-based virus detection technologies. As part of the Royal Society of Chemistry Detection Science Series, this volume brings together leading experts from academia and industry, covering a broad spectrum of topics—from classical detection principles to cutting-edge innovations such as clustered regularly interspaced short palindromic repeats (CRISPR)-based biosensors, wearable devices, and artificial intelligence (AI)-driven data analysis.

The book is structured to guide readers through a logical progression of knowledge. It begins with a foundational introduction to the field, highlighting the growing importance of virus detection and the transformative role of sensors. The second section delves into state-of-the-art sensor technologies, providing insights into optical, electrochemical, and mass-based detection methods. This is followed by an exploration of novel approaches and emerging platforms, emphasizing the integration of advanced materials, computation, and intelligent systems.

Detection Science Series No. 32
Advances in Virus Detection with Sensors
Edited by Adil Denizli
© The Royal Society of Chemistry 2026
Published by the Royal Society of Chemistry, www.rsc.org

The subsequent chapters address real-world applications in both clinical and environmental contexts, showcasing how sensor technologies are being leveraged to support medical diagnostics, public health surveillance, and ecosystem monitoring. Recognizing the multifaceted challenges inherent to virus detection—such as sensitivity, specificity, and data interpretation—the book also dedicates space to discussing obstacles and proposing actionable solutions. Case studies from successful implementations further bridge theory and practice, offering valuable lessons for future innovations.

Finally, the book concludes with a forward-looking perspective on the role of sensors in pandemic preparedness, international collaboration, and translational research.

We hope this volume serves as a valuable resource for researchers, engineers, clinicians, and policymakers working at the intersection of biosensing, virology, and public health. The field of sensor-based virus detection is rapidly evolving, and it is our aspiration that this book will both inform and inspire continued advancements in this critical area of science and technology.

Adil Denizli

Contents

Detection Science Series No. 32
Advances in Virus Detection with Sensors
Edited by Adil Denizli
© The Royal Society of Chemistry 2026
Published by the Royal Society of Chemistry, www.rsc.org

Section VI: Case Studies

Section VII: Future Directions

Section I: Introduction

CHAPTER 1

Introduction to Sensor-based Virus Detection Studies

YEŞEREN SAYLAN[a], SEROK DEVRIM AYDOĞAN[a], SERHAT ÜNAL[b] AND ADIL DENIZLI[a*]

[a]Department of Chemistry, Hacettepe University, Ankara, Turkey; [b]Department of Infectious Disease and Clinical Microbiology, Hacettepe University, Ankara, Turkey
*E-mail: denizli@hacettepe.edu.tr

1.1 Brief Overview of the Increasing Importance of Virus Detection

The terminological meaning of the word virus is defined as "poison" in Latin and "infectious disease" in English.[1] Viruses can be classified according to their various characteristics. Some viruses are encoded by RNA and some by DNA. Their ability to reproduce, such as their structure and inheritance characteristics, also forms the basis of some classifications. Basically, they are structures smaller than 200 nm that can reproduce themselves in a host cell.[2] Although they have small molecular structures, they have caused important events in the history of world public health. Humans, who interact with each other in large numbers instead of living in small groups, have caused disease-causing agents to be displaced, transmitted and, indirectly, mutated.[3]

For instance, smallpox was one of the most widely spread viral diseases in the world population. Smallpox is thought to date back to 3000–6000 years ago. In 1978, the epidemic was defeated in one of the most important events

Detection Science Series No. 32
Advances in Virus Detection with Sensors
Edited by Adil Denizli
© The Royal Society of Chemistry 2026
Published by the Royal Society of Chemistry, www.rsc.org

in the history of science. Knowledge and detection of smallpox, which is still deadly today, has enabled the production of vaccines and medicines for it.[4] Influenza was one of the largest epidemics in 1918, with more than 50 million people dying. Repeated outbreaks in 1957 and 1968 once again demonstrated the necessity of developing diagnostic techniques for viruses. Today, deaths due to influenza infection account for 1% of the US population per 2 million infected.[5,6] Bubonic plague killed almost killed one third of Europe (25 million people) during the fourteenth century, one of the greatest pandemics in human history. Although it is bacteria-based, it is like the corona virus disease (COVID)-19 pandemic that took place in the recent past and whose effects are still being experienced.[7]

With respect to the influenza virus, viral epidemics appeared in Spain in 1918, Asian flu in 1957, Hong Kong flu in 1968, and swine flu in 2009. The COVID-19 pandemic caused by the corona virus in the recent past is like these epidemics. The detection of the first cluster of cases once again demonstrated the importance of early detection of cases and the development of appropriate treatments in the fight against viruses.

Preventive studies will be possible with the isolation and detection of viruses to reduce the burden of pandemics and epidemics on public health and public health policies.[8] Viruses affect not only humans but also plants and animals, and it is known that each of these hosts can infect each other in different ways and spread the viruses. Today, the fight against viruses that rapidly threaten animal, human and plant life is critical for public health.[9] Viral diseases, especially in poor regions, cause many people to lose their lives. In these regions, the failure to isolate the first cluster of cases after the early detection of the first case causes an increase in transmission and a public health crisis.[10] Molecular detection of viruses that can change their characteristics rapidly can slow down transmission at an early stage. To control transmission, control points must be brought to the areas where transmission is high and made accessible. In fact, the detection of viruses is very important not only for epidemics, but also for food safety, medical laboratory safety, basic science studies, and non-specific disease determination.[11,12]

Today, nucleic acid detection and immunoassay methods are the most used methods for rapidly identifying viral infection.[13] One of the detection methods is to force virus particles to multiply by inserting them into a host cell. The virus is isolated within the host cell. Virus cell culture methods used in the past can take a long time due to the slow replication rates of the viruses.[13,14] Different methods have also been developed to accelerate the absorption of the virus into the host cell. The host cell and the virus are centrifuged at low speed in the same environment, shortening the absorption time and accelerating the process leading to diagnosis.[15] Shortening the time to detection will reduce the possible economic and administrative damage that may occur under the risk of epidemic and will save time for prophylactic treatment. The method of detecting viruses that replicate inside the host cell is an indirect method. In addition, direct detection of the relevant virus can be performed using a nucleic acid-based approach. The gold standard here is polymerase chain reaction (PCR) technology. The polymerase used as standard can synthesize DNA specifically, so RNA amplification requires the use of an enzyme with reverse transcription activity. Reverse transcription-PCR is a common method used for the detection of DNA and RNA viruses.[16]

The PCR method has been used during the COVID-19 pandemic that has spread all over the world in the recent past. According to the data of the World Health Organization (WHO), 775 million COVID-19 diagnoses have been documented. With 7 million lives lost, humanity faced the 21st century's greatest pandemic; throughout this crisis, our greatest strength in pandemic management was fast and effective testing. During this pandemic process, since corona virus and influenza are clinically close to each other, billions of PCR tests have been performed and to determine which of the viruses patients are infected with. Keeping the transmission of the two existing viruses under control is also facilitated through making the correct diagnoses. On January 13, 2022, the number of samples tested for severe acuter respiratory syndrome coronavirus 2 (SARS-CoV-2) was reported as 55 351 010 worldwide, while 109 670 tests were conducted for Influenza.[17,18]

Moreover, according to the data of the WHO in 2022, 39 million people continue their lives with human immunodeficiency virus (HIV) infection. Close to 4% of these people are under the age of 15. In 2022, there were 1.3 million new diagnoses. With the development of virus diagnosis and control methods, it was reported that in 2000, there were 54% fewer people who were newly diagnosed.[19] When examining the leading causes of death from 2000 to 2021, it is observed that deaths due to HIV infection dropped out of the top 10 as of 2019 (Figure 1.1). The effective fight against one of the most fundamental viral threats to public health has yielded positive results over time. However, in 2020, the COVID-19 pandemic emerged as one of the most significant global health crises, ranking third among the leading causes of death. By 2021, COVID-19-related fatalities had risen to the second position. This underscores the critical importance of public health protection, virus detection, diagnostics, isolation measures, and the development of related drugs, vaccines, and therapeutic approaches.[20]

1.2 The Importance of Virus Detection

Viruses cause diseases as pathogens, and these diseases can spread from a single case to a population. Although there are no specific treatment methods for most viruses, they have distinctive features. After clinical diagnosis, broad-spectrum anti-viral drugs can be used, and ideally, the characteristics of the relevant virus are expected to be known. In recent history, many different variants of the coronavirus have been faced in the COVID-19 epidemic. Although the clinical observations are similar, both vaccine studies and disease management have become much more effective thanks to the ability to isolate and detect viruses.[22]

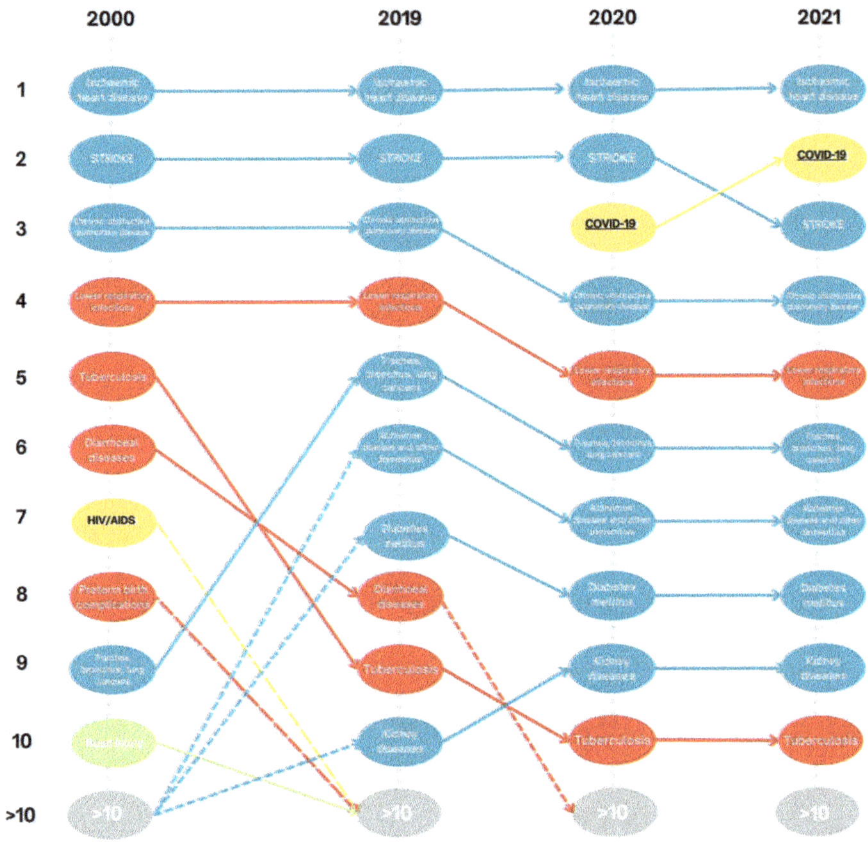

Figure 1.1 The top 10 causes of death globally in 2000–2021.[21]

On March 31, 2019, the China Country Office of the WHO reported that the virus causing pneumonia was 2019-nCoV, but the Coronavirus Research Group (CSG) of the International Committee classified it as SARS-CoV-2 on February 11, 2020.[23] This difference was not only necessary for classification, but also important for future data. Clinical symptoms and interventions may change depending on the occurrence of mutations, and classification has an important place in scientific, vaccine and drug studies.[22,24]

Similarly, variants of the hepatitis B virus (HBV) are clinically significant. Gao *et al.* point out that different variants of antiviral drugs that inhibit DNA replication by targeting HBV polymerase (P) may cause the drug to be ineffective. Some variants may cause clinical and preclinical negative effects, such as ineffective vaccines, failure of diagnosis, and progression of liver fibrosis. For example, it has been observed that the rate of X gene mutation or deletion is high in hepatocellular carcinoma (HCC) patients, and the presence of the mutation reveals the risks for the development of HCC. The recognition of the HBV X variant may have clinical significance for the development of HCC.[25,26]

1.3 The Role of Sensors in Revolutionizing Virus Detection Technologies

Viruses can be detected by many different methods, such as PCR and isolation. However, these techniques have several challenges. These include the challenge of the time elapsed during the testing process, such as the sample taken for the PCR test, and the difficulties of cell culture for isolation. In addition to technical difficulties like this, the consistency of the tests with each other can also vary. False-negative or false-positive results can be obtained in tests performed at the same centre. Sensors can make direct measurements with very low concentration sample loads.[27] By using different markers in optical and electrochemical methods, detection studies can be performed with different limit of detection (LOD) options.[28,29]

HIV infection and its related acquired immune deficiency syndrome (AIDS) can sometimes be symptomatic and sometimes asymptomatic. Therefore, detection of viral load in early diagnosis is important for preventive measures. An electro-chemiluminescence biosensor for HIV-1 gene detection has been developed. This achieved a very sensitive detection of the HIV gene in the range of 3.0 fM to 0.3 nM. The prepared biosensor demonstrated good selectivity for HIV detection compared with non-complementary sequences. Taking precise measurements at low concentrations is important and urgent for viruses and especially HIV.[30] Rozmysłowicz *et al.* used p24 as an HIV biomarker, which correlated with viral load. Piezoelectric biosensors can be used to detect the HIV virus at an early stage in a low-cost and rapid manner. Piezoelectric biosensors can detect HIV at ng ml^{-1} levels in a short time without using more than 10–100 ml of blood, which is sensitive, fast, portable, accessible and low-cost.[31]

Hepatitis virus, which also causes chronic organ failure, is a virus that can be found in human blood serum, which has been known since the 1940s. Most of the deaths from infection are caused by effects such as liver failure. Early detection of the virus and widespread screening of the population is one of the important stages in the fight against the virus. There are different diagnostic methods for hepatitis B with electrochemical and optical biosensors. In diagnostic methods, not only the virus itself, but also its DNA, surface antigens or antibody can be detected. This is one of the important strengths of biosensors.

With the use of gold nanoparticles, hepatitis B DNA can be tested in a dynamic range of 1–20 µg ml^{-1}. 5-Amino-2-mercapto-1,3,4-thidiazole (AMT) enriched with gold nanoparticles (AuNPs) was used to bind to the DNA surface of viruses, selective signals were obtained for the sequence related to the hepatitis B virus (HBV). The presence of AMT on the surface before the electrodes were used increased the performance of the AuNPs, allowing them to reach a lower limit of detection.[32]

Human papillomavirus (HPV) is the leading risk factor for cervical cancer worldwide. Early diagnosis plays a key role in treatment of cervical cancer, which can be fatal According to WHO data for 2022, there were approximately

500 000 cases of cervical cancer worldwide.[33] By 2030, the WHO has identified a number of steps to eliminate cervical cancer, one of which includes screening 70% of women by age 35 with high-performance tests. Sensors will be an important option in this regard, especially for low-income groups.[34] Tasoglu *et al.* presented alternatives to existing diagnostic mechanisms using microfluidics with nanotechnology. There are gold or silver nanoparticle-based sensors that can detect HPV-6, -11, -16, and-18 genotypes up to a concentration of 50 nM. The use of accurate biomarkers in the design of biosensors provides an analysis area in a wide spectrum of LODs.[35] Different LOD values and biomarkers will add diversity, reproducibility and reliability to the WHO's targeted screening efforts by 2030. Different instruments will provide extra opportunities to practitioners both in clinical trials and in the diagnosis–treatment process.

Considering Ebola, diagnosis is very difficult during its clinical course. Because the first symptoms are similar to the flu. It is vital to diagnose and intervene faster than the spread of the virus. In the later stage of the disease, there are dangers that can lead to multiple organ failure. Again, different approaches can provide different opportunities under different conditions. Saylan *et al.* drew attention to digital sensors, which can provide great convenience in field studies for Ebola. Netesan *et al.* produced sensors that captured specific antibodies with microarrayed recombinant antigens. Thanks to the combination with smartphones, it is possible to make sense of the data faster by using cloud communication.[36,37] The fact that it is not necessary to wait for clinical symptoms for diagnosis increases the importance of plasmonic, optic, and microfluidic sensors for groups with estimated prevalence in the population. The quick response time, cost and ease of applicability also indicate that the areas of usage will increase in the future.[38]

In epidemic situations, such as the coronavirus epidemic, the reliability and speed of results of traditional tests are negatively affected by the increase in the burden on all steps of the healthcare system. It is only possible to combat this burden with innovative diagnostic methods and fast and reproducible methods. Biosensors have allowed an important step forward in terms of being able to be designed quickly and to take measurements with high accuracy with low sample quantity.[39,40] Saylan *et al.* have identified 24 000 publications on the use of sensors in medical diagnostics in the Science Direct database since 1997. It has been observed that this number is increasing rapidly.[41]

At the bedside, point of care (POC) systems are prepared for diagnosis within health centres. These POC systems have reduced the need for large, complex laboratories and brought about a new stage in clinical diagnostic strategies in the last 10 years. Especially in cases where urgent action is required, POC devices make a great contribution to conventional diagnostic methods (Figure 1.2). Aslan *et al.* separated POC systems according to four basic techniques. (i) Colorimetric, (ii) electrochemical, (iii) fluorescent, and (iv) optical; While the high sensitivity of all methods is a common feature,

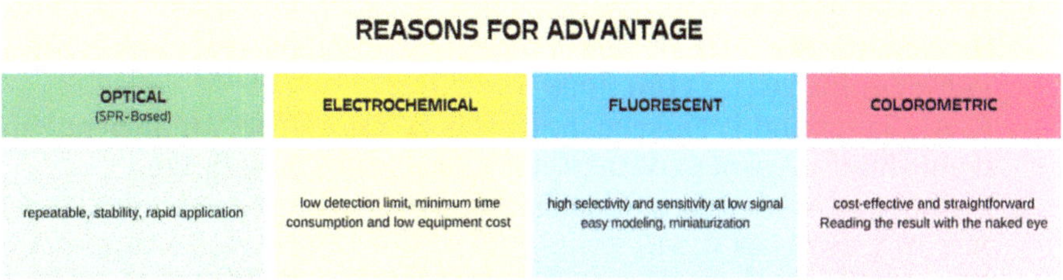

Figure 1.2 Main advantages of different POC systems.

the observation of measurements may vary. For example, although optical sensors are more expensive than others, they can measure in real-time.[42]

As in the cases of coronavirus, HIV and hepatitis, biosensors are an important step in ensuring that the screening of epidemics and chronic problems with different methods is accessible to the entire world population. In the past, the requirements for specialist technicians and large amounts of instrumental equipment were not achievable for low-income, developing populations. However, the ability of biosensors to obtain stable and fast results specific to the relevant virus and pathogen will alleviate the workload of clinics and increase the success rate. The fight against viruses by low-income and limited populations is not only a concern for their members, but also for all public health.

Acknowledgements

Yeşeren Saylan gratefully acknowledges the support from Scientific and Technological Research Council of Turkey (TÜBİTAK) Directorate of Science Fellowships and Grant (BİDEB) 2247-D National Young Researchers (Project No: 121C226) and also thanks the Young Scientist Awards Program (BAGEP) award from the Science Academy.

References

1. B. A. Lwoff, The concept of virus the third Marjory Stephenson memorial lecture, *J. Gen. Microbiol.*, 1957, **17**(1), 239–253.
2. J. Louten, Virus structure and classification, in *Essential Human Virology*, 2016, p. 19. Available at: https://pmc.ncbi.nlm.nih.gov/articles/PMC7150055/.
3. A. P. Dobson and E. Robin Carper, Infectious diseases and human population history, *Bioscience*, 1996, **46**(2), 115–126.
4. A. M. Geddes, The history of smallpox, *Clin. Dermatol.*, 2006, **24**(3), 152–157.
5. J. K. Taubenberger, J. C. Kash and D. M. Morens, The 1918 influenza pandemic: 100 years of questions answered and unanswered, *Sci. Transl. Med.*, 2019, **11**(502), eaau5485.

6. D. M. Morens, J. K. Taubenberger and A. S. Fauci, A centenary tale of two pandemics: The 1918 influenza pandemic and COVID-19, Part I, *Am. J. Public Health*, 2021, **111**(6), 1086–1094.
7. K. A. Glatter and P. Finkelman, History of the Plague: An Ancient Pandemic for the Age of COVID-19, *Am. J. Med.*, 2020, **134**(2), 176.
8. L. Akin and M. G. Gözel, Understanding dynamics of pandemics, *Turk J. Med. Sci.*, 2020, **50**(3), 515.
9. A. A. Malik, C. Nantasenamat and T. Piacham, Molecularly imprinted polymer for human viral pathogen detection, *Mater. Sci. Eng.: C.*, 2017, **77**, 1341–1348.
10. P. J. Hotez, Blue marble health and "the big three diseases": HIV/AIDS, tuberculosis, and malaria, *Microbes Infect.*, 2015, **17**(8), 539–541.
11. D. H. Choi, A. Thaxton, I. C. Jeong, K. Kim, P. R. Sosnay, G. R. Cutting and P. C. Searson, Sweat test for cystic fibrosis: Wearable sweat sensor vs. standard laboratory test, *J. Cystic Fibrosis*, 2018, **17**(4), e35–e38.
12. J. E. Chang, D. S. Lee, S. W. Ban, J. Oh, M. Y. Jung, S. H. Kim, S. J. Park, K. Persaud and S. Jheon, Analysis of volatile organic compounds in exhaled breath for lung cancer diagnosis using a sensor system, *Sens. Actuators, B*, 2018, **255**, 800–807.
13. A. Cassedy, A. Parle-McDermott and R. O'Kennedy, Virus detection: a review of the current and emerging molecular and immunological methods, *Front. Mol. Biosci.*, 2021, **8**, 637559.
14. L. Tonoyan, C. V. Olivieri, M. Chevalier, R. Marsault and A. Doglio, Detection of Epstein-Barr virus infection in primary junctional epithelial cell cultures, *J. Oral Microbiol.*, 2024, **16**(1), 2301199.
15. A. Hematian, N. Sadeghifard, R. Mohebi, M. Taherikalani, A. Nasrolahi and M. Amraei, *et al.*, Traditional and modern cell culture in virus diagnosis, *Osong Public Health Res. Perspect.*, 2016, **7**(2), 77–82.
16. A. A. Rabaan, R. Tirupathi, A. A. Sule, J. Aldali, A. Al Mutair and S. Alhumaid, *et al.*, Viral dynamics and real-time RT-PCR Ct values correlation with disease severity in COVID-19, *Diagnostics*, 2021, **11**, 1091.
17. COVID-19 circulation | WHO COVID-19 dashboard. Available at: https://data.who.int/dashboards/covid19/circulation?m49=001&n=o.
18. Microsoft Power BI. Available at: https://app.powerbi.com/view?r=eyJrIjoiNzc4YTIxZjQtM2E1My00YjYxLWIxMDItNzEzMjkyY2E1MzU1IiwidCI6ImY2MTBjMGI3LWJkMjQtNGIzOS04MTBiLTNkYzI4MGFmYjU5MCIsImMiOjh9.
19. *World Health Statistics 2023: Monitoring Health for the SDGs, Sustainable Development Goals*. Available at: https://iris.who.int/handle/10665/367912.
20. *World Health Statistics 2025: Monitoring Health for the SDGs, Sustainable Development Goals*, World Health Organization, Geneva, 2025.
21. *Global Health Estimates*. Available at: https://www.who.int/data/global-health-estimates.
22. K. Tao, P. L. Tzou, J. Nouhin, R. K. Gupta, T. de Oliveira and S. L. Kosakovsky Pond, *et al.*, The biological and clinical significance of emerging SARS-CoV-2 variants, *Nat. Rev. Genet.*, 2021, **22**(12), 757–773.

23. D. Vasireddy, R. Vanaparthy, G. Mohan, S. V. Malayala and P. Atluri, Review of COVID-19 Variants and COVID-19 Vaccine Efficacy: What the Clinician Should Know?, *J. Clin. Med. Res.*, 2021, **13**(6), 317.
24. Ö. Erdem, I. Eş, Y. Saylan and F. Inci, Unifying the efforts of medicine, chemistry, and engineering in biosensing technologies to tackle the challenges of the COVID-19 pandemic, *Anal. Chem.*, 2022, **94**(1), 3–25.
25. J. H. Lee, K. H. Han, J. M. Lee, J. H. Park and H. S. Kim, Impact of Hepatitis B Virus (HBV) X Gene Mutations on Hepatocellular Carcinoma Development in Chronic HBV Infection, *Clin. Vaccine Immunol.*, 2011, **18**(6), 914.
26. S. Gao, Z. P. Duan and C. S. Coffin, Clinical relevance of hepatitis B virus variants, *World J. Hepatol.*, 2015, **7**(8), 1086.
27. Y. Saylan, Ö Erdem, S. Ünal and A. Denizli, An alternative medical diagnosis method: biosensors for virus detection, *Biosensors*, 2019, **9**(2), 65.
28. A. Baskar, K. Madhivanan, R. Atchudan, S. Arya and A. K. Sundramoorthy, Nanoparticle electrochemical biosensors for virus detection, *Clin. Chim. Acta*, 2025, **566**, 120054.
29. M. J. Kim, I. Haizan, M. J. Ahn, D. H. Park and J. H. Choi, Recent advances in lateral flow assays for viral protein detection with nanomaterial-based optical sensors, *Biosensors*, 2024, **14**(4), 197.
30. B. Babamiri, A. Salimi and R. Hallaj, A molecularly imprinted electrochemiluminescence sensor for ultrasensitive HIV-1 gene detection using EuS nanocrystals as luminophore, *Biosens. Bioelectron.*, 2018, **117**, 332–339.
31. T. Rozmysłowicz, A. M. Sci, E. Ergezen, J. Desa, D. O. Conover, R. Lec, *et al.* Experimental research A low-cost diagnostic technology for HIV detection, 2022, DOI: 10.5114/aoms/143344.
32. M. Muti, M. Muti and A. Erdem, Impedimetric nanobiosensor for the detection of sequence-selective DNA hybridization, *Hacettepe J. Biol. Chem.*, 2018, **46**(4), 495–503.
33. Cancer Today. Available at: https://gco.iarc.fr/today/en/dataviz/tables-prevalence?mode=population&key=total&sort_by=value1&types=2&cancers=23&sexes=2&populations=903_904_905_908_909_935&multiple_populations=1.
34. Global strategy to accelerate the elimination of cervical cancer as a public health problem, 2020. Available at: http://apps.who.int/bookorders.
35. S. Tasoglu, H. Cumhur Tekin, F. Inci, S. Knowlton, S. Q. Wang and F. Wang-Johanning, *et al.*, Advances in nanotechnology and microfluidics for Human Papillomavirus diagnostics, *Proc. IEEE*, 2015, **103**(2), 161–178.
36. M. Natesan, S. W. Wu, C. I. Chen, S. M. R. Jensen, N. Karlovac and B. K. Dyas, *et al.*, A smartphone-based rapid telemonitoring system for Ebola and Marburg disease surveillance, *ACS Sens.*, 2019, **4**(1), 61–68.
37. Y. Saylan and A. Denizli, Virus detection using nanosensors, in *Nanosensors for Smart Cities*, 2020, pp. 501–511.
38. S. Asir, M. Bakhshpour, S. Unal and A. Denizli, Nanoparticle-based plasmonic devices for bacteria and virus recognition, in *Modern Practical Healthcare Issues in Biomedical Instrumentation*, 2022, pp. 167–183.

39. E. Vermisoglou, D. Panáček, K. Jayaramulu, M. Pykal, I. Frébort and M. Kolář, *et al.*, Human virus detection with graphene-based materials, *Biosens. Bioelectron.*, 2020, **166**, 112436.
40. J. U. Lee, A. H. Nguyen and S. J. Sim, A nanoplasmonic biosensor for label-free multiplex detection of cancer biomarkers, *Biosens. Bioelectron.*, 2015, **74**, 341–346.
41. Y. Saylan, E. Özgür and A. Denizli, Recent advances of medical biosensors for clinical applications, *Med. Devices Sens.*, 2021, **4**(1), e10129.
42. Y. Aslan, M. Atabay, H. K. Chowdhury, I. Göktürk, Y. Saylan and F. Inci, Aptamer-based point-of-care devices: emerging technologies and integration of computational methods, *Biosensors*, 2023, **13**(5), 569.

Section II: State of the Art Sensor Technologies

CHAPTER 2

Overview of Existing Sensor Technologies for Virus Detection

DIDEM AKSU[a*], TOLGA ERDOGAN[b], ECE IPEK[c], EMINE GULER CELIK[b] AND SUNA TIMUR[d]

[a]Central Research Testing and Analysis Laboratory Research and Application Center, Ege University Bornova, Izmir, 35100, Turkey; [b]Department of Bioengineering, Faculty of Engineering, Ege University, Izmir 35100, Turkey; [c]Department of Biotechnology, Institute of Natural Sciences, Ege University, Izmir 35040, Turkey; [d]Biochemistry Department, Faculty of Science, Ege University, 35100 Bornova, Izmir, Turkey
*E-mail: didem.eroglu@ege.edu.tr

2.1 Virus Detection and Its Importance

Viruses, which are responsible for numerous diseases, lead to the deaths of thousands of people every year. The rapid and accurate detection of viruses is crucial, especially in global health crises such as pandemics, to control the spread of infectious diseases. The world has experienced and continues to experience multiple pandemics, including severe acute respiratory syndrome (SARS), avian influenza virus (H5N1), swine influenza virus (H1N1), and Middle East respiratory syndrome (MERS), along with outbreaks of

Detection Science Series No. 32
Advances in Virus Detection with Sensors
Edited by Adil Denizli
© The Royal Society of Chemistry 2026
Published by the Royal Society of Chemistry, www.rsc.org

other deadly diseases such as human immunodeficiency virus (HIV), Ebola, and Zika. The coronavirus disease (COVID-19) pandemic, caused by severe acute respiratory syndrome coronavirus-2 (SARS-CoV-2) in 2019, once again demonstrated the critical importance of rapid and accurate diagnosis, particularly in situations where no specific vaccine or effective treatment is available.[1,2]

The most used viral detection methods in clinical settings involve molecular techniques that measure nucleic acids or protein biomarkers. However, these techniques have several disadvantages, including being time-consuming and costly. In this context, rapid test systems present a valuable alternative due to their speed, accessibility, and lower costs.[3] From a public health perspective, rapid and accurate diagnostic tests play a vital role in the early diagnosis of patients, the detection of asymptomatic carriers, who are significant sources of transmission, and the reduction of infection risk.[4] Early intervention enables more effective implementation of strategies such as contact tracing, quarantine measures, and targeted vaccination. These preventive measures help to alleviate the burden on healthcare systems and personnel during outbreaks, allowing for more effective crisis management.[5] High specificity and sensitivity are critical criteria for diagnostic tests. Errors in test results, including false positives and false negatives, can significantly affect the course of epidemics. False-negative results can lead to undetected infected individuals, increasing the risk of uncontrolled disease spread. In contrast, false-positive results can cause unnecessary treatments and isolation measures, placing an additional burden on healthcare systems. Rapid diagnostic tests with high specificity and sensitivity minimize the occurrence of false results, ensuring a safer approach to disease management.[6] Additionally, the widespread use of these tests enables real-time epidemiological monitoring, allowing researchers and policymakers to track disease progression instantly. Consequently, public health decisions can be updated quickly, facilitating the rapid adaptation of different strategies based on disease transmission patterns.[7]

2.2 Advancements in Sensor Technologies

Over the years, diagnostic systems have rapidly evolved, with recent developments in nanotechnology and technological integration. In clinical settings, molecular diagnostic methods such as virus isolation, enzyme-linked immunosorbent assay (ELISA), polymerase chain reaction (PCR), and hemagglutination/inhibition assays are used for virus identification and viral load detection.[3] Although these methods offer high accuracy, they have several limitations, including being time-consuming and costly, requiring laboratory infrastructure and specialized personnel, and being inaccessible in underdeveloped countries. These challenges have driven the demand for faster, more accessible, and cost-effective diagnostic systems and complementary approaches. In response to this need, point-of-care testing (POCT) has been developed. These tests meet essential criteria, such as affordability, sensitivity, specificity, user-friendliness, rapid processing, and equipment-free usability, making them a promising alternative for viral infection detection.[8,9]

The era of POCT began in the 1960s with the development of a rapid and effective method for diabetes patients and further advanced with the introduction of pregnancy tests in the 1970s, ultimately reaching its current widespread use.[5] Over time, advancements in materials science, nanotechnology, and microfluidic systems have led to the development of more sophisticated sensor platforms capable of detecting biomolecules, such as nucleic acids and viral antigens, with high sensitivity and speed. One of these advanced sensor platforms is nano-biosensors. These sensor technologies integrate nanomaterials, such as gold nanoparticles (AuNPs), silver nanoparticles (AgNPs), quantum dots, and carbon-based nanostructures, with sensors. Through this integration, the surfaces of nanoparticles can be modified with biomolecules, enabling them to selectively recognize target molecules with high efficiency. Furthermore, the optical, magnetic, and electrochemical properties of nanoparticles facilitate the design of different types of sensors for target molecules.[10]

These designs minimize false-positive and false-negative results, enhance sensitivity and specificity, shorten analysis times, and significantly improve surface–analyte interactions due to their high surface to volume ratio, allowing for the detection of even low concentrations of viral loads. These features are particularly advantageous in multiplex sensor systems, enabling the simultaneous detection of multiple viruses within a single sample. With multiplex systems, respiratory viruses causing similar symptoms can be detected concurrently, allowing for the rapid and accurate differentiation of viral infections, thereby accelerating the treatment process.[11]

Recent advancements in sensor technologies have also enabled the integration of these sensors with smartphones. Through such integrations, systems that previously provided only qualitative measurements can now perform quantitative analyses *via* analytical software, while also gaining additional capabilities, such as remote monitoring of sensor data. These enhanced features contribute to improved diagnostic accuracy and enable the early detection of diseases, ultimately strengthening healthcare management and public health responses.[12]

2.3 Comparing Different Sensor Technologies

Virus detection has become increasingly important in public health, particularly following multiple virus outbreaks. Various sensor technologies have been developed to improve virus detection sensitivity, specificity, and speed. This section compares several sensor technologies according to their principles of operation, advantages and disadvantages.

The molecularly imprinted polymer (MIP) sensor, which combines viral detection and capture on a single platform, is among the most promising technologies. Batra *et al.* showed that the effectiveness of virus detection from aerosol particles can be significantly increased by using a highly porous membrane or a metal–organic framework based molecular absorbent layer. By using an electrostatic mechanism to draw in charged particles, this sensor improves the efficiency of viral capture.[13] Because of their pre-established binding sites, MIPs also have the benefit of great specificity, which makes them perfect for identifying viruses in a complicated environment.

Another important technology is lab-on-a-chip (LoC) platforms that enable simultaneous electrochemical detection of both viral RNA and antibodies produced by the body against viruses. Najjar and colleagues demonstrated the potential of such platforms to provide rapid results from plasma and saliva samples.[14] The multiprocessing capability of these platforms allows for parallel detection of multiple viral targets, which will provide practicality and speed in clinical applications.

Fluorescent barcoded probes have also been of particular interest because of their multiplexing capability for the identification of multiple pathogens, including influenza viruses. Huang *et al.* described how this technology allows for sensitive detection and typing of several pathogens but has not yet been fully utilized in single-virus typing.[15] Multiplex capacity is a huge advantage when more than a single virus would probably be encountered as it eliminates time and cost of testing.

Waveguide mode sensors developed by Gopinath *et al.* are another sensitive method of detecting influenza viruses. Working on the principle of immobilization of antibodies on a detection plate, waveguide sensors can detect viruses at lower concentrations, thus allowing early diagnosis.[16] The high sensitivity of waveguide sensors makes them suitable for applications where early detection is important.

Silicon nanowire (SiNW) sensors are also a promising technology for detection of influenza viruses. Shen *et al.* showed that SiNW sensors can detect influenza viruses sensitively and rapidly.[17] They also noted in their study that SiNW sensors could be easily functionalized for the detection of viruses other than influenza viruses.

Point-of-care biosensors such as those developed by Baca *et al.* are designed to detect viruses like Ebola. These sensors use surface acoustic wave (SAW) technology to detect viral antigens in a reagent-free situation, this makes them suitable for use on site.[18] The most important advantage of these sensors is that they can be applied in areas with low resources without complex chemicals.

New wireless immunoassay technologies have also been created, as described by Li *et al.* The technology enables passive and automatic detection of respiratory virus aerosols, which enables on-site testing with minimal user interaction.[19] Wireless technology integration enhances the ease of virus detection in various settings, including public places.

Terahertz nano-gap metamaterials have been proposed as highly sensitive virus sensors capable of sensing a single virus. Park *et al.* identified the sensitivity of the sensors that can be achieved through optimized geometry and immobilized antibodies.[20] The technology is a new approach in the detection of viruses, but it is still being researched in practice.

Surface plasmon resonance imaging (SPR) imaging was also used for the real-time virus detection by Zybin *et al.* The results of their study indicated the potential of SPR sensors for the detection of single viral particles and for offering a powerful tool for virus detection for clinical and research applications.[21] The potential of SPR sensors for real-time monitoring is most beneficial for tracing viral infections over time.

In conjunction with such technologies, interferometric sensors have been used to provide instant detection of viruses. Ymeti *et al.* reported the detection of herpes simplex virus type 1 using an ultrasensitive Young interferometer sensor, identifying the capability of optical techniques in use for detecting viruses.[22] The interferometric techniques are mostly highly sensitive, which is an advantage in recognizing trace amounts of virus.

In conclusion, a number of sensor technologies exist or are under development nowadays, each has their own advantages and disadvantages when several parameters like accessibility, ease of use, cost *etc.* are considered. Molecularly imprinted polymers, LoC devices, fluorescent barcoded probes, waveguide sensors, SiNW sensors, point-of-care biosensors, wireless immunoassays, terahertz metamaterials, surface plasmon resonance imaging, and interferometric sensors all represent a range of methods for detecting viruses. Ongoing research will advance existing and developing sensor technologies and will raise public health to higher levels in the future. In addition, increasing the allocated resources and encouraging sensor studies will ensure that society is much more prepared for possible pandemics.

2.4 Application of Sensor Detection for Different Viruses

Sensors play a critical role in various fields, such as healthcare, biotechnology, and environmental monitoring, by enabling the rapid and accurate identification of viral pathogens.[23,24] Numerous viruses, including coronaviruses, influenza virus, tobacco mosaic virus, HIV, human norovirus (HuNoV), Ebola virus (EBOV), dengue virus (DENV), Zika virus (ZIKV), and hepatitis viruses, can be rapidly detected using various sensor technologies.[25–27]

Extensive biosensor research has been conducted following the emergence of the coronavirus in China in 2019. Some studies utilize magnetic nanoparticle-based biosensor platforms for the rapid detection of SARS-CoV-2 antibody responses in human serum.[28] Other studies focus on different biosensor technologies, such as clustered regularly interspaced short palindromic repeats (CRISPR)-based sensors and immunosensors.[29,30] This

Table 2.1 Sensor-based detection of viruses.

Virus type	Sensor type	Limit of detection (LOD)	References
SARS-CoV-2	Graphene-based electronic biosensor	0.9 pM	31
Covid-19	Au micropattern-based highly uniform electrochemical biosensor	0.276 pmol l^{-1}	32
Tobacco mosaic virus	Paper-based molecularly imprinted sensor	4.8 fM	33
Norovirus	Electrochemical biosensor	0.84 copies per ml	25
Rift Valley fever virus (RVFV)	Aptasensor	0.015 ng ml^{-1}	34
H5N1	Paper-based molecularly imprinted sensor	0.58 fM	35
SARS-CoV-2	Electrochemical biosensor	0.53–0.75 ng ml^{-1}	24
Rotavirus	Electrochemical biosensor using antibody	10^3 plaque forming units (pfu) ml^{-1}	36
Zika virus	RNA sensor	78.8–287.5 copies per µl	37
Hepatitis B virus (HBV)	Sequence-specific DNA sensor	3.00×10^{-13} M	38
H1N1	Electrochemical impedance spectroscopy (EIS)	0.004 ng in 6 ml	39
HIV	Imprinting-based electrochemiluminescence biosensor	0.3 fM	40
Human papilloma virus	Optical biosensor	2.87 ng ml^{-1}	41
Dengue virus (DENV)	Electrochemical biosensor	2.81 ng ml^{-1}	42
DENV and chikungunya virus (CHIKV)	Combines 3D printing technology with electrochemical biosensors	1.10^6–1.10^6 ng ml^{-1}	27
Monkeypox virus (Mpox)	Optical biosensor	200 pfu ml^{-1}	43
5 types of hepatitis virus antigens	Electrochemical immunosensor array	≤1.0 ng ml^{-1}	44
CHIKV	CaF_2/TiO_2 nanophotonic biosensor	795–1411 nm per refractive index unit (RIU)	45
DENV	Gold nanoparticle-based lateral flow	5.12×10^2 pfu	46
Respiratory syncytial virus A (RSV-A) Respiratory syncytial virus B (RSV-B)	Colorimetric toehold switch sensors in a paper-based cell-free system	RSV-A: 52 aM RSV-B: 91 aM	47

section explores various sensor applications for virus detection and presents examples of specific viruses identified using biosensors in Table 2.1.

As shown in Table 2.1, various biosensor technologies are used for detecting different viruses. For instance, electrochemical biosensors have been successfully utilized for detecting SARS-CoV-2 by targeting specific RNA sequences while optical biosensors have been employed for human papillomavirus identification.[24,41] Moreover, gold nanoparticle-enhanced sensors have demonstrated improved sensitivity in detecting dengue viruses.[46] These advancements highlight the critical role of biosensor technology in early virus detection, enabling effective disease control and prevention.

In conclusion, the continuous development of biosensor technologies offers promising opportunities for rapid, accurate, and cost-effective virus detection, contributing significantly to global health security.

2.5 Sensor Technologies in the Future

Today, thanks to rapidly advancing integrated sensor platforms, significant developments have been made in the detection and monitoring of infectious diseases. Compared with laboratory-based tests, POCT and systems like LoC, which offer many advantages, play a critical role in early diagnosis, infection tracking, and controlling the spread of disease, especially in urgent health situations like pandemics.

The technologies implemented on microfluidic chips at millimetre and centimetre scales, where traditional laboratory processes are carried out, are referred to as LoC systems. These allow samples taken from patients to move through microchannels on the chip, facilitating chemical, biological, and physical analyses. These systems yield rapid test results even with small amounts of samples, paving the way for the development of low-cost and portable diagnostic systems.[48] These capabilities are also of great importance in the detection of viral diseases. In addition to being cost-effective, the ability to operate with small sample volumes and to detect multiple viruses within the same testing system allows for the rapid identification of the infecting virus in cases where viruses such as influenza A/B, respiratory syncytial virus (RSV), and COVID-19 are indistinguishable due to similar clinical symptoms, thus accelerating the patient's treatment process.[49] Another advantage of LoC systems is their low contamination risk due to their closed system design and their ability to be integrated with different detection systems. The integration of these systems with PCR and loop-mediated isothermal amplification (LAMP) systems results in the development of more accessible and precise diagnostic systems.[50,51]

The internet of things (IoT) refers to network systems that enable sensors to share data with central systems over the internet. These network systems allow for real-time data flow between devices and facilitate quick analysis. Through this integration, especially in rural areas or situations where health services are inadequate, patients can have their symptoms monitored remotely without having to visit hospitals. Monitoring of symptoms is critical for stopping the spread of diseases. In particular, in the tracking of infectious diseases, the advantages of being able to identify infected individuals and monitor contacts play a significant role in controlling the spread and reducing the burden on the healthcare system. With the development of

IoT-based systems, it is anticipated that personalized health monitoring and early warning systems will become more widespread in the coming years, along with wearable sensor systems.[52,53]

Wearable sensors are one of the most important application areas of IoT. This application area has gained popularity and continues to make significant progress due to its advantages in terms of lightness, portability, and comfort. Looking at the historical development of wearable sensors, their first use for monitoring human health dates back to 1960. It began with the need to monitor astronauts' heart rates, body temperatures, and other physiological parameters during the Apollo program, and has since been used in many areas, including tracking athletes' physical performance. Today, wearable biosensors capable of conducting biochemical analyses from sweat, saliva, and interstitial fluid continue to be used for disease diagnosis and health monitoring.[54]

One of the greatest advantages of wearable sensors in the health field is their ability to be used to predict and diagnose dangerous situations or diseases early, thus allowing for faster treatment options. When examined from the perspective of viral diseases, it is anticipated that the development of multifunctional wearable sensors and systems capable of monitoring various complex physiological signals from patients could play a critical role in the early diagnosis of infections and the prevention of their spread. With these sensor technologies, it is possible to detect viral infection symptoms using non-invasive methods from body fluids (sweat, saliva, interstitial fluid) and to identify infections even before symptoms appear.[55] After the pandemic, the importance of these technologies and the need for fast, reliable diagnostic methods have been emphasized once again, accelerating research in this area.[56]

The integration of sensor technologies commonly used in personalized health services with advanced technologies like artificial intelligence provides a more efficient means of utilization. Through artificial intelligence, it is anticipated that connections can be established between the data obtained from the sensors and vital signs and other health measurements, allowing for the most accurate risk analyses for patients. Wearable sensors, viewed as a good alternative to traditional measurement methods, continue to be improved to obtain the maximum amounts of clinical information with minimal sensors.[57]

Research is being conducted in various fields, such as materials science, artificial intelligence, and data analytics, to advance wearable sensors. Self-healing materials and energy-independent systems are being developed to make sensors more precise and durable. Through these systems, data analysis and artificial intelligence-supported algorithms can be integrated into the sensors, transforming them into the most useful systems for patients.[58]

2.6 Future Perspectives and Research Gaps

Herein we have discussed the existing sensor technologies for virus detection by focusing on the recent advancements and various applications. Different sensing strategies have been compared and their POCT implementation has been evaluated in terms of multiprocessing capabilities, particularly in pandemic scenarios where rapid detection is essential. The currently available sensors utilize electrochemical, optical, or fluorescent methods for signal transduction; employ immunodiagnostics or molecular diagnostics for signal detection; and operate on microfluidic or paper-based platforms as their working principle. Even though molecular diagnostics based on recombinase polymerase amplification (RPA), LAMP, and PCR have high sensitivity, they are still expensive, equipment-dependent, and use reagents that are not easy to move and store. Additionally, the complexity of their implementation into portable POCT systems limits their capabilities for rapid testing settings. While nucleic acid amplification provides superior sensitivity in molecular diagnostics, its POCT feasibility still remains a challenge due to the complex workflows (such as requirement for nucleic acid extraction and amplification), long analysis time, and difficultly in miniaturization.[59] Electrochemical biosensors have shown rapid evolution by offering cost-effective, user-friendly and portable detection of viruses. Unfortunately, limited clinical studies or unmet clinical requirements, are the most important barrier to their commercialization. Moreover, other limitations are scalability and multiplexing difficulties. On the other hand, their sensitivity and successful integration into POCT settings as well as artificial intelligence makes them promising for the next generation sensor technologies. To pave the way for the commercialization of these diagnostic tools, the number of clinical studies needs to be increased and improvements in the simultaneous detection multiple targets and machine learning assisted enhancements are required.[60] Paper-based platforms have demonstrated significant advantages, including affordability, minimal reagent and sample requirements, mass production potential, fast response time and ease of transportation. Lateral flow assays (LFA) are the representative commercial products among paper-based platforms due to their properties, such as being fully portable, equipment free, rapid, user-friendly, and generating results that are easy to interpret using the naked eye. LFA formats require further developments in terms of sensitivity, specificity and quantitation. On the other hand, results published in the literature indicate that their combination with sensitive signal transduction methods, including electrochemical and fluorescent technologies, CRISPR–CRISPR-associated protein (Cas)-assisted detection, and smartphone integration for quantitative analysis, holds great promise for enhancing sensitivity and reliability.[61] To date, many sensing systems based on microfluidics have been reported. Microfluidic systems offer great potential for fast responses, compact devices, minimal sample and reagent requirements, and multiplexing. While these advantages make microfluidics sensitive and high-speed platforms, efforts continue to develop fully autonomous devices to overcome their drawbacks of reliance on external instruments, such as fluidic pumps and valves.[59] Despite extensive research efforts, commercialization of novel sensing technologies remains restricted due to key barriers, including difficulties in obtaining regulatory approval of innovative diagnostic technologies, cost and scalability, and ease of operation without experts. Future studies should focus on performing more clinical studies, effective multiplexing and miniaturization and artificial intelligence driven data analysis by developing highly sensitive, low-cost, and automated POCT systems that minimize laboratory dependency while maintaining high accuracy. By addressing these research gaps, next-generation sensor technologies can be produced that will play a pivotal role in advancing public healthcare and responses to emerging infectious diseases.[62]

2.7 Conclusion

In summary, the development of sensor technologies for virus detection has completely changed the approach to managing and overcoming infectious diseases. This chapter has reviewed various diagnostic methods, from traditional molecular techniques to point-of-care devices, and describes how developments in innovative fields, such as nanotechnology, LoC systems, and wearable sensor technologies, have changed the process of viral disease diagnosis.

Various sensor platforms have been compared and analysed based on examples, and each has its own advantages and disadvantages. Technologies such as molecularly imprinted polymers, SiNW sensors, and new generation optical and electrochemical biosensors have been shown to be promising in terms of increasing diagnostic accuracy, reducing time, reducing cost, and improving accessibility. In addition, the integration of these technologies with smartphones and wearable devices and IoT systems will increase their advantages in terms of time, cost and, especially, accessibility.

From a future perspective, it is an inevitable fact that researchers working in many fields, such as engineering, life sciences, software and artificial intelligence technologies, must work together to create a collective product in order to meet needs such as the need for more sensitive, scalable, and user-friendly designs. Developing materials science and data processing technologies will enable viral diagnostic technologies to take an important place in personalized medicine applications.

In conclusion, the rapid progress seen in sensor technologies for viral disease diagnoses, the encouragement of interdisciplinary work, studies carried out on personalized medicine and the integration of sensor technologies into them show that infectious diseases can be managed much more actively and easily globally.

References

1. A. Poghossian, M. Jablonski, D. Molinnus, C. Wege and M. J. Schöning, *Front. Plant Sci.*, 2020, **11**, 598103.
2. B. V. Ribeiro, T. A. R. Cordeiro, G. R. Oliveira, E. Freitas, L. F. Ferreira and D. L. Franco, *Talanta Open*, 2020, **2**, 100007.
3. G. Marrazza, M. Ramalingam, A. Jaisankar, L. Cheng, G. Selvolini and I. A. Vitale, *TrAC, Trends Anal. Chem.*, 2024, **172**, 117609.

4. M. Goh, C. Joy, A. N. Gillespie, Q. R. Soh, F. He and V. Sung, *Pediatr. Res.*, 2024, **95**, 508–531.
5. P. K. Drain, E. P. Hyle, F. Noubary, K. A. Freedberg, D. Wilson, W. R. Bishai, W. Rodriguez and I. V. Bassett, *Lancet Infect. Dis.*, 2014, **14**, 239–249.
6. D. S. Mouliou and K. I. Gourgoulianis, *Expert Rev. Respir. Med.*, 2021, **15**, 993–1002.
7. K. Lakshmanan and B. M. Liu, *Diagnostics*, 2025, **15**, 123.
8. M. L. Everitt, A. Tillery, M. G. David, N. Singh, A. Borison and I. M. White, *Anal. Chim. Acta*, 2021, **1146**, 184–199.
9. P. Yager, G. J. Domingo and J. Gerdes, *Annu. Rev. Biomed. Eng.*, 2008, **10**, 107–144.
10. C. Jianrong, M. Yuqing, H. Nongyue, W. Xiaohua and L. Sijiao, *Biotechnol. Adv.*, 2004, **22**, 505–518.
11. H. Xi, H. Jiang, M. Juhas and Y. Zhang, *ACS Omega*, 2021, **6**, 25846–25859.
12. R. E. Madrid, F. Ashur Ramallo, D. E. Barraza and R. E. Chaile, *Bioengineering*, 2022, **9**, 101.
13. J. S. Batra, T. Chi, M. Huang, D. Zhu, Z. Chen, D. Lee and J. Kameoka, *Biosensors*, 2023, **13**, 861.
14. D. Najjar, J. Rainbow, S. T. Sanjay, P. Jolly, H. D. Puig, M. Yafia and D. E. Ingber, *Nat. Biomed. Eng.*, 2022, **6**, 968–978.
15. B. Huang, N. P. West, J. Vider, A. J. Cox, T. Constantino, B. Harrower and D. Warrilow, *Sci. Rep.*, 2017, **7**, 18092.
16. S. C. B. Gopinath, K. Awazu and M. Fujimaki, *Anal. Methods*, 2010, **2**, 1880.
17. F. Shen, J. Wang, Z. Xu, Y. Wu, Q. Chen, X. Li and T. Zhu, *Nano Lett.*, 2012, **12**, 3722–3730.
18. J. T. Baca, V. Severns, D. M. Lovato, D. W. Branch and R. A. Larson, *Sensors*, 2015, **15**, 8605.
19. X. Li, R. Sun, J. Pan, Z. Shi, Z. An, C. Dai, J. Lv, G. Liu, H. Liang, J. Liu, Y. Lu, F. Zhang and Q. Liu, *Nat. Commun.*, 2024, **15**, 4035.
20. S. J. Park, S.H. Cha, G. Shin and Y. H. Ahn, *Biomed. Opt. Express*, 2017, **8**, 3551.
21. A. Zybin, Y. A. Kuritsyn, E. L. Gurevich, V. Temchura, K. Überla and K. Niemax, *Plasmonics*, 2009, **5**, 31.
22. A. Ymeti, J. Greve, P. Lambeck, T. Wink, V. Hövell, R. Wijn and J. Kanger, *Nano Lett.*, 2006, **7**, 394.
23. E. Cesewski and B. N. Johnson, *Biosens. Bioelectron.*, 2020, **159**, 112214.
24. C. Durmus, S. B. Hanoglu, D. Harmanci, H. Moulahoum, K. Tok, F. Ghorbanizamani, S. Sanli, F. Zihnioglu, S. Evran, C. Cicek, R. Sertoz, B. Arda, T. Goksel, K. Turhan and S. Timur, *Talanta*, 2022, **243**, 123356.
25. H. Zhao, W. Xie, R.-L. Zhang, X.-D. Wang, H.-F. Liu, J. Li, T. Sha, X.-S. Guo, J. Li, Q.-M. Sun, Y.-P. Zhang and C.-P. Li, *Talanta*, 2022, **237**, 122896.
26. X. Ma, Y. Shi, G. Gao, H. Zhang, Q. Zhao and J. Zhi, *J. Electroanal. Chem.*, 2023, **950**, 117867.
27. P. Sharma, M. R. Hasan, A. Yardim, E. G. Celik, S. Timur, M. Z. Abdin and J. Narang, *ACS Appl. Bio Mater.*, 2025, **8**, 1251.

28. D. Harmanci, S. Balaban Hanoglu, G. Akkus Kayali, E. Durgunlu, N. Ucar, C. Cicek and S. Timur, *Biosensors*, 2023, **13**, 851.
29. W. Heo, K. Lee, S. Park, K. A. Hyun and H. I. Jung, *Biosens. Bioelectron.*, 2022, **201**, 113960.
30. M. Drobysh, A. Ramanaviciene, R. Viter, C. F. Chen, U. Samukaite-Bubniene, V. Ratautaite and A. Ramanavicius, *Int. J. Mol. Sci.*, 2022, **23**, 666.
31. A. N. Herdina, A. Bozdogan, P. Aspermair, J. Dostalek, M. Klausberger, N. Lingg, M. Cserjan-Puschmann, P. P. Aguilar, S. Auer, H. Demirtas, J. Andersson, F. Lötsch, B. Holzer, A. Steinrigl, F. Thalhammer, J. Schellnegger, M. Breuer, W. Knoll and R. Strassl, *Biosens. Bioelectron.*, 2025, **267**, 116807.
32. W. A. El-Said, A. S. Al-Bogami, W. Alshitari, D. A. El-Hady, T. S. Saleh, M. A. El-Mokhtar and J. W. Choi, *Biochip J.*, 2021, **15**, 287.
33. H. Gong, T. Pang, X. Yang, F. Chen, N. Jiang, Y. Li, C. Chen and C. Cai, *Sens. Actuators, B*, 2025, **424**, 136909.
34. M. A. Alharbi, A. Rhouati and M. Zourob, *Sci. Rep.*, 2024, **14**, 23892.
35. H. Gong, L. Tang, C. Chen, F. Chen and C. Cai, *Chem. Eng. J.*, 2023, **477**, 146990.
36. F. Liu, Y. H. Kim, D. S. Cheon and T. S. Seo, *Sens. Actuators, B*, 2013, **186**, 252e257.
37. C. Cheng, J. J. Wu and J. Chen, *Anal. Chem.*, 2021, **93**, 11379.
38. J. Zheng, C. Chen and X. L. Wang, *Sens. Actuators, B*, 2014, **199**, 168.
39. H. Ravina, P. S. Gill and A. Kumar, *Int. J. Biol. Macromol.*, 2019, **130**, 720.
40. Babamiri, A. Salimi and R. Hallaj, *Biosens. Bioelectron.*, 2018, **117**, 332.
41. H. Inan, S. Wang and F. Inci, *Sci. Rep.*, 2017, **7**, 3322.
42. O. Y. Siew, E. L. Pang, H. S. Loh and M. T. Tan, *Biosens. Bioelectron.*, 2021, **176**, 112895.
43. M. Aslan, E. Seymour, H. Brickner, A. E. Clark, I. Celebi, M. B. Townsend, P. S. Satheshkumar, M. Riley, A. F. Carlin, M. S. Ünlü and P. Ray, *Biosens. Bioelectron.*, 2025, **269**, 116932.
44. D. Tang, J. Tang and B. Su, *Biosens. Bioelectron.*, 2010, **25**, 1658.
45. H. Arafa Aly, B. A. Mohamed and M. Al-Dossari, *Appl. Opt.*, 2024, **63**, 7909.
46. M. Liu, C. M. Williams, C. M. Acuña, N. L. Sepulveda, S. G. Huerta, D. M. Villanueva, M. A. Nolazco, D. P. Ibarra, K. G. Rodríguez, J. G. A. Juarez and R. A. M. Estilla, *Biosensors*, 2022, **12**, 495.
47. M. Cao, Q. Sun, X. Zhang, Y. Ma and J. Wang, *Biosens. Bioelectron.*, 2021, **182**, 113173.
48. L. C. Lim, A. Z. Kouzani and W. Duan, *Microsyst. Technol.*, 2010, **16**, 1995–2015.
49. L. Gandhi, D. Maisnam, D. Rathore, P. Chauhan, A. Bonagiri and M. Venkataramana, *Eur. J. Med. Res.*, 2022, **27**, 236.
50. K. T. L. Trinh, H. Zhang, D.-J. Kang, S.-H. Kahng, B. D. Tall and N. Y. Lee, *Int. Neurourol. J.*, 2016, **20**, S38–S48.
51. H. Zhang, Y. Xu, Z. Fohlerova, H. Chang, C. Iliescu and P. Neuzil, *TrAC, Trends Anal. Chem.*, 2019, **113**, 44–53.

52. S. M. T. Gharibzahedi and Z. Altintas, *Trends Anal. Chem.*, 2024, **170**, 117473.
53. M. S. Al-kahtani, F. Khan and W. Taekeun, *Sensors*, 2022, **22**, 5738.
54. S. Jia, H. Gao, Z. Xue and X. Meng, *Biosensors*, 2022, **12**, 1057.
55. K. Wang, W. Liu, J. Wu, H. Li, H. Peng, J. Zhang, K. Ding, X. Wang, C. Hou, H. Zhang and Y. Luo, *ACS Appl. Mater. Interfaces*, 2025, **17**(9), 13279–13301.
56. D. S. Temple, M. Hegarty-Craver, R. D. Furberg, E. A. Preble, E. Bergstrom, Z. Gardener, P. Dayananda, L. Taylor, N.-M. Lemm, L. Papargyris, M. T. McClain, B. P. Nicholson, A. Bowie, M. Miggs, E. Petzold, C. W. Woods, C. Chiu and K. H. Gilchrist, *J. Infect. Dis.*, 2023, **227**, 864.
57. J. Heikenfeld, A. Jajack, J. Rogers, P. Gutruf, L. Tian, T. Pan, R. Li, M. Khine, J. Kim, J. Wang and J. Kim, *Lab Chip*, 2018, **18**, 217.
58. H. C. Ates, P. Q. Nguyen, L. Gonzalez-Macia, E. Morales-Narváez, F. Güder, J. J. Collins and C. Dincer, *Nat. Rev. Mater.*, 2022, **7**, 887.
59. Y. Seok, M. G. Mauk, R. Li and C. Qian, *Anal. Chim. Acta*, 2023, **1264**, 341283.
60. B. Lou, Y. Liu, M. Shi, J. Chen, K. Li, Y. Tan, L. Chen, Y. Wu, T. Wang, X. Liu, T. Jiang, D. Peng and Z. Liu, *Trends Anal. Chem.*, 2022, **157**, 116738.
61. M. Sharafeldin and J. J. Davis, *Anal. Chem.*, 2021, **93**, 184–197.
62. X. Song, Z. Fredj, Y. Zheng, H. Zhang, G. Rong, S. Bian and M. Sawan, *J. Pharm. Anal.*, 2023, **13**(11), 1252–1268.

CHAPTER 3

Optical Sensors

DUYGU ÇIMEN[a], AYŞEGÜL BÜLBÜL[a] AND FATMA YıLMAZ[b*]

[a]Department of Chemistry, Hacettepe University, 06800, Ankara, Turkey;
[b]Department of Chemistry Technology, Bolu Abant Izzet Baysal University, 14900 Bolu, Turkey
*E-mail: yilmaz_f@ibu.edu.tr

3.1 Introduction

Medicine and public health services have improved significantly in the last few decades in most countries. As a result, the demand for cost-effective and user-friendly remote health monitoring systems for older adults is steadily increasing. Sensors, used in both medical and non-medical electronic equipment, function by converting various vital signs into electrical signals. Sensors are extensively used in the medical field, including applications like life-supporting implants, long-term patient monitoring, and preventive healthcare services.[1]

Optical sensors detect changes in biological, chemical, or physical parameters by measuring specific optical properties, such as wavelength, intensity, and phase of optical signals. Optical sensors offer exceptional sensitivity, a high signal-to-noise ratio, and immunity to electromagnetic radiation, allowing for the acquisition of more parameters. Due to their unique advantages over traditional wearable devices, they hold great potential in both clinical applications and pervasive medicine. However, achieving higher precision in optical sensor design generally results in higher volume levels and higher demodulation costs. Wearability, manufacturability, and sensing capabilities

Detection Science Series No. 32
Advances in Virus Detection with Sensors
Edited by Adil Denizli
© The Royal Society of Chemistry 2026
Published by the Royal Society of Chemistry, www.rsc.org

often result in a trade-off when considering usability.[2] Optical biosensors provide an alternative method for virus detection that is safe, easy to use, and cost-effective, eliminating the need for nucleic acid amplification.

The recent coronavirus disease (COVID)-19 outbreak has led to major global disruptions because of the shortage of testing centers and equipment. Over the past 15 years, the use of optical techniques for viral detection has flourished, offering more reliable, cost-effective, and accurate methods.[3] In this section, optical phenomena concepts are evaluated according to the carrier medium and operating principle. Classifications based on the carrier medium includes fibre optic sensors, free-space optical sensors, gas optical sensors, and liquid medium optical sensors. Detection of virus pathogens with plasmonic sensors, fluorescent sensors, refractive index sensors, and Raman sensors is examined according to their working principles.

In this context, the biomedical sensing applications of optical fibres are becoming increasingly important.[4] Optical fibre technology holds great potential to transform current polymerase chain reaction (PCR) systems by offering miniaturised sample requirements, flexible signal guidance and acquisition, cost efficiency, immunity to electromagnetic interference, and the ability for remote diagnostics. Jin *et al.* introduced an integrated fibre PCR (iF-PCR) system that overcomes all the above-mentioned challenges and enables quantitative PCR (qPCR), replacing PCR instruments with a single fibre system.[5]

Recent diagnostic research has primarily focused on the need for rapidly identifying potential diseases from biological specimens using compact, high-precision technology. Optofluidic sensors enable single-molecule detection through planar, waveguide-coupled multipoint fluorescence excitation. Amin *et al.* presented a novel approach for single-particle fluorescence sensing by applying free-space, top-down illumination to liquid-core antiresonant reflecting optical waveguide (ARROW) devices, combined with two distinct multi-spot excitation techniques.[6] The incorporation of various integrated photonic technologies into various precision schemes on lab-on-a-chip platforms has displayed significant potential in biosensors at the point of care. ARROW-based optofluidic sensors have demonstrated exceptional capabilities in detecting a wide range of target biomarkers with single-particle sensitivity, such as DNA, pathogenic proteins, and individual viruses, by harnessing optical signal guidance in a low-index liquid medium.[7]

With the rise in air pollution and the onset of the pandemic, there is a greater need for sensors capable of detecting gases and airborne pathogens with high accuracy. However, existing detection technologies are not suitable for real-time, on-site applications.[8] Pathogens in the air, like influenza and severe acute respiratory syndrome coronavirus 2 (SARS-CoV-2), present serious health threats. Although conventional diagnostic techniques have their drawbacks, optical biosensors provide a potential solution due to their speed, sensitivity, and affordability.[9] The COVID-19 pandemic has created an urgent need for quick and precise diagnostic methods. One innovative, non-invasive method involves using an electronic nose (GeNose C19) for COVID-19 detection through breath samples. GeNose C19 shows great potential as an effective tool in combating the virus.[10] Ongoing research and development will enhance its functionality and broaden the scope of its possible applications. A total of 615 breath samples were collected, including those from 43 COVID-19 patients and 40 non-infected individuals. The samples were tested with GeNose C19, a portable device featuring a set of metal oxide semiconductor gas sensors. Volatile organic compounds (VOCs) in human breath, which can be used as indicators for various diseases, including COVID-19, are detected by the device.[11] Preventing and controlling future epidemics and bioterrorism threats is possible through the application of rapid and sensitive virus detection techniques that can be rapidly deployed at multiple locations. A label-free optofluidic nanoplasmonic sensor was presented by Yanik *et al.*, which directly identifies intact viruses in biological samples at clinically significant concentrations, with minimal or no sample preparation required. Both small enveloped RNA viruses (vesicular stomatitis virus and pseudotyped Ebola) and large enveloped DNA viruses (vaccinia virus) were detected within a dynamic range spanning three orders of magnitude.[12]

3.2 Viruses

In recent years, many diseases have caused hundreds of thousands of deaths with the pollution of resources due to viruses around the world. Today, the spread of infectious diseases that have an effect on all living things such as animals, humans, and plants is increasing day by day.[13] Various infectious diseases such as tuberculosis, malaria, influenza (A, B, and C types), COVID, and human immunodeficiency virus have adversely affected many people in recent years and all over the world.[14] Viruses replicate their genetic material as intracellular parasites, requiring a host cell and complex protection mechanisms, and change very rapidly. Viruses have emerged that are adapted to break and manipulate the immune reaction of the host cell. Some outbreaks of influenza AH1N1 subtype (in 2019), Ebola (in 2014), and COVID (in 2021) spread all over the world. In order to detect viruses, the enzyme-related, reverse transcription–polymerase chain reaction, immunosorbent techniques, polymerase chain reaction, and different optical sensors are used.[15–18]

The most widely used optical sensor types for the detection of different viruses are summarised in Table 3.1.

3.3 Optical Sensors

Optical sensors are electronic devices that can detect the presence of objects using light energy. These sensors can be used in many different applications, such as motion detection, colour detection, distance measurement, position determination, and optical communication. The working principle of these sensors is based on light from a light emitting source being directed towards the target object and reflected from the object's surface back to the sensor. The sensor receives the reflected light and measures its properties to detect

Table 3.1 Optical sensors most commonly used for the detection of virus families.

Virus family	Example viruses	Most commonly used optical sensors	Advantages
Flaviviridae	Dengue, Zika, hepatitis C virus (HCV), West Nile virus	Surface plasmon resonance (SPR), fluorescence, Raman spectroscopy	Label-free, high sensitivity
Coronaviridae	SARS-CoV, Middle East respiratory syndrome (MERS)-CoV, SARS-CoV-2	SPR, fluorescence, enzyme linked immunosorbent assay (ELISA)-optical, localised surface plasmon resonance (LSPR)	High sensitivity, rapid response time
Herpesviridae	Herpes simplex virus (HSV), Varicella zoster virus (VZV), Epstein–Barr virus (EBV)	Fluorescence, SPR, Raman spectroscopy	Suitable for viral DNA or RNA detection
Orthomyxoviridae	Influenza A, B, C	LSPR, fluorescence	Ideal for hemagglutinin and neuraminidase analysis
Retroviridae	Human immunodeficiency virus (HIV)	SPR, fluorescence, LSPR	Can monitor viral protein interactions
Papillomaviridae	Human papillomavirus (HPV)	Fluorescence *in situ* hybridisation (FISH)	DNA- or RNA-based detection
Adenoviridae	Adenoviruses	Fluorescence, ELISA-optical	Suitable for gene expression analysis
Paramyxoviridae	Measles, mumps, respiratory syncytial virus (RSV)	Fluorescence, SPR,	Ideal for real-time detection
Filoviridae	Ebola, Marburg	LSPR, Raman spectroscopy	Critical for rapid diagnostics
Bunyaviridae	Hantavirus, Crimean–Congo haemorrhagic fever (CCHF)	SPR, fluorescence, LSPR	Low sample requirement
Poxviridae	Variola (Smallpox), Molluscum contagiosum	Fluorescence, SPR, Raman spectroscopy	Suitable for antibody-based detection

the presence or movement of the object. They react faster than other sensors and have a high accuracy rate. They are also less affected by environmental factors, such as temperature, humidity, and vibration. For this reason, they are used in many different application areas, such as the automotive industry, medical devices, industrial automation, security systems, and even smartphones. Optical sensors are divided into two classes according to the carrier medium and operating principle and applications.[19,20]

3.3.1 Classification by Carrier Medium

3.3.1.1 Fibre Optic Sensors

The rapid development of optical fibre communication technology since 1977 has led to optical fibre sensing technology becoming a standard measurement information technology in several countries. The optical signal propagated in the optical fibre is used by optical fibre sensing technology to detect the physical quantity to be measured. The principle of optical fibre sensing is depicted in Figure 3.1 using a schematic diagram. In fibre optic sensors, the light emitted from the light source first reaches the sensing region of interest. The physical quantity to be measured will change if the light source remains unchanged. The characteristic parameters of the light wave are impacted by this through a mechanism that affects the light in the optical fibre. The optical spectrum analyser (OSA) receives the modified light and demodulates the useful signal to detect it. The most important advantages of fibre sensor technology are fast response, high sensitivity, miniaturisation, and freedom from electromagnetic interference. Due to these important features, they are preferred in medical applications.[21–23]

The COVID-19 outbreak, which poses a grave threat to human health, is diagnosed and identified by using the acute respiratory syndrome coronavirus-2 nucleocapsid protein (N-protein). Xia *et al.* designed label-free and sensitive biosensors based on optical fibre cylindrical micro-resonators for COVID-19 detection.[24] The surface of the whispering gallery mode (WGM)

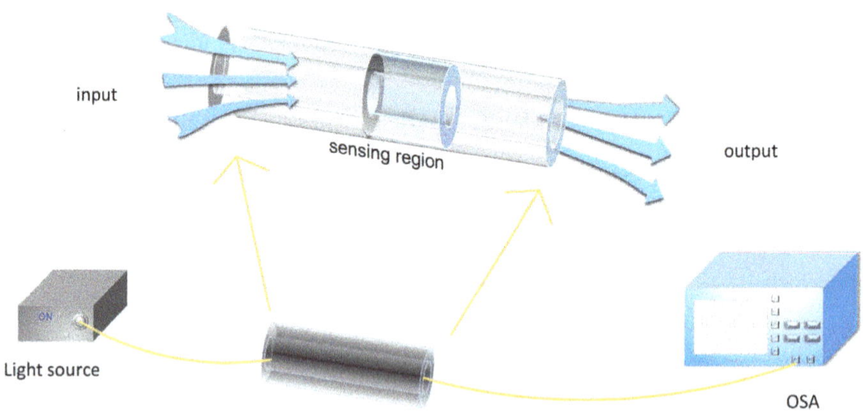

Figure 3.1 Schematic representation of the optical fibre sensor working principle. Abbreviation: optical spectral analyser, OSA. Adapted from ref. 23, https://doi.org/10.3390/coatings14020173, under the terms of the CC BY 4.0 license, https://creativecommons.org/licenses/by/4.0/.

micro-resonator was modified with N-protein monoclonal antibody to evaluate the biokinetic process of the prepared biosensor and to detect N-proteins. With the observation of a decline in WGM resonance, an increased red-shift in the spectral features and transmission spectra was observed. The observed spectral feature of WGM red shift was due to the antibody–antigen reaction. It was shown that the wavelength shift is proportional to the N-protein concentration, varying between 0.1 and 100 $\mu g\ ml^{-1}$ concentration range. The lowest detection limit was calculated as 0.23 $\mu g\ ml^{-1}$. The specificity of the optical fibre sensor was also studied in the mixed solution of microresonator tumor necrosis factor (TNF), spike protein (S-protein) solution, and N-protein prepared in human saliva and normal human serum.

It is very important to quickly detect epidemic diseases, such as SARS-Cov-2 and influenza A, caused by viruses. Yang *et al.* prepared a reusable dual-channel optical fibre immunosensor that gives results in 10 min by combining the tandem detection mode with the transient wave detection properties of the mobile phase (Figure 3.2).[25] The prepared immunosensors can detect the nucleoproteins of influenza A (FluA) virus strains, such as H1N1, H3N2, and H7N9, and the spike proteins of SARS-CoV-2 omicron and delta variants with the lowest detection limit with a wider concentration range than tests

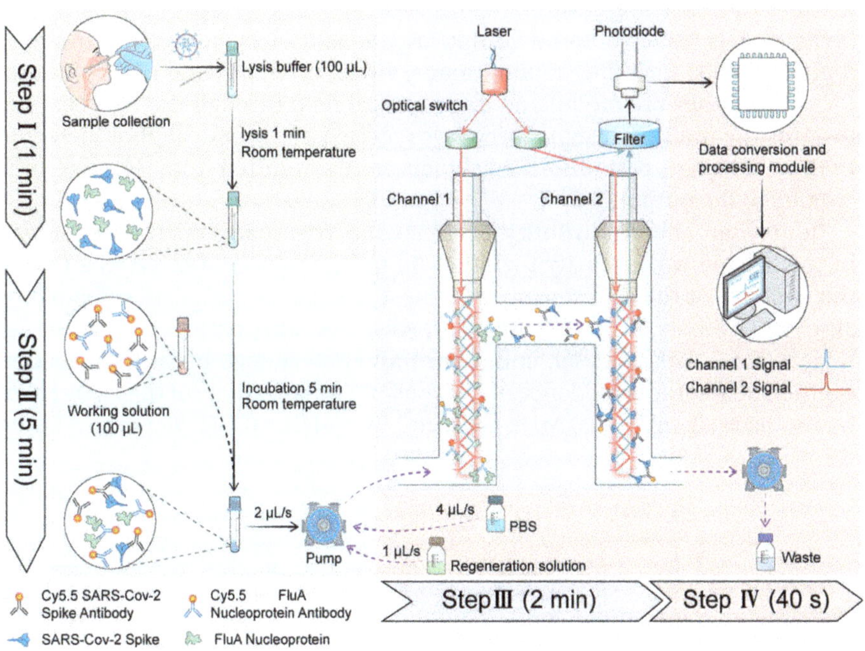

Figure 3.2 Schematic representation of the detection of influenza A (FluA) and SARS-CoV-2 viruses by optical fibre immunosensor. Adapted from ref. 25 with permission from John Wiley & Sons, Copyright 2022, The Authors. *J. Med. Virol.* published by Wiley Periodicals LLC.

with enzyme-linked immunosorbent assay (ELISA). When the SARS-CoV-2 detection results in 26 clinical samples were compared with the commercial colloidal gold test strip, 35% sensitivity, 100% specificity, and 94% sensitivity were observed. The prepared optical fibre immunosensor can be reused more than 80 times, which both saves costs and facilitates real-time detection of pathogenic viruses. The lowest detection limits for the determination of SARS-CoV-2 and FluA nucleoprotein attached protein were 0.2 ng ml^{-1} and 64.9 ng ml^{-1}, respectively.

3.3.1.2 *Free-Space Optical Sensors*

Free space optical communication (FSO) is wireless optical transmission from one point to another point in an undirected propagation medium of visible light (VL), infrared (IR) or ultraviolet (UV) spectra using visible light and infrared bands. As a general operating principle, outdoor FSO links operate in the UV and VL bands, while indoor and underwater communications use the near VL and IR bands. While FSO operates at extremely high frequencies and photodetectors are immune to multipath signal fluctuations, radio frequency networks are highly susceptible to multipath fading. Figure 3.3 schematically shows a typical free-space optical communication system, including the transmission medium, optical receivers, light sources, electrical-to-optical conversion, and *vice versa* in the FSO operating principle. Light signals are created by an optical transmitter by converting multiple digital signals, including audio, image, video, internet data, and computer files. Then, they are modulated in an optical field through the FSO channel towards the multipath signal, which is received at the optical photodetector. In the final stage, demultiplexed signals are transmitted from the electronic circuits to the output.[26–28]

Tabib-Azar and Chowdhury fabricated a free-space microwave sensor in the X band with a cavity region of 30 radial wires connected to a hub at the center.[29] At the FSO sensor centre is a circular disk capacitor divided by an electric dipole at 10 GHz. Viruses, soot, and dust particles in the cavity region, a network analyser, and a horn antenna change both the reflection coefficient and the impedance of the sensor. The sensitivity of the FSO sensor was 85.02 MHz ml^{-1}, 89.5 MHz ml^{-1}, and 94.6 MHz ml^{-1} for deionised water,

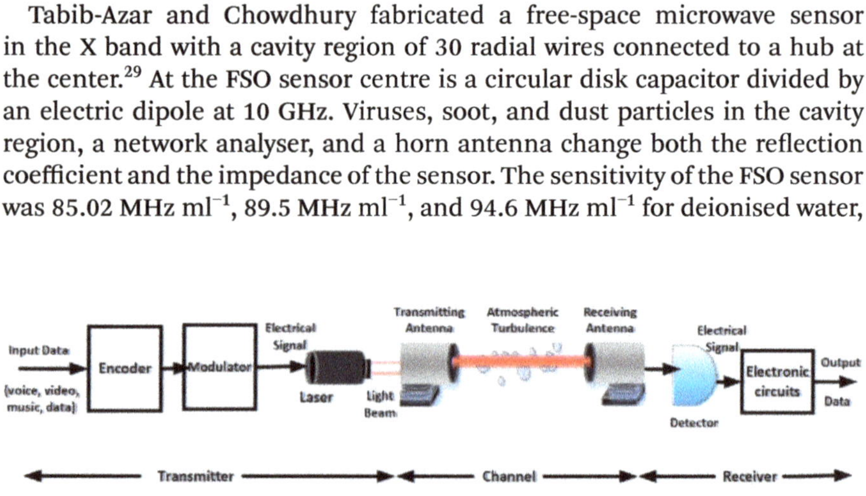

Figure 3.3 Schematic representation of the operating principle of an FSO system. Adapted from ref. 28 with permission from Elsevier, Copyright 2022.

uninfected saliva, and infected saliva solutions with SARS-COV-2 added at 103 viruses per µl respectively. Furthermore, the sensitivity of the FSO sensor to the dielectric sample was 3.23 MHz mm^{-3}, while for iron particles it was 16.25 MHz mm^{-3}. All these analysed samples were smaller than $\lambda/30$ at 10 GHz and therefore not detectable on metallic substrates.

3.3.1.3 *Gas Optical Sensors*

In recent years, with the development of optical technology, optical gas sensors, which have an important contribution to human health, are among the most important areas of use for protection from hazardous gases, detecting and minimising those that cause air pollution. In 2012, according to studies conducted by the World Health Organization, approximately seven million people died due to exposure to air pollution. In addition, optical gas sensors are frequently used in the health sector, food production and control, indoor air quality control, and fire and explosion prevention. With the rapid advancement of technology, there are optical gas sensors that can detect a large number of gases, sometimes from a kilometre or more away, and even be used to detect and visualise the gas clouds in real time. In the detection of optical gas sensors, factors such as gas types, resolution, operating environment, concentration range, sensitivity, analysis time, size, and cost are important.[30]

Hepatitis B virus (HBV), a hepatotropic virus, causes a chronic and permanent infection with immune effects. Wang *et al.* developed a resonance light scatter sensor using a specific gas sensitive molecular imprinted method for HBV determination (see Figure 3.4).[31] In the prepared sensors, the binding energy of the virus to be detected as a result of changing N_2 and CO_2 according to the electrical properties of the amines in the suppressed cavities is important to obtain both rapid binding and release. The changes in electrical properties are used to determine the sensitivity and specificity of the virus to be detected. In this study, the imprinted factor was 6.7 and the lowest limit of detection was 1.9 pM. The results of cyclical experiments indicated that the gas response sensor is stable and can be reused repeatedly. These features indicate that the sensor is an important method in controlling virus epidemics for the determination of viral infections.

3.3.1.4 *Liquid Medium Optical Sensors*

Liquid medium optical sensors are devices that use light to detect properties or changes in liquid media. These sensors often rely on the interaction of light with the liquid, measuring factors like concentration, turbidity, refractive index, and the presence of specific substances. They are widely used in applications such as environmental monitoring, industrial processes, and medical diagnostics. The most important advantages of liquid media optical sensors are non-invasiveness, high sensitivity, real-time monitoring, and a

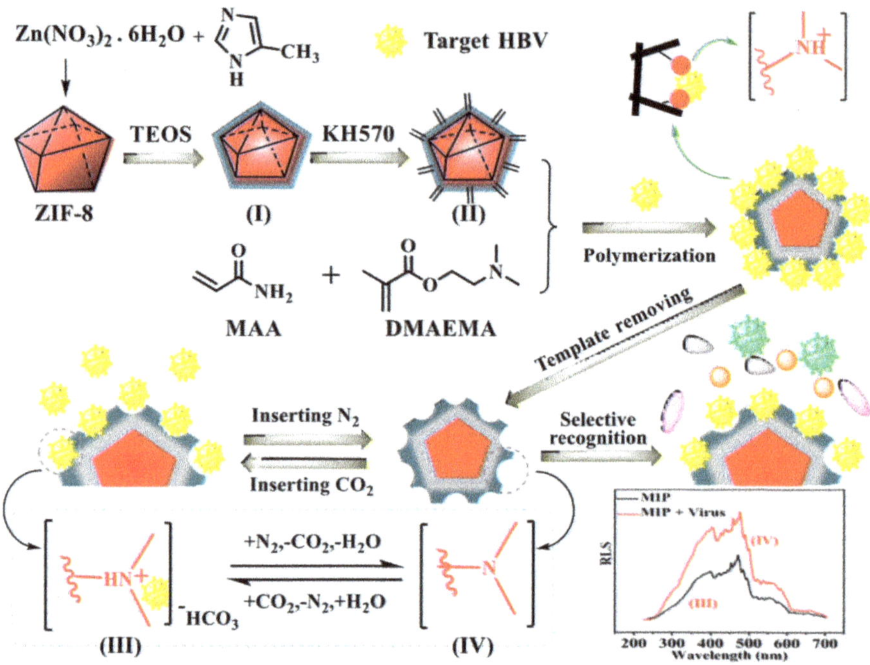

Figure 3.4 Schematic representation of the preparation of molecularly imprinted polymers for virus determination. Adapted from ref. 31 with permission from Elsevier, Copyright 2022.

wide range of applications. The most widely used types of optical sensors for liquid media are fluorescence sensors and pH sensors.[32,33]

Viral hepatitis viruses (types A, B, C, and D) are a group of infectious diseases whose clinical symptoms cause liver function damage. These virus types can interact with each other and coexist in the patient's body. This causes difficulties in differentiating between these virus types.[34] Luo *et al.* designed fluorescence sensors using a molecular imprinting method for the simultaneous detection of hepatitis A (HAV) and hepatitis B (HBV) viruses (Figure 3.5).[35] It utilized changes in the colour of red and green quantum dots according to the changing concentrations of hepatitis viruses to be detected. The imprinting factors of hepatitis A and hepatitis B viruses with the molecularly imprinted fluorescence sensor were 3.70 and 3.35, respectively. With 20 min analysis time, the lowest detection limits of hepatitis A and hepatitis B viruses were determined with a sensitivity of 3.4 and 5.3 pmol l^{-1}, respectively. The high recoveries of the prepared fluorescence sensor with simultaneous and single detection modes prove that it can simultaneously detect a large number of viruses in real clinical samples for the determination of hepatitis A and hepatitis B in human serum samples.

Flores *et al.* prepared pH sensors for the detection of white spot shrimp virus (WSSV) using nucleic acid based methods.[36] In the detection of WSSV virus, hydrogen ions are produced during the amplification process. In this

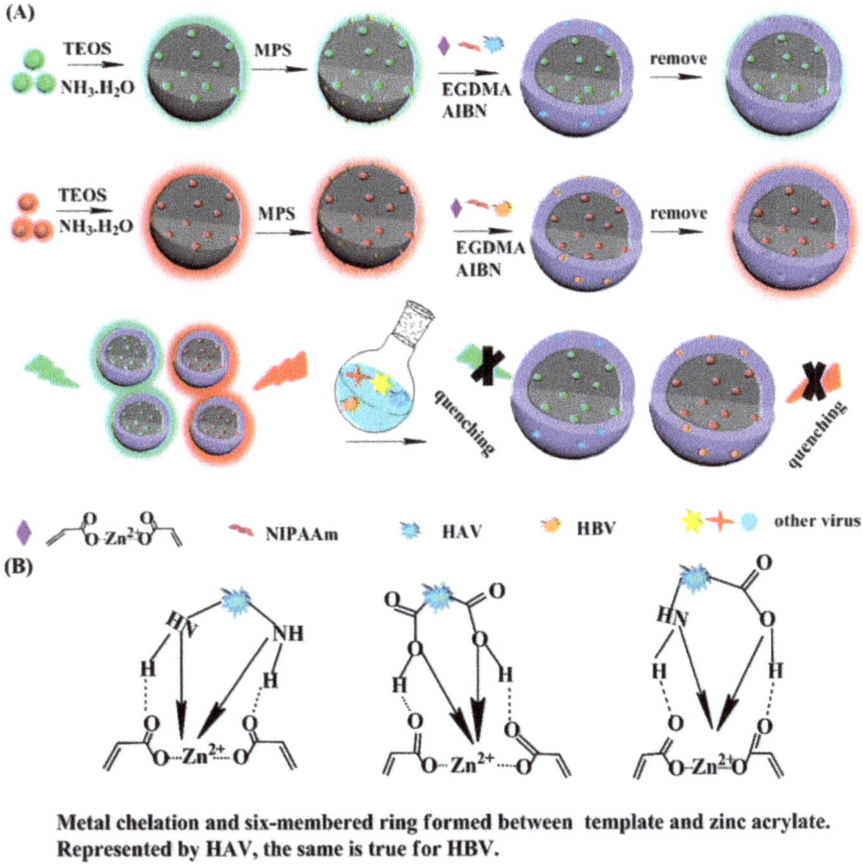

Figure 3.5 Schematic representation of the preparation of the fluorescence sensor prepared by the molecular imprinting method (a) and representation of the ring structure formed between the template molecule and zinc acrylate by metal chelation (b). Adapted from ref. 35 with permission from American Chemical Society, Copyright 2019.

study, the pH sensor was designed as a micro probe potentiometric sensor using the loop-mediated isothermal amplification (LAMP) technique. The pH microprobe was modified by galvanostatic electroplating from pencil graphite with polypyrrole nanotubes containing methyl orange. The prepared probe was characterised by Fourier transform infrared spectroscopy and field emission scanning electron microscopy. The potentiometric measurement with the highest sensitivity in the range of pH 2–8 and the lowest detection limit of 5.19×10^{-8} M was found. Potentiometric measurements showed a low potential shift of ±1.00 mV at pH 8 with a sub-Nernstian response of 45.5 mV per pH. The prepared potentiometric sensor showed a high diagnostic sensitivity of 93.8% and a positive LAMP reaction to pH measurements with a specificity of 100%. Furthermore, WSSV was detected in shrimp samples.

Table 3.2 Summary of studies on sensors classified by carrier medium used in virus detection.

Sensor type	Analyte	Linear range	LOD	Real sample	References
Optical fibre sensor	Vaccinia virus	1 fM–20 pM	—	Serum	37
Optical fibre sensor	COVID-19	200–500 μm	—	—	38
Fibre-optic label-free biosensor	SARS-CoV-2	1 pg ml^{-1}–100 μg ml^{-1}	—	Transport medium	39
Optical fibre aptasensor	SARS-CoV-2	—	37 nM	Human serum	40
Ratiometric fluorescent sensors	SARS-CoV-2 RNA-dependent RNA polymerase (RdRp) gene	—	30 fM	Saliva sample	41
Fluorescence aptasensor	H5N1 influenza virus	0–100 ng ml^{-1}	3.5 ng ml^{-1}	Human serum	42
Fluorescence sensor	Hepatitis A and B viruses	6.0–88.5 pM	1.97 pM	—	43
Fluorescence immuno-sensor	Influenza A (H1N1)	1.0×10^{-12}–1.0×10^{-8} g ml^{-1}	1.0×10^{-13} g ml^{-1}	—	44

Table 3.2 summarises studies on sensors classified by carrier medium used in virus detection in the literature.

3.3.2 Classification by Operating Principle and Detector Types

3.3.2.1 Plasmonic Sensors

Plasmonic sensors, particularly surface plasmon resonance (SPR) systems, have emerged as cutting-edge tools for real-time, label-free, and highly sensitive detection of viral biomarkers, making them indispensable for diagnostics, vaccine development, and epidemiological monitoring. SPR operates by detecting shifts in the resonance angle caused by biomolecular interactions on a metallic surface, typically gold, offering unparalleled precision and sensitivity.[45] Recent advancements, such as the integration of gold nanoparticles (AuNPs) and molecularly imprinted polymers (MIPs), have further enhanced the sensitivity of SPR biosensors, enabling detection at femtomolar (fM) and attomolar (aM) concentrations.[46] These modifications also facilitate the selective recognition of target analytes, including viral proteins, antibodies, and nucleic acids, through functionalised biorecognition elements. Additionally, SPR biosensors' multiplexing capabilities allow simultaneous detection of multiple analytes, enabling their application in tracking viral mutations and co-infections. This versatility, combined with their ability to eliminate complex sample preparation, establishes SPR systems as a cornerstone in advancing global health security and pandemic preparedness.[47]

SPR based biosensors have proven to be effective tools in viral diagnostics due to their high sensitivity and real-time detection capabilities. Maddali *et al.* developed a SPR-based biosensor for the detection of SARS-CoV-2, employing a gold-coated sensor chip functionalised with viral proteins to facilitate label-free virus–host interaction analysis. Their study demonstrated high sensitivity, achieving a limit of detection of 200 ng ml^{-1} with a rapid response time of 10 min. The biosensor exploited the strong affinity between the SARS-CoV-2 spike (S) protein and the angiotensin-converting enzyme 2 (ACE2) receptor, enabling precise viral detection. Furthermore, the study highlighted its potential application in evaluating immune responses in vaccinated individuals, offering real-time antibody profiling capabilities. To enhance detection efficiency, the surface was functionalised using gold-binding polypeptides (GBPs), ensuring stable antigen immobilisation and improved sensitivity. These findings underscore the potential of SPR technology as a rapid and reliable alternative to conventional diagnostic methods, particularly in pandemic preparedness and clinical virology applications. As illustrated in Figure 3.6, the SPR system measures shifts in resonance angle caused by molecular interactions on the gold sensor surface, enabling real-time monitoring with exceptional accuracy.[48]

Viral diagnostics have increasingly relied on advanced biosensing technologies to improve detection speed and accuracy. In this regard, Anand *et al.* explored the application of SPR and LSPR-based biosensors for identifying influenza (H1N1, H3N2, H5N1, and H6N1), dengue (DENV), hepatitis B (HBV), hepatitis C (HCV), SARS-CoV-2, norovirus (NoV), and Ebola. These biosensors were engineered with specific surface modifications, including antibody immobilisation, aptamer coatings, gold nanoparticles (AuNPs), gold nanorods (AuNRs), and gold-binding polypeptides (GBPs), to enhance sensitivity and target specificity. The study reported limits of detection (LOD) for various viruses: influenza [0.128–1.28 HAU, 5.14×10^5 50% egg infectious dose (EID50) per 0.1 ml], dengue (0.0001–10 nM, 100% selectivity in human serum), HBV (5 copies per 25 µl, 1 pg ml^{-1}), SARS-CoV-2 (200 ng ml^{-1}, 0.22 pM, 0.18 ng μl^{-1}), Ebola (pM-level, 20-minute detection), and norovirus (high accuracy in serum samples). To further improve detection stability, modifications such as streptavidin–biotin binding and self-assembled monolayers (SAMs) were utilised. Additionally, microfluidic integration enabled real-time, portable, and cost-effective diagnostics, reinforcing these biosensors as essential tools for viral surveillance and clinical applications. As shown in Figure 3.7, plasmonic biosensors detect viral interactions by measuring resonance angle shifts, enabling real-time and precise diagnostics.[49]

SPR biosensors have shown high sensitivity in detecting viruses such as influenza and Ebola, beyond their established role in COVID-19 diagnostics. While challenges like sensor stability and high fabrication costs persist, advancements in alternative materials and artificial intelligence (AI)-driven

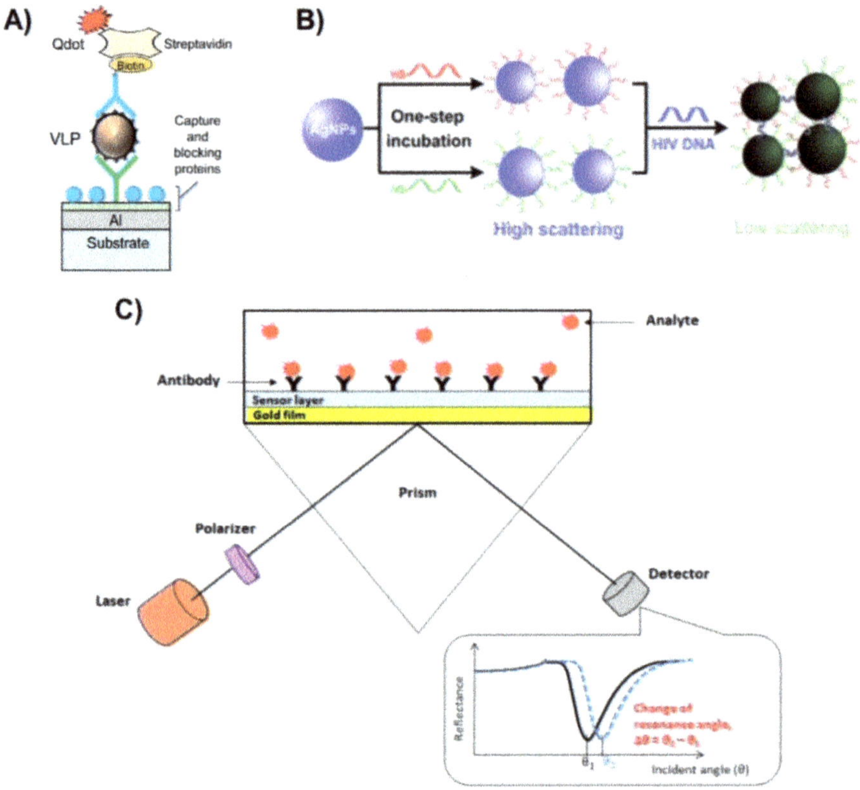

Figure 3.6 SPR biosensors and applications. (a) Sensor surface functionalisation for analyte capture. Abbreviations: quantum dot (Qdot); virus-like particle (VLP). (b) AgNP-based HIV DNA detection with scattering signal changes. (c) Prism-based SPR setup showing resonance angle shifts during biomolecular binding. Adapted from ref. 48, https://doi.org/10.1002/cbic.202000744, under the terms of the CC BY 4.0 license, https://creativecommons.org/licenses/by/4.0/.

analytics aim to improve performance. The integration of portable and smartphone-based SPR systems enhances real-time viral screening, positioning SPR as a critical tool for infectious disease surveillance and global health security.[49]

3.3.2.2 Fluorescence Sensors [Förster Resonance Energy Transfer (FRET)]

Fluorescence biosensors have emerged as powerful tools for highly sensitive, rapid, and real-time viral detection, providing optical signal generation through interactions between fluorescent molecules and viral biomarkers. These biosensors facilitate the detection of viral nucleic acids, proteins, or whole viral particles with minimal sample preparation, rendering them

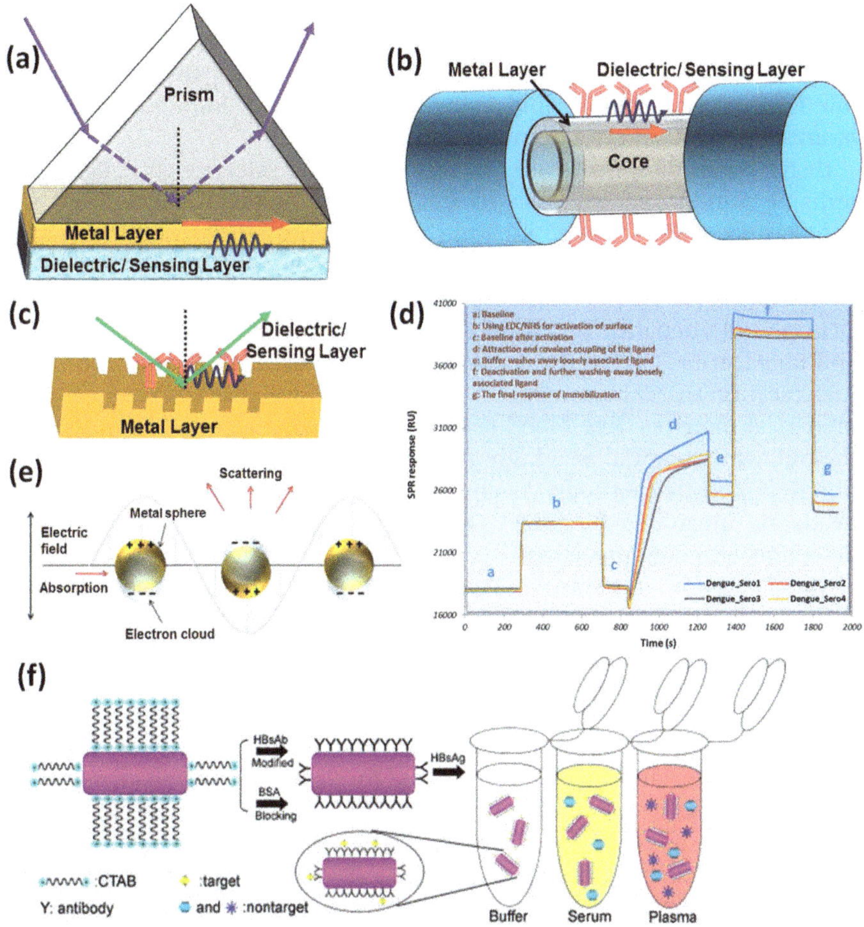

Figure 3.7 SPR biosensor configurations and applications. (a) Prism-based SPR, (b) waveguide-based SPR, (c) nanostructured SPR, (d) dynamic signal response during ligand immobilisation, [abbreviations: *N*-hydroxysuccinimide (NHS), *N*-ethyl-*N*-(dimethylaminopropyl) carbodiimide (EDC)] (e) nanomaterial-enhanced plasmonic effects, and (f) detection of viral antigens in biological samples (buffer, serum, plasma) [abbreviations: cetyltrimethylammonium bromide (CTAB), hepatitis B surface antibody (HBsAb), bovine serum albumin (BSA), hepatitis B surface antigen (HBsAg)]. Adapted from ref. 49, https://doi.org/10.1038/s42003-020-01615-8, under the terms of the CC BY 4.0 license, http://creativecommons.org/licenses/by/4.0/.

particularly suitable for point-of-care (PoC) diagnostics and epidemiological surveillance. Their operational principle is based on photon excitation and subsequent emission, wherein fluorophores absorb light at specific wavelengths and emit fluorescence upon returning to their ground state. Advanced fluorescence-based techniques, including Förster resonance energy transfer (FRET), fluorescence quenching, and single-molecule counting, significantly enhance detection specificity and sensitivity. A notable innovation in this field is the implementation of upconversion luminescence nanoparticles (UCNPs), which enable the conversion of near-infrared (NIR) light into visible fluorescence, thereby improving the detection of low-abundance viral biomarkers with superior accuracy and efficiency.[50]

UCNP-based biosensors offer ultrasensitive and selective virus detection, utilising anti-Stokes emission to minimise background interference and enhance accuracy. Ma *et al.* integrated UCNPs into FRET-based biosensing, achieving precise differentiation of viral RNA sequences. Their study reported exceptionally low LODs, including for human immunodeficiency virus (HIV) DNA (15 pM), hepatitis B virus (0.1 IU ml^{-1}), influenza A virus (60.9 pg ml^{-1}), and Ebola virus (300 fM). By converting near-infrared (NIR) light into visible fluorescence, UCNP biosensors enable real-time viral diagnostics in complex biological matrices, like blood serum and saliva. These findings highlight their clinical potential as a high-throughput, reliable alternative to conventional diagnostic methods. As shown in Figure 3.8, UCNPs and magnetic beads are integrated for target DNA capture and fluorescence-based viral detection with high precision.[51]

Metal–organic framework (MOF)-based fluorescence biosensors have emerged as powerful tools for virus detection due to their unique fluorescence quenching and recovery mechanisms. In this context, Wang *et al.* utilised MIL-101 MOFs to selectively bind fluorophore-labelled DNA probes *via* π–π stacking and hydrogen bonding. Upon interaction with complementary viral RNA or DNA, fluorescence was restored, allowing for real-time and highly sensitive detection. LOD of 160 pM for Ebola virus RNA and 73 pM for Sudan virus RNA were reported, demonstrating the high specificity and practicality of this method for selective virus diagnostics. Figure 3.9 illustrates how MOF-based biosensors enable fluorescence quenching and recovery through selective interactions with nucleic acids, demonstrating their efficiency in real-time viral detection.[52]

Fluorescence biosensors face challenges like photobleaching, autofluorescence, and reliance on complex optics. Solutions such as plasmon-enhanced fluorescence, AI-driven imaging, and clustered regularly interspaced short palindromic repeats (CRISPR)-based detection are improving sensitivity. Advances in microfluidics and self-amplifying probes aim to enable single-virus detection, strengthening their role in virology and public health.[52]

3.3.2.3 *Refractive Index Sensors*

Refractive index (RI) biosensors detect viral particles by measuring variations in the refractive index induced by biomolecular interactions at the sensor surface. When viral antigens, nucleic acids, or whole viral particles bind to a functionalised sensing layer, they alter the local refractive index, thereby modifying the propagation of incident light. This shift is quantitatively analysed through interferometric, waveguide-based, or reflectance imaging techniques, allowing for highly sensitive, real-time virus detection. Unlike

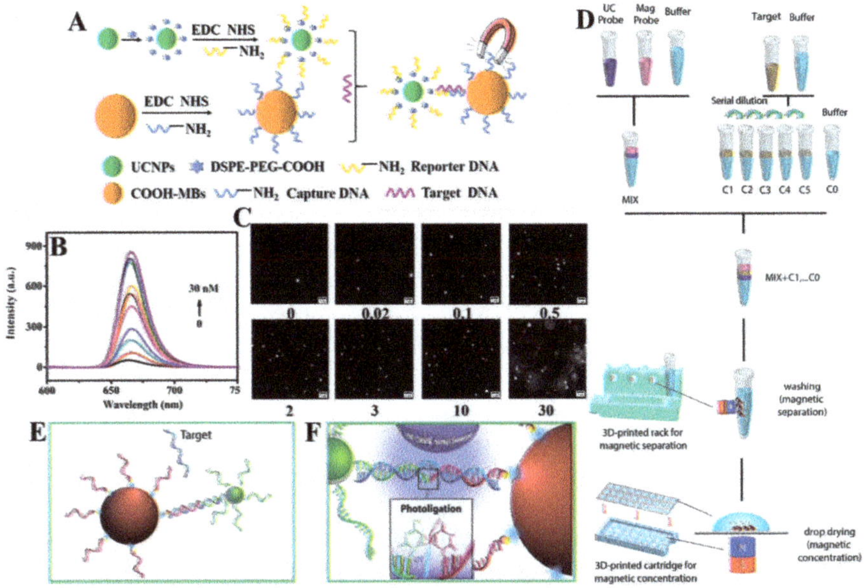

Figure 3.8 (a) Schematic representation of UCNP-based fluorescence biosensor platform, detailing target DNA binding, magnetic separation, and fluorescence signal generation. Abbreviations: Magnetic beads (MBs). (b) Fluorescence intensity profiles across varying DNA concentrations. (c) Fluorescence imaging for sensitivity analysis. (d) Workflow showcasing magnetic separation and 3D-printed cartridge integration for enhanced viral detection. (e) The target is captured with the UCNPs probe. (f) The relative photoligation reaction between ssDNA strands under UV irradiation. Adapted from ref. 51, https://doi.org/10.1002/EXP.20210216, under the terms of the CC BY 4.0 license, https://creativecommons.org/licenses/by/4.0/.

fluorescence-based methods, RI biosensors offer label-free detection, minimal sample preparation, and rapid response times, making them highly suitable for pandemic surveillance and clinical diagnostics.[53]

RI biosensors are increasingly employed for precise viral detection and antibody characterisation. Seymour *et al.* introduced the single-particle interferometric reflectance imaging sensor (SP-IRIS) as a high-throughput platform for monoclonal antibody screening and viral antigen analysis (Figure 3.10). Utilising recombinant vesicular stomatitis virus (rVSV) models expressing Ebola (EBOV), Sudan (SUDV), and Lassa (LASV) glycoproteins, the study demonstrated real-time virus–antibody interaction profiling. The SP-IRIS system enhanced nanoparticle contrast, facilitating highly sensitive viral detection and epitope mapping. The 8.9F monoclonal antibody exhibited the strongest affinity for LASV, while 13F6 and 6D8 antibodies demonstrated robust binding to EBOV glycoproteins. Furthermore, an antibody competition assay revealed synergistic interactions among ZMapp's therapeutic antibodies (13C6, 2G4, and 4G7), elucidating their enhanced efficacy against Ebola. The platform achieved a LOD of 10^4–5×10^4 plaque forming

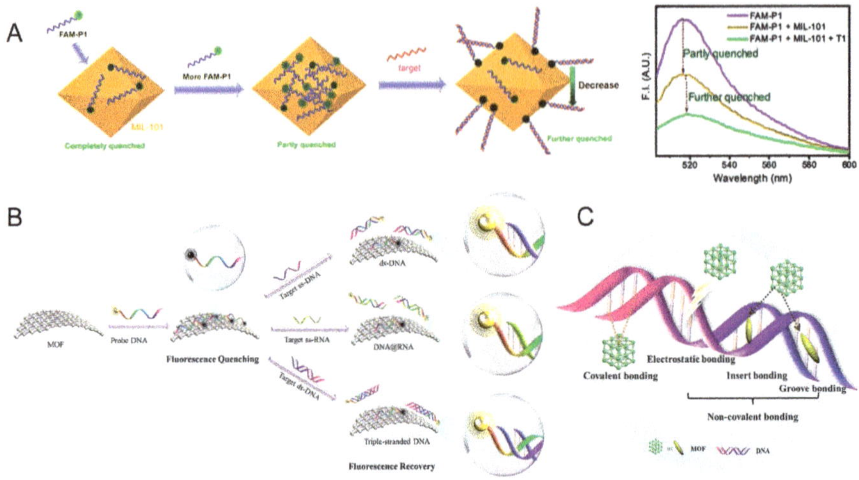

Figure 3.9 (a) The fluorescence quenching and recovery mechanism of MOF-based biosensors, where fluorophore-labelled probes (FAM-P1) bind to MIL-101 MOFs, quenching fluorescence, and recover upon hybridisation with target nucleic acids. (b) Different molecular interactions, including single-stranded DNA (ssDNA), single-stranded RNA (ssRNA), and double-stranded DNA (dsDNA), showcase the versatility of MOFs in detecting various nucleic acid structures. (c) Molecular binding mechanisms, including covalent, electrostatic, groove, and insertion bonding, highlight the adaptability of MOF-based biosensors in viral diagnostics. Adapted from ref. 52 with permission from Elsevier, Copyright 2020.

units (pfu) ml^{-1} for LASV, underscoring its potential for rapid diagnostics and therapeutic antibody development.[54]

Refractive index-based biosensing enables highly specific and real-time viral detection. Troyer *et al.* introduced SPIRFISH, combining SP-IRIS with single-molecule fluorescence *in situ* hybridisation (smFISH) for simultaneous protein and RNA analysis at the single-virion level. This approach differentiates HIV-1 virions from extracellular vesicles (EVs), which have a similar size and composition. SPIRFISH successfully detected HIV-1 envelope protein (gp120) and genomic RNA (gRNA), facilitating high-resolution characterisation of infectious particles. Additionally, it enabled co-localisation analysis of viral glycoproteins with host-derived markers such as cluster of differentiation 9 (CD9) and CD63, providing insights into viral assembly and antibody interactions. With applications in therapeutic antibody evaluation, vaccine development, and high-throughput screening, SPIRFISH offers exceptional specificity for distinguishing virions from host-derived vesicles as illustrated in Figure 3.11, SPIRFISH effectively differentiates HIV-1 virions from EVs by analysing surface markers such as CD9, CD63, and CD81, enabling precise characterisation of viral particles.[55]

RI biosensors, particularly SP-IRIS, facilitate monoclonal antibody screening and vaccine development by enabling precise analysis of virus–antibody

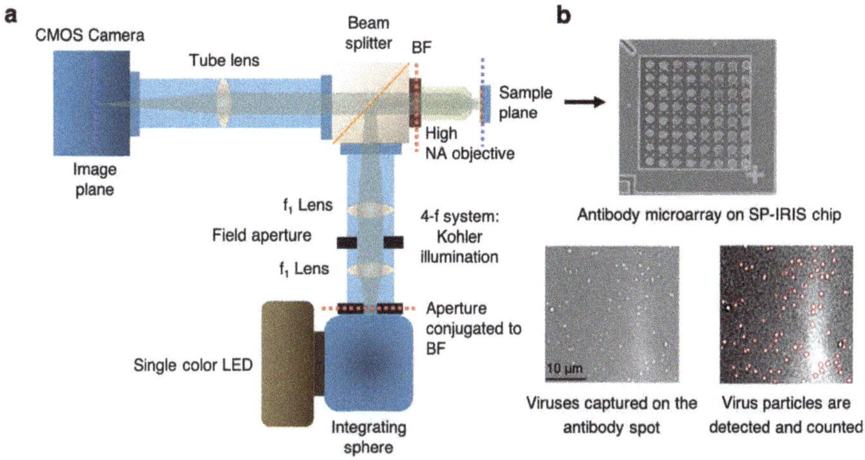

Figure 3.10 (a) Optical setup of the SP-IRIS system. The diagram demonstrates key components, including a complementary metal oxide semiconductor (CMOS) camera, beam splitter, high numerical aperture (NA) objective, and Kohler illumination system, designed to maximise the detection of virus particles (abbreviation: back focal plane, BF). The integrating sphere and LED source ensure uniform illumination, enhancing signal clarity. (b) Antibody microarray on the SP-IRIS chip used to capture virus particles. The captured particles are visualised and counted using the SP-IRIS system, demonstrating its capabilities for high-precision viral diagnostics. Adapted from ref. 54, https://doi.org/10.1038/s41598-022-27281-w, under the terms of the CC BY 4.0 license, http://creativecommons.org/licenses/by/4.0/.

interactions. Challenges such as non-specific binding and environmental instability are being mitigated through surface functionalisation, AI integration, and nanomaterial enhancements.[55]

3.3.2.4 Raman Sensors

Raman spectroscopy is an advanced optical technique that enables label-free, real-time detection of biological molecules, including viruses, by analysing inelastic light scattering to generate molecular fingerprints. Its non-invasive nature and minimal sample preparation make it a promising alternative to conventional diagnostic methods. However, traditional Raman spectroscopy has limitations due to weak signal intensity, reducing its effectiveness in detecting low-concentration biomolecules. To enhance sensitivity, surface-enhanced Raman spectroscopy (SERS) was developed, significantly amplifying Raman signals and enabling ultrasensitive virus detection at femtomolar or attomolar concentrations. SERS also allows for multiplexed virus detection, facilitating the simultaneous identification of multiple viral strains within a single sample. These advancements make

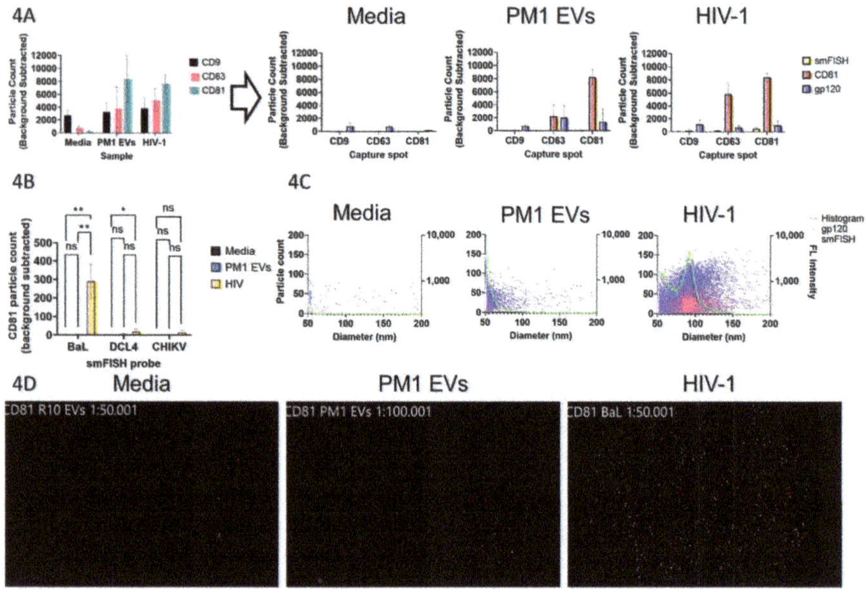

Figure 3.11 Analysis of media, PM1 extracellular vesicles (EVs), and HIV-1 particles using a biosensing platform. (a) Quantitative detection of surface proteins (CD9, CD63, and CD81) across different samples. (b) Co-localisation of surface proteins and gp120, illustrating the specific detection of HIV-1 particles. (c) Particle size and density comparison between media, PM1 EVs, and HIV-1. (d) High-resolution imaging of CD81-labeled particles in three sample groups, showcasing the system's capacity for precise virus–vesicle differentiation. Adapted from ref. 55 with permission from American Chemical Society, Copyright 2024.

Raman-based biosensors efficient and scalable tools for clinical diagnostics, virology research, and epidemiological surveillance.[56]

The integration of engineered plasmonic nanostructures with SERS-based biosensing has significantly enhanced the sensitivity and specificity of viral diagnostics. Lin *et al.* developed a SERS-based biosensing platform utilising precisely designed plasmonic nanostructures for highly selective and ultrasensitive viral detection, including detection of SARS-CoV-2. The sensor was functionalised with virus-specific ligand coatings to enhance binding specificity and integrated SiO_2 or polymer layers to improve stability and prevent nanoparticle aggregation. The study achieved a LOD in the femtomolar to attomolar range, demonstrating exceptional sensitivity. Additionally, the platform facilitated serological diagnostics by detecting host antibodies, providing critical insights for pandemic surveillance and immune response monitoring. Figure 3.12 illustrates a SERS-based biosensor for rapid viral detection, from capture to real-time classification.[58]

The integration of AI-driven machine learning algorithms with SERS-based biosensing has significantly enhanced the accuracy and efficiency of viral diagnostics. Yang *et al.* developed a SiO_2-coated silver nanorod SERS

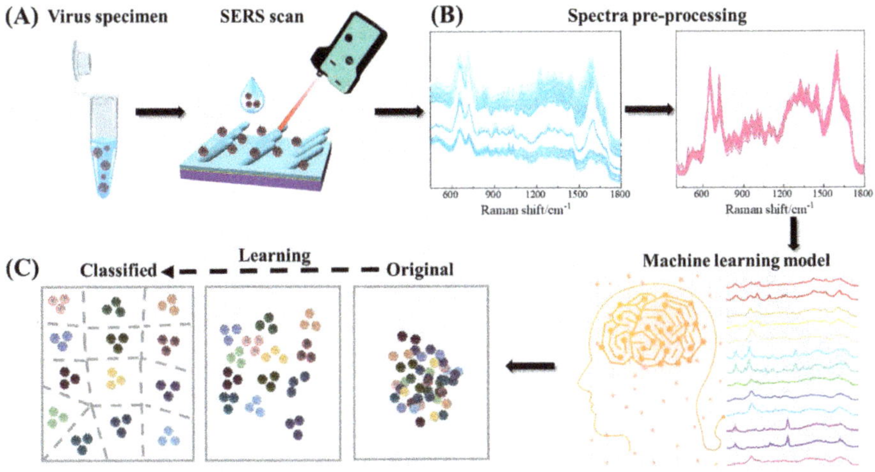

Figure 3.12 Schematic representation of a SERS-based biosensing platform for viral detection. (a) Viral lysis releases nucleic acids and proteins, which are selectively captured by functionalised nanoparticle-based aptasensors, enhancing Raman signal intensity through localised surface plasmon resonance. (b) A portable Raman spectrometer generates molecular fingerprints, enabling the rapid differentiation of virus-positive and virus-negative samples through unique spectral features. (c) The integration of a lateral flow-based Raman sensor demonstrates the platform's potential for real-time, point-of-care diagnostics with high sensitivity and specificity. Adapted from ref. 58, https://doi.org/10.1186/s12951-023-01890-7, under the terms of the CC BY 4.0 license, http://creativecommons.org/licenses/by/4.0/.

platform combined with support vector machines (SVM), K-nearest neighbour (KNN), and random forest algorithms, achieving over 99% classification accuracy across 13 respiratory viruses, including SARS-CoV-2, influenza, and RSV. Utilising convolutional neural networks (CNNs) to identify unique spectral signatures, the system enabled quantitative virus detection in buffer and saliva samples, with a LOD of 195 pfu ml^{-1} for H1N1 and adenovirus (Ad5) and 391 pfu ml^{-1} for other respiratory viruses. The AI-enhanced analysis streamlined diagnostics, delivering results within 30 minutes, demonstrating its potential for rapid, label-free virus identification in clinical and field settings. Figure 3.13 depicts the SERS-AI integration workflow, from Raman spectral acquisition to machine learning-based classification, enabling rapid and accurate virus detection in clinical settings.[58]

Raman and SERS-based biosensors face challenges in signal variability, background noise, and substrate reproducibility, limiting their clinical application. The lack of standardised detection protocols and scalability issues further hinder widespread adoption. Research efforts are focussed on AI-driven signal processing, hybrid biosensing strategies, and portable Raman spectrometers to enhance diagnostic accuracy. These advancements aim to improve reliability and applicability in clinical and field settings,

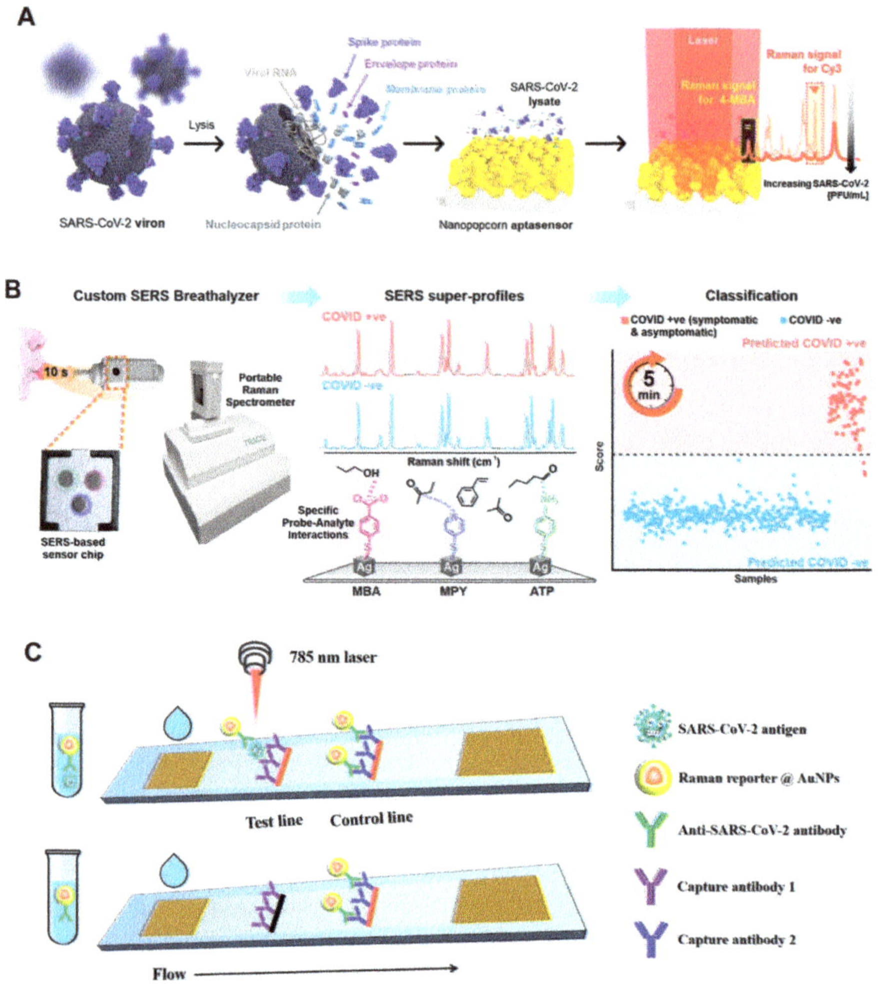

Figure 3.13 SERS-based biosensing platform integrated with machine learning for viral detection and classification. (a) Viral specimens are applied to SERS substrates, where Raman scans capture molecular signals. (b) Spectral data undergo pre-processing to extract distinct Raman features, which are then used to create molecular fingerprints. (c) A machine learning model processes these features for classification, distinguishing between different viral strains on the basis of spectral patterns. The workflow demonstrates the integration of advanced AI tools with Raman spectroscopy for real-time, high-accuracy viral diagnostics. Adapted from ref. 57 with permission from Elsevier, Copyright 2022.

positioning Raman-based biosensors as a valuable tool for precision diagnostics and epidemiological surveillance.[44–46] Table 3.3 summarises recent studies on sensors classified by detector types used in virus detection.

Table 3.3 Summary of studies on sensors classified by detector types used in virus detection.

Method/sensor type	Target virus	Key findings	Advantages	Limit of detection (LOD)	References
Reusable SPR with ferromagnetic particles	H1N1 Influenza	Developed a reusable SPR biosensor capable of multiple uses (seven times) with minimal signal degradation.	Cost-effective, reusable design for multiple detections.	0.22 pM	59
CRISPR–SPR fibre tip biosensor	Monkeypox	CRISPR–SPR system enables amplification-free, highly sensitive detection of Monkeypox virus.	Integration of CRISPR technology with SPR for enhanced sensitivity.	8.251 pM	60
Graphene-based SPR biosensor	Dengue	SPR sensor optimised for detecting Dengue E-proteins with low false-positive rates.	Enhanced detection sensitivity using graphene oxide.	0.08 pM	61
SPR-based biosensor using monoclonal antibodies	Ebola	Monoclonal antibody (mAb3) based SPR detection of Ebola virus with high affinity.	Biosensor application for biodefense and rapid viral detection.	0.5 pg ml^{-1}	62
Graphene-coated SPR biosensor	SARS-CoV-2	Optimised SPR sensor for detecting viral RNA, and immunoglobulins IgM, and IgG for COVID-19 diagnostics.	Graphene-enhanced SPR sensitivity for improved viral diagnostics.	183.33° per refractive index unit (RIU)	63
5G-enabled ultra-sensitive fluorescence sensor	SARS-CoV-2	Enabled real-time, ultra-sensitive viral detection through 5G connectivity for internet of medical things (IoMT) integration.	Internet of things (IoT)-compatible, long-distance usability, ultra-sensitive performance.	1.6 ng ml^{-1} (S-protein), 2.2 ng ml^{-1} (N-protein)	64
Paper-based fluorescent DNA sensor	Hepatitis C virus (HCV)	Developed a smartphone-compatible, portable DNA sensor for rapid viral RNA detection.	Low-cost, field-deployable, smartphone-integrated design.	5 pmol	65

(continued)

Table 3.3 (*Continued*)

Method/sensor type	Target virus	Key findings	Advantages	Limit of detection (LOD)	References
Non-autofluorescence aptamer sensor	H5N1 influenza virus	Designed an aptamer-based sensor with minimal autofluorescence for high specificity in serum samples.	High specificity, low background noise, effective in complex matrices.	0.0128 HAU ml^{-1}	66
CRISPR–CRISPR-associated 12a (Cas12a) fluorescence-based sensor	African swine fever virus (ASFV)	Achieved rapid viral detection with high specificity using CRISPR–Cas12a's fluorescence signal amplification.	Highly specific, rapid detection with stable fluorescence output.	100 fM	67
Fluorescent silver nanoclusters (AgNCs)	SARS-CoV-2, HIV, Influenza	DNA-templated AgNCs allow for tuneable fluorescence to detect viral DNA or RNA.	High photostability, biocompatibility, and cost-effectiveness; enables label-free, highly sensitive detection.	Down to 18 nM for norovirus RNA; 500 pM for H5N1 DNA	68
Optical interferometric detection biosensor	SARS-CoV-2	Anti-SARS-CoV-2 immunoglobulins detected in serum and saliva using recombinant spike proteins immobilised on photonic transducers.	Cost-effective, high-throughput screening, saliva compatibility, and ELISA correlation.	Not explicitly specified	69
Interferometric reflectance imaging sensor (IRIS)	General viruses (*e.g.*, DNA, protein)	Real-time, label-free detection of DNA–protein and protein–protein interactions with broad applicability in viral diagnostics.	Label-free, high-throughput, highly sensitive, and versatile for clinical use.	1 ng ml^{-1} (for DNA hybridisation)	70
Asymmetric Mach–Zehnder interferometric biosensor	SARS-CoV-2	Sensitive multiplex detection of anti-SARS-CoV-2 antibodies in human plasma using spike, receptor-binding domain (RBD), and nucleocapsid proteins as targets.	High sensitivity, cost-effective, multiplexing capability, miniaturisation for point-of-care testing.	0.3 IU ml^{-1} (anti-SARS-CoV-2 antibodies)	71

Michelson interferometric optical fibre biosensor	Canine distemper virus (CDV)	Achieved trace-level antigen detection using multilayer composite films for immobilisation of CDV antibodies.	High specificity, repeatability, cost-effective, and practical for clinical samples.	0.1236 pg ml^{-1}	72
Higher refractive index antibody-coated microbeads	HIV-1	Point-of-care HIV-1 viral load quantification using microfluidic-integrated antibody-coated microbeads and computational lensless imaging.	Low-cost, portable, field-deployable, eliminates need for PCR, highly specific virus quantification.	1569 virions per ml	73
SERS-based detection	Monkeypox Virus (MPXV)	Successfully detected MPXV genomic DNA and antigen proteins using SERS-enhanced nanoparticles.	Rapid, sensitive, and label-free detection for early-stage viral diagnosis.	0.45 fg ml^{-1} (genomic DNA)	74
Magnetic-responsive SERS	General viral nucleic acids	Developed a tuneable 'hot spot' platform to enhance Raman signals for nucleic acid detection.	Highly sensitive, adaptable for multiple virus strains, and minimises background noise.	1.2 aM (viral RNA)	75
Machine learning assisted SERS	Influenza A and B	Applied SERS combined with ML algorithms to accurately classify and detect influenza strains.	High specificity, automation capability, and rapid analysis in clinical samples.	7.8×10^{-15} M	76
CRISPR/Cas12a-SERS	Multiple viral genomes	Designed a CRISPR-Cas12a-mediated AuNP aggregation SERS platform for genetic virus detection.	Universal detection method applicable to different viral strains.	1.5 fM (viral genome)	77
SERS-based wastewater surveillance	SARS-CoV-2	Used SERS to detect SARS-CoV-2 in wastewater samples for epidemiological monitoring.	Onsite detection, real-time pandemic surveillance, and ultra-low detection limits.	2.3 copies per ml	78

3.4 Conclusion and Future Outlook

With the world's population expanding and aging, healthcare professionals around the globe are adopting advanced biomedical technologies to enhance the effectiveness of patient diagnosis, monitoring, and treatment. So, early detection of infectious viral diseases is a critical concern for public health, national security, and military forces. Recent viral outbreaks, such as H1N1 flu, H5N1 flu, and SARS, have heightened concerns that these viruses could spread quickly and potentially become a pandemic like the 1918 Spanish flu, which claimed over 50 million lives. An essential factor in preventing and managing future epidemics will be the creation of fast and highly sensitive diagnostic methods that can be rapidly implemented at various locations.[12]

Biosensors are used in a wide range of fields, including medical diagnostics, environmental monitoring, pharmaceutical testing, food safety, and defence and security applications. Their advancement integrates various fields, such as chemistry, physics, materials science, nanotechnology, and engineering. The importance of biosensors is steadily increasing as a key element in comprehensive health monitoring and assessment of solutions. Additionally, numerous emerging sensor applications are being developed to cover extensive areas while maintaining flexibility and comfort. Optical biosensors have brought about major progress in the detection of viruses and the enhancement of imaging technologies. Coupled with cutting-edge instrumentation, they have enhanced sensitivity and boosted the speed of sample testing. These methods have the potential to be transformed into point-of-care (POC) diagnostics for virus detection and offer promising alternatives for identifying COVID-19.[3]

Optical sensors detect biological, chemical, or physical changes by measuring parameters like wavelength, intensity, and phase of optical signals. In optical sensing, light is typically emitted from a source, interacts with physical objects, and the resulting light is then detected using a photodiode, demodulator, or other types of light detectors. Various optical sensing principles have been employed according to the application scenarios. This chapter began by examining the principles of optical sensing based on the types of carrier mediums, followed by a brief overview of the various detector types used and their different operating principles. The growing significance of optical sensors in clinical and biomedical applications for virus detection was also addressed. This emphasises how optical sensors employ spectroscopic techniques, such as fluorescence, refractive index, or Raman spectroscopy, to detect specific molecules and identify or quantify various virus types of interest.

This chapter has examined various topics related to sensing technologies, sensor systems, and their use in health and wellness monitoring. This chapter has also examined recent research and advancements in the application of various biosensor platforms for detecting viral infections. After providing an overview of viruses and the significance of virus detection in medical fields, the following sections cover the properties, design, characterisation, applications, and recent advancements in optical sensor technologies. The chapter also explores cutting-edge technological advancements that integrate platforms like microfluidic systems, lab-on-a-chip devices, and molecular imprinting techniques, while encompassing essential principles from chemistry, physics, biology, medicine, and nanotechnology. Each biosensor's application and potential for viral detection were discussed, with detailed explanations provided for their use in detecting viruses such as Coronavirus, HIV, Hepatitis, Ebola, Zika, Norovirus, Influenza, SARS, and more. This chapter is aimed at clinical and technical researchers, engineers, students, and the general public interested in gaining insight into the latest developments in sensor applications across the discussed fields. It is hoped that the reader will gain a comprehensive understanding of the critical technical and non-technical challenges that must be overcome to develop effective end-to-end sensor applications.

References

1. G. V. Angelov, D. P. Nikolakov, I. N. Ruskova, E. E. Gieva and M. L. Spasova, Healthcare sensing and monitoring, in *Enhanced Living Environments*, ed. I. Ganchev, N. Garcia, C. Dobre, C. Mavromoustakis and R. Goleva, 2019, vol. 11369, pp. 226–262.
2. J. Zhao, S. Zhang, Y. Sun, N. Zhou, H. Yu, H. Zhang and D. Jia, Wearable optical sensing in the medical internet of things (miot) for pervasive medicine: opportunities and challenges, *ACS Photonics*, 2022, **9**, 2579–2599.
3. H. Maddli, C. E. Miles, J. Kohm and D. M. O'Carroll, Optical biosensors for virus detection: prospects for SARS-CoV-2/COVID-19, *ChemBioChem*, 2021, **22**, 1176–1189.
4. A. Leal-Junior, J. S. Santos, L. Macedo, A. Marchesi, S. Morau, J. Valentino, F. Valentim and M. Costa, Wearable optical sensing in the medical internet of things (MIoT) for pervasive medicine: opportunities and challenges, *Sens. Diagn.*, 2024, **3**, 1135–1149.
5. F. Jin, X. He, J. Zhang, L. Chen, Y. Zhang, Q. Yang, Z. Liu, M. Li, Y. Ran and B.-O. Guan, A simple and integrated fiber-optic real-time qPCR platform for remote and distributed detection of epidemic virus infection, *Biosens. Bioelectron.*, 2025, **273**, 117145.
6. M. N. Amin, V. Ganjalizadeh, M. Hamblin, A. R. Hawkins and H. Schmidt, Free-space excitation of optofluidic devices for pattern-based single particle detection, *IEEE Photonics Technol. Lett.*, 2021, **33**, 884–887.
7. D. Ozcelik, A. Jain, A. Stambaugh, M. A. Stott, J. W. Parks, A. Hawkins and H. Schmidt, Scalable spatial-spectral multiplexing of single-virus detection using multimode interference waveguides, *Sci. Rep.*, 2017, **7**, 12199.
8. S. Kaaliveetil, J. Yang, S. Alssaidy, Z. Li, Y.-H. Cheng, N. H. Menon, C. Chande and S. Basuray, Microfluidic gas sensors: detection principle and applications, *Micromachines*, 2022, **13**, 1716.

9. R. W. Raut, H. S. Naik, P. M. Sah, P. Golińska and A. Gade, A comparative analysis of optical biosensors for rapid detection of SARS-CoV-2 and influenza, *BioNanoScience*, 2026, **16**, 109.
10. B. A. Zayed, A. N. Ali, A. A. Elgebaly, N. M. Talaia, M. Hamed and F. R. Mansour, Smartphone-based point-of-care testing of the SARS-CoV-2: a systematic review, *Sci. Afr.*, 2023, **21**, e01757.
11. D. K. Nirwati, A. K. Kusumaatmaja, M. S. Hakimi, N. S. Nur, T. J. S. Budi, Y. Mahendradhata, A. M. I. Saktiawati, H. S. Kusuma and K. Triyana, Fast and noninvasive electronic nose for sniffing out COVID-19 based on exhaled breath-print recognition, *Digit. Med.*, 2022, **5**, 115.
12. A. A. Yanik, M. Huang, O. Kamohara, A. Artar, T. W. Geisbert, J. H. Connor and H. Altug, An optofluidic nanoplasmonic biosensor for direct detection of live viruses from biological media, *Nano Lett.*, 2010, **10**, 4962–4969.
13. Y. K. Chan and M. U. Gack, Viral evasion of intracellular DNA and RNA sensing, *Nat. Rev. Microbiol.*, 2016, **14**, 360–373.
14. A. Afzal, A. Mujahid, R. Schirhagl, S. Z. Bajwa, U. Latif and S. Feroz, Gravimetric viral diagnostics: QCM based biosensors for early detection of viruses, *Chemosensors*, 2017, **5**, 7.
15. H. Maddali, C. E. Miles, J. Kohn and D. M. O'Carroll, Optical biosensors for virus detection: prospects for SARS-CoV-2/COVID-19, *ChemBioChem*, 2021, **22**, 1176–1189.
16. Y. Sun, Z. Li, P. Y. Liu, B. T. T. Nguyen, W. Wang, Q. Zhang, L. K. Chin, M. Wu, P. H. Yap, X. Zhang, H. Zhang, D. Yuan, D. P. Tsai and A. Q. Liu, On-chip optical detection of viruses: a review, *Adv. Photonics Res.*, 2021, **2**, 2000150.
17. C. L. Wong, M. Clyne, H. M. H. Chan, L. X. Chen, H. Q. Lam and M. Olivo, A phase-intensity surface plasmon resonance biosensor for avian influenza A (H5N1) detection, *Sensors*, 2017, **17**, 2363.
18. M. Khater, A. Escosura-Muñiz and A. Merkoçi, Biosensors for plant pathogen detection, *Biosens. Bioelectron.*, 2017, **93**, 72–86.
19. M. Elsherif, A. E. Salih, M. G. Muñoz, F. Alam, B. AlQattan, D. S. Antonysamy, M. F. Zayed, Y. Ali, S. Park, T. D. Wilkinson and H. Butt, Optical fiber sensors: working principle, applications, and limitations, *Adv. Photonics Res.*, 2022, **3**, 2100371.
20. D. Çimen, N. Bereli and A. Denizli, Advanced plasmonic nanosensors for monitoring of environmental pollutants, *Curr. Anal. Chem.*, 2023, **19**, 2–17.
21. J. Chen, Q. Liu and Z. He, Time-domain multiplexed high resolution fiber optics strain sensor system based on temporal response of fiber Fabry-Perot interferometers, *Opt. Express*, 2017, **25**, 21914–21925.
22. C. Pendão and I. Silva, Optical fiber sensors and sensing networks: overview of the main principles and applications, *Sensors*, 2022, **22**, 7554.
23. W. Qu, Y. Chen, C. Ma, D. Peng, X. Bai, J. Zhao, S. Liu and L. Luo, Application of optical fiber sensing technology and coating technology in blood component detection and monitoring, *Coatings*, 2024, **14**, 173.

24. R.-L. Xia, B. Liu, Y. Hu, J. Liu, Y. Fu, X.-D. He, P. Lu, G. Farrell, J. Yuan and Q. Wu, Rapid detection of SARS-CoV-2 nucleocapsid protein by a label-free biosensor based on optical fiber cylindrical micro-resonator, *IEEE Sens. J.*, 2023, **23**, 12511.
25. Y. Yang, R. Zhao, Y. Wang, D. Song, B. Jiang, X. Guo, W. Liu, F. Long, H. Song and R. Hao, Rapid and universal detection of SARS-CoV-2 and influenza A virus using a reusable dual-channel optic fiber immunosensor, *J. Med. Virol.*, 2022, **94**, 5325–5335.
26. S. A. H. Mohsan, M. A. Khan and H. Amjad, Hybrid FSO/RF networks: a review of practical constraints, applications and challenges, *Opt. Switch. Netw.*, 2023, **47**, 100697.
27. Z. Saima and K. Hira, *Wireless Pers. Commun.*, 2021, **121**, 429.
28. A. Jahid, M. H. Alsharif and T. J. Hall, A contemporary survey on free space optical communication: Potentials, technical challenges, recent advances and research direction, *J. Netw. Comput. Appl.*, 2022, **200**, 103311.
29. M. Tabib-Azar and M. A. H. Chowdhury, A 10 GHz metamaterial sensor to detect SARS COV-2 and dust particles in free space, *IEEE Sens. J.*, 2022, **22**, 12846–12851.
30. R. Bogue, Detecting gases with light: a review of optical gas sensor technologies, *Sens. Rev.*, 2015, **35**, 133–140.
31. L. Wang, J. Yang, S. He, H. Gong, C. Chen and C. Cai, A mild and safe gas-responsive molecularly imprinted sensor for highly specific recognition of hepatitis B virus, *Sens. Actuators, B*, 2022, **366**, 131990.
32. A. U. Alam, Y. Qin, S. Nambiar, J. T. W. Yeow, M. M. R. Howlader, N.-X. Hu and M. J. Deen, Polymers and organic materials-based pH sensors for healthcare applications, *Prog. Mater. Sci.*, 2018, **96**, 174–216.
33. Y.-H. Shin, T. M. Gutierrez-Wing and J.-W. Choi, Review—recent progress in portable fluorescence sensors, *J. Electrochem. Soc.*, 2021, **168**, 017502.
34. J. Perez-Vargas, F. Amirache, B. Boson, C. Mialon, N. Freitas, C. Sureau, F. Fusil and F.-L. Cosset, Enveloped viruses distinct from HBV induce dissemination of hepatitis D virus in vivo, *Nat. Commun.*, 2019, **10**, 2098.
35. L. Luo, F. Zhang, C. Chen and C. Cai, Visual simultaneous detection of hepatitis A and B viruses based on a multifunctional molecularly imprinted fluorescence sensor, *Anal. Chem.*, 2019, **91**, 15748–15756.
36. J. M. Flores, M. B. B. Maningas and F. B. Sevilla III, Micro-probe potentiometric ph sensor for detection of amplification in the LAMP assay for white spot syndrome virus (WSSV) in shrimps, *IEEE Sens. J.*, 2022, **22**, 9289–9296.
37. K. Seitkamal, B. Kenges, A. Tleuzhanova, M. Sypabekova, Y. Amanova, A. Amantayeva, A. Gonzalez, C. Caucheteur, D. Tosi and L. Vangelista, Proof of principle for a sensitive, real time and label-free detection of poxviruses using optical fiber biosensors, *Optik*, 2023, **288**, 171195.
38. M. U. Hadi and M. Khurshid, SARS-CoV-2 detection using optical fiber based sensor method, *Sensors*, 2022, **22**, 751.

39. S.-L. Lee, J. Kim, S. Choi, J. Han, G. Seo and Y. W. Lee, Fiber-optic label-free biosensor for SARS-CoV-2 spike protein detection using biofunctionalized long-period fiber grating, *Talanta*, 2021, **235**, 122801.
40. N. Cennamo, L. Pasquardini, F. Arcadio, L. Lunelli, L. Vanzetti, V. Carafa, L. Altucci and L. Zeni, SARS-CoV-2 spike protein detection through a plasmonic D-shaped plastic optical fiber aptasensor, *Talanta*, 2021, **233**, 122532.
41. W. Wang, L. Luo, Y. Li, B. Hong, Y. Ma, K. Kang and J. Wang, Detection of SARS-CoV-2 using machine learning-enabled paper-assisted ratiometric fluorescent sensors based on target-induced magnetic DNAzyme, *Biosens. Bioelectron.*, 2024, **255**, 116272.
42. Y. Pang, Z. Rong, J. Wang, R. Xiao and S. Wang, A fluorescent aptasensor for H5N1 influenza virus detection based-on the core-shell nanoparticles metal-enhanced fluorescence (MEF), *Biosens. Bioelectron.*, 2015, **66**, 527–532.
43. H. Gong, G. Cai, Y. Li, N. Jiang, C. Chen, F. Chen and C. Cai, Portable dual-function ratio-type triple-emission molecularly imprinted fluorescence sensor for the simultaneous visual detection of hepatitis A and B viruses, *Anal. Chim. Acta*, 2025, **1336**, 343451.
44. Y. Li, M. Hong, B. Qiu, Z. Lin, Y. Chen, Z. Cai and G. Chen, Highly sensitive fluorescent immunosensor for detection of influenza virus based on Ag autocatalysis, *Biosens. Bioelectron.*, 2014, **54**, 358–364.
45. D. Çimen, S. Ünal and A. Denizli, Nanoparticle-assisted plasmonic sensors: recent developments in clinical applications, *Anal. Biochem.*, 2025, **698**, 115753.
46. S. Akgönüllü, H. Yavuz and A. Denizli, SPR nanosensor based on molecularly imprinted polymer film with gold nanoparticles for sensitive detection of aflatoxin B1, *Talanta*, 2020, **219**, 121219.
47. D. D. Türkmen, M. Bakhshpour, I. Göktürk, S. Aşır, F. Yılmaz and A. Denizli, Selective dopamine detection by SPR sensor signal amplification using gold nanoparticles, *New J. Chem.*, 2021, **45**, 18296–18306.
48. H. Maddali, C. E. Miles, J. Kohn and D. M. O'Carroll, Optical biosensors for virus detection: prospects for SARS-CoV-2/COVID-19, *ChemBioChem*, 2021, **22**, 1176–1189.
49. A. M. Shrivastav, U. Cvelbar and I. Abdulhalim, A comprehensive review on plasmonic-based biosensors used in viral diagnostics, *Commun. Biol.*, 2021, **4**, 70.
50. F. Ma, Y. Li, B. Tang and C.-Y. Zhang, Fluorescent biosensors based on single-molecule counting, *Acc. Chem. Res.*, 2016, **49**, 1722–1730.
51. Y. Ma, M. Song, L. Li, X. Lao, M.-C. Wong and J. Hao, Advances in upconversion luminescence nanomaterial-based biosensor for virus diagnosis, *Exploration*, 2022, **2**, 20210216.
52. Y. Wang, Y. He, Q. Hou, J. Yuan, H. Xie, N. Wang, S. Cheng, D. Peng, Y. Liu and Z. Li, A label-free and sensitive optical fiber biosensor based on an optofluidic chip for the detection of the SARS-CoV-2 spike protein, *Biosens. Bioelectron.*, 2020, **169**, 112604.

53. A. Panda and P. P. Devi, Sensitivity enhancement of a surface plasmon resonance based optical fiber biosensor using blue phosphorene and MoS_2: a theoretical study, *Opt. Fiber Technol.*, 2020, **54**, 102123.
54. E. Seymour, F. E. Kanik, S. D. Gür, M. Bakhshpour-Yucel, A. Araz, N. L. Ünlü and M. S. Ünlü, Solid-phase optical sensing techniques for sensitive virus detection, *Sensors*, 2023, **23**, 5018.
55. Z. Troyer, O. Gololobova, A. Koppula, Z. Liao, F. Horns, M. B. Elowitz, J. Pablo Tosar, M. Batish and K. W. Witwer, Simultaneous protein and RNA analysis in single extracellular vesicles, including viruses, *ACS Nano*, 2024, **18**(39), 26568–26584.
56. K. V. Serebrennikova, A. N. Berlina, D. V. Sotnikov, A. V. Zherdev and B. B. Dzantiev, Raman scattering-based biosensing: new prospects and opportunities, *Biosensors*, 2021, **11**, 512.
57. Y. Yang, B. Xu, J. Murray, J. Haverstick, X. Chen, R. A. Tripp and Y. Zhao, Rapid and quantitative detection of respiratory viruses using surface-enhanced Raman spectroscopy and machine learning, *Biosens. Bioelectron.*, 2022, **217**, 114721.
58. C. Lin, Y. Li, Y. Peng, S. Zhao, M. Xu, L. Zhang, Z. Huang, J. Shi and Y. Yang, Recent development of surface-enhanced Raman scattering for biosensing, *J. Nanobiotechnol.*, 2023, **21**, 149.
59. H. Yoo, J. Shin, J. Sim, H. Cho and S. Hong, Reusable surface plasmon resonance biosensor chip for the detection of H1N1 influenza virus, *Biosens. Bioelectron.*, 2020, **168**, 112561.
60. Y. Chen, Z. Chen, T. Li, M. Qiu, J. Zhang, Y. Wang, W. Yuan, A. Ho-Pui Ho, O. Al-Hartomy, S. Wageh, A. G. Al-Sehemi, X. Shi, J. Li, Z. Xie, L. Xuejin and H. Zhang, Ultrasensitive and specific clustered regularly interspaced short palindromic repeats empowered a plasmonic fiber tip system for amplification-free Monkeypox virus detection and genotyping, *ACS Nano*, 2023, **17**, 12903.
61. N. A. S. Omar, Y. W. Fen, J. Abdullah, A. R. Sadrolhosseini, Y. M. Kamil, N. M. Fauzi, H. S. Hashim and M. A. Mahd, Quantitative and selective surface plasmon resonance response based on a reduced graphene oxide–polyamidoamine nanocomposite for detection of dengue virus E-proteins, *Nanomaterials*, 2020, **10**, 569.
62. P. K. Sharma, J. S. Kumar, V. V. Singh, U. Biswas, S. S. Sarkar, S. I. Alam, P. K. Dash, M. Boopathi, K. Ganesan and R. Jain, Surface plasmon resonance sensing of Ebola virus: a biological threat, *Anal. Bioanal. Chem.*, 2020, **412**, 4101–4112.
63. T. B. A. Akib, S. F. Mou, M. M. Rahman, M. M. Rana, M. R. Islam, I. M. Mehedi, M. A. P. Mahmud and A. Z. Kouzan, Design and numerical analysis of a graphene-coated SPR biosensor for rapid detection of the novel coronavirus, *Sensors*, 2021, **21**, 3491.
64. J. Guo, S. Chen, S. Tian, K. Liu, J. Ni, M. Zhao, Y. Kang, X. Ma and J. Guo, 5G-enabled ultra-sensitive fluorescence sensor for proactive prognosis of COVID-19, *Biosens. Bioelectron.*, 2021, **181**, 113160.

65. P. Teengam, N. Nisab, N. Chuaypen, P. Tangkijvanich, T. Vilavaivan and O. Chailapakul, Fluorescent paper-based DNA sensor using pyrrolidinyl peptide nucleic acids for hepatitis C virus detection, *Biosens. Bioelectron.*, 2021, **189**, 113381.
66. S. Chen, G. Cai, X. Gong, L. Wang, C. Cai and H. Gong, Non-autofluorescence detection of H5N1 virus using photochemical aptamer sensors based on persistent luminescent nanoparticles, *ACS Appl. Mater. Interfaces*, 2022, **14**, 46964–46971.
67. Q. He, D. Yu, M. Bao, G. Korensky, J. Chen, M. Shin, J. Kim, M. Park, P. Qin and K. Du, High-throughput and all-solution phase African swine fever virus (ASFV) detection using CRISPR-Cas12a and fluorescence based point-of-care system, *Biosens. Bioelectron.*, 2020, **154**, 112068.
68. D. Li, H. Chen, X. Gao, X. Mei and L. Yang, Development of general methods for detection of virus by engineering fluorescent silver nanoclusters, *ACS Sensors*, 2021, **6**, 613–627.
69. A. M. M. Murillo, J. Tomé-Amat, Y. Ramírez, M. Garrido-Arandia, L. G. Valle, G. Hernández-Ramírez, L. Tramarin, P. Herreros, B. Santamaría, A. Diaz-Perales and M. Holgado, Developing an optical interferometric detection method based biosensor for detecting specific SARS-CoV-2 immunoglobulins in Serum and Saliva, and their corresponding ELISA correlation, *Sens. Actuat. B-Chem.*, 2021, **345**, 130394.
70. M. Bakhshpour-Yucel, N. Lortlar Unlu, E. Seymour and A. Denizli, A comprehensive review of interferometric reflectance imaging sensor as a sensitive detection platform and its application areas, *Biosens. Bioelectron. X*, 2025, **22**, 100574.
71. G. Besselink, A. Schütz-Trilling, J. Veerbeek, M. Verbruggen, A. van der Meer, R. Schonenberg, H. Dam, K. Evers, E. Lindhout, A. Garritsen, A. van Amerongen, W. Knoben and L. Scheres, 2022, Asymmetric Mach–Zehnder interferometric biosensing for quantitative and sensitive multiplex detection of anti-SARS-CoV-2 antibodies in human plasma, *Biosensors*, 2022, **12**, 553.
72. X. Zhang, X. Hou and W. Feng, Trace detection of canine distemper virus based on Michelson-interferometer sensing probe, *J. Biophotonics*, 2023, **17**, e202300329.
73. M. Sher, B. Coleman, M. Caputi and W. Asgha, Development of a point-of-care assay for HIV-1 viral load using higher refractive index antibody-coated microbeads, *Sensors*, 2021, **21**, 1819.
74. D. Zhang, X. Zhang, R. Ma, S. Deng, X. Wang, X. Wang, X. Zhang, X. Huang, Y. Liu, G. Li, J. Qu, Y. Zhu and J. Li, Ultra-fast and onsite interrogation of severe acute respiratory syndrome coronavirus 2 (SARS-CoV-2) in waters via surface enhanced Raman scattering (SERS), *Water Res.*, 2021, **200**, 117243.
75. A. Su, Y. Liu, X. Cao, W. Xu, C. Liang and S. Xu, A universal CRISPR/Cas12a-mediated AuNPs aggregation-based surface-enhanced Raman scattering (CRISPR/Cas-SERS) platform for virus gene detection, *Sens. Actuators, B*, 2022, **369**, 132295.

76. A. Tabarov, V. Vitkin, O. Andreeva, A. Shemanaeva, E. Popov, A. Dobroslavin, V. Kurikova, O. Kuznetsova, K. Grigorenko, I. Tzibizov, A. Kovalev, V. Savchenko, A. Zheltuhina, A. Gorshkov and D. Danilenko, Detection of A and B influenza viruses by surface-enhanced raman scattering spectroscopy and machine learning, *Biosensors*, 2022, **12**, 1065.
77. B. Yin, W. K. Hei Ho, Q. Zhang, C. Li, Y. Huang, J. Yan, H. Yang, J. Hao, S. H. Dexter Wong and M. Yang, Magnetic-responsive surface-enhanced Raman scattering platform with tunable hot spot for ultrasensitive virus nucleic acid detection, *ACS Appl. Mater. Interfaces*, 2022, **14**, 4714–4724.
78. Z. Zhang, H. Jiang, S. Jiang, T. Dong, X. Wang, Y. Wang and Y. Li, Rapid detection of the Monkeypox virus genome and antigen proteins based on surface-enhanced Raman spectroscopy, *ACS Appl. Mater. Interfaces*, 2023, **15**, 34419–34426.

CHAPTER 4

Electrochemical Sensors

ÇAĞAN TOMUR[a], VILDAN SANKO[a,b] AND FİLİZ KURALAY[a,*]

[a]Department of Chemistry, Faculty of Science, Hacettepe University, 06800 Ankara, Turkey; [b]Middle East Technical University, Micro-Electro-Mechanical Systems (MEMS) Center, 06530 Ankara, Turkey
*E-mail: filizkur@hacettepe.edu.tr, kuralay.filiz@gmail.com

4.1 Introduction

Infections by viruses have caused many people's deaths owing to the fact that they can easily reproduce themselves in the human body's cells. These small pathogens are nano-sized intracellular parasites that are dependent on the cell for their replication. Viruses not only affect human beings but also affect many living organisms since they need a host cell for multiplication. The mechanisms by which the viruses affect the organism are as follows. First, they enter the living body and find a region that they can infect, then they select the cells in that region as host cells and infect them, and realize their proliferation within the host cell.[1,2] There have been many classifications made of viruses to date. The Baltimore classification system is one of the most common classifications used to better understand viruses. According to the Baltimore classification system, viruses are divided into seven different groups based on their viral genome: double stranded DNA (dsDNA), single stranded DNA (ssDNA), double stranded RNA (dsRNA), positive strand (+) RNA, negative strand (−) RNA, reverse RNA, and dsDNA retroviruses. The classification, which was developed for better identification by using evolving

Detection Science Series No. 32
Advances in Virus Detection with Sensors
Edited by Adil Denizli
© The Royal Society of Chemistry 2026
Published by the Royal Society of Chemistry, www.rsc.org

new technologies, can now be used as a reference and is very important for the diagnosis of viral diseases.[3,4] When this story began, there was little information about viruses. Rapid development of biotechnology has made these nano-sized pathogens more understandable and visible.

Viruses are dangerous organisms because of their ability to easily transmit from one living organism to another and infect their host cells. When we look at the impact of viruses, which can use most living organisms in the ecosystem as hosts, on the human body and where they are hosted, it has been observed that they infect different parts of the body, including mainly the respiratory tract, the intestinal system, the skin, the urogenital system, and even the intrauterine environment, or spread out to the blood vessels and nerves.[5,6] For instance, some papillomaviruses are poxviruses that cause skin lesions because they infect the skin. Paramyxoviruses and influenza viruses cause respiratory problems because they infect the respiratory system. Human papillomavirus (HPV) and human immunodeficiency virus (HIV) infect the genital system and cause many types of cancer, such as cervical cancer and penile cancer.[5,7] Looking at viruses in general, it has been observed that some viruses have an incredibly fast ability to mutate and reproduce. For example, RNA viruses are classified as much more dangerous because they reproduce faster than DNA viruses and their mutation cycles are faster. In addition, the degree of danger is commonly related to the adaptation of the virus to environmental conditions and reproduction *via* host cell infection.[8] Some viruses can renew themselves every year, and these viruses are considered particularly dangerous. For instance, many people get infected by the influenza virus every year because the influenza virus has the ability to change very quickly. For this reason, it may become critical to get the influenza virus vaccine every year. At this time, some methods (drugs, rapid diagnostic kits, *etc.*) that can be used to protect people from viruses are being developed with rapid emerging technology. Rapid diagnosis is particularly important for people infected with the virus, as it makes treatment more effective. Considering this, new systems have to be developed to overcome the disadvantages of traditional methods.[4,9,10] Western blot, the polymerase chain reaction (PCR) technique, and enzyme-linked immunosorbent assay (ELISA) are among the technologies that have been used for a long time and are still the most widely used today. In general, these techniques have some disadvantages, such as their high-cost, long operational time, and need for trained staff. However, in real world conditions, it is better to detect viruses with rapid, sensitive, low-cost, and powerful analytical devices.[11] In this case, electrochemistry, a popular field, may have the power to create a change in our lives.

Electrochemical sensors, one of the subjects studied in the field of electrochemistry, are among the most important systems used in the determination of viruses, being fast, simple, portable, precise, accurate, and specific determination systems.[4,5] Specifically, electrochemical biosensor systems have played major role in virus detection. These biosensors have three major components: the biorecognition element or receptor, the transducer, and the signal processing system or signal output. In the case of detection, the biorecognition element is usually first placed on the surface and interacts directly with the analyte to be determined. The second part of the electrochemical sensor is a transducer. A transducer is responsible for converting this biorecognition event into a signal that can be measured. The last part of the electrochemical sensor is signal output, and this output system generates electroanalytical usable data.[5,12] It is well known that with the addition of nanomaterials or micromaterials amplification of the electrode response can be achieved.

Electrochemical approaches to developing virus detection systems are very promising for real-world applications. These detection systems, such as point-of-care diagnostics and wearable sensors, provide helpful insights in terms of bedside testing and near-patient testing. In this chapter, we highlight the power of electrochemical virus detection systems and their potential to contribute to healthcare applications. To reveal these situations, types of electrochemical techniques used for virus detection, virus biomarkers and their use in electrochemical detection, nanomaterials and micromaterials utilized for electrochemical virus detection and their functionalization strategies, design of electrochemical virus sensors and the applications of these sensors are discussed. Finally, the chapter ends with conclusions and discussion of the future prospects of these sensors. The studies in diagnosing various viral infections, such as respiratory viruses and several other clinically significant virus groups, are taken into account and these discussed studies reveal the great potential of virus sensors based on electrochemistry.

4.2 Fundamentals of Electrochemical Sensors

Chemical sensor technologies, one of the greatest benefits of growing technology, are among the important fields that continue to develop day by day. In general, a chemical sensor is an analytical device that measures chemical signals when an interaction occurs between the sensing material and the analyte of interest. The trend of the resulting signal depends directly or indirectly on the changing concentration of the analyte. Among different types of sensors, such as optical, thermal, and piezoelectric, electrochemical sensors are one of the most frequently used. This sensor type allows accurate, rapid, effective, sensitive, and selective analyte detection with superior characteristics.[13,14] Electrochemical sensors are used for the detection of specific chemical constituents through electrochemical reactions by using electrochemical techniques. Thus, they are mainly related to redox reactions between the electrode materials used and the target analyte.[15,16] In these sensors, attempts are made to improve electron transfer capability by modifying the electrode surface with conductive, functional, and chemically stable materials, such as nanoparticles or microparticles, including carbon nanotubes, graphene, quantum dots, metal or metal oxide nanoparticles, and polymeric structures. These modifications not only improve the electroactive surface area of the working electrode but also can provide a good immobilization platform for biomolecules in the case of biosensing processes.[15,17]

If the first electrochemical sensor technology is compared with today's sensors, it is seen that there has been a great deal of development. This field, which began with an oxygen electrode, followed by an enzyme electrode for glucose detection, has evolved to include some features such as flexible, wearable and mobile sensors and providing point-of-care diagnostics.[18–20] Additionally, electrochemical sensors have many advantages over the traditional methods. For example, if the traditional methods in the field of chemistry such as mass spectrometry (MS), gas chromatography (GC), nuclear magnetic resonance (NMR), spectroscopy *etc*., are compared with electrochemical techniques, it is seen that these conventional methods are high-cost, time consuming, complex to work, and need pretreatment processes and trained personnel.[16]

In electrochemical sensing methodology, to facilitate a specific electrochemical detection mechanism, three fundamental components are required in terms of electrodes. The working electrode is the primary component, responsible for measuring the analyte in the medium or sample matrix. The reference electrode is an electrode with a known electrode potential, which helps to provide a constant potential. The counter or auxiliary electrode, the third component, is a chemically stable electrode that facilitates the conduction of current to the working electrode in solution. As a result, three-electrode (working, counter and reference electrodes) or two-electrode (working and reference electrodes) setups are utilized in electrochemical sensing based on the preferred technique.[21,22] In some cases, it is known that a biological component is integrated into the sensor system in order to maintain biorecognition of the analyte by the component in order to achieve more specific detection. These biological components, such as nucleic acids, proteins, enzymes, antibodies, peptides, cells, and tissues, determine the specificity, selectivity, and stability of the sensors. The areas of use of electrochemical sensors are quite wide. The following fields can be given as examples for these areas: biomedicine, food monitoring, agricultural, environmental, and industrial.[23–25]

Looking particularly at the electrochemical sensors used for detecting viruses, it is clearly seen that the detection mechanisms supported by electrochemical techniques provide great benefits.[26] Within the field of electrochemistry there are many techniques that can ensure a good understanding of electrochemical responses. In this context, cyclic voltammetry (CV), differential pulse voltammetry (DPV), square wave voltammetry (SWV), amperometry (AMP), and electrochemical impedance spectroscopy (EIS) are commonly used.[4,27,28] One of the common techniques in electrochemistry that is used in electrochemical biosensors for virus determination is the CV technique. The CV technique is used to understand the electron transfer mechanisms by measuring the current that is obtained during the cycling of the potential of a working electrode between two potential values. This technique is mostly used for the characterization of modified electrodes at a fixed concentration of the analyte not for a precise detection in virus-related sensor studies. Another technique, DPV, is one of the most useful techniques with high precision for the detection of viruses. This technique can be used to understand the relationship between peak currents obtained with different target analyte concentrations by applying a series of regular voltage pulses.[29] One of the effective techniques for the detection of viruses at low concentrations is SWV, which depends on square wave and staircase potential applied to the working electrode.[21,30] AMP is also utilized to detect viruses electrochemically, and this technique depends on the subsequent measurement of the steady-state current when applying constant potential to the working electrode.[31] EIS is one of the popular techniques of recent times and helps in understanding the interfacial properties related to recognition of the analyte at the working electrode.[32] These techniques are the most commonly used electrochemical techniques for virus detection and they can be used together with other electrochemical techniques.

4.3 Viral Biomarkers and Their Detection

Since the corona virus disease (COVID)-19 pandemic emerged in 2019, rapid virus detection has become more critical than ever in protecting global health.[33] This chapter explores the advancements in electrochemical sensor technologies and their application in diagnosing various viral infections, such as respiratory viruses and several other clinically significant virus groups. These groups are divided into six categories: Respiratory viruses, gastrointestinal viruses, hepatitis viruses, vector-borne viruses, blood-borne viruses, and nervous system viruses.

1. Respiratory viruses: viruses like influenza and coronaviruses can lead to severe respiratory illnesses, especially in young children, older adults, and individuals with weakened immune systems. Rapid detection is essential for preventing outbreaks and managing acute respiratory infections.[34,35]
2. Gastrointestinal viruses: viruses like norovirus, rotavirus, and adenovirus lead to severe gastroenteritis and cause infections that mainly affect the stomach and intestines, causing nausea, vomiting, diarrhoea, abdominal pain, and fever. Detecting them is essential for preventing outbreaks and managing acute diarrheal illnesses.[36]
3. Hepatitis viruses: these viruses, which include hepatitis A, B, C, D, and E, primarily affect the liver and lead to significant morbidity and mortality worldwide.[37] Rapid and accurate detections through advanced sensor platforms could improve early intervention strategies.
4. Vector-borne viruses: Viruses transmitted by insect vectors, such as dengue, Zika, chikungunya virus (CHIKV), and West Nile virus, remain a significant challenge to public health, particularly in tropical and subtropical regions.[38] In these circumstances, electrochemical sensors present the potential for prompt detection in areas prone to outbreaks.

5. Blood-borne viruses: HIV and Ebola are transmitted through blood or bodily fluids. Their high fatality rates and complex transmission dynamics require the creation of reliable, field-deployable diagnostic tools.[39]
6. Nervous system viruses: pathogens such as herpes simplex viruses, rabies virus, and poliovirus can affect the central nervous system, leading to severe neurological complications.[40] The sensitive detection of these viruses is paramount for early therapeutic intervention.

Common viral biomarkers are the primary indicators utilized in diagnosing viral infections. Rapid and accurate detection of these biomarkers is essential for early diagnosis, effective monitoring, and timely intervention. This emphasizes the capabilities of advanced electrochemical sensor platforms in virus detection.

Identification of specific biomarkers is valuable for diagnosing viral diseases and their complications. When a patient presents with an infection, one of the first steps is to determine its origin to enable appropriate treatment; it is essential to differentiate between viral infections.[8] Viral biomarkers are biological indicators that provide critical information about viral infections' presence, progression, and severity. These biomarkers are usually viral nucleic acids (DNA or RNA), viral proteins, and host response markers of antibodies.[41] For this purpose, forming a covalent bond between gold and the thiol group is a widely used and effective method of immobilizing nucleic acid-based probes.[42] One of the most commonly used viral biomarkers is the hepatitis B surface antigen (HBsAg), essential for diagnosing hepatitis B (HBV) infection.[43] Also, in the case of HIV, the p24 antigen and HIV RNA act as early diagnostic indicators before antibody production occurs.[44,45] In another technique, severe acute respiratory syndrome corona virus 2 (SARS-CoV-2) RNA viral load measurement has become widely accepted for diagnosing and monitoring COVID-19.[46] Concurrently, spike (S) and nucleocapsid (N) proteins and specific immunoglobulin IgG are vital for understanding immune responses.[47] In addition to direct viral components, host response biomarkers provide valuable insights. C-reactive protein (CRP) and interleukin-6 (IL-6) are often elevated in severe viral infections, such as influenza and COVID-19, indicating systemic inflammation.[48,49] Accurately and rapidly detecting viral biomarkers is essential for timely medical intervention and effective disease control. While traditional diagnostic methods, such as PCR and ELISA, are commonly employed, they require specialized laboratories and prolonged processing times as mentioned above.[50,51] Conversely, electrochemical sensors have surfaced as promising alternatives due to their high sensitivity, quick response times, and potential for point-of-care applications.[52] By integrating various materials, particularly nanomaterials, with advanced surface modifications, electrochemical biosensors can enhance the detection of viral biomarkers and act as valuable tools for real-time disease monitoring and epidemic prevention.[11]

However, identifying specific viral biomarkers presents several challenges that complicate the development of reliable diagnostic assays. The high mutation rates of viruses can lead to significant genetic variability, making it difficult to design tests that accurately detect all viral strains. Furthermore, the levels of these biomarkers may be very low during the early stages of infection, requiring highly sensitive detection methods. The complexity of biological samples also adds to the challenge, as interfering substances can obscure the signals of the target biomarkers. Additionally, the expression of particular biomarkers is temporary, occurring only during specific phases of the viral lifecycle. Variations in individual immune responses further contribute to inconsistencies in biomarker levels. Addressing these challenges is crucial for advancing robust electrochemical sensor technologies in virus detection.[53,54]

4.4 Material Innovations in Electrochemical Virus Sensors

Improving the characteristics of sensors is an important step for the enhancement of their analytical performance. In this sense, certain actions need to be taken, such as increasing the electron transfer on the electrode surface *via* material integration, immobilizing biomolecules for specific and/or selective recognition, and functionalization of the sensing materials. Therefore, it is necessary to have an understanding of the types of materials used for the effectiveness and improvement of diagnosis. Nanomaterials or micromaterials, with their good physical and chemical properties, are widely used in the electrochemical detection of viruses. These materials provide increased electroactive surface area, improved electron transfer, high electrical conductivity, rapid diagnosis, and good functionalization, stability, sensitivity, and selectivity. In some cases, they also ensure biocompatibility and enhance flexibility. Carbon nanotubes, graphene, graphene oxide, graphene-like structures, quantum dots, metallic or metal-based particles (such as gold and platinum nanoparticles), microsphere materials, silicon-based nanoparticles, metal–organic frameworks (MOFs), covalent organic frameworks (COFs), and polymeric materials have been used to construct effective electrochemical virus detection platforms.[55–57]

Figure 4.1 illustrates a schematic presentation of the materials that can be used for detection of different types of viruses. In virus detection, biomolecules are used to provide biorecognition for sustaining specificity or, mostly, selectivity. Therefore, the sensing system should also be a good host for the integration of bioreceptor while guaranteeing the excellent electrical properties of the transducer. It is important to note that the amount of bioreceptor also affects the availability of binding sites for the analyte.[11] In general, nanomaterials and/or micromaterials, such as shown in Figure 4.1, have been very useful for the preparation of electrochemical sensors for virus detection. Nanomaterials and micromaterials, mostly nanomaterials, have

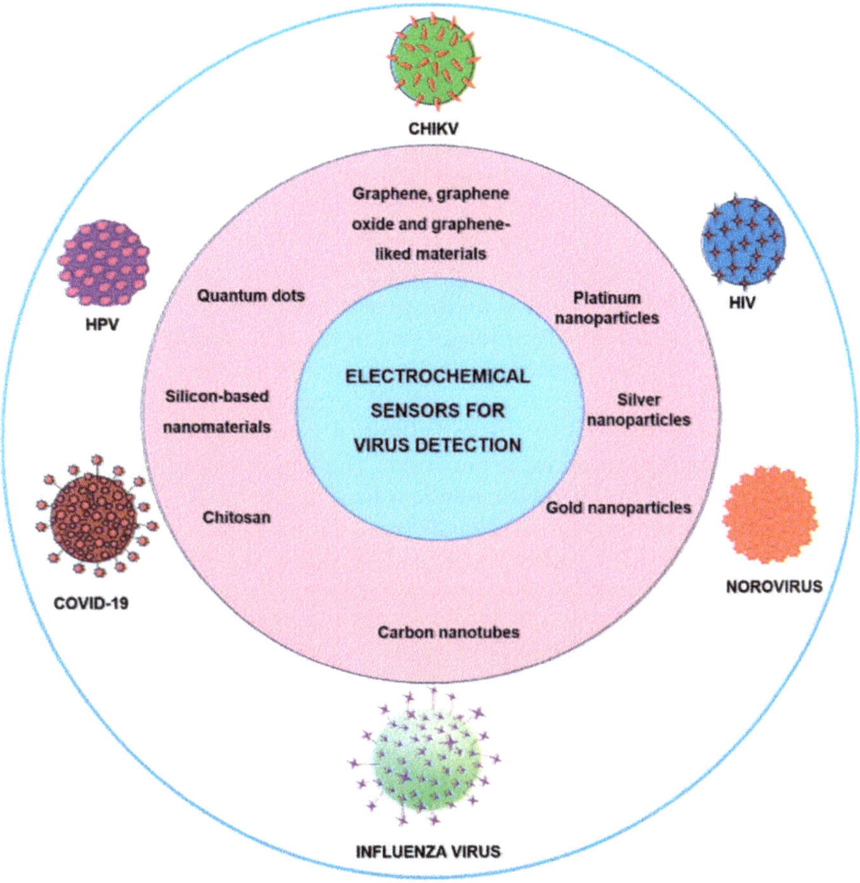

Figure 4.1 Schematic presentation of materials used in electrochemical-based detection of different viruses.

attracted great attention in virus-related diagnostics since they meet challenges in terms of accuracy, sensitivity, cost-effectiveness, rapidity, and early diagnosis.

Nanomaterials are said to be the most commonly helpful class of these materials. The synthesis of novel and functional nano-sized materials increases its contribution day by day. Since detection of low concentrations of biomarkers is valuable, nanoparticles provide the amplification of the electrochemical signals that is needed for high sensitivity. Another perspective is that functionalized nanoparticles can be used as labelling agents or bioreceptor immobilization or conjugation interfaces. As an example, gold nanoparticles (AuNPs) can be used for bioreceptor attachment through gold–thiol interactions, cross-linking, and physical adsorption.[55,58] Another example is carbon-based nanomaterials being used in electrochemical sensors for virus detection. Carbon nanotubes, graphene and its derivatives, and carbon dots can be considered in this context. Graphene-like structures can be also preferred. In these material groups, adsorption plays a particularly important role in biomolecule incorporation.[11,59,60]

Micromaterials are also utilized in virus detection due to the high surface area, reliability, and sensitivity they provide to electrochemical sensors. Microfluidics, combined with nanomaterials, microelectrode arrays, and micro-sized polymeric structures (such as conducting polymer microspheres), are good examples for this situation.[61,62] For example, in the case of conducting polymer-based microspheres, biomolecules can be employed as dopants or can be immobilized on the polymer surface.[62]

Another polymer-based technology that needs to be considered separately is molecular imprinting polymer (MIP) technology. One of the important points that makes this technique stand out is the high selectivity it provides. In this approach, high selectivity and affinity depends on the creation of template-based cavities in polymer networks. Thus, as a result, highly selective receptors are created for electrochemical virus detection. MIP is a promising technology for the design of artificially created binding sites (synthetic receptors). In this design, the appropriate functional monomer that has affinity or interaction with the target is chosen. After polymerization, the template is removed from the polymeric structure and therefore binding sites for the target analyte are created.[32,63]

In terms of materials-based approaches, several current examples can be given as follows: Lu *et al.* designed flexible paper-based nickel metal–organic framework composite, gold nanoparticles, carbon nanotubes, and polyvinylalcohol film (Ni–Au composite–CNT–PVA) electrodes for the determination of HIV DNA through DNA hybridization by using methylene blue (MB) redox indicator. The linear concentration response of this flexible substrate was between 10 nM and 1 µM, with a detection limit of 0.13 nM. It has been suggested that the film electrode can be an effective immobilization matrix for probe DNA by providing good interaction between the MOF and probe DNA *via* hydrogen bonds. The flexible material substrate was a cellulose membrane in the study. Firstly, a mixture of CNTs and PVA was modified on the membrane by vacuum filtration. Then, Au–Ni composite (including AuNPs and Ni-MOF) was dropped onto the CNT–PVA film modified cellulose membrane. Finally, the platform thus formed was used as an immobilization matrix for probe DNA. DNA hybridization was monitored by using the DPV technique.[64] Carbon nanotubes are often used in electrochemical sensing platforms for virus detection since they improve the electrode's stability and electron transfer and ensure a highly specific surface area. AuNPs are also one of the most preferred materials due to their excellent properties, such as enhancement of sensitivity and selectivity, good conductivity, biocompatibility, and providing an easy surface or interface for biorecognition elements (for instance antigen, antibodies, nucleic acids *etc.*).[55,58] In the study of Pusomjit *et al.*, Pt nanoparticles on single-walled carbon nanotubes (SWCNTs) on paper-based screen-printed graphene electrodes were used for antibody immobilization to perform hepatitis C virus (HCV) detection.

This detection protocol based on immunosensing displayed a good linearity to the target [hepatitis C virus core antigen (HCV-cAg)] in the range of 0.05 to 1000 pg ml^{-1} with a detection limit of 0.015 pg ml^{-1}. The DPV technique was used for detection in the $[Fe(CN)_6]^{3-/4-}$ redox probe containing solution. The application of the electrode was presented in human serum samples.[65] Very recently, Khorshed and colleagues developed monkeypox (MPX) virus detection, *via* an A29 protein, immunosensor based on a functionalized monoclonal antibody at the working gold electrode. The DPV technique was used for the detection of A29 protein in the presence of $[Fe(CN)_6]^{3-/4-}$ redox probe. The proposed electrode showed a linearity to A29 protein in the concentration range of 1.8 to 100 ng ml^{-1}. The performance of the immunosensor was tested by spiking the protein in saliva samples.[66] In another very recent study, MIP-based technology was utilized to detect Zika virus (ZIKV) in urine. The study demonstrated an acrylamide–methacrylic acid–methyl methacrylate–vinylpyrrolidone copolymer and graphene oxide (GO)-based composite for the preparation of molecularly imprinted polymer. In order to fabricate the sensing platform, firstly pre-polymer–graphene oxide mixture was dropped on screen-printed gold electrodes (gold SPE). Then, ZIKV suspension was dispersed on the composite film. After exposure to UV light and polymerization step in oven, the ZIKV template was removed from the composite polymer by acetic acid washing. Electrochemical performance of the electrode was shown with linear sweep voltammetry (LSV) in the presence of redox probe {$[Fe(CN)_6]^{3-/4-}$ in phosphate buffered saline (PBS) medium}. A machine learning method, support vector machine (SVM), was applied to sort the electrochemical responses.[67]

These electrochemical virus detection sensors based on various materials indicate the benefits of integrating materials science to create new, practical, and powerful analytical devices.

4.5 Design and Fabrication of Electrochemical Sensors

Creating electrochemical sensors for detecting viruses demands careful attention to sensitivity, specificity, and stability for accurate, reliable readings. These sensors generally comprise three key elements: a working electrode, a reference electrode, and a counter electrode. Popular electrode materials include gold, carbon-based options, and screen-printed electrodes, facilitating miniaturization and portability.[52,68,69] The fabrication process for these sensors incorporates various techniques, including drop casting, electrodeposition, and chemical functionalization, to immobilize viral bioreceptors, such as antibodies, aptamers, or nucleic acid probes, on the electrode surface.[70,71] Additionally, advanced nanomaterials like gold nanoparticles, graphene oxide, and MOFs are frequently integrated to enhance electron transfer and signal amplification.[55,64]

As has been mentioned earlier in this chapter, various electrochemical techniques achieve high sensitivity and real-time monitoring, including CV, LSV, DPV, SWV, EIS, and AMP. These methods enable rapid, label-free, and cost-effective detection of viral biomarkers, making them ideal for point-of-care applications. Moreover, integrating microfluidic platforms and wearable technologies has improved the portability and usability of electrochemical virus sensors. Ongoing advancements in nanotechnology, bioengineering, and microfabrication techniques have paved the way for developing highly efficient and reliable electrochemical sensors for virus detection, essential for early diagnosis and outbreak control.[72–74]

Alternatively, hybrid sensor design involves altering electrode surfaces with fluorescent or plasmonic nanomaterials, such as quantum dots, gold nanoparticles, and graphene-based materials. These nanomaterials enhance electron transfer for electrochemical sensing and serve as optical reporters for surface-enhanced Raman scattering (SERS) detection. They combine electrochemical and spectroscopic techniques, enabling detection of viral biomarkers with enhanced accuracy. For instance, hybrid sensors can employ SWV to monitor charge transfer in viral RNA detection while utilizing fluorescent-labelled probes to confirm hybridization events.[75] As research in nanotechnology and biosensors progresses, hybrid electrochemical–optical sensors are expected to play a vital role in next-generation virus detection platforms, leading to more efficient and accessible diagnostic solutions.

4.6 Design and Fabrication of Electrochemical Sensors

In virus detection, electrochemical (bio)sensors mainly convert biological interactions into measurable electrical signals, enabling the detection of viral components, such as nucleic acids, proteins, antigens, or whole-virus particles. Their abilities to work with minimal sample preparation and to be integrated into portable, point-of-care diagnostic devices make them particularly valuable for real-time virus detection. Recent advancements in nanomaterials, biofunctionalization strategies, and microfluidic integration have further improved the performance of electrochemical sensors, enhancing their sensitivity, specificity, and multiplexing capabilities. Electrochemical sensors offer immense potential to meet demand by delivering real-time, cost-effective solutions for virus detection. Their continued development and integration with digital healthcare systems, such as smartphone-based readout platforms and internet of things (IoT) technologies, will further revolutionize infectious disease diagnostics, making them essential tools for future pandemic preparedness and global health security.[76,77]

Table 4.1 provides an overview of various electrochemical sensors designed for virus detection categorized by virus groups, such as respiratory viruses, gastrointestinal viruses, hepatitis viruses, vector-borne viruses, blood-borne viruses, and nervous system viruses. It outlines the types of

Table 4.1 Overview of electrochemical biosensors developed for the detection of various viruses.[a]

Group	Virus	Sensor type	Electrode material	Detection technique	Linear range	LOD	Reference
Respiratory viruses	H1N1	Aptasensor	Fe_3O_4–NH_2@Au	DPV	1 pg ml^{-1}–1 μg ml^{-1}	0.51 pg ml^{-1}	56
	RSV	Immunosensor	GOx/$Mo_2Ti_2C_3T_x$/RSV	SWV	0.01 pg ml^{-1}–10 μg ml^{-1}	0.015 pg ml^{-1}	30
	SARS-CoV-2	Aptasensor	Truncated multi-thiol aptamer	CV/CC	8.0 fg ml^{-1}–38 ng ml^{-1}	8.0 fg ml^{-1}	84
	SARS-CoV-2	Aptasensor	$Mn_3[Fe(CN)_6]_2$@Pt	DPV	8.83 fg ml^{-1}–88.25 pg ml^{-1}	0.088 fg ml^{-1}	78
	H1N1	Aptasensor	TAPT-TP COFs/MWCNTs	DPV	10 fM–1 nM	1.01 fM	85
	SARS-CoV-2	Genosensor	Aldehyde-modified cellulose paper	AMP	0.05–100 nM	0.12 pM	86
	H1N1				0.05–1000 nM	0.35 pM	
	RSV				10–1000 nM	0.36 pM	
	SARS-CoV-2	Genosensor	AuNPs/Bi_2Se_3	DPV	1.0×10^{-15}–1.0×10^{-11} M	3.46×10^{-16} M	87
	COVID-19	Immunosensor	Ag-MOF	DPV	100 fM–10 nM	60 fM	29
	Influenza A	Genosensor	TPB–DVA COFs	DPV	10 fM–1000 nM	5.42 fM	88
Gastrointestinal viruses	HAdV	eIP	Acrylic acid, BIS, APMA polymer	PE	0.2 fM–20 pM	0.2 fM	57
	Norovirus	Aptasensor	AuNPs	SWV	10^{-9} μg ml^{-1}–10^{-2} μg ml^{-1}	1.365×10^{-10} μg ml^{-1}	79
	Norovirus	MIP	NoV-LPs	EIS	1 pg ml^{-1}–1 μg ml^{-1}	6.5 pg ml^{-1}	32
	Rotavirus	Immunosensor	MXenes@CNTs@AuNPs	SWV	0.08 pg ml^{-1}–31.25 ng ml^{-1}	0.0207 pg ml^{-1}	89
	Norovirus	Aptasensor	Black phosphorous nanosheet@Ti_3C_2-Mxene	DPV	10^{-2}–10^5 copies per ml	0.003 copies per ml	80
	Rotavirus	Genosensor	ssDNA/HFGNs/MXene/PPY	DPV	10^{-18}–10^{-7} M	0.8 aM	90
	Norovirus	Aptasensor	MB@Apt@WP5A@Au@COF@Fe_3O_4	DPV	2.5–2.5×10^5 copies per ml	0.84 copies per ml	91

(continued)

Table 4.1 (*Continued*)

Group	Virus	Sensor type	Electrode material	Detection technique	Linear range	LOD	Reference
Hepatitis viruses	Hepatitis B	Genosensor	CeO_2/AuPt	AMP	1 fmol l^{-1}–1 nmol L^{-1}	0.36 fmol l^{-1}	31
	Hepatitis B	Immunosensor	Au@Co_3O_4NPS/NiCo	SWV	1 pg ml^{-1}–100 ng ml^{-1}	15 fg ml^{-1}	92
	Hepatitis B	Genosensor	$Ti_3C_2NH_2$ MXene@Au	DPV	1.0×10^{-17}–1.0×10^{-7} M	1.05×10^{-14}	93
	Hepatitis C	Genosensor	Gold nanoparticle-labelled reporter probe	DPASV	1 fM–1 μM	43 ng μl^{-1}	94
	Hepatitis C	Immunosensor	Fe-BDC-CA3	DPV	0–1000 ng ml^{-1}	0.75 ng ml^{-1}	95
	Hepatitis B	Immunosensor	Thionine and ferrocene-carboxylic acid	SWV	0.01–10.000 ng ml^{-1}	1.66 pg ml^{-1}	96
	Hepatitis C	Immunosensor	GOx-CTAB-thiophene	SWV	0.2–8 ng ml^{-1}	0.07 ng ml^{-1}	97
Vector-borne viruses	Malaria	Immunosensor	f-CNF + PEDOT@Th	CV	0.25–100 ng ml^{-1}	0.5 ng ml^{-1}	81
	Dengue virus	Immunosensor	HSPC	DPV	1.0 pg ml^{-1}–100 μg ml^{-1}	0.665 pg ml^{-1}	69
	Dengue virus	Immunosensor	Au/ZnO/GOx	CV	100 ng ml^{-1}–100 μg ml^{-1}	100 ng ml^{-1}	98
	Malaria	Immunosensor	hPG/PEDOT/MB	SWV	0.064–5000 ng ml^{-1}	0.026 pg ml^{-1}	99
	Dengue virus	Immunosensor	S40	DPV	0.001–10 ng ml^{-1}	0.932 pg ml^{-1}	100
	West Nile virus	Aptasensor	$Ti_3C_2T_x$	SWV	10 pM–1 μM	1.06 pM	101
Blood-borne viruses	HIV	Immunosensor	PdPtCu mesoporous nanocubes/AuNPs@GOx	DPV	0.04 pg ml^{-1}–100 ng ml^{-1}	20 fg ml^{-1}	82
	HIV	Aptasensor	GNR	DPV	0–80 aM	2.3 aM	102
	HIV	Genosensor	cys-GQDs/AuNps	DPV	10 μg ml^{-1}–1 fg ml^{-1}	1.0 fg ml^{-1}	59
	HIV	Aptasensor	FMMA	SWV	0.1 fM–10 nM	0.037 fM	103
Nervous system viruses	HSV-1	Genosensor	MB@MOF-PC	DPV	10^{-18}–10^{-8} M	0.79 aM	83
	Rabies virus	Genosensor	Ly-rGO	CA	0.145–25.39 ng μl^{-1}	0.104 ng μl^{-1}	60

[a]Abbreviations: graphene oxide (GOx), respiratory syncytial virus (RSV), chronocoulometric (CC), 4,4′,4″-(1,3,5-triazine-2,4,6-trial) trianiline (TAPT), 1,3,5-triformylphloroglucinol (TP), multiwalled carbon nanotubes (MWCNTs), 1,3,5-tri(4-aminophenyl)benzene (TPB), 1,4-benzenedicarboxaldehyde (DVA), epitope-imprinted polymer (eIP), *N*,*N*′-methylenebis(acrylamide) (BIS), *N*-(3-aminopropyl)methacrylamide (APMA), molecular imprinting polymer (MIP), norovirus virus-like particles (NoV-LPs), transition metal–carbon/nitrogen two-dimensional inorganic compounds (MXenes), hierarchical flower-like gold nanostructures (HFGNs), polypyrrole (PPY), magnetic covalent organic framework/pillararene heterosupramolecular nanocomposites (MB@Apt@WP5A@Au@COF@Fe_3O_4), nanoparticles (NPs), differential pulse anodic stripping voltammetry (DPASV), benzenedicarboxylic acid (BDC), cetyltrimethylammonium bromide (CTAB), poly(3,4-ethylenedioxythiophene) (PEDOT), high-surface porous carbon (HSPC), highly porous gold (hPG), surface terminations (Tx), graphene nanoribbon (GNR), graphene quantum dots (GQDs), ferrocenylmethyl methacrylate (FMMA), lysozyme-reduced graphene oxide (Ly-rGO).

electrochemical sensors used, including aptasensors, immunosensors, nanosensors, and molecularly imprinted polymer (MIP) sensors. It also discusses electrode materials, ranging from materials like gold AuNPs, transition metal–carbon/nitrogen two-dimensional inorganic compounds (MXenes), and MOFs to advanced hybrid structures. Detection techniques mainly include voltametric, amperometric, and impedimetric methods, which have been observed to provide high sensitivity and selectivity for viral biomarkers. In addition, the table highlights the ability of these sensors to detect target viruses at low concentrations by listing linear detection ranges and limits of detection (LOD) reported in studies, which make them valuable tools for rapid and reliable virus diagnosis.

Xiong *et al.* presented a label-free electrochemical DNA aptamer biosensor for rapidly and sensitively detecting H1N1 hemagglutinin (HA) protein, a major surface glycoprotein of influenza virus (Figure 4.2A). The developed biosensor uses a specific DNA aptamer that selectively binds to HA protein, providing high specificity without additional labelling steps. The detection mechanism is based on electrochemical signal changes upon target binding, enabling real-time monitoring with minimal sample preparation. DPV was used as the electrochemical detection technique, offering a broad linear detection range (1 pg ml^{-1}–1 μg ml^{-1}) and a low LOD (0.51 pg ml^{-1}), providing remarkable results for early-stage infection diagnosis. An electrochemical biosensor was developed by immobilizing a sulfhydryl-modified single-stranded DNA aptamer probe on the surface of magnetic nanoparticles (MNPs) Fe_3O_4–NH_2@Au based on the HA gene sequence. This new surface further enhances the sensor's sensitivity. This H1N1 biosensor may provide a promising alternative to traditional diagnostic methods and enable portable virus detection, making it a valuable tool for point-of-care applications and pandemic preparation.[56]

Zhu *et al.* introduced a novel electrochemical DNA sensor based on nanocubic $Mn_3[Fe(CN)_6]_2$@Pt designed to rapidly and accurately detect SARS-CoV-2. The biosensor employs a DNA probe that specifically hybridizes with the viral genetic sequence, enabling selective and precise detection (Figure 4.2B). The incorporation of platinum (Pt)-decorated $Mn_3[Fe(CN)_6]_2$ nanocubes enhances the electrochemical properties of the sensor, resulting in improved signal transduction and sensitivity. DPV is the technique that was utilized. The sensor allowed for quick response times and an exceptionally low detection limit (0.088 fg ml^{-1}), making it suitable for early-stage and low-concentration viral detection. The sensor features an easy fabrication process, high conductivity, and is ultrasensitive and selective, making it an effective tool for real-time virus monitoring. This work merges nanomaterial-enhanced electrochemical properties with a straightforward yet efficient sensing strategy, contributing to advancing next-generation biosensing platforms for infectious disease detection.[78]

Li *et al.* introduced a novel dual-mode aptasensor that combines colorimetric and electrochemical detection methods for sensitive and specific norovirus (NoV) identification in faecal samples. This biosensor was designed to

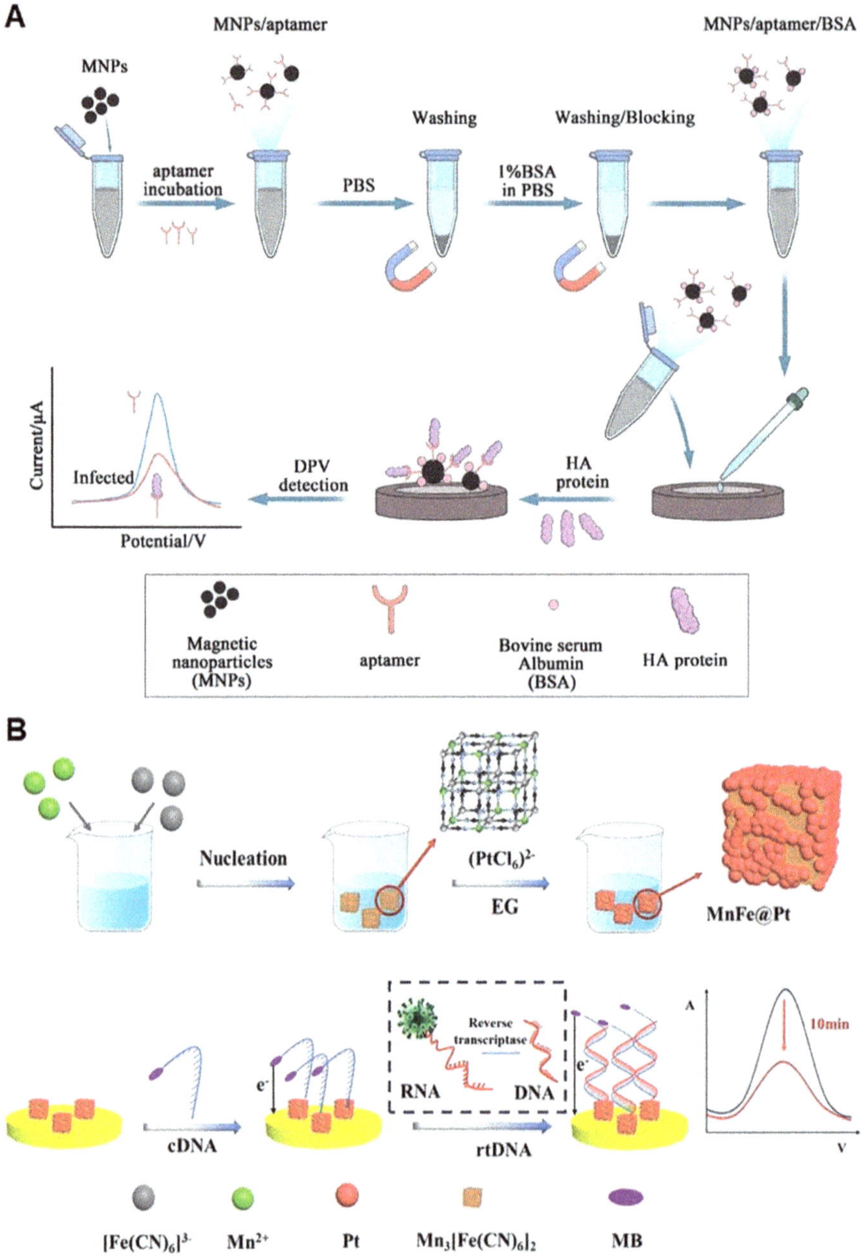

Figure 4.2 (A) Flow diagram of construction of an electrochemical biosensor that uses MNPs and aptamers to identify H1N1–HA proteins. Reproduced from ref. 56 with permission from Elsevier, Copyright 2025. (B) Flow diagram of MnFe@Pt nanocube-based DNA biosensors and their detection approach for the reverse transcribed DNA (rtDNA) of SARS-CoV-2. Reproduced from ref. 78 with permission from Elsevier, Copyright 2024. Abbreviations: ethylene glycol (EG), methylene blue (MB).

utilize a highly selective aptamer that binds to norovirus, ensuring sensitive recognition. Colorimetric detection offers a visible colour change, providing a simple and rapid visual screening method, while the electrochemical approach enables highly sensitive and quantitative detection (Figure 4.3A). Combining these two detection strategies enhances the sensor's reliability and broadens its potential applications in clinical diagnostics and point-of-care testing. The sensor platform was built using a screen-printed carbon

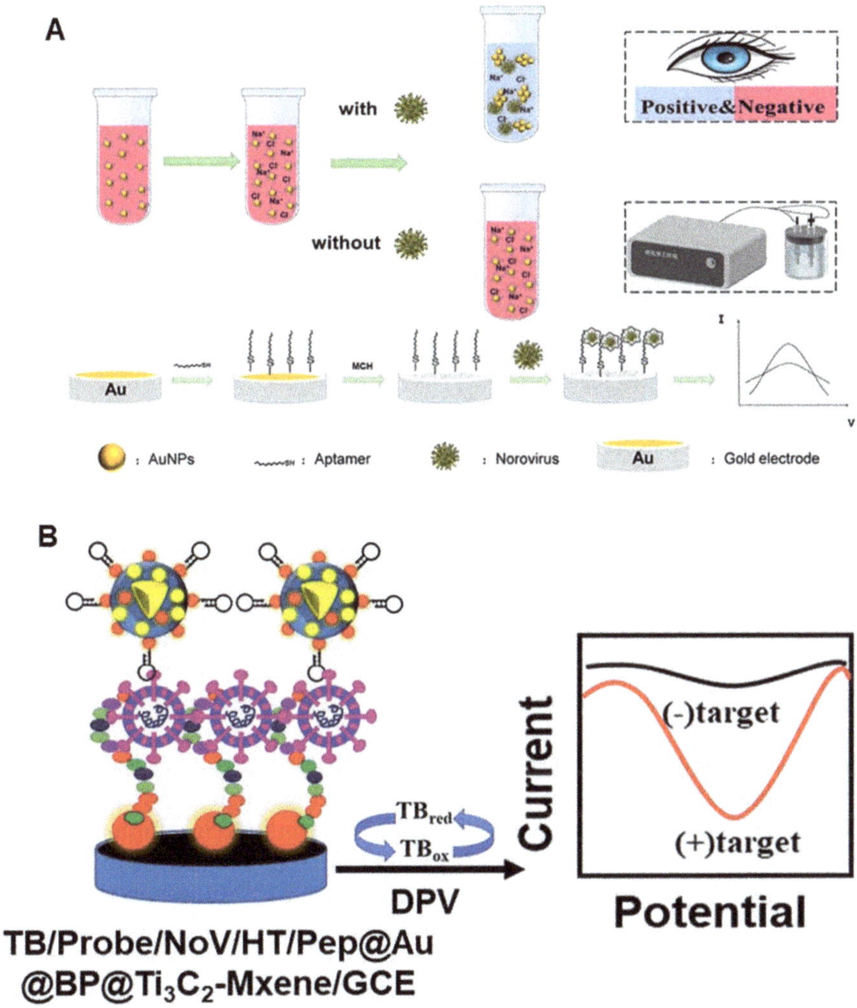

Figure 4.3 (A) Detection mode of NoV by dual-mode aptasensor. Reproduced from ref. 79 with permission from Elsevier, Copyright 2024. (B) Assembly of the NoV electrochemical Apt@Au@$ZnFe_2O_4$@COF biosensor. Abbreviations: toluidine blue (TB). 1-Hexaththiol (HT), peptide (Pep), black phosphorous (BP), glassy carbon electrode (GCE). Reproduced from ref. 80 with permission from Elsevier, Copyright 2023.

electrode modified with AuNPs. Data revealed a low detection limit of 1.365 × 10^{-10} μg ml^{-1} and a wide dynamic range from 10^{-9} to 10^{-2} μg ml^{-1} using SWV, making the sensor suitable for diagnosing early-stage infections. With its dual-mode functionality, high specificity, and user-friendly design, this aptasensor presents a promising alternative to traditional norovirus detection techniques.[79]

Liu *et al.* developed an advanced electrochemical biosensor for sensitively detecting NoV using a peptide-target–aptamer recognition strategy. The biosensor was designed with a nanohybrid material composed of black phosphorus (BP) nanosheets and Ti_3C_2-MXene, which enhances conductivity and signal transduction. Additionally, a magnetic COF was incorporated into the system to facilitate target capture and improve sensitivity. The peptide-target–aptamer system ensures selective binding to NoV, leading to accurate detection. DPV was employed as the detection method in the developed sensor, achieving a linear working range of 10^{-2} to 10^{-5} copies per ml and LOD value of 0.003 copies per ml. This study demonstrates that combining nanomaterials with biofunctionalization strategies offers a robust and innovative approach for NoV diagnosis by enhancing detection efficiency (Figure 4.3B).[80]

Mao *et al.* presented the development of a sandwich-type electrochemical DNA sensor to detect HBV DNA sensitively. This sensor used cerium oxide/gold–platinum nanoparticles (CeO_2/AuPt NPs) nanoprobes to improve the performance (Figure 4.4). The biosensor has a probe recognition strategy in which a capture probe is immobilized on the electrode surface and specifically binds to the target HBV DNA. These nanoprobes significantly improve the electrochemical signal and increase the sensitivity due to their excellent conductivity, catalytic activity, and large surface area. The CeO_2/AuPt nanocomposite utilizes the catalytic properties of H_2O_2 formed in the system. The sensor performance was evaluated using AMP, where a glassy carbon electrode was used as the working electrode, giving a wide linear range (1 fmol l^{-1}–1 nmol l^{-1}) and a LOD value of 0.36 fmol l^{-1}. Thanks to its specificity, stability, and reproducibility, this biosensor has become an effective tool for early and accurate diagnosis of HBV, contributing to better management and monitoring of HBV.[31]

In a study by Monisha *et al.* a rapid and highly sensitive electrochemical immunosensor was developed to detect a malaria protein biomarker using a novel thionine-functionalized carbon nanofiber (CNF) surface for effective antibody immobilization. Integrating redox-active thionine into the operating mechanism enhanced electron transfer, improving electrochemical performance and signal stability. The carbon nanofiber structure offered a large surface area for increased antibody loading, which enhanced biomarker capture and improved sensitivity. The developed immunosensor functions with a label-free detection strategy, where the antigen–antibody interaction generates measurable electrochemical signals. CV served as the detection technique, providing rapid and accurate identification of malaria biomarkers. With a detection range of 250 pg ml^{-1} to 100 ng ml^{-1} and a LOD value of

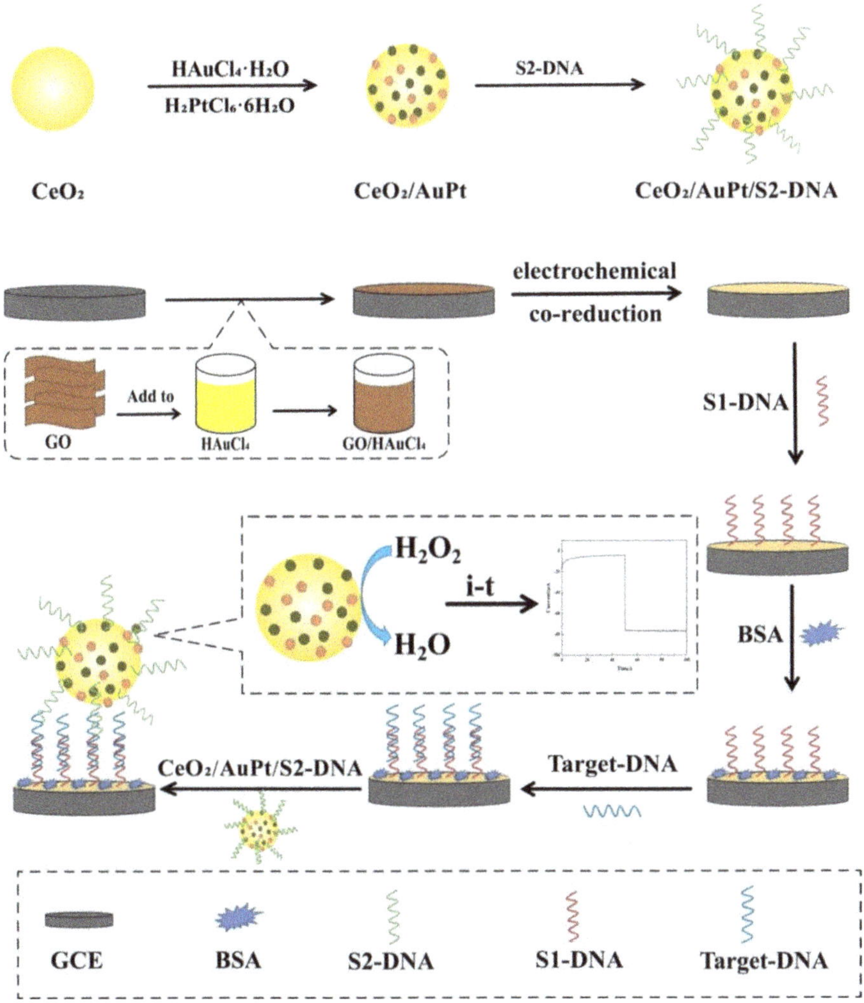

Figure 4.4 The development procedure of an electrochemical DNA biosensor for HBV DNA detection. Reproduced from ref. 31 with permission from Elsevier, Copyright 2025.

0.5 ng ml^{-1}, this sensor has shown very effective results for early-stage malaria diagnosis with good selectivity properties.[81]

Li *et al.* reported a highly sensitive electrochemical sandwich immunosensor for the detection of HIV-p24, an important viral protein biomarker used for early HIV diagnosis. The sensor comprised PdPtCu mesoporous nanocubes, which act as an advanced nanomaterial for signal amplification due to their high surface area, excellent catalytic activity, and superior conductivity (Figure 4.5A). The study employed a sandwich detection strategy in which a capture antibody immobilized on the electrode surface of the immunosensor specifically binds to HIV-p24 and then interacts with a secondary antibody

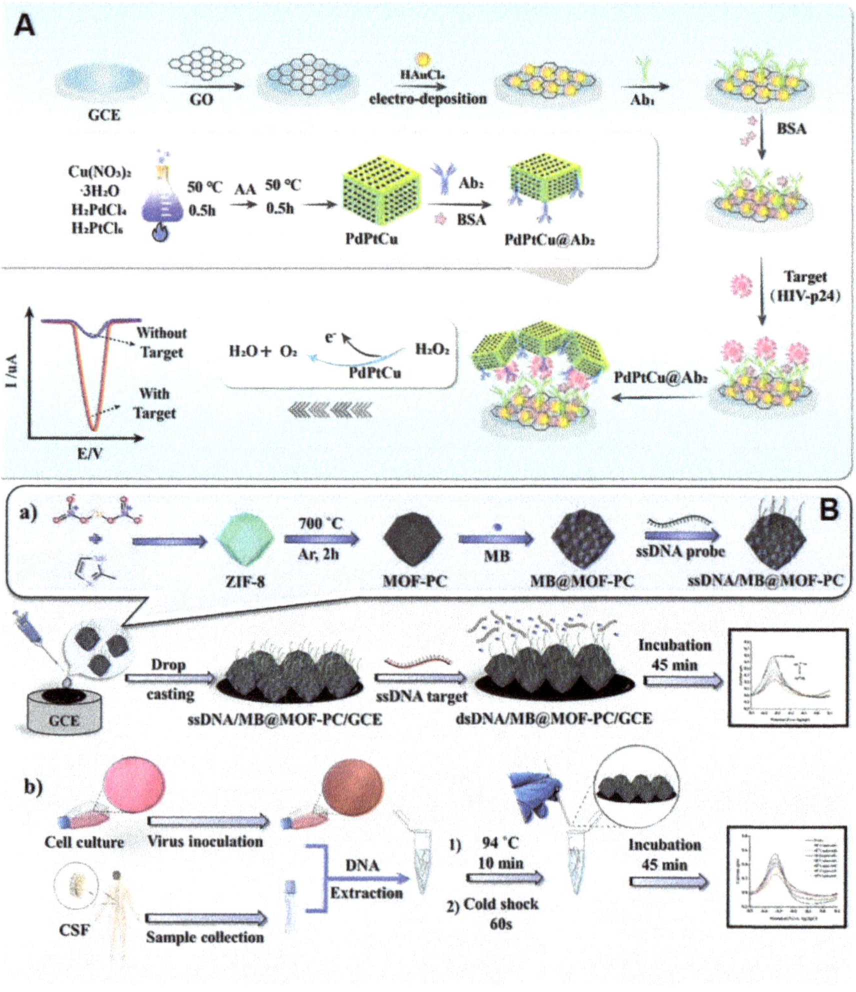

Figure 4.5 (A) Layer-by-layer assembly of a sandwich-format electrochemical immunosensor for HIV-p24 detection. Abbreviations: graphene oxide (GO). Reproduced from ref. 82 with permission from Elsevier, Copyright 2025. (B) Genosensor-based HSV-1 detection: (a) fabrication of ssDNA/MB@metal–organic framework-derived porous carbon (MOF-PC)/GCE. Abbreviations: zeolitic imidazolate framework 8 (ZIF-8). (b) Electrochemical sensing procedure. Abbreviations: cerebrospinal fluid (CSF). Reproduced from ref. 83 with permission from Elsevier, Copyright 2024.

conjugated to PdPtCu nanocubes to enhance the electrochemical signal output. The DPV technique was used for detection, and it was shown that the sensor, which has a working range of 0.04 pg ml^{-1}–100 ng ml^{-1} and LOD of 20 fg ml^{-1}, also provides excellent specificity against non-target proteins.[82]

In another study, a highly specific and sensitive electrochemical biosensor was introduced for detecting herpes simplex virus type 1 (HSV-1). This biosensor employed a DNA-gated nanometre-scale porous carbon bioplatform derived from a MOF. It utilized a target-specific DNA probe that selectively hybridizes with HSV-1 genetic sequences, ensuring high specificity in detection (Figure 4.5B). The nanometre-scale porous carbon structure significantly enhanced the electrochemical performance of the sensor by providing a large surface area, excellent conductivity, and improved signal transmission. The DNA-gating mechanism allowed for controlled molecular recognition, resulting in a sensitive and stable detection process. The sensor, utilizing DPV, achieves a wide linear detection range (10^{-18}–10^{-8} M) and a low LOD (0.79 aM), yielding highly suitable results for early and accurate virus identification. With its high stability and excellent selectivity, this biosensor appears to be efficient for effectively managing and controlling HSV-1 infections.[83]

According to literature reviews, strategies involving aptamers, immunoreactive elements, and DNA hybridization techniques have been shown to enhance the specificity of these sensors further, enabling sensitive virus detection even at ultra-low concentrations. Furthermore, the development of dual-modality and label-free detection methods has broadened the practical applications of these biosensors in clinical diagnostics and point-of-care environments. Despite these advancements, challenges such as sensor miniaturization, multiplex detection, and real-world validation remain essential research areas. Nevertheless, electrochemical biosensors have significant potential to revolutionize virus diagnostics by offering cost-effective, portable, and real-time monitoring solutions essential for timely disease management and outbreak control.

4.7 Conclusions and Future Perspectives

Viruses can have serious effect on health and the environment due to the numerous infections they cause. Infections caused by viruses are one of the leading causes of mortality worldwide. Viruses infect all organisms and are found in pretty much all Earth's ecosystems. Therefore, it's important and essential to detect these pathogens before they caused serious problems in any life. While aiming to prevent the lethal effects of viruses by developing systems, it is very important to detect them more rapidly and accurately. There are effective techniques used besides electrochemistry for early detection of viruses as has been mentioned in this chapter. However, compared with electrochemical techniques, most of them are time consuming, expensive, not portable, require long times to process, and need trained staff. With electrochemistry, it is possible and easy to modify the sensing platforms to gain desired properties, resulting a high sensitivity, selectivity, and stability. Additionally, with the developing technology, electrochemical sensors provide great advantages, such as quick and efficient integration to flexible or wearable technologies, thus promising adaptation to point-of-care diagnostics.

It is clear that the electrode modifications and immobilization strategies for bioreceptors are the key factors to improve the electrochemical sensors' performance. Mostly, for the immobilization of bioreceptors strong and oriented interactions, such as covalent or crosslinking, should be preferred compared with adsorption strategies to sustain the greater availability of analyte binding sites, and high repeatability and stability. Investigation of the development of novel and functional materials to contribute to electroanalytical performance is an important research field. Various properties in the resulting technology, such as faster response time, improved electroactive surface area, accuracy, precision, cost-effectiveness, sensitivity, and selectivity, show the contribution of the material used. Moreover, biocompatibility, being environmentally friendly, and flexibility are among the prominent developments of recent years. In virus detection, it is also important to make a diagnosis without having to go to the hospital. This situation is also important in terms of controlling a possible epidemic. Therefore, more efforts on these topics will bring about more advanced electroanalytical tools.

The progress in label-free detection and dual-modality has expanded the practical applications of electrochemical sensors in clinical and point-of-care diagnostics. There are still challenges including miniaturization, multiplex detection, complexity of biological samples, and variations in individual immune responses. However, electrochemical sensors meet the conditions of being promising and fascinating detection tools for virus detection at very good levels with their unique properties. It is anticipated that more effective and practical commercialized products will be produced through the joint application of electroanalytical chemistry, materials science, and engineering technologies.

Acknowledgements

F. Kuralay acknowledges the Turkish Academy of Sciences (TÜBA) as an associate member.

References

1. D. Sumbria, E. Berber, M. Mathayan and B. T. Rouse, *Front. Immunol.*, 2021, **11**, 594963.
2. C. J. Hurst and N. J. Adcock, Relationship Between Humans and Their Viruses, in *Viral Ecology*, ed. C. J. Hurst, Academic Press, USA, 2000, pp. 519–548.
3. L. Yuwen, S. Zhang and J. Chao, *Biosensors*, 2023, **13**, 822.
4. N. Manring, M. M. Ahmed, N. Tenhoff, J. L. Smeltz and P. Pathirathna, *Anal. Chem.*, 2022, **94**, 7149.

5. X. Ma, Y. Shi, G. Gao, H. Zhang, Q. Zhao and J. Zhi, *J. Electroanal. Chem.*, 2023, **950**, 117867.
6. L. E. Ahangar and M. A. Mehrgardi, *Bioelectrochemistry*, 2017, **117**, 83.
7. F. Fenner, P. A. Bachmann, E. P. J. Gibbs, F. A. Murphy, M. J. Studdert and D. O. White, *Veterinary Virology*, Academic Press, USA, 1987.
8. A. Cassedy, A. Parle-McDermott and R. O'Kennedy, *Front. Mol. Biosci.*, 2021, **8**, 637559.
9. J. H. Strauss and E. G. Strauss, *Viruses and Human Disease*, Academic Press, UK, 2nd edn, 2008.
10. M. A. Abdul Ghani, A. N. Nordin, M. Zulhairee, A. C. M. Nor, M. S. A. Noorden, M. K. F. M. Atan, R. Ab Rahim and Z. M. Zain, *Biosensors*, 2022, **12**, 666.
11. L. C. Brazaca, P. L. dos Santos, P. R. de Oliveira, D. P. Rocha, J. S. Stefano, C. Kalinke, R. A. A. Muñoz, J. A. Bonacin, B. C. Janegitz and E. Carrilho, *Anal. Chim. Acta*, 2021, **1159**, 338384.
12. G. Kabay, J. DeCastro, A. Altay, K. Smith, H. W. Lu, A. M. Capossela, M. Moarefian, K. Aran and C. Dincer, *Adv. Mater.*, 2022, **34**, 2201085.
13. J. Wang, *Electroanalysis*, 1991, **3**, 255.
14. H. Karadeniz, A. Erdem, F. Kuralay and F. Jelen, *Talanta*, 2009, **78**, 187.
15. K. Mahato and J. Wang, *Sens. Actuators, B*, 2021, **344**, 130178.
16. K. Beaver, A. Dantanarayana and S. D. Minteer, *J. Phys. Chem. B*, 2021, **125**, 11820.
17. J. Kang, A. Tahir, H. Wang and J. Chang, *Wiley Interdiscip. Rev.: Nanomed. Nanobiotechnol.*, 2021, **13**, e1700.
18. A. P. F. Turner, Biosensors: Sense and sensibility, *Chem. Soc. Rev.*, 2013, **42**, 3184.
19. F. Beck, C. Horn and A. J. Baeumner, *Anal. Bioanal. Chem.*, 2022, **414**, 475.
20. L. F. de Lima, P. C. R. Corsato, M. A. Beluomini, A. L. Ferreira, L. Ester-dos Santos and P. P. Barbosa, *ACS Appl. Electron. Mater.*, 2025, 7, 2882.
21. A. J. Bard, L. R. Faulkner and H. S. White, *Electrochemical Methods: Fundamentals and Applications*, John Wiley & Sons, Inc., USA, 3rd edn, 2022.
22. R. B. Channon, Y. Yang, K. M. Feibelman, B. J. Geiss, D. S. Dandy and C. S. Henry, *Anal. Chem.*, 2018, **90**, 7777.
23. F. Kuralay, H. Özyörük and A. Yıldız, *Sens. Actuators, B*, 2005, **109**(14), 194.
24. N. J. Ronkainen, H. B. Halsall and W. R. Heineman, *Chem. Soc. Rev.*, 2010, **39**, 1747.
25. G. Marrazza, M. Ramalingam, A. Jaisankar, L. Cheng, G. Selvolini and I. A. Vitale, *TrAC, Trends Anal. Chem.*, 2024, **172**, 117609.
26. S. B. Hanoglu, D. Harmanci, S. Evran and S. Timur, *Bioelectrochemistry*, 2024, **160**, 108784.
27. M. Muti, F. Kuralay, A. Erdem, S. Abaci, T. Yumak and A. Sinağ, *Talanta*, 2010, **82**, 1680.
28. H. Ilkhani and S. Farhad, *Anal. Biochem.*, 2018, **557**, 151.

29. S. Adel, A. Firoozbakhtian, H. Rabbani, M. Hosseini, A. B. Pebdeni, N. Sadeghi, J. Gilnezhad and M. R. Ganjali, *Anal. Biochem.*, 2024, **689**, 115500.
30. A. Rhouati, A. Thakur, B. Anasori and M. Zourob, *Nanoscale*, 2025, **17**, 5161.
31. J. Mao, J. Wang, H. Chen and Q. Yan, *Bioelectrochemistry*, 2025, **163**, 108901.
32. S. Kaur, P. Singla, A. J. Dann, J. McClements, M. V. Sullivan, M. Kim, S. Stoufer, J. A. Dawson, R. D. Crapnell, C. E. Banks, N. W. Turner, M. D. Moore, I. Kaur and M. Peeters, *ACS Appl. Mater. Interfaces*, 2024, **16**, 51397.
33. T. Acter, N. Uddin, J. Das, A. Akhter, T. R. Choudhury and S. Kim, *Sci. Total Environ.*, 2020, **730**, 138996.
34. N. H. L. Leung, *Nat. Rev. Microbiol.*, 2021, **19**, 528.
35. W. P. Hanage and W. Schaffner, *Infect. Dis. Ther.*, 2024, **14**, 5.
36. N. Maison, A. Peck, S. Illi, M. Meyer-Buehn, E. von Mutius, J. Hübner and U. Von Both, *Infection*, 2022, **50**, 519.
37. M. A. Odenwald and S. Paul, *World J. Gastroenterol.*, 2022, **28**, 1405.
38. W. Socha, M. Kwasnik, M. Larska, J. Rola and W. Rozek, *J. Clin. Med.*, 2022, **11**, 3026.
39. G. S. Orf, A. Olivo, B. Harris, S. L. Weiss, A. Achari, G. Yu, S. Federman, D. Mbanya, L. James, S. Mampunza, C. Y. Chui, M. A. Rodgers, G. A. Cloherty and M. G. Berg, *Viruses*, 2023, **15**, 1022.
40. L. Feige, L. M. Zaeck, J. Sehl-Ewert, S. Finke and H. Bourhy, *Viruses*, 2021, **13**, 2364.
41. B. M. Gunasekaran, S. Srinivasan, M. Ezhilan and N. Nesakumar, *Clin. Chim. Acta*, 2024, **559**, 119715.
42. M. Manzano, S. Viezzi, S. Mazerat, R. S. Marks and J. Vidic, *Biosens. Bioelectron.*, 2018, **100**, 89.
43. P. Teengam, P. Tangkijvanich, N. Chuaypen and O. Chailapakul, *Sci. Rep.*, 2023, **13**, 3523.
44. J. L. Gogola, G. Martins, A. Gevaerd, L. Blanes, J. Cardoso, F. K. Marchini, C. E. Banks, M. F. Bergamini and L. H. Marcolino-Junior, *Anal. Chim. Acta*, 2021, **1166**, 338548.
45. A. C. R. Moço, J. A. S. Neto, D. D. de Moraes, P. H. Guedes, J. G. Brussasco, J. M. R. Flauzino, L. F. G. Luz, M. M. C. N. Soares, J. M. Madurro and A. G. Brito-Madurro, *Microchem. J.*, 2021, **170**, 106739.
46. H. Zhao, F. Liu, W. Xie, T.-C. Zhou, J. OuYang, L. Jin, H. Li, C.-Y. Zhao, L. Zhang, J. Wei, Y.-P. Zhang and C.-P. Li, *Sens. Actuators, B*, 2021, **327**, 128899.
47. N. Kumar, N. P. Shetti, S. Jagannath and T. M. Aminabhavi, *Chem. Eng. J.*, 2022, **430**, 132966.
48. S. Madhurantakam, Z. J. Lee, A. Naqvi, J. B. Karnam, S. Muthukumar and S. Prasad, *Biosens. Bioelectron., X*, 2023, **13**, 100307.
49. C. Adrover-Jaume, A. Alba-Patiño, A. Clemente, G. Santopolo, A. Vaquer, S. M. Russell, E. Barón, M. del Mar González del Campo, J. M. Ferrer,

M. Berman-Riu, M. García-Gasalla, M. Aranda, M. Borges and R. de la Rica, *Sens. Actuators, B*, 2021, **330**, 129333.
50. N. R. Blumenfeld, M. A. E. Bolene, M. Jaspan, A. G. Ayers, S. Zarrandikoetxea, J. Freudman, N. Shah, A. M. Tolwani, Y. Hu, T. L. Chern, J. Rogot, V. Behnam, A. Sekhar, X. Liu, B. Onalir, R. Kasumi, A. Sanogo, K. Human, K. Murakami, G. S. Totapally, M. Fasciano and S. K. Sia, *Nat. Nanotechnol.*, 2022, **17**, 984.
51. M. Di Domenico, A. De Rosa and M. Boccellino, *Diagnostics*, 2021, **11**, 698.
52. V. Sanko and F. Kuralay, *Biosensors*, 2023, **13**, 333.
53. F. H. N. Howard, A. Kwan, N. Winder, A. Mughal, C. Collado-Rojas and M. Muthana, *Viruses*, 2022, **14**, 1493.
54. Z. Zhang, P. Ma, R. Ahmed, J. Wang, D. Akin, F. Soto, B.-F. Liu, P. Li and U. Demirci, *Adv. Mater.*, 2022, **34**, 2103646.
55. S. Homayoonnia, Y. Lee, D. Andalib, Md. S. Rahman, J. Shin, K. Kim and S. Kim, *Biomed. Eng. Lett.*, 2021, **11**, 335.
56. C.-r. Xiong, Q.-C. Yu, Y. Liu, B.-C. Wang, X.-M. Lu, L.-C. Wang and X. Chen, *Microchem. J.*, 2025, **210**, 112998.
57. E. Sehit, G. Yao, G. Battocchio, R. Radfar, J. Trimpert, M. A. Mroginski, R. Süssmuth and Z. Altintas, *ACS Sens.*, 2024, **9**, 1831.
58. F. Nasrollahi, R. Haghniaz, V. Hosseini, E. Davoodi, M. Mahmoodi, S. Karamikamkar, M. A. Darabi, Y. Zhu, J. Lee, S. Emir Diltemiz, H. Montazerian, S. Sangabathuni, M. Tavafoghi, V. Jucaud, W. Sun, H.-J. Kim, S. Ahadian and A. Khademhosseini, *Small*, 2021, **17**, 2100692.
59. A. C. R. Moço, J. A. L. Gomide, J. M. R. Flauzino, J. G. Brussasco, L. F. G. Luz, M. M. C. N. Soares, J. M. Madurro and A. G. Brito-Madurro, *J. Pharm. Biomed. Anal.*, 2024, **242**, 116025.
60. R. Challhua, L. Akashi, J. Zuñiga, H. B. de C. Batista, R. Moratelli and A. Champi, *Biosens. Bioelectron.*, 2023, **232**, 115291.
61. Z. Wu, W.-J. Guo, Y.-Y. Bai, L. Zhang, J. Hu, D.-W. Pang and Z.-L. Zhang, *Anal. Chem.*, 2018, **90**, 1683.
62. V. V. Tran, N. H. T. Tran, H. S. Hwang and M. Chang, *Biosens. Bioelectron.*, 2021, **182**, 113192.
63. M. Antipchik, J. Reut, A. G. Ayankojo, A. Öpik and V. Syritski, *Talanta*, 2022, **250**, 123737.
64. Q. Lu, T. Su, Z. Shang, D. Jin, Y. Shu, Q. Xu and X. Hu, Flexible paper-based Ni-MOF composite/AuNPs/CNTs film electrode for HIV DNA detection, *Biosens. Bioelectron.*, 2021, **184**, 113229.
65. P. Pusomjit, P. Teengam, N. Chuaypen, P. Tangkijvanich, N. Thepsuparungsikul and O. Chailapakul, *Microchim. Acta*, 2022, **189**, 339.
66. A. A. Khorshed, T. Jiang and J. Chen, *Biosens. Bioelectron.*, 2025, **278**, 117337.
67. W. Sukjee, P. Sirisangsawang, C. Thepparit, P. Auewarakul, T. Puttasakul and C. Sangma, *Anal. Biochem.*, 2025, **702**, 115854.
68. M. Yadav, M. Dhanda, R. Arora, G. Singh, H. Mohan and S. Lata, *Microchem. Journal*, 2024, **197**, 109831.

69. S. S. Hegde, Shivakumar, B. R. Bhat, P. Mishra, U. Dalimba, M. U. Ahmed and G. N. Santos, *Biosens. Bioelectron. X*, 2024, **20**, 100525.
70. Y. Liu, Y. Zhang, Z. Li, Z. Li, S. Zhou, Y. Xia and L. Mou, *Microchem. Journal*, 2024, **207**, 112064.
71. Y. Tian and L. Dong, *ACS Sensors*, 2024, **9**, 1978.
72. B. Hosnedlova, J. Werle, J. Cepova, V. H. B. Narayanan, L. Vyslouzilova, C. Fernandez, A. A. Parikesit, M. Kepinska, E. Klapkova, K. Kotaska, O. Stepankova, G. Bjorklund, R. Prusa and R. Kizek, *Crit. Rev. Anal. Chem.*, 2025, **55**(6), 1203.
73. T. Lehnert and M. A. M. Gijs, *Lab Chip*, 2024, **24**, 1441.
74. R. Godin, S. Hejazi and N. F. Reuel, *Adv. Sens. Res.*, 2024, **3**, 2300061.
75. W. A. El-Said, S. Al-Bogami and W. Alshitari, *Spectrochim. Acta, Part A*, 2022, **264**, 120237.
76. M. Xiao, F. Tian, X. Liu, Q. Zhou, J. Pan, Z. Luo, M. Yang and C. Yi, *Adv. Sci.*, 2022, **9**, 2105094.
77. M. Nasajpour, S. Pouriyeh, R. M. Parizi, M. Dorodchi, M. Valero and H. R. Arabnia, *J. Healthcare Inf. Res.*, 2020, **4**, 325.
78. M. Zhu, Y. Liu, M. Wang, T. Liu, Z. Chu and W. Jin, *Bioelectrochemistry*, 2024, **156**, 108598.
79. X. Li, Y. Zhao, W. Gu, Y. Qian, Q. Huang, X. Hu and H. Xing, *Anal. Biochem.*, 2024, **687**, 115444.
80. H. Liu, S. Ma, G. Ning, R. Zhang, H. Liang, F. Liu, L. Xiao, L. Guo, Y. Zhang, C.-P. Li and H. Zhao, *Talanta*, 2023, **258**, 124433.
81. S. Monisha, N. S. Jayaprakash and A. Senthil Kumar, *Langmuir*, 2025, **41**, 4446.
82. X. Li, D. Han, X. Li, C. Zhou, B. Shen, H. Wei, Q. Lou, C. Liu and T. Chen, *Bioelectrochemistry*, 2025, **161**, 108819.
83. A. Vojdani, H. Asaadi, Z. Meshkat, E. Aryan, S. B. Tavakoly Sany, H. Farsiani, J. Dorazehi, B. Hatamluyi and S. Abolbashari, *Sens. Actuators, B*, 2024, **411**, 135707.
84. S. R. M. Ramirez, N. Samiseresht, M. A. Martínez-Roque, F. Catania, K. Graef, M. Rabe, A. Offenhäusser, D. Mayer and G. Figueroa-Miranda, *Biosensors*, 2025, **15**, 24.
85. J. Yan, C. Zhang, Q. Cheng, H. Liu, S. Wang, M. Zhang and L. Wang, *ACS Appl. Nano Mater.*, 2024, 7, 13241.
86. A. Lomae, K. Teekayupak, P. Preechakasedkit, E. Pasomsub, T. Ozer, C. S. Henry, D. Citterio, T. Vilaivan, O. Chailapaku and N. Ruecha, *Talanta*, 2024, **279**, 126613.
87. J. Bai, Y. Jiang, F. Tan, P. Zhu, X. Li, X. Xiong, Z. Wang, T. Song, B. Xie, Y. Yang and J. Han, *Bioelectrochemistry*, 2024, **159**, 108748.
88. J. Yan, Q. Cheng, H. Liu, L. Wang and K. Yu, *Bioelectrochemistry*, 2023, **153**, 108497.
89. W. Hu, S. Yang, X. Wang, X. Li, L. Lei, H. Lin, Q. Yuan, D. Mao and Y. Luo, *Anal. Chem.*, 2024, **96**, 19678.
90. K. Ketabi, H. Soleimanjahi, A. Teimoori, B. Hatamluyi, M. Rezayi and Z. Meshkat, *Microchim. Acta*, 2023, **190**, 293.

91. H. Zhao, W. Xie, R.-L. Zhang, X.-D. Wang, H.-F. Liu, J. Li, T. Sha, X.-S. Guo, J. Li, Q.-M. Sun, Y.-P. Zhang and C.-P. Li, *Talanta*, 2022, **237**, 122896.
92. Y. Li, W. Wang, W. Yue, Q. Lei, Z. Zhao, Y. Sun, H. Xu, W. Zhang, L. Chen, J. K. Kim and J. Hu, *Chem. Eng. J.*, 2024, **483**, 149087.
93. P. Fan, E. Ma, C. Liu, Y. Zhao, X. Wen, L. Wang, L. Li and Q. Qu, *Ionics*, 2024, **30**, 541.
94. T. Chaibun, S. Karunaithas, T. Ngamdee, R. Wasitthankasem, S. Lapchai, Y. Poovorawan, L. S. Yin and B. Lertanantawong, *Sci. Rep.*, 2024, **14**, 23792.
95. M. S. Annas, M. T. Ulhakim, N. L. W. Septiani, S. A. Abrori, A. Hermawan, O. Floweri, G. Gumilar and B. Nugraha, *Appl. Phys. A*, 2024, **130**, 956.
96. S. Traipop, S. Jampasa, P. Tangkijvanich, N. Chuaypen and O. Chailapakul, *Sens. Actuators, B*, 2023, **387**, 133769.
97. G. M. Santana, A. K. S. Silva, P. A. B. Ferreira, S. R. Benjamin, E. K. G. Trindade and R. F. Dutra, *Sens. Bio-Sens. Res.*, 2023, **42**, 100594.
98. M. R. Hasan, S. Singh, P. Sharma, C. Rawat, M. Khanuja, R. Pilloton and J. Narang, *Sensors*, 2024, **24**, 801.
99. J. C. Glória, D. S. Oliveira, A. D. Gandarilla, Y. R. Barcelay, L. A. M. Mariúba, P. A. Nogueira, W. R. Brito and F. T. C. Moreira, *ACS Omega*, 2024, **9**, 33130.
100. K. K. Dewi, N. L. W. Septiani, S. Wustoni, Nugraha, S. N. A. Jenie, R. V. Manurung and B. Yuliarto, *ACS Omega*, 2024, **9**, 1454.
101. H. Park, N. Kwon, G. Park, M. Jang, Y. Kwon, Y. Yoon, J. An, J. Min and T. Lee, *Bioelectrochemistry*, 2023, **154**, 108540.
102. N. K. Swamy, K. N. S. Mohana, A. M. Madhusudana, J. G. Manjunatha and H. M. Manukumar, *Results Chem.*, 2024, **7**, 101393.
103. M. Ma, L. He, X. Shi, Y. Wang, H. Hai and X. Wei, *Int. J. Electrochem. Sci.*, 2023, **18**, 100286.

CHAPTER 5

Mass Sensors

SEMRA AKGÖNÜLLÜ*, ILGIM GÖKTÜRK AND HANDAN YAVUZ

Department of Chemistry, Hacettepe University, Ankara, Turkey
*E-mail: semraakgonullu@hacettepe.edu.tr

5.1 Introduction

Real-time and label-free monitoring of living cells,[1] toxins,[2] viruses,[3] bacteria,[4] glucose,[5] proteins,[6] and various chemical gases[7] is crucial in food control,[8] homeland security,[9] microbiological detection,[10] pharmaceutical research,[11] environmental monitoring,[12] and so on.[13] Today, pandemics are a major threat to global health and quality of life. The formation and spread of viruses, which can lead to serious illness, outbreaks, and death, is a cause for concern.[14–16] It is still difficult to effectively diagnose pathogens before they spread to other organisms.[17] Identification of disease early and accurately is critical for effective early diagnosis and successful patient survival.[18] Analytical techniques for the detection of viruses and pathogenic microorganism require substantially lower detection limits.[19] Therefore, biological molecular recognition designs, which depend on antibodies or enzymes, are often needed for selective binding or enrichment of the target. These are usually produced following complicated protocols and need special processing conditions because of their restricted stability, thus resulting in higher costs of production. As a result, cost-effective, robust and reusable synthetic receptors that mimic the recognition features of natural antibodies will be especially beneficial for the detection and analysis of viruses in a wide range of

Detection Science Series No. 32
Advances in Virus Detection with Sensors
Edited by Adil Denizli
© The Royal Society of Chemistry 2026
Published by the Royal Society of Chemistry, www.rsc.org

health-related scenarios.[20] The mimicking of specific immunological interactions has considerably contributed to the advancement of molecularly imprinted technologies.[21,22] Although originally designed for low molecular weight components (*i.e.* small molecules), the molecularly imprinted polymer technique was then extended to macro biomolecules and even whole cells, bacteria, and viruses.[23] New recognition molecules and their integration are very attractive for clinical application due to the need to improve the detection capabilities of sensors and to enable fast analysis.[24] The development of sensor systems is an evolving, ever-expanding, and diversifying field of study. Sensors for the identification of therapeutically relevant bacteria and viruses have attracted great interest due to their being fast, cheap, and portable.[25] They are potential alternative devices that can provide fast, accurate, sensitive, and early detection.[26] They have the advantages of cheapness, simplicity, portability, ease-of-use, and mass production. In the literature, there have been reports of diverse spectroscopic, optical, piezoelectric, and electrochemical methods for the design of sensors.[27] A sensor is an instrument for detecting changes from a physical stimulus and transforming the changes into a measurable signal that makes it possible to evaluate and record them. It obtains the physical amount and transforms it into a signal that is proportional for operation. The amount of the target compound or analyte can therefore be analyzed. Nowadays, almost all sensors transform the measurement of a physical phenomenon into a measurable electrical signal. A transducer is the active part of a sensor. They can be utilized to convert one kind of energy into a different one.[28]

Viruses are small fragments that are made up of nucleic acids (DNA, RNA).[29] Viruses depend on the host cell to reproduce and replicate genetic material. Their sophisticated defense mechanisms can change rapidly.[30] Viruses can be rapidly spread between people through droplets, aerosols, and even contact with fecal substances. They are between 20 and 400 nm in diameter.[31,32] Many infectious diseases are caused by pathogenic viruses. Viruses can quickly and easily spread from infected people to healthy people. Many can live and replicate in the human body.[33] They cause a lot of infectious including hepatitis, influenza, meningitis, and acquired immune deficiency syndrome (AIDS), *etc.* The timely detection of disease-causing viruses enables the monitoring of the early medical treatment and prevention of the unwanted spread of infectious diseases. Ebola virus, Human immunodeficiency virus (HIV), avian influenza, and coronavirus severe acute respiratory syndrome (SARS) are viruses that originated in wild animals. These are all examples of zoonoses that increasingly pose a serious threat to public health and world economic growth.[34] Influenza, commonly known as “the flu”, is a rapidly wide spreading pandemic illness that has caused the deaths of between 290 000 to 650 000 people per year worldwide, according to the World Health Organization (WHO).[35] There is a novel coronavirus known as corona virus disease (Covid)-19, an infectious virus with a very high fatality risk and rate, causing the ongoing pandemic. Since December 2019, there have been more than 760 million cases and 6.9 million deaths recorded in the world, however, the actual number is estimated to be significantly higher.[36] Bacteria and viruses cause serious health problems and diseases caused by these pathogens can be a major cause of death. It is critical to prevent the spread of these easily transmissible infectious diseases at an early stage. Accurate and rapid methods that can detect these pathogens with good sensitivity are being developed to provide appropriate treatment to patients, treating the patient with appropriate therapies at the appropriate time. Currently, such techniques are needed to counter the global threat posed by these pathogens. Traditional methods, such as specific culture methods, colony counting methods, polymerase chain reaction (PCR)-based methods, and enzyme-linked immunosorbent assays (ELISAs), have been developed and tested to detect and identify these infectious viruses and bacteria in the field. These techniques are incredibly efficient and sensitive. They are also time-consuming, labour-intensive, and require expensive equipment.[37]

This chapter describes the combination of specific recognition material models with new techniques with mass-based sensor technology and methods for selective detection of viruses.

5.2 Mass Sensors

5.2.1 Acoustic Wave Biosensors

5.2.1.1 Bulk Acoustic Wave Sensors

Bulk acoustic wave (BAW) sensors are based on the inverse piezoelectric effect, in which an electrical signal induces mechanical vibrations within a piezoelectric substrate. Gravimetric quantification in rigid film implementation is enabled through a classical Sauerbrey equation.[38–40] BAW sensor devices, which are mainly operated in thickness shear mode (TSM), do not suffer from the energy dissipating overly in fluid media.

5.2.1.2 Surface Acoustic Wave (SAW) Sensors

Surface acoustic wave (SAW) sensors have emerged as a class of sensitive detection tools for monitoring biomolecular interactions. Due to their potential for sensitive, rapid, and low-cost detection, applications include medical diagnostics, environmental monitoring, food safety, and pathogen detection.[41] SAW sensors utilize the piezoelectric property to monitor mass loading and changes of interfacial properties in real-time.[42] SAW sensors function by inducing surface acoustic waves using interdigitated transducers (IDT) on piezoelectric material.[43] The waves travel through the sensor surface and interact with the biomolecules, which causes a change in the properties of the waves, like frequency, amplitude, and wave velocity.[41] Depending on the type of molecular recognition process involved in the biosensing, the SAW-based sensor can be modified and accommodates various sensors for targets ranging from cancer biomarkers to foodborne pathogens.[44]

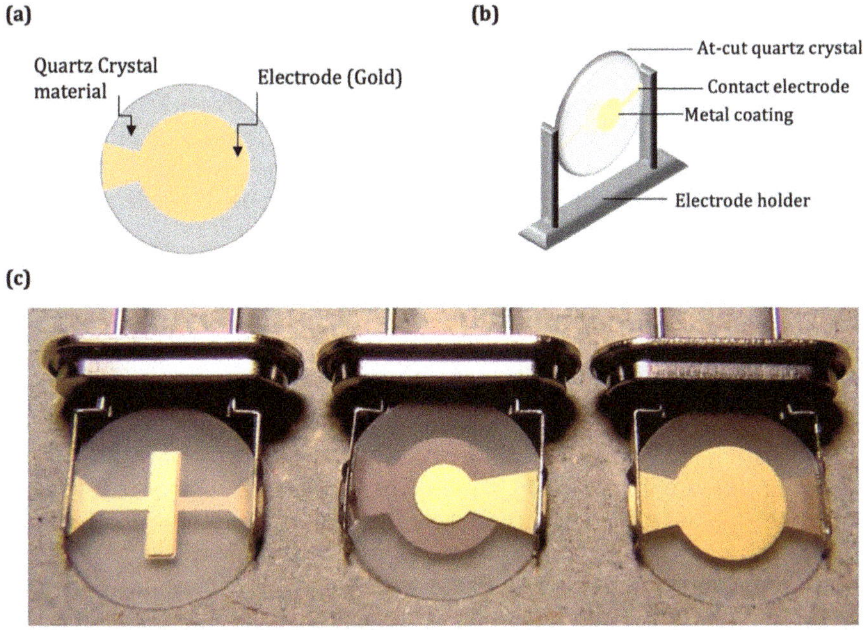

Figure 5.1 (a) Top image of the quartz crystal gold electrode;[46] (b) AT&T (AT)-cut quartz crystal;[46] (c) quartz crystal photographs.[47] (a) and (b) reproduced from ref. 46, https://doi.org/10.3390/chemosensors10030106, under the terms of the CC BY 4.0 license, https://creativecommons.org/licenses/by/4.0/. (c) reproduced from ref. 47 with permission from Elsevier, Copyright 2007.

5.2.1.3 Piezoelectric Sensors

Piezoelectric sensors are particularly popular because of their high sensitivity, low-cost, light weight, and easy application to many environments.[45] These sensors work according to the piezoelectric effect, the property some materials possess in which mechanical stress produces an electrical charge, first discovered by the Curie brothers in 1880. Harnessing the inverse piezoelectric effect governed by electrical input to facilitate mechanical deformation expands the application of these sensors.[45] When the charges are pulled out of their centers by extra forces, they build up charge on top of the material's surface and form an electrical potential. This principle has been exploited in piezoelectric sensors—to capture mechanical stimuli—making them appropriate for uses from human-motion tracking to artificial intelligence systems.

A top view of a quartz crystal electrode is shown in Figure 5.1a. The top and bottom piezoelectric gold electrodes run the oscillation of the piezoelectric sensor (Figure 5.1b).[46] Three different quartz crystals (AT-cut 10 MHz) are presented in Figure 5.1c.[47]

Piezoelectric sensors have played an important role in upgrading structural health monitoring and providing highly sensitive, lightweight, and multi-functional sensing solutions. Continued advancements in piezoelectric materials, transducer designs, and AI-driven diagnostics will continue to escalate their performance and expand their utility. The focus of the future researchers should be on self-powered, adaptive, and intelligent sensor networks to make structural health monitoring systems more efficient, reliable, and cost-effective.

5.3 Virus Detection

5.3.1 Dengue Virus

Dengue virus (DENV) is the most important mosquito-borne viral infection in humans in the tropics worldwide. Tai and co-workers[48] reported a molecularly imprinted polymer (MIP)-based quartz crystal microbalance sensor for the detection of DENV. For this purpose, the nonstructural protein 1 of flavivirus was imprinted using an epitope imprinting technique. The MIP-coated quartz crystal microbalance (QCM) showed high sensitivity for the detection of dengue virus and had a linear concentration range of 1–10 ng ml^{-1}. Analysis times in the range of 20–30 min per sample were reported. It was reported that the developed QCM sensor system and the MIP technique have the advantages of high sensitivity (approximately μg l^{-1} scale), low technical demand, easy interpretation of data, low cost, and short working time for the detection of dengue fever virus.

Lim and co-workers[49] developed a MIP-based QCM sensor with a biomimetic recognition element for quantitatively detecting DENV non-structural 1 (NS1) antigen as an early disease biomarker (Figure 5.2). The synthetic peptide mimicking a DENV NS1 epitope was used as a substitute template for the synthesis of an epitope-imprinted polydopamine (EMIPDA) sensing film on the sensor chip surface. The designed EMIPDA–QCM sensor showed a lower limit of detection and limit of quantitation calculated as 0.091 μg ml^{-1} and 0.436 μg ml^{-1}, respectively.

5.3.2 Human Immunodeficiency Virus Type 1 (HIV-1)

Human immunodeficiency virus type 1 (HIV-1) glycoprotein 41 (HIV-1 gp41) is a protein found in the virus coat of HIV-1, causative agent of acquired immunodeficiency syndrome (AIDS).[50] Lu and co-workers[51] developed an epitope imprinting polymer film based QCM sensor for the detection of HIV-1 gp41 (Figure 5.3). Dopamine was selected as a functional monomer and crosslinking agent. The HIV gp41 imprinted polymer film was characterized using Fourier transform infrared spectroscopy (FTIR) and X-ray photoelectron spectroscopy (XPS). The limit of detection for gp41 was found to be 2 ng ml^{-1}.

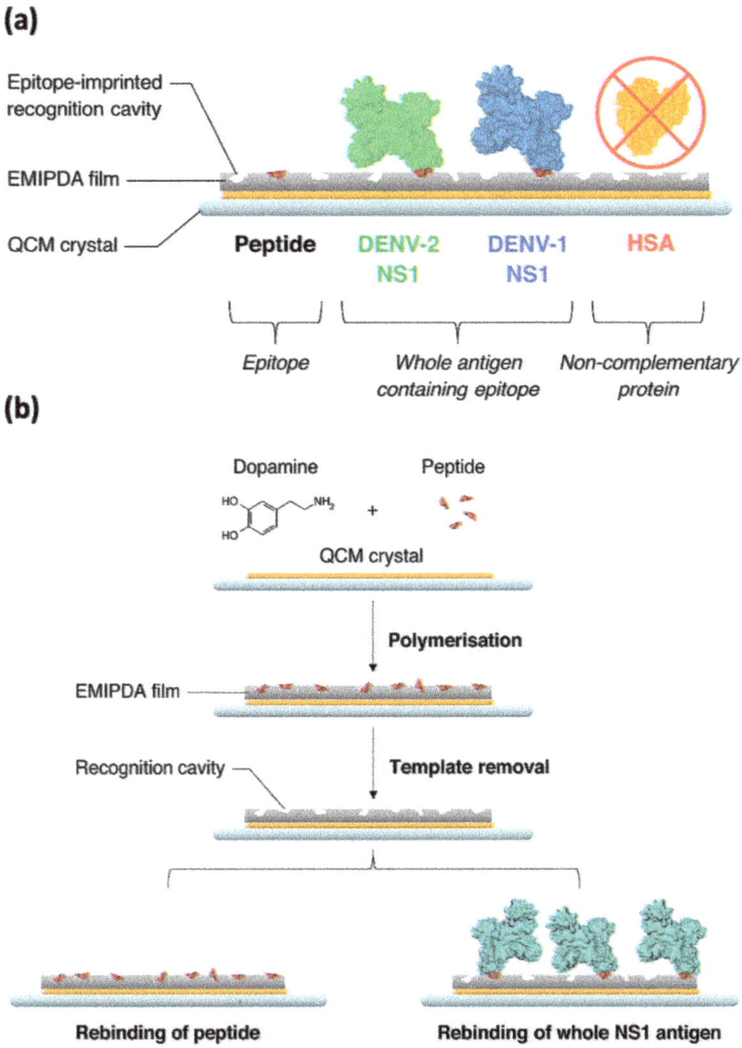

Figure 5.2 (a) Schematic image of epitope imprinting surface; (b) synthesis of EMIPDA polymer film.[49] Reproduced from ref. 49, https://doi.org/10.1016/j.talanta.2023.125376, under the terms of the CC BY 4.0 license, https://creativecommons.org/licenses/by/4.0/.

5.3.3 Human Adenovirus, Picornaviruses, and Human Papillomavirus (HPV)

Sehit and co-workers[52] reported another epitope-mediated imprinted polymer for detecting human adenovirus in biological fluid and buffer solution samples. They synthesized epitope imprinting polymers using the solid-phase synthesis technique (Figure 5.4). In which the template epitope was immobilized on a glass support prior to polymerization. The adenovirus

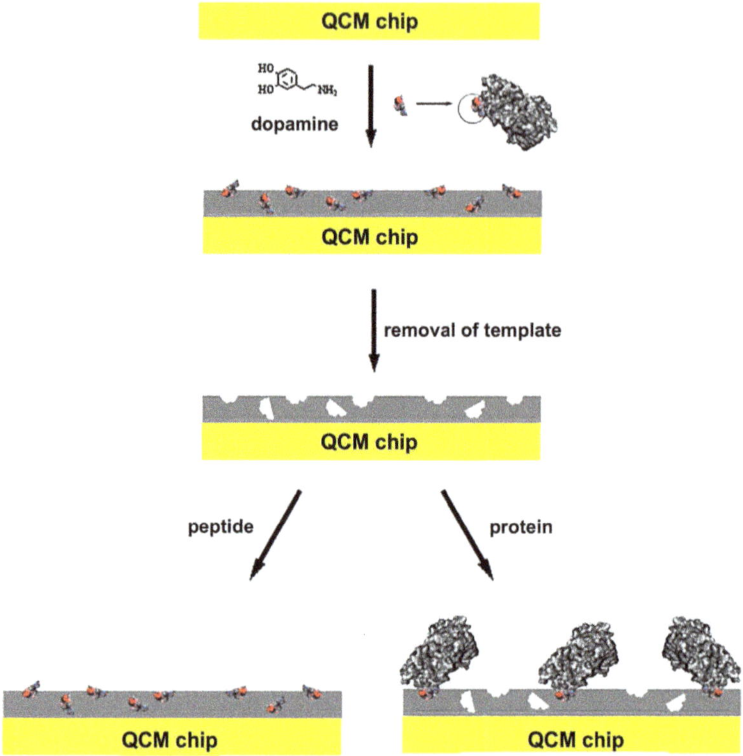

Figure 5.3 Design of QCM chip surface using epitope imprinting technique.[51] Reproduced from ref. 51 with permission from Elsevier, Copyright 2012.

epitope with 22 residues was chosen as a surrogate template. The limit of detection was 10^2 plaque forming units (PFU) ml^{-1} within the concentration range of 10^2–10^7 PFU ml^{-1}.

Picornaviruses are considered to be among the most significant human and animal pathogens and the human rhinovirus (HRV) and foot-and-mouth disease virus (FMDV) are typical models. Jenik and co-workers[53] reported a MIP-based QCM sensor for the detection of HRV and FMDV. Porcine circovirus type 2 (PCV2) is an important epidemic agricultural pathogen. Klangprapan and co-workers[54] reported a nano-MIP modified QCM sensor for the detection of PCV2. Fluorescence measurements showed a limit of detection of 47 μg ml^{-1} for the binding of MIP particles to PCV2.

Recent advancements in biosensor technology have led to the development of mass-sensitive sensors for detecting human papillomavirus (HPV). One notable example is the use of piezoelectric biosensors, which detect mass changes through variations in oscillation frequency. The developed loop-mediated isothermal amplification (LAMP)–QCM assay represents a significant advancement in the diagnostic detection of HPV-58.[55] Compared with conventional LAMP and nested PCR methods, the LAMP–QCM system offers several notable advantages. It enables rapid detection with an assay time of

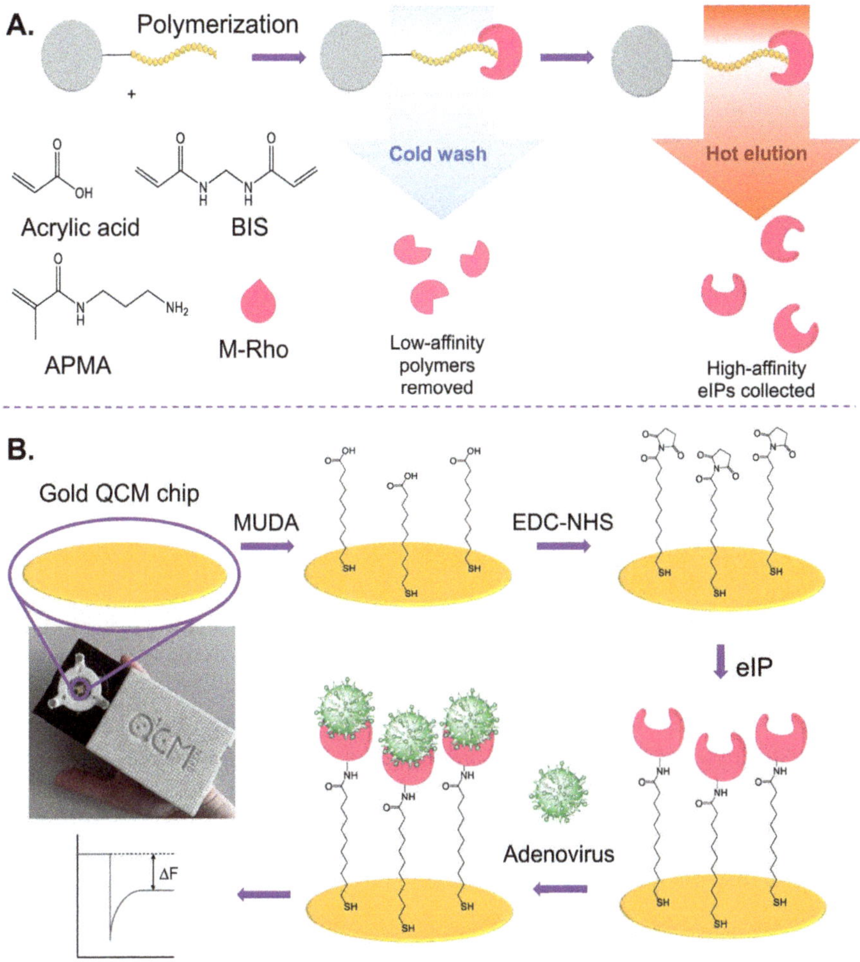

Figure 5.4 (a) Solid-phase synthesis process of epitope imprinting polymers; (b) portable epitope imprinting polymers QCM sensor surface.[52] Reproduced from ref. 52, https://pubs.acs.org/doi/10.1021/acssensors.3c02374, under the terms of the CC BY 4.0 license, https://creativecommons.org/licenses/by/4.0/.

approximately 30 minutes, in contrast to 1 hour for conventional LAMP and 6–8 hours for nested PCR. Additionally, the QCM sensor demonstrates versatility, as it can be readily adapted for the detection of various DNA targets. Overall, the LAMP–QCM assay exhibits enhanced sensitivity, improved efficiency, and broader applicability for nucleic acid-based diagnostics.

The system incorporated a peristaltic pump and temperature control unit enclosing the QCM chamber, as illustrated in Figure 5.5a. A heating block was specifically designed to maintain a constant temperature within the QCM chamber during LAMP–QCM reactions. Figure 5.5b shows the steps for QCM electrode preparation. LAMP reaction conditions were optimized to

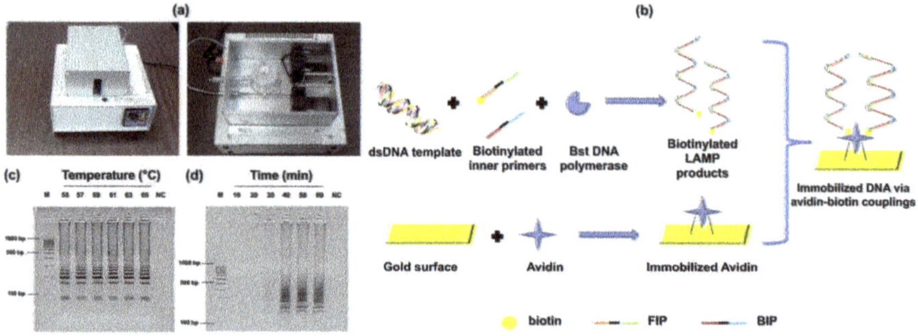

Figure 5.5 The in-house temperature control box housed the QCM sensor within a flow cell and was linked to a frequency counter and the optimal conditions of LAMP–QCM assay: (a) it featured a bottom heating block and a proportional-integral-derivative (PID) controller for precise temperature regulation. Two internal fans ensure even air circulation, maintaining a uniform temperature throughout the box. (b) Diagram of enhancement and detection of biotinylated LAMP on QCM. *Bacillus stearothermophilus* (Bst) polymerase was used. (c) Reaction temperature optimization. (d) The reactions were conducted at 57 °C for varying durations. LAMP products began to appear faintly at 30 minutes. Lane M contains HyperLadder IV DNA markers, and NC is the negative control.[55] Reproduced from ref. 55 with permission from Elsevier, Copyright 2013.

balance QCM signal stability and amplification efficiency for HPV-58 detection. Biotinylated LAMP products generated at various reaction temperatures (55–65 °C) were analyzed by gel electrophoresis (Figure 5.5c). A reaction temperature of 57 °C was selected on the basis of gel electrophoresis results, with detectable amplification achieved within 30 minutes (Figure 5.5d).

5.3.4 Ebola Virus

Acoustic-based biosensing systems, particularly QCM, show great promise for the rapid and effective detection of diseases across various applications.[56] Ebola virus (EBOV) epidemics can be severe, with mortality rates of between 50 and 90 percent. The incubation period lasts 2 to 21 days, and individuals become contagious only after symptoms appear. Deaths typically occur 7 to 10 days after symptoms start, which may include hemorrhages, coagulopathy, pain, and neurological issues like seizures. The virus is believed to amplify early in mononuclear phagocytes, leading to serious infections in the liver and lungs, endothelial leakage in the spleen, and potentially death.[57] In light of the advancements in QCM-based sensors, Yu and colleagues proposed the use of QCM immunosensors to detect EBOV glycoproteins (GP).[58] Due to their inherent sensitivity and the fact that they do not require labeling, QCM transducers are employed to create sensors for a wide range of analytes. This frequency is affected by changes in the conditions surrounding the crystal. Sauerbery's equation describes the factors that influence the

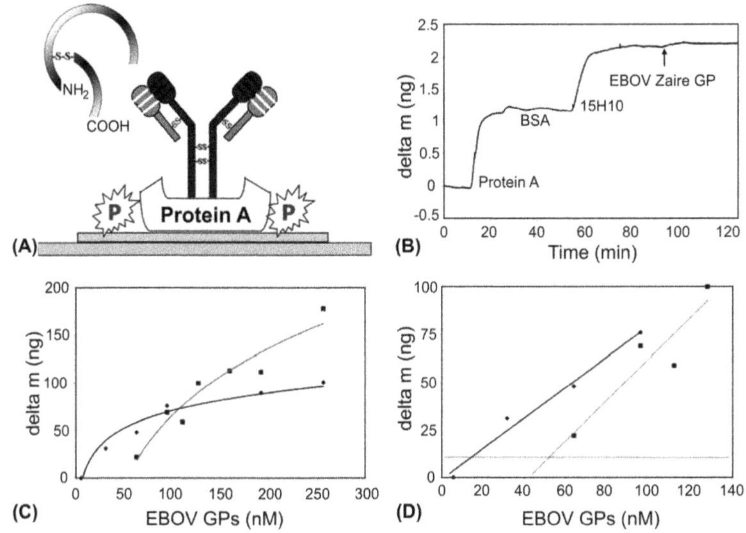

(E)

Antibody	Capture Agent Used	Capture Agent Δm (μg)	Antibody Δm (μg)
Rabbit antiserum	Protein A	1.6± 0.3	1.8± 0.2
6D11	Protein G	0.9± 0.2	0.56± 0.09
15H10	Protein A	1.8± 0.2	0.6± 0.1
17A3	Protein L	0.36± 0.04	1.3± 0.1

Figure 5.6 Detection of EBOV glycoprotein (GP) using MAb 15H10 and QCM. (a) Schematic illustration of immunosensor assembly and detection of EBOV Zaire GP. (b) QCM frequency shifts corresponding to stepwise immunosensor construction and target detection. (c) Binding of EBOV GPs from Zaire (filled diamonds) and Sudan-Gulu (filled squares) to immobilized MAb 15H10, fitted with a logarithmic model. (d) Linear response region at low concentrations used to determine limits of detection; (e) immobilization steps of capture agent and antibody.[58] Reproduced from ref. 58 with permission from Elsevier, Copyright 2006.

readout frequency.[58] The QCM system has been effectively employed for the detection of low concentrations of EBOV GP.

The QCM immunosensor was composed of a quartz crystal with a gold electrode, functionalized with a capture functional group (protein A, G, or L) as illustrated in Figure 5.6a, and real-time binding was monitored over 12 minutes (Figure 5.6b). EBOV GP detection from Zaire or Sudan-Gulu strains using rabbit antiserum, monoclonal antibody (MAb) 6D11, or MAb 17A3 yielded limited binding responses. In contrast, MAb 15H10 exhibited clear, concentration-dependent binding to both EBOV Zaire and Sudan-Gulu GPs as represented in Figure 5.6c. The lowest concentrations of EBOV glycoproteins that could be reliably detected were 14 nM for the Zaire strain and 56 nM for the Sudan-Gulu strain (Figure 5.6d). Immobilization was performed at low pH to enhance adsorption near the isoelectric points of the proteins. Protein-A was utilized for rabbit sera and MAb 15H10, while protein-G and protein-L were employed for MAbs 6D11 and 17A3, respectively (Figure 5.6e).

5.3.5 Influenza Virus

Influenza viruses are a global threat to people. The early recognition of influenza is the most efficient method of protection against flu. Typically, flu viruses are spherical, enveloped, and have a diameter of 80 to 120 nm. The three main species of influenza virus are influenza A, B, and C. Influenza A often spreads between birds and mammals, often causing death.[59] Erofeev and co-workers[60] have reported fast detection of Influenza A virus using the first radial oscillation mode of lead zirconium titanate piezoelectric (PZT) disks. PZT disks are made of a piezoelectric membrane with a radius of 2 mm and a thickness of 100 μm (Figure 5.7a). For influenza A detection, PZT disks coated with a synthetic sialylglycopolymer receptor layer were placed in a flowing virus suspension. Label-free and fast detection of Influenza A viruses at concentrations below 105 virus particles per ml has been reported. Resonance curves of the PZT disk in water and air are shown in Figure 5.7b. The 30SL-poly(acrylic acid) (PAA) polymer immobilized with a sialyloligosaccharide group on the surface of the PZT disc was selected as the virus detection sensor layer (Figure 7c).

Avian influenza virus (AIV) was first recognized in infected turkeys in 1996. Due to its high contagiousness and mortality rate, it has attracted intense attention all over the world. H9N2 can be spread between animals and humans, according to previous reports, which indicates that H9N2 is a potential threat. Zhang and co-workers[61] have developed an array of microcantilever sensors, a "reading system" that operates without an additional driver and generates an external self-sensing signal. This system has been successfully used for the label-free detection of the AIV H9N2 virus. Figure 5.7d shows a schematic diagram of the prepared microcantilever sensor surface. The immunosensor exhibits a linear detection range of 100–1000 ng ml^{-1} for AIV H9N2 virus, with a reported limit of detection of 2.9 ng ml^{-1}. This approach may enable detection of other analytes through antigen–antibody responses.

5.3.6 SARS-CoV

Comprehensive analysis of antigen–antibody interaction kinetics and immune complex formation is fundamental to the advancement of medical diagnostics and bioanalytical methodologies. These insights are crucial for the design of highly sensitive and rapid immunosensors capable of continuous monitoring. A combined spectroscopic ellipsometry and QCM with dissipation monitoring (SE/QCM-D) approach was employed by Plikusiene *et al.*, to investigate the immobilization of SARS coronavirus 2 (SARS-CoV-2) nucleocapsid protein (SCoV2-N) and its interaction with specific polyclonal antibodies (Figure 5.8a).[62] QCM-D measurements across multiple overtones (3rd, 5th, and 7th) were used to monitor real-time changes in frequency (ΔF) and dissipation (ΔD), reflecting the viscoelastic properties of the sensor surface. As shown in Figure 5.8b, the low ΔD value (1.09×10^{-6}) following covalent

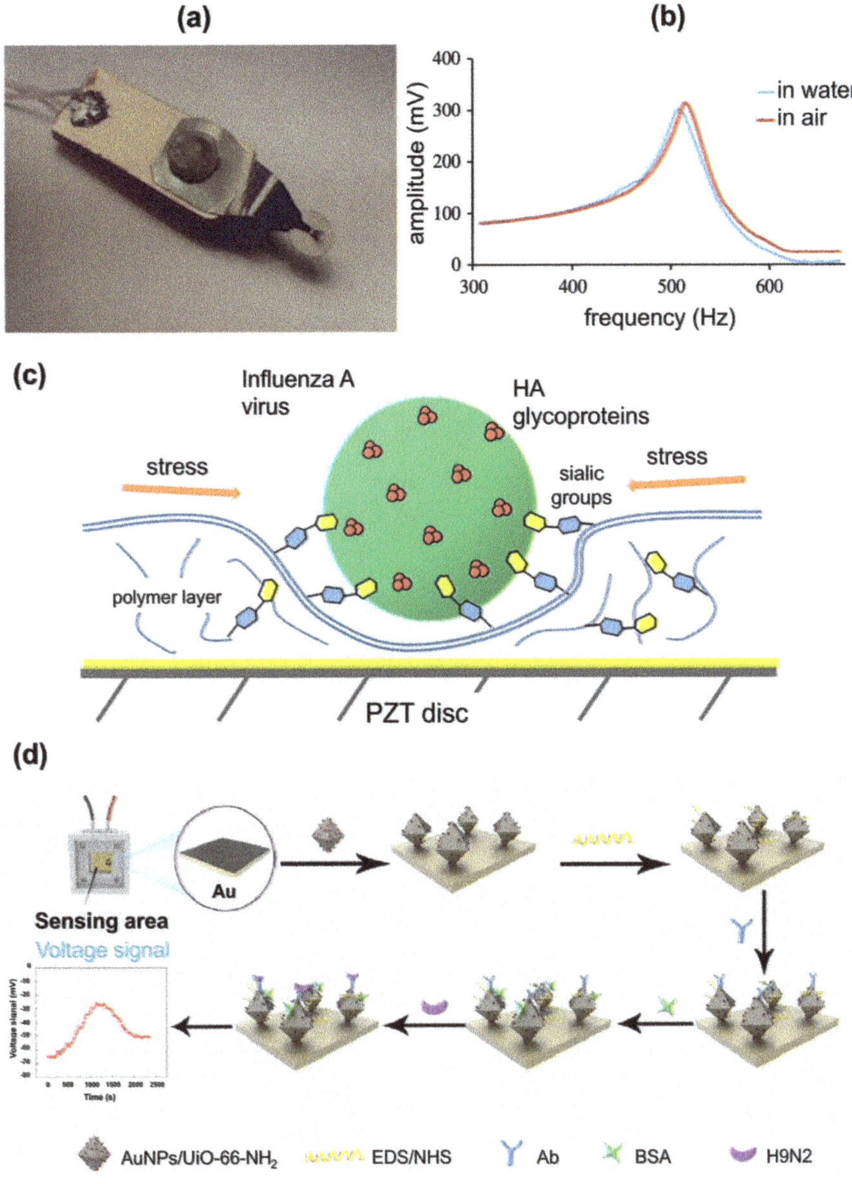

Figure 5.7 (a) Disc holder. (b) Resonance peaks of piezo transducers in air and water solution. (c) Virus particles captured on polymer coated PZT disc.[60] (d) Design of micro cantilever immunosensor surface.[61] (a)–(c) reproduced from ref. 60, https://doi.org/10.1098/rsos.190255, under the terms of the CC BY 4.0 license, http://creativecommons.org/licenses/by/4.0/. (d) Reproduced from ref. 61 with permission from Elsevier, Copyright 2023.

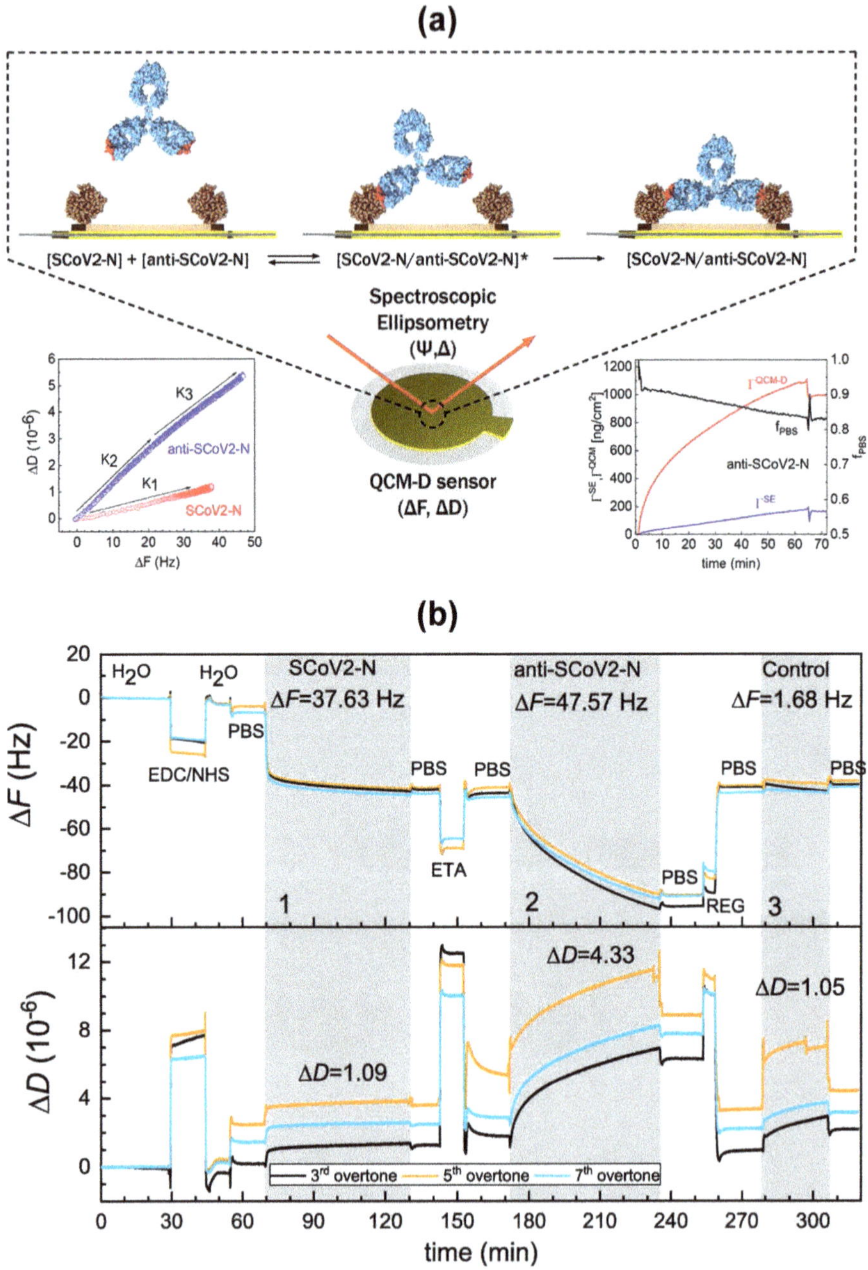

Figure 5.8 (a) Illustration of the interaction between covalently immobilized SARS-CoV2-N and *anti*-SARS-CoV2-N. (b) Real-time monitoring of frequency shifts (ΔF) and energy dissipation (ΔD) was performed for: (1) covalent immobilization of SARS-CoV2-N; (2) specific affinity interaction with *anti*-SARS-CoV2-N antibodies; and (3) nonspecific binding by polyclonal antibodies derived from non-immunized mice.[62] Reproduced from ref. 62 with permission from Elsevier, Copyright 2022.

immobilization of SCoV2-N indicates the formation of a rigid monolayer, suitable for analysis using the Sauerbrey equation.[63] Upon antibody binding, ΔD increased to 4.33×10^{-6}, indicating the formation of a more viscoelastic immune complex layer.

As a result, the SCoV2-N monolayer exhibited a rigid profile, while antibody binding formed a soft, viscoelastic immune complex layer. The spatial distribution of immobilized antigen and the structural flexibility of antibody binding fragments (Fab) facilitated bivalent binding. This study demonstrates the utility of SE–QCM-D for quantitatively characterizing the kinetics and conformational dynamics of surface-bound immune interactions.

In another study, Saleh *et al.* developed a QCM biosensor utilizing polyethylene glycol (PEG)-based surface functionalization to enhance sensitivity and specificity.[64] The platform demonstrated a detection limit of 53.3 50% tissue culture infectious dose ($TCID_{50}$) per ml and a sensitivity of 0.263 Hz per $TCID_{50}$ per ml, with assay results obtainable within approximately 15 minutes. Cross-reactivity assessments with influenza A confirmed the biosensor's high specificity toward SARS-CoV-2.

According to the kinetics of change of frequency (Hz) represented in Figures 5.9b and c, the sensor demonstrated a strong and specific response to SARS-CoV-2, evidenced by a marked frequency decrease upon viral binding, while showing minimal response to Influenza A, indicating high specificity and absence of cross-reactivity. This specificity is critical for accurate SARS-CoV-2 diagnostics, reducing false positives. Stability testing under controlled

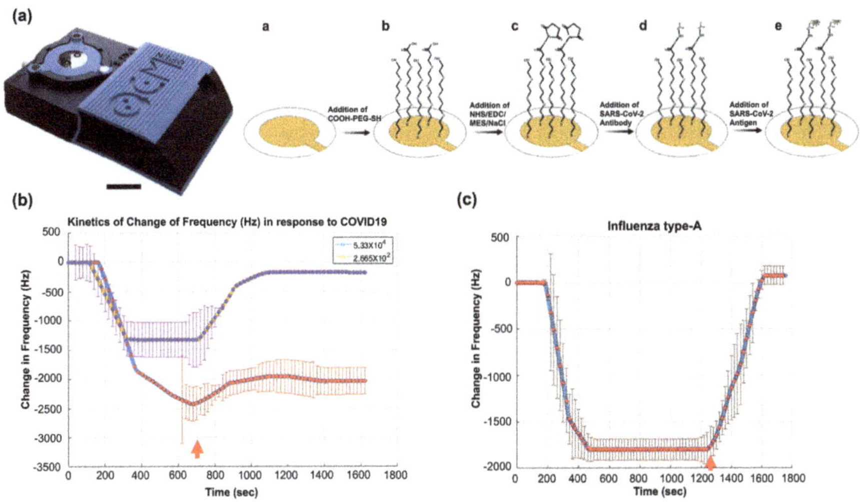

Figure 5.9 (a) OpenQCM device and the surface functionalization steps [sensor surface cleaning, thiol modification, attachment of SARS-CoV-2 nucleocapsid protein (E8R1L) mAb]. (b) The change in frequency (Hz) over time in response to COVID-19 viral solution. (c) The change in frequency (Hz) over time in response to Influenza Type-A.[64] Reproduced from ref. 64 with permission from Elsevier, Copyright 2025.

environmental conditions further supports the sensor's reliability. However, broader cross-reactivity assessments with additional respiratory viruses are recommended to fully validate clinical performance.

5.4 Conclusions

This chapter has gathered the fundamentals of recognition surfaces for the design of mass sensors for the detection of viruses. In conclusion, the mass-sensitive sensor technologies represent significant progress in virus detection, providing rapid, sensitive, and specific diagnostic tools that could enhance early detection and treatment strategies for virus-related conditions. They play a crucial role in the detection of viruses by measuring changes in mass associated with virus binding to the sensor surface. These sensors achieve high sensitivity through the use of specially designed antibodies or aptamers for virus recognition, allowing for the detection of a variety of viruses. With the ability to provide rapid results within minutes, QCM sensors offer significant advantages over traditional virus detection methods in terms of speed and efficiency.

QCM sensors have the potential to revolutionize virus detection and improve public health outcomes. Mass-based biosensors have allowed a new approach to biomolecular detection with real-time, label-free, and ultra-sensitive analytical methodology. As research continues, mass-based biosensors will probably see broader applications in the fields of personalized medicine, environmental safety, and rapid diagnostics of infectious diseases.

List of Abbreviations

BAW	Bulk acoustic wave biosensors
FTIR	Fourier transform Infrared spectroscopy
HIV	Human immunodeficiency virus
LAMP	Loop-mediated isothermal amplification
MIP	Molecularly imprinted polymer
QCM	Quartz crystal microbalance
SARS	Severe acute respiratory syndrome
SAW	Surface acoustic wave
XPS	X-ray photoelectron spectroscopic

References

1. X. Zhao, Y.-C. Tsao, F.-J. Lee, W.-H. Tsai, C.-H. Wang, T.-L. Chuang, M.-S. Wu and C.-W. Lin, Optical fiber sensor based on surface plasmon resonance for rapid detection of avian influenza virus subtype H6: Initial studies, *J. Virol. Methods*, 2016, **233**, 15–22.
2. S. Akgönüllü, H. Yavuz and A. Denizli, SPR nanosensor based on molecularly imprinted polymer film with gold nanoparticles for sensitive detection of aflatoxin B1, *Talanta*, 2020, 121219.

3. Z. Rahmati and M. Roushani, SARS-CoV-2 virus label-free electrochemical nanohybrid MIP-aptasensor based on $Ni_3(BTC)_2$ MOF as a high-performance surface substrate, *Microchim. Acta*, 2022, **189**, 287.
4. J. Shen, X. Zhou, Y. Shan, H. Yue, R. Huang, J. Hu and D. Xing, Sensitive detection of a bacterial pathogen using allosteric probe-initiated catalysis and CRISPR-Cas13a amplification reaction, *Nat. Commun.*, 2020, **11**, 267.
5. H. Teymourian, C. Moonla, F. Tehrani, E. Vargas, R. Aghavali, A. Barfidokht, T. Tangkuaram, P. P. Mercier, E. Dassau and J. Wang, Microneedle-based detection of ketone bodies along with glucose and lactate: toward real-time continuous interstitial fluid monitoring of diabetic ketosis and ketoacidosis, *Anal. Chem.*, 2020, **92**, 2291–2300.
6. F. Arcadio, L. Zeni, C. Perri, G. D'Agostino, G. Chiaretti, G. Porto, A. Minardo and N. Cennamo, Bovine serum albumin protein detection by a removable spr chip combined with a specific mip receptor, *Chemosensors*, 2021, **9**, 218.
7. R. K. Jha, J. V. D'Costa, N. Sakhuja and N. Bhat, MoSe2 nanoflakes based chemiresistive sensors for ppb-level hydrogen sulfide gas detection, *Sens. Actuators, B*, 2019, **297**, 126687.
8. H. Li, W. Ahmad, Y. Rong, Q. Chen, M. Zuo, Q. Ouyang and Z. Guo, Designing an aptamer based magnetic and upconversion nanoparticles conjugated fluorescence sensor for screening Escherichia coli in food, *Food Control*, 2020, **107**, 106761.
9. E. Aznar-Gadea, P. J. Rodríguez-Canto, J. P. Martínez-Pastor, A. Lopatynskyi, V. Chegel and R. Abargues, Molecularly Imprinted Silver Nanocomposites for Explosive Taggant Sensing, *ACS Appl. Polym. Mater.*, 2021, **3**, 2960–2970.
10. R. Gupta, N. Raza, S. K. Bhardwaj, K. Vikrant, K.-H. Kim and N. Bhardwaj, Advances in nanomaterial-based electrochemical biosensors for the detection of microbial toxins, pathogenic bacteria in food matrices, *J. Hazard. Mater.*, 2021, **401**, 123379.
11. L. Qian, S. Durairaj, S. Prins and A. Chen, Nanomaterial-based electrochemical sensors and biosensors for the detection of pharmaceutical compounds, *Biosens. Bioelectron.*, 2021, **175**, 112836.
12. O. Çakır and Z. Baysal, Pesticide analysis with molecularly imprinted nanofilms using surface plasmon resonance sensor and LC-MS/MS: Comparative study for environmental water samples, *Sens. Actuators, B*, 2019, **297**, 126764.
13. M. Wang, H. Li, T. Xu, M. Yu, G. Li, H. Zheng, J. Wu and J. Xu, Sensing and Manipulation of Bianisotropic Biomolecules Using a Surface Plasmon Resonance Based Optical Fiber Sensor, *J. Lightwave Technol.*, 2018, **36**, 5927–5934.
14. M. Hellmér, N. Paxéus, L. Magnius, L. Enache, B. Arnholm, A. Johansson, T. Bergström and H. Norder, Detection of pathogenic viruses in sewage provided early warnings of hepatitis A virus and norovirus outbreaks, *Appl. Environ. Microbiol.*, 2014, **80**, 6771–6781.

15. N. E. Moore, J. Wang, J. Hewitt, D. Croucher, D. A. Williamson, S. Paine, S. Yen, G. E. Greening and R. J. Hall, Metagenomic Analysis of Viruses in Feces from Unsolved Outbreaks of Gastroenteritis in Humans, *J. Clin. Microbiol.*, 2015, **53**, 15–21.
16. C. Herrington, P. Coates and W. Duprex, Viruses and disease: emerging concepts for prevention, diagnosis and treatment, *J. Pathol.*, 2015, **235**, 149–152.
17. N. Boudaud, C. Machinal, F. David, A. Fréval-Le Bourdonnec, J. Jossent, F. Bakanga, C. Arnal, M. P. Jaffrezic, S. Oberti and C. Gantzer, Removal of MS2, Qβ and GA bacteriophages during drinking water treatment at pilot scale, *Water Res.*, 2012, **46**, 2651–2664.
18. S. Asir, M. Bakhshpour, S. Unal and A. Denizli, *Nanoparticle-based plasmonic devices for bacteria and virus recognition*, Elsevier Inc., 2021.
19. J. T. Connelly and A. J. Baeumner, Biosensors for the detection of waterborne pathogens, *Anal. Bioanal. Chem.*, 2012, **402**, 117–127.
20. M. Gast, H. Sobek and B. Mizaikoff, Advances in imprinting strategies for selective virus recognition a review, *TrAC, Trends Anal. Chem.*, 2019, **114**, 218–232.
21. K. Mosbach and O. Ramström, The Emerging Technique of Molecular Imprinting and Its Future Impact on Biotechnology, *Nat. Biotechnol.*, 1996, **14**, 163–170.
22. G. Wulff, Molecular Imprinting in Cross-Linked Materials with the Aid of Molecular, *Angew Chem., Int. Ed. Engl.*, 1995, **34**, 1812–1832.
23. M. Gast, H. Sobek and B. Mizaikoff, Advances in imprinting strategies for selective virus recognition a review, *TrAC, Trends Anal. Chem.*, 2019, **114**, 218–232.
24. Y. Saylan and A. Denizli, Virus detection using nanosensors, in *Nanosensors for Smart Cities*, 2020, pp. 501–511.
25. S. A. Piletsky, N. W. Turner and P. Laitenberger, Molecularly imprinted polymers in clinical diagnostics—Future potential and existing problems, *Med. Eng. Phys.*, 2006, **28**, 971–977.
26. X. Huang, Y. Zhu and E. Kianfar, Nano Biosensors: Properties, applications and electrochemical techniques, *J. Mater. Res. Technol.*, 2021, **12**, 1649–1672.
27. B. D. Gupta, A. Pathak and V. Semwal, Carbon-Based Nanomaterials for Plasmonic Sensors: A Review, *Sensors*, 2019, **19**, 3536.
28. S. Kurbanoglu and S. A. Ozkan, Electrochemical carbon based nanosensors: A promising tool in pharmaceutical and biomedical analysis, *J. Pharm. Biomed. Anal.*, 2018, **147**, 439–457.
29. S. Akgönüllü, M. Bakhshpour, Y. Saylan and A. Denizli, Virus-Based Nanocarriers for Targeted Drug Delivery, *Viral and Antiviral Nanomaterials*, 2021, 173–191.
30. Y. Saylan, Ö. Erdem, S. Ünal and A. Denizli, An alternative medical diagnosis method: Biosensors for virus detection, *Biosensors (Basel)*, 2019, **9**, 65.
31. Y. Saylan and A. Denizli, *Biosensors for Virus Detection*, IOP Publishing, Bristol, UK, 2021, pp. 1–21.

32. A. Denizli and Y. Saylan, *Principles of Plasma Spectroscopy*, IOP Publishing, Bristol, UK, 2021.
33. S. Akgönüllü and A. Denizli, *Biosensors for Virus Detection*, IOP Publishing, Bristol, UK, 2021, pp. 4–18.
34. S. Akgönüllü and A. Denizli, Nano-sensors and nano-devices for biological disaster monitoring (virus/disease epidemics/animal plagues detections), in *Nanotechnology-Based Smart Remote Sensing Networks for Disaster Prevention*, 2022, pp. 43–57.
35. WHO, Influenza (Seasonal), https://www.who.int/news-room/fact-sheets/detail/influenza-.
36. WHO, Coronavirus Disease, https://www.who.int/news-room/fact-sheets/detail/coronavirus-disease-(covid-19).
37. A. K. Yadav, D. Verma, N. Dalal, A. Kumar and P. R. Solanki, Molecularly imprinted polymer-based nanodiagnostics for clinically pertinent bacteria and virus detection for future pandemics, *Biosens. Bioelectron.: X*, 2022, **12**, 100257.
38. G. Sauerbrey, The use of quartz oscillators for weighing thin layers and for microweighing, *Z. Med. Phys.*, 1959, **155**(2), 206–222.
39. M. C. Marco de Lucas, B. Jacquier, E. Lebrasseur, J. Y. Rivoire and B. Varrel, New setup for end-fire coupling excitation of planar waveguides at low temperatures, *Rev. Sci. Instrum.*, 1997, **68**, 2275–2278.
40. R. Fogel, J. Limson and A. A. Seshia, Acoustic biosensors, *Essays Biochem.*, 2016, **60**, 101–110.
41. S. I. Zida, Y. Der Lin and Y. L. Khung, Current trends on surface acoustic wave biosensors, *Adv. Mater. Technol.*, 2021, **6**, 2001018.
42. A. Janshoff, H. J. Galla and C. Steinem, Piezoelectric mass-sensing devices as biosensors—an alternative to optical biosensors?, *Angew. Chem., Int. Ed.*, 2000, **39**(22), 4004–4032.
43. R. M. White and F. W. Voltmer, Direct piezoelectric coupling to surface elastic waves, *Appl. Phys. Lett.*, 1965, **7**, 314–316.
44. D. S. Ballantine, S. J. Martin, A. J. Ricco, G. C. Frye, H. Wohltjen, R. M. White and E. T. Zellers, *Acoustic Wave Sensors*, Elsevier, 1997, pp. 1–9.
45. M. Ju, Z. Dou, J. W. Li, X. Qiu, B. Shen, D. Zhang, F. Z. Yao, W. Gong and K. Wang, Piezoelectric materials and sensors for structural health monitoring: fundamental aspects, current status, and future perspectives, *Sensors*, 2023, **23**(1), 543.
46. S. Akgönüllü, E. Özgür and A. Denizli, Recent advances in quartz crystal microbalance biosensors based on the molecular imprinting technique for disease-related biomarkers, *Chemosensors*, 2022, **10**(3), 106.
47. H. Anderson, M. Jönsson, L. Vestling, U. Lindberg and T. Aastrup, Quartz crystal microbalance sensor design. I. Experimental study of sensor response and performance, *Sens. Actuators, B*, 2007, **123**, 27–34.
48. D.-F. Tai, C.-Y. Lin, T.-Z. Wu, J.-H. Huang and P.-Y. Shu, Artificial Receptors in Serologic Tests for the Early Diagnosis of Dengue Virus Infection, *Clin. Chem.*, 2006, **52**, 1486–1491.

49. H. J. Lim, T. Saha and C. W. Ooi, Site-specific imprinting of dengue virus non-structural 1 antigen on a polydopamine-based sensing film for early detection and prognosis of dengue, *Talanta*, 2024, **268**, 125376.
50. X. Wang, M. Cao, Y. Wu, W. Xu, Q. Wang, T. Ying, L. Lu and S. Jiang, Synergistic Effect by Combining a gp120-Binding Protein and a gp41-Binding Antibody to Inactivate HIV-1 Virions and Inhibit HIV-1 Infection, *Molecules*, 2021, **26**, 1964.
51. C. H. Lu, Y. Zhang, S. F. Tang, Z. Bin Fang, H. H. Yang, X. Chen and G. N. Chen, Sensing HIV related protein using epitope imprinted hydrophilic polymer coated quartz crystal microbalance, *Biosens. Bioelectron.*, 2012, **31**, 439–444.
52. E. Sehit, G. Yao, G. Battocchio, R. Radfar, J. Trimpert, M. A. Mroginski, R. Süssmuth and Z. Altintas, Computationally Designed Epitope-Mediated Imprinted Polymers versus Conventional Epitope Imprints for the Detection of Human Adenovirus in Water and Human Serum Samples, *ACS Sens.*, 2024, **9**, 1831–1841.
53. M. Jenik, R. Schirhagl, C. Schirk, O. Hayden, P. Lieberzeit, D. Blaas, G. Paul and F. L. Dickert, Sensing Picornaviruses Using Molecular Imprinting Techniques on a Quartz Crystal Microbalance, *Anal. Chem.*, 2009, **81**, 5320–5326.
54. J. Klangprapan, W. A. A. Sudjarwo, P. A. Lieberzeit and K. Choowongkomon, Synthesis and characterization of molecularly imprinted polymer nanoparticles against porcine circovirus type 2 viral-like particles, *Anal. Bioanal. Chem.*, 2024, **416**, 7357–7368.
55. P. Prakrankamanant, C. Leelayuwat, C. Promptmas, T. Limpaiboon, S. Wanram, P. Prasongdee, C. Pientong, J. Daduang and P. Jearanaikoon, The development of DNA-based quartz crystal microbalance integrated with isothermal DNA amplification system for human papillomavirus type 58 detection, *Biosens. Bioelectron.*, 2013, **40**, 252–257.
56. N. Karaseva and T. Ermolaeva, A regenerable piezoelectric immunosensor on the basis of electropolymerized polypyrrole for highly selective detection of Staphylococcal Enterotoxin A in foodstuffs, *Microchim. Acta*, 2015, **182**, 1329–1335.
57. P. Sharma, S. Suleman, A. Farooqui, W. Ali, J. Narang, S. J. Malode and N. P. Shetti, Analytical methods for Ebola virus detection, *Microchem. J.*, 2022, **178**, 107333.
58. J. S. Yu, H. X. Liao, A. E. Gerdon, B. Huffman, R. M. Scearce, M. McAdams, S. M. Alam, P. M. Popernack, N. J. Sullivan, D. Wright, D. E. Cliffel, G. J. Nabel and B. F. Haynes, Detection of Ebola virus envelope using monoclonal and polyclonal antibodies in ELISA, surface plasmon resonance and a quartz crystal microbalance immunosensor, *J. Virol. Methods*, 2006, **137**, 219–228.
59. S. C. B. Gopinath, T. H. Tang, Y. Chen, M. Citartan, J. Tominaga and T. Lakshmipriya, Sensing strategies for influenza surveillance, *Biosens. Bioelectron.*, 2014, **61**, 357–369.

60. A. S. Erofeev, P. V. Gorelkin, D. V. Kolesov, G. A. Kiselev, E. V. Dubrovin and I. V. Yaminsky, Label-free sensitive detection of influenza virus using PZT discs with a synthetic sialylglycopolymer receptor layer, *R. Soc. Open. Sci.*, 2019, **6**(9), 190255.
61. Y. Zhang, F. Shi, C. Zhang, X. Sheng, Y. Zhong, H. Chong, Z. Yang and C. Wang, Detection of avian influenza virus H9N2 based on self-driving and self-sensing microcantilever piezoelectric sensor, *Chin. Chem. Lett.*, 2023, **34**(4), 107700.
62. I. Plikusiene, V. Maciulis, S. Juciute, A. Ramanavicius and Z. Balevicius, Investigation of SARS-CoV-2 nucleocapsid protein interaction with a specific antibody by combined spectroscopic ellipsometry and quartz crystal microbalance with dissipation, *J. Colloid Interface Sci.*, 2022, **626**, 113–122.
63. P. Hampitak, D. Melendrez, M. Iliut, M. Fresquet, N. Parsons, B. Spencer, T. A. Jowitt and A. Vijayaraghavan, Protein interactions and conformations on graphene-based materials mapped using a quartz-crystal microbalance with dissipation monitoring (QCM-D), *Carbon*, 2020, **165**, 317–327.
64. S. Saleh, H. Alkalamouni, K. Antar, J. Rahme, M. Kazan, P. Karam, J. Muthuswamy, H. Zaraket and M. L. Khraiche, Quartz crystal microbalance-based biosensor for rapid and ultrasensitive SARS-CoV-2 detection, *J. Pharm. Biomed. Anal. Open*, 2025, **5**, 100071.

CHAPTER 6

Comparative Analysis of Different Sensor Types

S. IREM KAYA[a], AHMET CETINKAYA[a] AND SIBEL A. OZKAN[b*]

[a]Department of Analytical Chemistry, Gülhane Faculty of Pharmacy, University of Health Sciences, Ankara, Turkey; [b]Department of Analytical Chemistry, Faculty of Pharmacy, Ankara University, Ankara, Turkey
*E-mail: ozkan@pharmacy.ankara.edu.tr

6.1 Introduction

The tiniest microbes, viruses, exist in various shapes and sizes and seriously harm human health.[1] As obligatory intracellular parasites, viruses are unable to reproduce without a host, and adapt by dismantling and modifying the host's immune response. When compared to bacteria, viruses are substantially tiny (20–500 nm). Compared to other microbes, the nanoscale size poses a barrier for viral identification and bio-imaging. Because viruses are so contagious, there are risks to human health. One of the main causes of illnesses that result in hundreds of thousands of fatalities annually is virus infection. There has been a notable increase in infectious diseases that directly affect all living things, including humans, animals, and plants.[2] Especially in many nations and impoverished areas of contemporary society, infectious diseases, like the human immunodeficiency virus, impact a large number of individuals and continue to cause serious health issues.[3]

Detection Science Series No. 32
Advances in Virus Detection with Sensors
Edited by Adil Denizli
© The Royal Society of Chemistry 2026
Published by the Royal Society of Chemistry, www.rsc.org

Significant medical obstacles have yet to be overcome in treating millions of people afflicted by various ailments. Viral infections like severe acute respiratory syndrome coronavirus 2 (SARS-CoV-2), dengue, hepatitis B, and influenza have been responsible for several recent outbreaks. The World Health Organization (WHO) identified SARS-CoV-2 as a global public health concern and pandemic in March 2020.[4] Another virus, which spreads rapidly through mosquitoes, is dengue, which has grown to be a serious public health issue, particularly in tropical nations like Bangladesh. Dengue primarily affects children, adolescents, and young adults globally; newborns are most commonly infected.[5] A major global health concern, the hepatitis B virus has a high transmission capacity. Hepatitis B problems impact between 257 and 291 million people despite the existence of an effective vaccine.[6] Last but not least, influenza is an acute respiratory illness that spreads easily. Typically, the disease manifests as a seasonal influenza outbreak. Influenza can be lethal if improperly handled. Globally, influenza is thought to cause 35 million severe cases and between 290 000 and 650 000 fatalities annually. Since the turn of the century, there have been four influenza pandemics. The influenza virus pandemics were H1N1 Spanish (1918), H2N2 Asian (1957), H3N2 Hong Kong (1968), and H1N1 swine (2009).[7]

The main cause of the increasing health effects caused by viruses is the absence of effective detection techniques. Depending on the type of virus, several basic methods are available for virus identification. Conventional virus detection techniques usually involve long-drawn-out laboratory procedures utilizing sophisticated analytical equipment.[8] While classical methods may identify viruses with high accuracy, their main disadvantages are their time-consuming, extensive procedures, costly consumables, complex sample preparation, and requirements for skilled personnel.[9] Thus, creating quick, low-cost, and highly accurate virus detection assays is critical. In this context, it is a hopeful sign that biosensors containing bio-recognition units can detect viruses with high sensitivity and selectivity.[10,11] Numerous biosensors have been proposed for the detection of viruses, including lateral flow immunoassay (LFIA) and surface-enhanced Raman scattering (SERS) using nanoscale materials,[12] colorimetric,[13] fluorometric,[14] electrochemical,[15] photoluminescence,[16] and quartz crystal microbalance (QCM).[17]

Nanotechnology and nanoscience have recently been shown to be effective diagnostic and detection tools for infectious viruses. Biosensors based on nanomaterials have emerged as one of the important topics in the field of virus detection and diagnostics. The high sensitivity, selectivity, and bio-imaging applications of nanomaterial-based sensors are well-known. Additionally, sensors based on nanomaterials can exhibit quick responses and a low detection limit. Biosensors with nanomaterials are attractive for viral detection because of these advantages.[18] Current research has led to new biosensor designs based on various nanomaterials, including magnetic nanoparticles (MNPs),[19] organic molecule based nanostructures,[20] upconversion nanomaterials (UCNPs),[21] gold nanoparticles (AuNPs),[22,23] and quantum dots (QDs).[24] Despite being a significant contributor to increased sensitivity, nanomaterial-based sensors are expensive and complicated, which may limit their widespread application.

In this chapter, the advantages and disadvantages of sensors using different sensing methods (electrochemical, optical, and mass-based) for detecting viruses are discussed in detail. The basic parameters to be considered in developing the sensor, such as sensitivity, specificity, linear range, limit of detection (LOD), speed, cost-effectiveness, and accuracy and applicability for real sample analysis, are summarized. Within this framework, recent reported uses of mass-based, optical, and electrochemical sensors in the detection of viruses, including influenza, hepatitis B, dengue fever, and SARS-CoV-2, are examined. Additionally, each sensor's unique benefits and drawbacks are assessed, along with the applications for which it works best. Lastly, sensor performance challenges and future perspectives in virus analysis are emphasized.

6.2 Key Parameters for Comparative Analysis

Developing a new analytical technique requires validation. The productivity of analytical and/or electroanalytical methods must be validated and evaluated to ascertain the degree of predicted error resulting from imprecision and inaccuracy and confirm that the degree of error satisfies anticipated laboratory or clinical criteria. The types of tests and expected applications determine the technique recommended for method validation. In particular, the different analytical methods used to detect viruses and important validation parameters are examined in detail. Although different analytical methods used to detect viruses have advantages and disadvantages with respect to each other, they should be comparable in terms of reliability, sensitivity, selectivity, specificity, and robustness. Therefore, before being accepted, the analytical measurements that will be used need to be confirmed and validated.[25]

6.2.1 Selectivity

Selectivity is the primary factor to consider while choosing an analytical method. Detection of the substance under analysis may be affected by its capacity to be assessed independently of other elements in the sample. A high degree of selectivity is necessary for both qualitative and quantitative aims. Insufficient selectivity casts doubt on the accuracy and precision of the procedure. Selectivity assessment for the specific methodology may vary with the potential interfering compounds. If a compound is present in the sample or matrices, the selectivity is verified using spiking analytes and compared with a blank matrix from many sources. However, samples with and without analytes are examined to test the selectivity, along with any potential interferences. The assessment of greater analyte concentrations is advised in relation to selectivity.[25]

6.2.2 Sensitivity and Specificity

Another crucial method for identifying departures from linearity is sensitivity. It is the rate at which a measuring device's response changes to a change in concentration. The response curve's slope indicates a method's sensitivity. It can be measured during linearity experiments and is referred to as the calibration line's slope. Sensitivity is frequently combined with the limit of detection (LOD) and limit of quantification (LOQ) levels of the approach. There are two complementary methods to define sensitivity. Analyte detection and quantification in the application environment is one of them. The other is the ability to distinguish between similar concentrations of an analyte in the application setting. Moreover, very low analyte concentrations must be detected in a sample to confirm the presence of trace amounts.[26–28]

Specificity, which characterizes the capacity of an analytical method to distinguish between targeted and untargeted biological entities in a sample, is considered a most important property. The capacity of an analytical technique to identify and measure analytes in a sample when endogenous matrix components, metabolites, degradation products, and other components are present is known as specificity.[26–28]

6.2.3 Detection Ranges and Limits

The function of the analyte concentration to the analytical methodology's response is referred to as linearity. The range of analyte concentrations at which the compound response changes linearly with concentration is known as the working range, and it is another phrase related to linearity.[29] The property known as linearity, which is expressed mathematically as $y = mc + n$, where c is the analyte concentration, y is the output signal, n is the intercept, and m is the analyte sensitivity, determines how accurately the measured response to a straight line is accurate. The study results are determined using statistical computations if the investigation reveals a linear relationship. The expression of linearity studies the linearity's slope (m), cut (y), correlation coefficient (r), and determination coefficient (r^2). The r and r^2 coefficients should have a mathematical value that is near 1.00.

Additionally, 0.99 is a suitable number to illustrate the linearity of the responses. Relative standard deviation (RSD%) or standard error (S.E.) values of slope and intersection are computed and provided to control linearity and associated validation parameters. The link between the concentration and the analytical response is expressed numerically by the correlation coefficient (r). The target concentration of the substance under study should be included in the concentration range. It should have at least the expected or necessary range of analytical results, the latter of which is directly tied to the specification's acceptance limits or the analytical method style's target test concentration. The analyte, process type, and application environment affect the working range. According to the suggested analytical procedure, the range is often given in the same units as the test results.[25]

The lowest quantity of analyte that can be identified is known as the detection limit. This parameter becomes particularly significant when the threshold limit is extremely low. The phrases "detection limit" and "sensitivity" are interchangeable. It also identifies the lowest concentration of the examined sample, which falls within the experiment's parameters but is impossible to measure exactly. Moreover, the LOD—also known as matrix sensitivity—can be changed within the matrix. Crucially, that ought to be ascertained in a matrix corresponding to the sample matrix. Depending on whether tools are used or not, several methods are used to calculate LOD. Both official guidelines and pharmacopeias endorse the most common method of calculation. By comparing the detected signals from samples known to have low analyte concentrations with those of blank samples, the signal to noise ratio (S : N) is calculated to establish the lowest concentration at which the analysis may be trusted. The lowest concentration at which the S : N ratio of 3 : 1 is achieved is used to calculate LOD. In studies performed using analytical methods, determining the LOD using the standard deviation of the response and the slope of the associated calibration curve is another frequently used computational technique. An analyte concentration around the LOD level is present in the samples used to analyse a particular calibration equation. The calculation for this method, for instance, can be expressed using the following equation:

$$\mathrm{LOD} = \frac{3.3 \times \sigma}{m}$$

where σ is the standard deviation of the signal and m is the slope of the corresponding calibration graph.[26,28] The lowest analyte concentration in a sample that can be quantified with the proper accuracy and precision is known as the limit of quantification, or LOQ. Less than or equal to the lower limit of its range, the LOQ is always greater than the LOD. If the result is found between LOD and LOQ, it can be reported that the analyte has been detected but at a level below the LOQ, and methods for calculating both LOD and LOQ are the same. The LOQ is the lowest concentration level of calibration at which the analyte may be quantified with a satisfactory degree of accuracy and precision when employing the visual evaluation method. LOQ can also be calculated by multiplying LOD by 3. The LOQ is the lowest measurement concentration at which the analyte may be accurately measured at 10 : 1 S : N, which is the primary distinction between the LOQ and LOD in terms of the S : N ratio. By comparing the observed signals from samples with known low analyte concentrations with blank samples that are similar to LOD, the S : N ratio is calculated using the approach used for LOQ calculation. According to the International Council for Harmonisation of Technical Requirements for Pharmaceuticals for Human Use (ICH) recommendations, the most common and suitable method of determining LOQ is to apply the following equation:

$$\mathrm{LOQ} = \frac{10 \times \sigma}{m}$$

The slope of the appropriate calibration line is denoted by m in the equation above, whereas σ represents the standard deviation.[26,28]

6.2.4 Accuracy and Precision

One of the most important criteria for validating analytical methods is accuracy. It describes how closely the mean test results produced by the suggested methodology match the analyte's nominal concentration. It also measures the discrepancy between the expected test result and the recognized reference value due to experimental and methodological errors. Lastly, accuracy is a qualitative term that describes how closely analytical measurement data match the reference value, which, depending on the definition, maybe the actual or predicted value. In addition, the recovery percentage using the suggested analytical method for a known amount of working compound included in the actual sample or the difference between the mean and the true value (Bias%) with confidence intervals are the two most common ways to report accuracy.[28,30–33]

The slope value of the calibration line's equation indicates the precision. Deviations from linearity can be found by examining precision. Under actual operating conditions, it is challenging to get good reproducibility and repeatability results from the analytical approach. The percentage of the coefficient of variation or RSD% indicates how accurate the analytical approach is. Three steps make up precision: reproducibility, intermediate precision, and repeatability. At least three concentrations and five repeated analyses per level should be used for precision.[26,28,34–36]

6.3 Performance Comparison of Currently Available Sensors for Virus Detection

Sensors developed for virus detection can be grouped as optical, electrochemical, and mass-based sensors. Since one of the most important purposes of virus detection is to be used as a diagnostic tool, some key points that stand out in sensor performance are rapidity, accuracy, sensitivity, and reliability. As mentioned in the previous section, these are associated with evaluation of various validation parameters. In this section, the application performances of different techniques for virus detection [such as voltammetry, QCM, and surface plasmon resonance (SPR)] are examined. Table 6.1 compares these studies with respect to technique, sensitivity, linear range, LOD, sample type, and recovery.

As experienced in the global coronavirus pandemic crisis, the most critical point in preventing the spread of the virus is the development of platforms that provide faster and more reliable results instead of traditional diagnostic tests. The main method used in the laboratory environment to diagnose SARS-CoV-2 is reverse transcriptase polymerase chain reaction (RT-PCR). However, it is impossible to get results quickly from a large number of samples with

Table 6.1 Performance comparison of currently available sensors for virus detection.[a]

		Method		
Virus	Parameter	Electrochemical	Optical	Mass
SARS-CoV-2	Technique	Chronoamperometry[15] (viral spike protein S1)	SPR[37] (virus spike receptor-binding domain)	QCM[38] (virus spike receptor-binding domain)
	Sensitivity	0.2478 mA cm^{-2} ml μg^{-1}	146.75556° RIU^{-1}	−10.331 Hz ml pg^{-1}
	Linear range	0.5–2 μg ml^{-1}	1.20–1.35 RIU	1 pg ml^{-1}–0.1 μg ml^{-1}
	LOD	0.074 fg ml^{-1}	NA	0.07 pg ml^{-1}
	Sample type	Saliva samples	Nasal and throat swabs	Saliva and plasma
	Recovery%	NA	NA	NA
SARS-CoV-2	Technique	DPV[39] (rtDNA)	Photoluminescence[16] (antibodies of recombinant N- and S-proteins)	NA
	Sensitivity	NA	NA	
	Linear range	8.83 fg ml^{-1}–88.25 pg ml^{-1}	0.025–0.5 nM (S-protein) 0.3–1 nM (N-protein)	
	LOD	0.088 pg ml^{-1}	0.01 nM (S-protein) 0.3 nM (N-protein)	
	Sample type	Serum samples	Nasopharyngeal swab specimens	
	Recovery%	99.00–102.31	NA	
SARS-CoV-2	Technique	SWV[40] (viral spike protein S1)	SERS[41] (spike receptor-binding domain S1)	NA
	Sensitivity	NA	NA	
	Linear range	26.7–194 fM (in PBS) 50–400 fM (in nasopharyngeal swap sample)	NA	
	LOD	15 fM (in PBS) 64 fM (in nasopharyngeal swap sample)	2.4 pg ml^{-1}	
	Sample type	Nasopharyngeal swab sample	Nasopharyngeal swab samples	
	Recovery%	NA	NA	

Table 6.1 (*Continued*)

			Method	
Virus	Parameter	Electrochemical	Optical	Mass
SARS-CoV-2	Technique	DPV[42] (SARS-CoV-2 nucleoprotein)	Colorimetric[43] (viral spike protein S1)	NA
	Sensitivity	0.002 μA fM^{-1}	0.0037 A ml ng^{-1}	
	Linear range	2.22–111 fM	130 ng ml^{-1}–1.3 pg ml^{-1}	
	LOD	15 fM	130 fg ml^{-1}	
	Sample type	Nasopharynx swab specimens	Saliva sample	
	Recovery%	NA	NA	
Dengue type 1 virus	Technique	DPV[44] (non-structural dengue protein NS1)	SPR[45] (dengue virus envelope protein)	QCM[17]
	Sensitivity	1.78 ng ml^{-1} mA l^{-}	39.96° nM^{-1}	189.61 Hz ml pfu^{-1}
	Linear range	1 ng ml^{-1}–100 μg ml^{-1}	0.0001–10 nM	10^{0}–10^{3} pfu ml^{-1}
	LOD	1 ng ml^{-1}	NA	0.58 pfu ml^{-1}
	Sample type	NA	NA	NA
	Recovery%	NA	NA	NA
Hepatitis B virus	Technique	DPV[46] (Hepatitis B virus DNA)	Fluorescence[47]	QCM[48] (Hepatitis B surface antigen)
	Sensitivity	1.41×10^{-10} μA ml per copy	0.032 FL pM^{-1}	0.6534 A mM^{-1}
	Linear range	1.1×10^{3}–1.21×10^{5} copies per ml	6–88.5 pM	1–1×10^{3} ng ml^{-1}
	LOD	1100 copies per ml	1.97 pM	3.14 ng ml^{-1}
	Sample type	Human blood samples	Human blood samples	Human blood samples
	Recovery%	NA	97.15–104.10	86.40–95.81
Hepatitis B virus	Technique	DPV[49] (Hepatitis B virus DNA)	Fluorescence[50] (Hepatitis B virus DNA)	QCM[51] (Hepatitis B virus core antigen)
	Sensitivity	173.8 μA M^{-1}	9.31 abs u μM^{-1}	1.662 Hz ml μg^{-1}
	Linear range	1 fM–1 nM	0.200–100 μM	1.56–25 μg ml^{-1}
	LOD	0.36 fM	0.05 μM	0.88 μg ml^{-1}
	Sample type	Serum sample	Human serum samples	Human serum samples
	Recovery%	98.7–102.7	NA	88.80–106.90

(*continued*)

Table 6.1 (*Continued*)

		Method		
Virus	Parameter	Electrochemical	Optical	Mass
Influenza A H1N1 virus	Technique	SWV[52] (H1N1 virus and viral DNA proteins)	UV-Vis spectroscopy[53] (H1N1 virus)	SAW[54] (H1N1 virus DNA)
	Sensitivity	3.42 μA ml pfu^{-1}	NA	11.95 KHz pM^{-1}
	Linear range	10^1–10^6 pfu ml^{-1}	NA	50 pM–10 nM
	LOD	1.52 pfu ml^{-1}	10 nM	14.4 pM
	Sample type	NA	NA	NA
	Recovery%	NA	NA	NA
Influenza A H5N1 virus	Technique	CV[55]	Fluorescence[56]	QCM[57]
	Sensitivity	-2×10^{-8} μA pM^{-1}	87.11 FL nm^{-1}	51.3 Hz 50 μl per HAU
	Linear range	1 pM–100 nM	0.4–290 fM	2^{-4}–2^4 HAUs per 50 μl
	LOD	9.43 pM	0.58 fM	2^{-4} HAUs 50 μl
	Sample type	NA	Human serum samples	Chicken tracheal swab samples
	Recovery%	NA	92.66–104.6	NA

[a]Abbreviations: surface plasmon resonance (SPR), refractive index unit (RIU), not available (NA), differential pulse voltammetry (DPV), reverse transcribed DNA (rtDNA), spike protein (S-protein), nucleocapsid protein (N-protein), square wave voltammetry (SWV), phosphate buffered saline (PBS), plaque-forming unit (pfu),: fluorescence units (FL), absorbance units (abs u), surface acoustic wave (SAW), cyclic voltammetry (CV), hemagglutination units (HAUs).

this method because the results need to be evaluated by experienced personnel using special equipment. Portable and easy-to-use devices are being worked on to replace RT-PCR. Nasopharyngeal and oropharyngeal swabs are the most preferred samples for RT-PCR analysis. However, the most significant disadvantages are the difficulties experienced by the patient during the sample collection phase, the possibility of obtaining false-negative results due to insufficient sample collection, and the risk to healthcare personnel working at this stage. Considering all these, Ramanujam *et al.*[15] developed an electrochemical sensor for the determination of COVID-19 from a neat saliva sample suitable for rapid analysis of a large number of samples. With this sensor, where the chronoamperometry method is used, it is possible to diagnose COVID-19 and perform selective analysis in the presence of HIV by determining the viral spike protein S1 in saliva (Figure 6.1). A miniature electrode system was used with a sample volume as low as 2 ml. The activation step was performed by cyclic voltammetry, followed by chronoamperometric measurements. The linear range was between 0.5 and 2 μg ml^{-1}, with an r^2 value of 0.89. The LOD was calculated as 0.074 fg ml^{-1}. The sensor's selectivity, an important parameter if a patient is co-infected, was evaluated in the presence of the HIV negative regulatory factor (NEF) protein. The sensor could

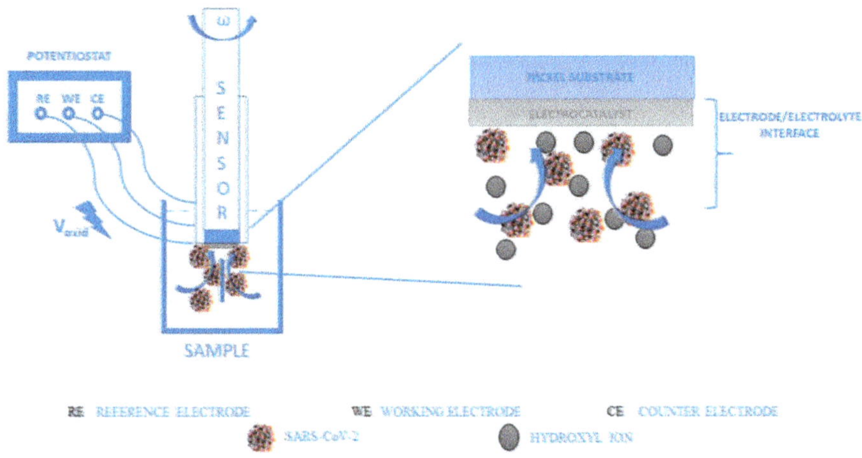

Figure 6.1 Schematic representation of the viral spike protein S1 determination in saliva using the electrochemical sensor. Reproduced from ref. 15, https://doi.org/10.3390/pr9071236, under the terms of the CC BY 4.0 license, https://creativecommons.org/licenses/by/4.0/.

clearly distinguish the two viral proteins, even when present at the same concentration. The real sample application was performed using saliva, which can be used for symptomatic and asymptomatic COVID-19 cases. As a result, this sensor offers a reliable option for clinical applications by developing a fast point of care (POC) device where users can collect and use their samples, thanks to the use of saliva samples.

In another work by Emon *et al.*,[37] an optical sensor using surface plasmon resonance (SPR) was developed to detect the SARS-CoV-2 spike receptor-binding domain. The composite layer of the sensor creates a 2D hybrid material, and the sensor detects the changes in surface plasmon polariton (SPP) and refractive index. Monoclonal antibodies (mAbs) are used as probe ligands to bind the target virus spike receptor-binding domain (Figure 6.2). Hence, the concentration in the sensing region is increased, thanks to the specific binding, which results in refractive index changes and SPR angle shifts. In general, SPR sensors are highly advantageous in terms of their rapidity and high sensitivity for analysing biological molecules. The produced mAbs are expressed as immunoglobulin IgG or IgM for this sensor design. The presence of the hybrid layer provides high charge transfer energy and stability for analyte collection. As a result, the sensor's angular sensitivity was the highest among those described in the literature. Besides, thanks to the broad refractive index range, it has the potential to be applied to various biological analytes.

Nemcekova *et al.*[38] developed a QCM aptasensor for detecting the SARS-CoV-2 virus spike receptor-binding domain (Figure 6.3). The QCM technique has the advantage of being a label-free and non-invasive approach, allowing real-time analysis based on aptamer–target interaction. The sensing mechanism is based on the oscillation of the quartz crystal and changes in its

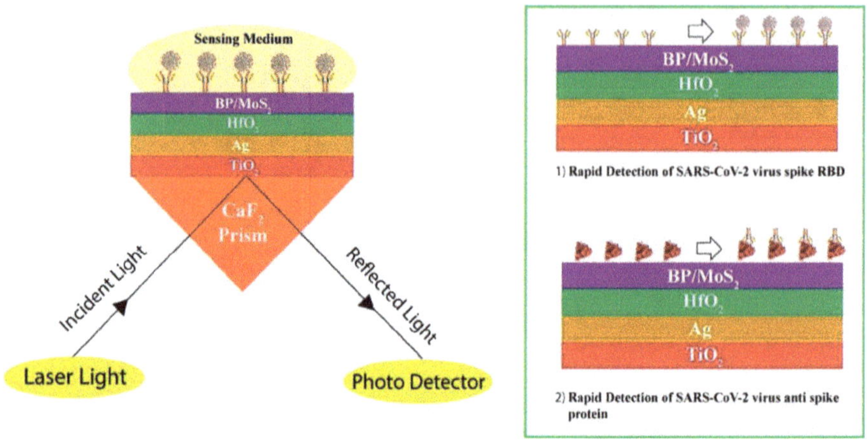

Figure 6.2 Schematic illustration of the SPR sensor principle. Abbreviations: blue phosphorous (BP). Reproduced from ref. 37 with permission from Elsevier, Copyright 2024.

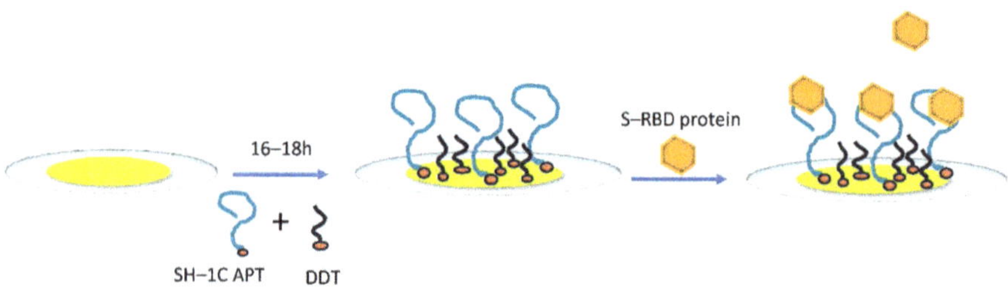

Figure 6.3 Schematic representation of the preparation of QCM aptasensor. Abbreviations: sulfhydryl aptamer 1C (SH-1C APT), dodecanethiol (DDT), spike receptor-binding domain (S-RBD) Reproduced from ref. 38, https://doi.org/10.3390/bios14090431, under the terms of the CC BY 4.0 license, https://creativecommons.org/licenses/by/4.0/.

frequency related to molecular reactions and interactions. Aptamers are popular 3D structures with high binding properties for viruses. Thanks to the mass-sensitive changes pertaining to the virus–aptamer interaction, it is possible to perform label-free virus detection with QCM sensors. For the design of this QCM aptasensor, an optimization procedure was performed, including buffer type, concentration, DNA sequence, and temperature. The optimized QCM aptasensor showed a linear response in the 1 pg ml^{-1}–0.1 μg ml^{-1} concentration range. The sensor's functionality and stability in real biological samples were evaluated using human plasma and saliva. Recovery values confirmed the sensor's accurate performance. It has good potential for practical use for clinical applications.

Developing sensor devices for effectively analysing SARS-CoV-2 is a popular topic. Among electrochemical sensors, DNA-based sensors aim to identify and detect the DNA of the target virus. However, the major challenge in virus detection is the presence of DNA and/or RNA in serum at very low concentrations. As a solution to this sensitivity problem, modification of electrochemical sensors with nanomaterials that have high electrocatalytic and conductive properties is employed. Zhu *et al.*[39] proposed an electrochemical DNA sensor modified with MnFe@Pt nanocubes for the ultrasensitive and ultrafast detection of the SARS-CoV-2 virus. The sensing strategy is based on immobilizing a specific DNA strand on nanocubes as capture and report probes. In 10 minutes, the reverse transcribed DNA (rtDNA) of the target virus can be detected, resulting in a decrease in peak current (Figure 6.4). The differential pulse voltammetry (DPV) method had a linear concentration range between 8.83 fg ml^{-1} and 88.25 pg ml^{-1} with an LOD value of 0.088 pg ml^{-1}. In order to emphasize the sensor's selectivity in practical use, carcinoembryonic antigen, alpha-fetoprotein, human chorionic gonadotropin, and prostate-specific antigen were tested at high concentrations and confirmed. Real serum samples were analysed, and good recovery results indicated the sensor's accuracy and applicability.

Viter *et al.*[16] proposed a photoluminescence-based biosensor for detecting antibodies to the SARS-CoV-2 virus proteins. Detection of antibodies (IgG and IgM) is an indirect approach that allows the interpretation of viral infection as a result of humoral response. Spike protein (S-protein) is responsible for the attachment and entry into cells, while nucleocapsid protein (N-protein) is associated with the synthesis and translation of RNA. These proteins. Target the specific antibodies caused by the viral infection. ZnO nanotetrapods were utilized as transducers thanks to their stable room temperature

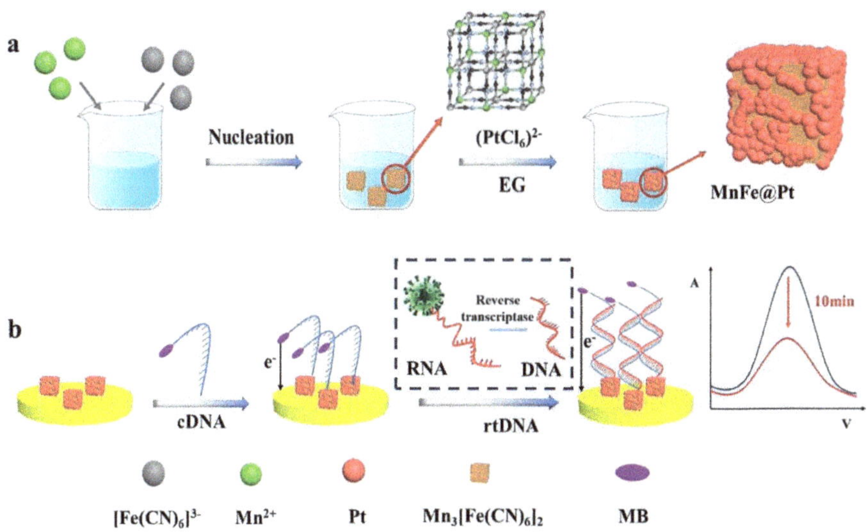

Figure 6.4 Schematic of the (a) preparation and (b) determination strategy for electrochemical DNA sensor. (EG: ethylene glycol, MB: methylene blue). Reproduced from ref. 39 with permission from Elsevier, Copyright 2024.

photoluminescence and high surface area. Photoluminescence measurements were performed based on the real-time interactions between virus proteins and antibodies. The linear range values for S- and N-proteins were 0.025–0.5 nM and 0.3–1 nM, respectively, with LOD values of 0.01 nM and 0.3 nM, respectively. Sensor performance was tested in blood serum samples. The results indicated that the sensor did not display increased photoluminescence intensity in control samples. The sensor's detection time was also short (15–20 min).

As in the study by Ramanujam *et al.*,[15] an electrochemical sensor for the SARS-CoV-2 viral spike protein S1 was developed by Ayankojo *et al.*[41] Their sensor was molecularly imprinted polymer (MIP) based, which provides high specificity as a synthetic biological receptor with specific recognition areas (Figure 6.5). The functional monomer was selected as 3-aminophenylboronic acid to give covalent interactions to the glycosylated protein. MIP formation was performed *via* electropolymerization, and the electrochemical measurements were performed using cyclic voltammetry and square wave voltammetry (SWV) methods. The sensor's detection performance was evaluated in phosphate buffered saline (PBS) solution and nasopharyngeal swap

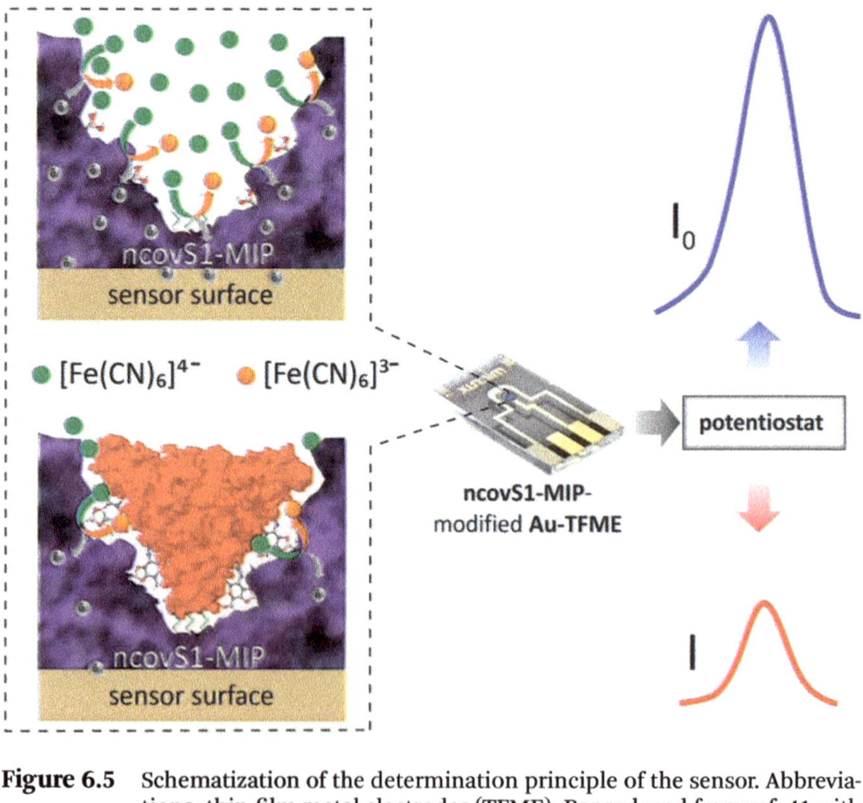

Figure 6.5 Schematization of the determination principle of the sensor. Abbreviations: thin-film metal electrodes (TFME). Reproduced from ref. 41 with permission from Elsevier, Copyright 2021.

samples in the linear concentration ranges of 26.7–194 fM and 50–400 fM, respectively. The LOD values were 15 fM and 64 fM for buffer and real samples, respectively. SARS-CoV-2 nucleoprotein was chosen to test the sensor's selectivity, and no response was obtained for the nucleoprotein. The sensor's response time was 15 min, which is the required time for incubating the patient's sample. This sensor offers on-site measurement, thanks to the portable device, and a good advantage in terms of POC analysis.

Another study for the determination of SARS-CoV-2 spike protein S1 was performed by Gholami *et al.*[43] using a colorimetric detection method. They designed a paper-based sensor with a target-specific extractor chip. This colorimetric paper-based sensor design consists of a nanopillar chip whose surface has attached mAb fragments. The extractor chip can selectively bind and purify the target protein from the biological sample. After that, the purified protein is reduced into biothiol form. The functional groups of this structure can react with the azo dye (blue coloured), resulting in a colour change to pink. The naked eye can observe this colour change. A linear relationship was observed between the colour intensity (pink colour) and the logarithm of the reduced viral protein concentration (130 ng ml^{-1}–1.3 pg ml^{-1}). The LOD value was calculated as 130 fg ml^{-1}. The sensor's performance in clinical samples was tested using spiked saliva and patients' nasopharyngeal swab samples and compared with the results of quantitative RT-PCR (qRT-PCR). The consistent results of the paper-based sensor platform demonstrated its high sensitivity, specificity, and potential for application in clinical settings.

Dengue is a serious disease caused by a mosquito-borne virus (dengue virus) that has the risk of causing epidemic and endemic infections. Dengue infection can affect many people in more than 100 countries, and it does not have very specific symptoms (generally being similar to malaria, typhoid, chikungunya, *etc.*), causing a challenging diagnosis process. Therefore, rapid and accurate diagnosis is the key to treatment and reduced mortality rate. Generally, enzyme-linked immunosorbent assay (ELISA) and chemiluminescence-based methods are preferred for detecting non-structural dengue protein NS1. On the other hand, electrochemical biosensors offer cost-effective, easy-to-use, and sensitive analysis options. Chaturvedi *et al.*[44] developed a bioinspired nanocomposite-based electrochemical immunosensor for detecting non-structural dengue protein NS1, using anti-NS1 antibodies as the recognition element. The liner range and LOD values were 1 ng ml^{-1}–100 μg ml^{-1} and 1 ng ml^{-1}, respectively, with the DPV method. It has excellent potential for clinical applications as a rapid, affordable, and sensitive electrochemical immunosensor.

Omar *et al.*[45] also focused on diagnosing the dengue virus, but they detected the dengue virus envelope protein using an SPR-based sensor. They used IgM antibodies as the sensing layer by immobilizing them and using some modification agents, such as magnetite nanoparticles, nanocellulose crystalline, *etc.*, to enhance the sensitivity and stability of the sensor. The spin coating technique was utilized for the immobilization of the antibodies. After the various concentrations of dengue virus envelope protein solutions are applied into the cell, they will contact the immobilized antibodies, resulting in resonance angle shifts. The linear response of the sensor was obtained in the range between 0.0001 and 10 nM. The sensitivity of the sensor was calculated as 39.96°nM^{-1}. Further studies on improving the selectivity and sensitivity of this sensor will pave the way for it to be an effective option for dengue virus diagnosis.

Navakul *et al.*[17] developed a QCM-based sensor using graphene oxide-MIP composites to detect the dengue virus (Figure 6.6). The MIP structure was designed as the copolymer of acrylamide, methacrylic acid, methylmethacrylate, and *N*-vinylpyrrolidone. Graphene oxide (GO) was incorporated into this structure to improve electrical properties and MIP performance. A cross-linker and an initiator were also included in the MIP structure, and the polymerization was performed at 55 °C for 18 h using an oven. Template removal was performed by stirring in a solution containing 10% acetic acid (v/v) and 0.1% sodium dodecyl sulphate (w/v) (1:1) for 30 min and in deionized water for 30 min. The linear working range was 10^0 and 10^3 pfu ml^{-1} with the sensitivity and LOD values of 189.61 Hz ml pfu^{-1} and 0.58 pfu ml^{-1}, respectively.

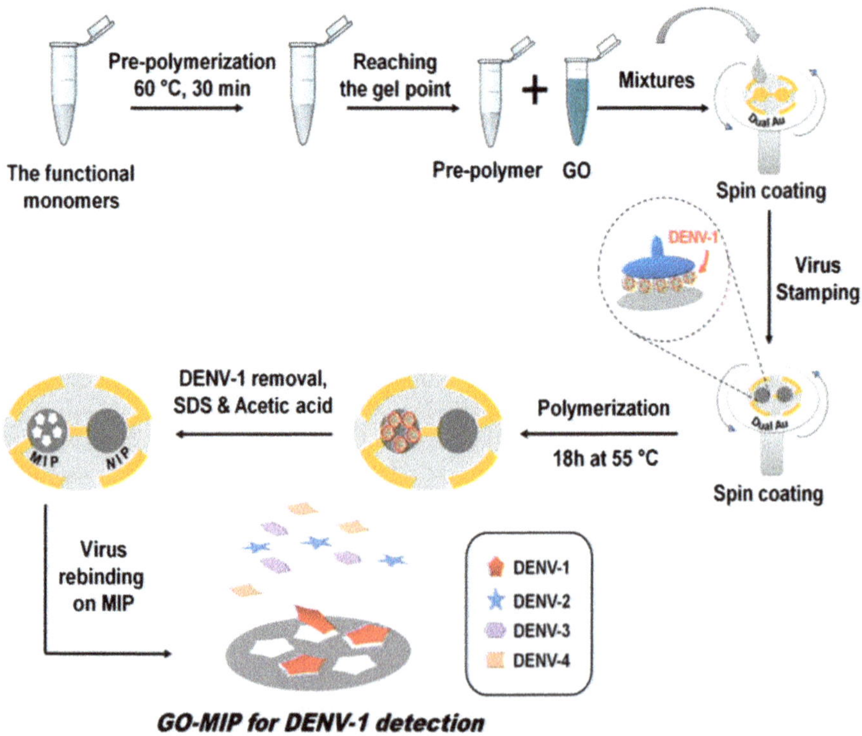

Figure 6.6 Schematic representation of the preparation of QCM-based sensor for dengue virus detection. Reproduced from ref. 17, https://doi.org/10.1007/s00216-021-03410-8, under the terms of the CC BY 4.0 license, http://creativecommons.org/licenses/by/4.0/.

The selectivity of the sensor was assessed against other serotypes of dengue virus 2, 3, and 4. Even though they are 65–70% homologous, other serotypes did not significantly affect the sensor response. Another advantage of this sensor is its short response time of 20 min. Consequently, this is a fast and practical method for dengue virus analysis, allowing mass production.

6.3.1 Comparison of Sensitivity and Specificity

Sensitivity and specificity are the most important parameters in evaluating and comparing sensor performance. In virus analysis, especially when evaluated in terms of early and accurate diagnosis, sensors with high sensitivity that are both selective and able to determine the analyte, even in low amounts, are needed. When some of the sensors used in the determination of SARS-CoV-2 are examined in terms of these parameters, optical (colorimetric)[43] and electrochemical (voltametric and chronoamperometric)[15,41] techniques have been used for the determination of viral spike protein S1. When compared, the highest sensitivity is observed for the electrochemical method based on SWV using MIPs,[41] and the least sensitive method is the chronoamperometric sensor.[15] The sensitivity of the MIP-based electrochemical sensor is highly competitive with the classical PCR method.

On the other hand, even though the optical (colorimetric)[43] method's sensitivity is not as high as the MIP sensor, it is still similar to that for antigen-based analysis. When evaluated in terms of specificity, it was seen that all three sensors performed successfully in the presence of other viral proteins and non-target proteins and did not give false results. However, when evaluated in more detail, the outstanding features, such as the superior selectivity obtained especially thanks to the integration with the MIP structure, both analysis in the presence of 10 times higher concentration of SARS-CoV-2 nucleocapsid protein and discrimination between other variants of SARS-CoV-2 can be performed using the MIP-based sensor. Researchers who chose the virus spike receptor-binding domain as an analyte for the same purpose have used SPR,[37] QCM,[38] and SERS[40] as techniques. Among these, the method with the highest sensitivity in determining the spike receptor-binding domain is QCM,[38] making detecting virus particles at ultra-low concentration levels possible. Regarding specificity, SERS[40] can distinguish variants of SARS-CoV-2 that are very similar to each other, thanks to its molecule fingerprinting advantage. The selectivity of SPR and QCM mostly depends on the presence of structures such as MIP or aptamer on the surface.

When three different sensors (electrochemistry, SPR, and QCM)[17,44,45] used in the determination of dengue type 1 virus are compared in terms of sensitivity and specificity, SPR[45] stands out with its high sensitivity and is similar to PCR in this respect. While using antibodies in electrochemical immunosensors provides enhanced binding specificity, there is potential interference from other proteins in the blood. The most important advantage of the QCM[17] sensor is that it can distinguish different dengue virus serotypes.

When the performance of two nanomaterial-based electrochemical DNA sensors[46,49] and a fluorescence DNA sensor[50] used for the determination of hepatitis B virus DNA is compared, it is observed that the sensor developed by Mao *et al.*[49] provides ultra-high sensitivity with gold–platinum nanoprobes and has similar sensitivity to the traditional PCR method. When the selectivity evaluation is made, it can be said that the sensor with the highest sensitivity[49] provides high selectivity even against similarly structured viruses thanks to specific DNA hybridization, whereas the fluorescence sensor[50] can be affected by other enzyme biomarkers.

There are voltametric,[52] spectroscopic,[53] and SAW-based[54] sensors for detecting influenza A H1N1 virus. These detect the virus itself, viral DNA proteins, and viral DNA. When evaluated from different perspectives, it was determined that the sensitivity of the SAW biosensor[54] was very high (similar to PCR). At the same time, the electrochemical sensor,[55] although capable of detecting whole virus particles, had low sensitivity. In addition, the SAW biosensor[54] can target specific sequences with its DNA probe and provides high specificity. Since the electrochemical sensor offers a peptide-based approach, similar proteins may affect the selectivity. The optical sensor is unsuitable for mixed samples because colour shifts can be affected.

When the CV,[55] fluorescence,[56] and QCM-aptamer-based[57] sensors were compared to detect the influenza A H5N1 virus, the highest sensitivity was found to belong to the fluorescence sensor,[56] which corresponds to a better performance than most commercial test kits. However, colorimetric results may be affected by light conditions. On the other hand, the QCM aptasensor[57] surpasses other sensors in selectivity performance thanks to the aptamer providing high specificity against H5N1 and preventing other influenza subtypes from affecting the analysis.

6.3.2 Comparison of Detection Ranges and Limits

The wide detection range of the sensor provides an advantage when working with various samples containing different concentrations of analyte, and the LOD value is another parameter that can be related to sensitivity. As in the previous section, when the optical (colorimetric)[43] and electrochemical (voltametric and chronoamperometric)[15,41] sensors used in the determination of SARS-CoV-2 viral spike protein S1 are compared, with respect to linear range and LOD values, it is seen that the linear range of the SWV-based voltametric sensor[41] is wider and the LOD value is lower. A wide working concentration range provides advantages in general diagnostic applications, and the low LOD value allows diagnosis at the beginning of the infection when the viral load is still low. In addition, SPR,[37] QCM,[38] and SERS[41] techniques for determining the SARS-CoV-2 virus spike receptor-binding domain were compared. Accordingly, the QCM sensor[38] has a wide detection range from pg per ml to relatively high concentrations. In this way, it can be used both in the initial stages of infection and in stages where the viral load is high in clinical applications. This sensor also has the lowest LOD value. Among the electrochemistry,[44] SPR,[45] and QCM-based[17] sensors used in determining dengue type 1 virus, the sensor with the widest linear range is the electrochemical immunosensor. The SPR sensor[45] stands out with its ultra-low LOD value. When the performances of two nanomaterial-based electrochemical DNA sensors[46,49] and a fluorescence DNA sensor[50] used for the determination of hepatitis B virus DNA are compared, the electrochemical sensor developed by Mao *et al.*[49] has the advantage of high sensitivity due to its linear range being both extensive and covering very low concentrations. Voltametric,[52] spectroscopic,[53] and SAW-based[54] sensors based on the determination of virus, viral DNA proteins, and viral DNA were evaluated to diagnose the influenza A H1N1 virus. While the SAW sensor,[54] which has the lowest LOD value, determines virus DNA, the voltametric sensor, which determines both virus and viral DNA protein, has the widest linear range. However, reaching very low or high concentration values with the optical sensor is impossible. When the CV,[55] fluorescence,[56] and QCM-aptamer-based[57] sensors were compared for detecting and quantifying the influenza A H5N1 virus, the electrochemical DNA sensor[55] offered a versatile approach with the widest linear range. The lower LOD value of the fluorescence-based sensor[56] leaves the other two methods behind regarding sensitivity.

6.3.3 Comparison of Real Sample Analysis

High-sensitivity sensors that can identify the analyte in various real samples are required to assess the accuracy and dependability of methods in virus analysis. When some of the sensors used in the determination of SARS-CoV-2 are examined in terms of these parameters, it is seen that different techniques, such as optical (colorimetric), electrochemical (voltametric and chronoamperometric), and QCM, are used. When electrochemical studies for the determination of SARS-CoV-2 were examined, recombinant viral proteins were used instead of the real virus in the sensor developed by Ramanujam *et al.*[15] After determining the basic level with medical-grade saliva, two different concentrations of S1 protein (10 and 30 μg) were mixed with saliva to prepare the samples. According to the results obtained, it was observed that saliva components caused no interference. In the technique used by Zhu *et al.*,[39] similarly to the nasopharyngeal swab test, serological testing offered a valuable method for accurately estimating the infection rate and assessing the severity of SARS-CoV-2 infection in clinical settings. As the target detection system, actual serum samples made by varying the concentration of rtDNA in human serum were utilized to assess the practical efficacy of these DNA sensors. The concentrations measured following the manual addition of rtDNA agreed with those obtained by conventional methods. The recovery was between 99.00% and 102.31%, while the RSD was between 2.00% and 2.65%. These results indicated that the sensor could detect SARS-CoV-2 viral RNA in real samples accurately and rapidly. In another study, the functionalization of SERS substrates with specific labels to precisely target viral proteins, and nucleic acids was developed by Daoudi *et al.*[40] It aims to improve the direct detection of the SARS-CoV-2 virus at the protein level [spike receptor binding domain (S-RBD)]; further improvements are needed to upgrade it to a real-time monitoring diagnostic device. Additionally, no recovery data showing the applicability of the developed sensor was reported in the publication. The clinical diagnostic applicability of a MIP-based electrochemical sensor created by Raziq *et al.*[42] for detecting the SARS-CoV-2 antigen was assessed by analysing samples taken from patient nasopharyngeal swabs. The sensor was first calibrated using COVID-19-negative samples to which known amounts of SARS-CoV-2 nucleoprotein (ncovNP) had been added. Additionally, the sensor's reaction was nearly insensitive to the addition of S1 to the COVID-19 negative sample, but it rapidly rose upon the addition of ncovNP, demonstrating high selectivity against ncovNP. Optical and QCM methods were also used to detect the SARS-CoV-2 virus. Emon *et al.*[37] presented different approaches for rapidly detecting the SARS-CoV-2 virus responsible for the COVID-19 pandemic by designing a surface plasmon optical refractometric sensor. Without getting false-positive findings, the presence or absence of the SARS-CoV-2 virus was ascertained within minutes based on the test results. Thus, the choice made will aid in detecting novel coronaviruses in the future, not just in lab settings. A microfluidic technology and a novel photoluminescence platform were created to identify anti-SARS-CoV-2 antibodies in blood probe samples by Viter *et al.*[16] As optical sensor templates, novel ZnO nanostructures with a high surface area and a steady photoluminescence signal were employed. Control blood samples without anti-SARS-CoV-2 were used for control measurements using ZnO/SARS-CoV recombinant spike protein (rS)-based immunosensors. Ayankajo *et al.*[41] studied the sensor's performance on patient nasopharyngeal swab samples to assess the clinical applicability of the sensor. For this purpose, an initial analysis of COVID-19-negative samples spiked with increasing ncovS1 concentrations was performed. Moreover, the variation observed in the sensor response to the different samples might be due to a number of factors. Prominent among these could be the difference in viral load, which strongly correlates to the viral protein concentration at different stages of the life cycle of a SARS-CoV-2 infection. A paper-based optical sensor was used by Gholami *et al.*[43] to screen for viruses completely resuspended in a viral transport medium (VTM) and then incubated with target-specific extractor chips. The colour change was visible to the unaided eye after the purified and reduced protein from the swabs was placed into paper-based optical sensors. While the sample from the infected person spontaneously changed from blue to pink, the sample from the healthy person did not produce any colour change on the optical sensor, indicating that the SARS-CoV-2 virus's S1 protein was present in the sample. A comparative analysis of QCM and electrochemical aptasensors for SARS-CoV-2 detection was performed by Nemčeková *et al.*[38] The aptasensors' ability to function in intricate biological settings was confirmed using diluted human plasma and saliva. The sensors were cleaned with diluted plasma or saliva to prevent interference from sample components in order to detect S-RBD in actual samples. Following stabilization, the sensor was filled with a diluted plasma solution that had been enhanced with S-RBD protein. The sensor's stability and functionality are demonstrated by its nonspecific contact with 1 $\mu g\ ml^{-1}$ N-protein and its selective response to 0.1 $\mu g\ ml^{-1}$ S-RBD protein.

Chaturvedi *et al.*[44] created an electrochemical immunosensor using a graphene nanocomposite to detect and measure the dengue virus NS1 protein. Early diagnosis of dengue virus infection has been shown to be possible using this immunosensor, which has an extraordinarily low detection limit. In another study, on the application of an SPR optical sensor in dengue virus diagnosis was presented by Omar *et al.*[45] Gold/Fe–3-mercaptopropionic acid (MPA)–nanocellulose crystalline (NCC)–hexadecyltrimethylammonium bromide (CTAB)/*N*-ethyl-*N*-3-(dimethylaminopropyl)carbodiimide (EDC)–*N*-hydroxysuccinimide (NHS) thin film immobilized monoclonal antibody (IgM) was made by spin coating. The presence of the dengue E-protein bound to the IgM surface was verified by recording the sensor chip's surface characteristics using atomic force microscopy. To increase its sensitivity in detecting the dangerous dengue disease pathogen, dengue type 1 virus, a MIP-based sensor recognition material was created and combined with GO by Navakul *et al.*[17] In order to compare them with "pure" MIP, or copolymer without GO, polymer composites with GO integrated into acrylamide (AAM)/methacrylic acid (MAA)/methyl methacrylate (MMA)/*N*-vinylpyrrolidone (VP) copolymer were created. Compared with pure MIP-based sensors, GO-MIP coated atop 10 MHz QCM sensors demonstrated double the sensitivity.

Furthermore, GO-polymer composites demonstrated a suppression factor of up to 21 in contrast to 3 for pure MIP. However, in the above-described studies, there were no studies on real sample applications for detecting dengue virus using electrochemical, optical, and QCM techniques. Therefore, the applicability of the sensor could not be compared.

A novel electrochemical DNA (E-DNA) sensor based on a multifunctional sandwich structure for the detection of hepatitis B-DNA replication was presented by Zhao *et al.*[46] By modifying the electrode surface, $Cu_3(PO_4)_2$–bovine serum albumin (BSA)–GO nanoflowers and AuNPs were formed to provide significant aptamer binding sites. Two aptamers were created to further capture and immobilize target DNA. Aptamer-1 captured DNA and was attached to the electrode *via* a thiol–gold bond. Aptamer-2 was utilized to supply and enhance the meagre electrochemical signal. Both hepatitis B-free and hepatitis B-positive human blood samples were used to examine detection of hepatitis B-DNA using the suggested E-DNA sensor, which was then compared to alternative techniques (ELISA). Another electrochemical study was aimed at developing a sandwich-type electrochemical DNA sensor using CeO_2/AuPt nanoprobes for highly sensitive detection of hepatitis B virus DNA by Mao *et al.*[49] In actual spiked serum sample recovery trials, recoveries ranged from 98.7% to 102.7%, and all relative standard deviations were less than 4.77%. A novel technique for hepatitis B virus DNA detection with possible clinical uses was presented in this work. A portable dual-function ratio-type triple-emission molecularly imprinted fluorescence sensor for simultaneous visual detection of hepatitis A and B viruses was designed by Gong *et al.*[47] The sensor was utilized to identify hepatitis A and B virus particles in human serum that had been diluted 100-fold to assess its suitability for identifying viruses in real blood samples. The RSD was calculated after the experiments were carried out in triplicate, and the recovery rates of the two viruses were compared with the amounts of virus spiked into the samples. Recovery rates between 101.1% and 113.0% for hepatitis A and 97.15% and 104.1% for hepatitis B were obtained, and on the basis of these results, the application of the developed sensor to real samples was validated. A proposed self-powered hepatitis B-DNA biosensor was created by Imbriano *et al.*[50] and was tested in practice using fluorescence measurements in human serum. Given its promising analytical characteristics, the suggested approach might open the door for creating innovative two-level logic biosensors. A biosensor consisting of a QCM crystal freshly coated with a molecularly imprinted polydopamine (MIPDA) layer to evaluate spiked human serum samples was designed by Lim *et al.*[51] The results were consistent with the frequency measurements obtained for spiked human serum samples. The MIPDA–QCM biosensor's applicability for quantifying hepatitis B surface antigen from a complex sample matrix and the reproducibility of results across different MIPDA–QCM crystals was demonstrated by the low RSD (6.2%) between the sensitivities of the two MIPDA–QCM crystals towards Hepatitis B surface antigen samples within the detection range. Additionally, the MIPDA–QCM biosensor recovered 88.8% to 106.9% of the hepatitis B surface antigen from the human serum samples, and the readings agreed well with the ELISA results.

In another study, Saffari *et al.*[48] used two different antibody-conjugated nanomaterials instead of horseradish peroxidase (HRP) as a catalyst: colloidal CuO nanoparticles and CuO/multiwall carbon nanotubes (MWCNT) nanocomposite. The method's recovery percentage was found to be between 89.0% and 103.6%. Comparatively speaking, their design necessitates using two monoclonal antibodies to detect antigens, which is a drawback for the ELISA methodology. Furthermore, compared with the QCM biosensor produced, the amplification stage of their design was more intricate. As a result, the QCM biosensor may be a more promising instrument for clinical laboratory measurement of a hepatitis B virus biomarker.

A label-free, peptide-based electrochemical sensor was developed by Kim *et al.*[52] to identify whole particles of the H1N1 virus with high sensitivity and selectivity. The performance of these peptide sensors was confirmed by electrochemical measurement (SWV), and the peptides that were easily immobilized on a gold chip were synthesized based on the exact sequences found through phage display. Using many peptides instead of just one enhanced the sensor's performance, and the detection capability was verified for real H1N1 and other virus samples. An electrochemical (EC) technique was used to construct a label-free avian influenza virus (AIV) H5N1 biosensor made of a multifunctional DNA construct on an electrode fabricated with porous Au nanoparticles (pAuNPs) by Lee *et al.*[55] Three-way junctions (3WJ) DNA was introduced as a multipurpose bioprobe. Every DNA fragment 3WJ was linked to the immobilization part (thiol group), EC signal generating part [horseradish peroxidase (HRP)-mimicked DNAzyme], and recognition part [hemagglutinin (HA) protein detection aptamer]. Each fragment was put together to create the DNA 3WJ for AI detection, and native-tris boric acid magnesium polyacrylamide gel electrophoresis (TBM–PAGE) was used to verify the constructed structure. The study demonstrated label-free, easily tailored, and straightforwardly fabricated detection components for AIV. This biosensor may be a strong contender for several virus detection systems. In both electrochemical studies conducted to determine influenza A, real sample applications were not performed. Therefore, no comparison could be made regarding the results obtained for the developed sensors. In order to create an effective bifunctional AuNP-based biosensor for the detection of influenza A and influenza B, the surface of AuNPs was functionalized simultaneously with oligonucleotide probes complementary to influenza A and influenza B by Zhang *et al.*[53] The special optical properties of AuNPs enabled the results to be obtained with the naked eye in a short time. In this study, no real sample application studies were performed. Gong *et al.*[56] developed a portable paper-based molecularly imprinted sensor and used 100-fold diluted human serum as a test sample in three different studies to detect H5N1. The constructed sensor's recovery rate for target detection ranged from 92.66% to 104.6%. These findings demonstrated the potential of the portable paper-based sensor to identify the target virus, H5N1, in samples from of actual patients. A highly sensitive, label-free, low-cost surface acoustic wave (SAW) biosensor that can rapidly and *in situ* detect target DNA sequences associated with H1N1 was fabricated by Sun *et al.*[54] The 128° YX-$LiNbO_3$ substrate was used to manufacture the SAW biosensor. Then, using aminopropyltriethoxysilane (APTES) for chemical functionalization, DNA probes with phosphate groups were immobilized on the SAW sensing region. The resonance frequency shift of the SAW biosensor was observed to identify the precise binding of the target DNA sequence to the immobilized DNA probes. However, there is no data regarding the application of the developed sensor to real samples.

The performance of the nanowell-based aptasensor developed by Wang *et al.*[57] for the detection of AIV H5N1 with a complex background was experimentally tested using chicken tracheal swabs spiked with inactivated AIV H5N1. A chicken tracheal swab sample devoid of AIV was employed as a negative control. The same protocol was used to detect inactivated AIV H5N1 for chicken swab samples. Each measured titre in the 2–4 to 20 hemagglutination units (HAUs) per 50 µl ranges was tested in triplicate. According to these test results, AIV H5N1 could be found in chicken tracheal swab samples using the nanowell-based aptasensor. However, bigger error bars were seen when comparing the detection of swab samples with the stock virus sample. This could be the result of impurity noise or interference with the detection signal from the tracheal swab samples.

6.4 General Overview of Advantages and Disadvantages of Different Sensor Types

It is possible to evaluate the advantages and disadvantages of sensors in virus detection and their superiority over each other using various evaluation parameters. However, it should be kept in mind that this provides a general perspective; it is not possible to offer a single best option, and when an analysis method is to be selected, the performance parameters should be evaluated in detail in terms of analysis procedure, analyte, and sample. When determining viruses, diagnosis of infections, and related clinical applications, low LOD values and high sensitivity are required for early-stage detection, where the viral load is low at the beginning of the infection or for the asymptomatic carriers. In addition, the method to be applied should compete with or outperform well-established traditional methods such as PCR and ELISA regarding sensitivity. Table 6.2 provides an overview of the sensors used for virus detection, their advantages and disadvantages, and which applications they are best for.

6.5 Future Perspectives and Challenges

Throughout this chapter, when we look at the comparative analysis of different sensor types used in virus detection and the studies in the literature, it is undeniable that promising progress has been made in this field. Viral sensor development studies, which are directly related to the accurate, precise, and rapid diagnosis of viral diseases, will continue to maintain their clinical importance in the future. In this context, if the issues open to development are to be addressed, the first is determining multiple analytes. Especially in the determination of rapidly mutating viruses, it may be possible to identify variants of the same virus or which virus it is in a short time, instead of a long and staged analysis procedure, using a sensor that can simultaneously determine more than one analyte. To improve sensor performance and obtain sensors with more sensitivity and a wider linear range, the integration of different materials, such as nanoparticles, is already being used and is expected to continue to increase in the future. In addition, the evolution of a sensor with effective results in selectivity, sensitivity, working range, and speed into affordable, reliable, and reproducible POC test devices is the most critical point to be worked on in terms of clinical value.[58–60] Another parameter related to simultaneous multiple measurements is the ability of the sensor to be used in diagnosis to adapt to rapidly developing viral mutations and increase diagnostic efficiency. The key aspects expected from the sensor developed with all these advantages are low cost and accessibility. Especially in critical situations such as global pandemics, it is important for global health that societies with limited resources can use it.[58,61,62] Considering all of these and the dynamic variability of viral infections, the major expectation in the future is that the sensors used in virus detection will be constantly evolving and up-to-date.

Table 6.2 General overview of the sensors used for virus detection.

Sensor	Advantages	Disadvantages	Best for when required
Electrochemical	• Simple to use • Low-cost • Portable, suitable for POC applications • High precision • High sensitivity • Fast analysis, rapid results	❖ Electrode fabrication and modification is a multi-stage process. ❖ Difficult to integrate as a routine clinical tool ❖ Relatively low robustness and reproducibility ❖ Potential cross-reactivity and selectivity problems ❖ Not suitable for simultaneous multiple analyte detection	✓ Rapid on-site screening ✓ Low-cost detection ✓ Ultra-sensitive analysis in resource-limited settings
Optical	• User-friendly and practical thanks to visual interpretation of results (*e.g.*, colour changes) • Easy to integrate into rapid self-testing kits • High specificity • Suitability for simultaneous multiple analyte detection	❖ Special equipment and experienced personnel are required for SPR sensors. ❖ Portability is not available for SPR sensors. ❖ Susceptibility to environmental factors	✓ Differentiation between virus strains ✓ Rapid mass-screening (Paper-based optical sensors)
Mass-based	• Label-free • Reusable • Easy and minimal sample preparation	❖ Suitability for portability is very low. ❖ Prone to false positive results	✓ Clinical settings (laboratory-based detection) ✓ Detecting early-stage infections

6.6 Conclusion

Thanks to the rapid change and development of sensor platforms used in virus detection, diagnosis, and identification of viral infections, they are carried out at lower costs, faster, more accurately, and more reliably. This chapter discusses the comparative analysis of currently developed sensor platforms for virus detection in terms of sensitivity, specificity, detection ranges and limits, speed, cost-effectiveness, and real sample analysis, which are critical parameters in performance evaluation. In this context, it is impossible to say that a single sensor platform or method provides great superiority over others. Instead, it would be more accurate to select by comparing and evaluating the advantages and disadvantages of sensor types according to the targeted specific application and purpose of the analysis.

Sensitivity and specificity are among the most essential performance aspects of virus detection, as they directly affect the reliability of diagnostics. Optical and electrochemical sensors are usually characterized by their high sensitivity, resulting from their ability to identify various viral biomarkers at very low concentrations. Nonetheless, the specificity of the sensor differs according to the kind of biorecognition element. For example, while cross-reactivity issues can be encountered in antibody-based assays, specificity is much higher in nucleic acid-based sensors such as clustered regularly interspaced short palindromic repeats (CRISPR). Although nanomaterial-based sensors are a major factor in improving sensitivity, they are complex and costly, which may hinder their broad use. Methods like ELISA and PCR, which are performed in a laboratory environment, yield great precision but come with high costs and lengthy analysis processes. Conversely, microfluidic biosensors and label-free sensors have shown the ability to perform rapid detection, providing outcomes in a matter of minutes rather than hours. Nonetheless, weighing the benefit:loss ratio between a quick and accurate analysis is crucial, given that rapid tests can be disadvantageous in terms of sensitivity. Although paper-based sensors and lateral flow assays provide inexpensive options, they need enhancement regarding accuracy and sensitivity compared with advanced biosensor platforms. Another important parameter is the performance of the developed sensor in real sample analysis. While many sensors give effective results under laboratory conditions during the development phase, sufficient performance may not be observed when applied to real samples. This may be due to the different properties of the sample matrix and the presence of other molecules in the environment, which greatly affect accuracy and reliability. Therefore, the real sample performance of the sensor for viral detection must be tested to confirm accuracy.

References

1. M. Breitbart and F. Rohwer, *Trends Microbiol.*, 2005, **13**, 278–284.
2. M. M. Alvarez, J. Aizenberg, M. Analoui, A. M. Andrews, G. Bisker, E. S. Boyden, R. D. Kamm, J. M. Karp, D. J. Mooney, R. Oklu, D. Peer, M. Stolzoff, M. S. Strano, G. Trujillo-De Santiago, T. J. Webster, P. S. Weiss and A. Khademhosseini, *ACS Nano*, 2017, **11**, 5195–5214.
3. A. A. Malik, C. Nantasenamat and T. Piacham, *Mater. Sci. Eng.: C*, 2017, **77**, 1341–1348.
4. C. W. Blackwell, *J. Nurse Pract.*, 2024, **20**, 104962.

5. K. Ghosh, S. R. Tony, K. Islam, R. Mobarak, Md. J. Alam, Md. Kamruzzaman, S. Alam, M. Talha, A. Al Faisal, N. Islam, M. M. Hossain, Mst. N. Begum and M. Rahman, *J. Infect. Public Health*, 2025, **18**, 102684.
6. A. Khatami, S. Bahadory, S. Ghorbani, H. Saadati, M. Zarei, A. Soleimani and M. Zibaei, *Microb. Pathog.*, 2021, **154**, 104828.
7. J. Lu, Z. Liu, Z. Li, J. Su, H. Zhen, Y. Qu, P. Herdewijn, H. Liu, Y. Liu and Z. Wang, *Bioorg. Med. Chem. Lett.*, 2025, **120**, 130151.
8. Z. Zhao, C. Huang, Z. Huang, F. Lin, Q. He, D. Tao, N. Jaffrezic-Renault and Z. Guo, *TrAC – Trends Anal. Chem.*, 2021, **139**, 116253.
9. J. A. Goode, J. V. H. Rushworth and P. A. Millner, *Am. Chem. Soc.*, 2015, DOI: 10.1021/la503533g.
10. Y. Saylan and A. Denizli, *Nanosensors for Smart Cities*, Elsevier, 2020, pp. 501–511.
11. J. Yoon, M. Shin, T. Lee and J.-W. Choi, *Materials*, 2020, **13**, 299.
12. H. Xi, M. Juhas and Y. Zhang, *Biosens. Bioelectron.*, 2020, **167**, 112494.
13. X. Qian and B. Städler, *Adv. Funct. Mater.*, 2020, **30**, 2004605.
14. F. Nasrin, A. D. Chowdhury, K. Takemura, I. Kozaki, H. Honda, O. Adegoke and E. Y. Park, *Anal. Chim. Acta*, 2020, **1109**, 148–157.
15. A. Ramanujam, S. Almodovar and G. G. Botte, *Processes*, 2021, **9**, 1236.
16. R. Viter, I. Tepliakova, M. Drobysh, V. Zbolotnii, S. Rackauskas, S. Ramanavicius, K. Grundsteins, V. Liustrovaite, A. Ramanaviciene, V. Ratautaite, E. Brazys, C. F. Chen, U. Prentice and A. Ramanavicius, *Sci. Total Environ.*, 2024, **939**, 173333.
17. K. Navakul, C. Sangma, P.-t. Yenchitsomanus, S. Chunta and P. A. Lieberzeit, *Anal. Bioanal. Chem.*, 2021, **413**, 6191–6198.
18. M. Holzinger, A. Le Goff and S. Cosnier, *Front. Chem.*, 2014, **2**, 63.
19. Y. Wang, J. Dostalek and W. Knoll, *Anal. Chem.*, 2011, **83**, 6202–6207.
20. O. R. Miranda, X. Li, L. Garcia-Gonzalez, Z. J. Zhu, B. Yan, U. H. F. Bunz and V. M. Rotello, *J. Am. Chem. Soc.*, 2011, **133**, 9650–9653.
21. M. H. Chan, C. Y. Lai, Y. C. Chan, M. Hsiao, R. J. Chung, X. Chen and R. S. Liu, *Nanomedicine*, 2019, **14**, 1791–1804.
22. H. Aldewachi, T. Chalati, M. N. Woodroofe, N. Bricklebank, B. Sharrack and P. Gardiner, *Nanoscale*, 2018, **10**, 18–33.
23. N. Elahi, M. Kamali and M. H. Baghersad, *Talanta*, 2018, **184**, 537–556.
24. F. Ma, C. Li and C. Zhang, *J. Mater. Chem. B*, 2018, **6**, 6173–6190.
25. M. Gumustas and S. A. Ozkan, *Open Anal. Chem. J.*, 2011, **5**, 1–21.
26. ICH Q2 (R1) Validation of analytical procedures: text and methodology|European Medicines Agency, https://www.ema.europa.eu/en/ich-q2-r1-validation-analytical-procedures-text-methodology. Accessed 28 Oct 2021.
27. N. Bhalla, P. Jolly, N. Formisano and P. Estrela, *Essays Biochem.*, 2016, **60**, 1.
28. *Development and Validation of Analytical Methods*, ed. C. M. Riley and T. W. Rosanske, Pergamon, Tarrytown, N.Y., 1996.
29. I. S. Krull and M. Swartz, *Anal. Lett.*, 1999, **32**, 1067–1080.

30. S. A. Ozkan, J.-M. Kauffmann, and P. Zuman, *Electroanalysis in Biomedical and Pharmaceutical Sciences, Chapter 8: Electroanalytical Method Validation in Pharmaceutical Analysis and Their Applications*, Springer, 2015, pp. 235–266.
31. S. A. Ozkan, J.-M. Kauffmann, and P. Zuman, Springer, 2015, pp. 1–5. *Electroanalysis in Biomedical and Pharmaceutical Sciences, Chapter 1: Introduction*
32. S. Park, *Analyst*, 2019, **144**, 5223–5231.
33. ICH M10 on bioanalytical method validatio|European Medicines Agency, https://www.ema.europa.eu/en/ich-m10-bioanalytical-method-validation, Accessed 1 November 2021.
34. J. Ermer and J. H. John, *Method Validation Pharm. Anal.*, 2005, 1–403.
35. S. Cárdenas and M. Valcárcel, *TrAC, Trends Anal. Chem.*, 2005, **24**, 477–487.
36. E. Trullols, I. Ruisánchez and F. X. Rius, *TrAC, Trends Anal. Chem.*, 2004, **23**, 137–145.
37. W. Emon, A. A. Rafsan, M. F. Nayan, R. R. Mahmud, N. H. Hai and M. A. Hossain, *Optik*, 2025, **322**, 172191.
38. K. Nemčeková, J. Korčeková, V. Svitková, D. Baraniak, M. Domšicová, E. Melníková, M. Hornychová, V. Szebellaiová, M. Gál and A. Poturnayová, *Biosensors*, 2024, **14**, 431.
39. M. Zhu, Y. Liu, M. Wang, T. Liu, Z. Chu and W. Jin, *Bioelectrochemistry*, 2024, **156**, 108598.
40. K. Daoudi, K. Ramachandran, H. Alawadhi, R. Boukherroub, E. Dogheche, M. A. El Khakani and M. Gaidi, *Surf. Interfaces*, 2021, **27**, 101454.
41. A. G. Ayankojo, R. Boroznjak, J. Reut, A. Öpik and V. Syritski, *Sens. Actuators, B*, 2022, **353**, 131160.
42. A. Raziq, A. Kidakova, R. Boroznjak, J. Reut, A. Öpik and V. Syritski, *Biosens. Bioelectron.*, 2021, **178**, 113029.
43. M. D. Gholami, K. Guppy-Coles, S. Nihal, D. Langguth, P. Sonar, G. A. Ayoko, C. Punyadeera and E. L. Izake, *Talanta*, 2022, **248**, 123630.
44. M. Chaturvedi, M. Patel, D. P. Mondal, A. K. Srivastava, N. Dwivedi and C. Dhand, *Electrochim. Acta*, 2024, **475**, 143630.
45. N. A. S. Omar, Y. W. Fen, J. Abdullah, C. E. N. C. E. Chik and M. A. Mahdi, *Sens Biosensing Res*, 2018, **20**, 16–21.
46. F. Zhao, Y. Bai, L. Cao, G. Han, C. Fang, S. Wei and Z. Chen, *J. Electroanal. Chem.*, 2020, **867**, 114184.
47. H. Gong, G. Cai, Y. Li, N. Jiang, C. Chen, F. Chen and C. Cai, *Anal. Chim. Acta*, 2025, **1336**, 343451.
48. Z. Saffari, M. Sepahi, R. Ahangari-Cohan, M. Khoobi, M. Hamidi-Fard, A. Ghavidel, M. R. Aghasadeghi and D. Norouzian, *Anal. Biochem.*, 2023, **661**, 114981.
49. J. Mao, J. Wang, H. Chen and Q. Yan, *Bioelectrochemistry*, 2025, **163**, 108901.
50. A. Imbriano, A. Tricase, E. Macchia, L. Torsi and P. Bollella, *Anal. Chim. Acta*, 2023, **1252**, 341037.

51. H. J. Lim, T. Saha, B. T. Tey, W. S. Tan, S. S. Hassan and C. W. Ooi, *Talanta*, 2022, **249**, 123659.
52. J. H. Kim, J. H. Shin, C. H. Cho, J. Hwang, D. H. Kweon, T. J. Park, C. H. Choi and J. P. Park, *Talanta*, 2022, **248**, 123613.
53. Y. Zhang, S. Zhang, H. Wang, M. Charconnet, J. Jiang, Y. Peng, L. Zhang and C. H. Lawrie, *Sens. Biosens. Res.*, 2024, **46**, 100689.
54. H. Sun, Y. Tao, Y. Li, M. Wang and H. Wang, *IEEE Sens. J.*, 2024, **24**, 17570–17578.
55. T. Lee, S. Y. Park, H. Jang, G. H. Kim, Y. Lee, C. Park, M. Mohammadniaei, M. H. Lee and J. Min, *Mater. Sci. Eng. C*, 2019, **99**, 511–519.
56. H. Gong, L. Tang, C. Chen, F. Chen and C. Cai, *Chem. Eng. J.*, 2023, **477**, 146990.
57. R. Wang, L. Wang, Z. T. Callaway, H. Lu, T. J. Huang and Y. Li, *Sens. Actuators, B*, 2017, **240**, 934–940.
58. M. Pal, V. Das, H. P. Deka Boruah and C. Chikkaputtaiah, in *Advanced Biosensors for Virus Detection*, Elsevier, 2022, pp. 219–239
59. S. Deumić, A. Lavić, N. Crnčević, E. Pramenković and A. Deumić, Biosensor for rapid methods for the detection of viruses, in *Mediterranean Conference on Medical and Biological Engineering and Computing*, 2024, pp. 423–429.
60. A. Poghossian, M. Jablonski, D. Molinnus, C. Wege and M. J. Schöning, *Front. Plant Sci.*, 2020, **11**, 1–14.
61. B. Lou, Y. Liu, M. Shi, J. Chen, K. Li, Y. Tan, L. Chen, Y. Wu, T. Wang, X. Liu, T. Jiang, D. Peng and Z. Liu, *TrAC, Trends Anal. Chem.*, 2022, **157**, 116738.
62. R. I. Stefan-Van Staden, *ECS Sens. Plus*, 2022, **1**, 011603.

Section III: Novel Approaches and Emerging Technologies

CHAPTER 7

Advances in CRISPR-based Biosensors for Virus Disease Detection: Across Viral Families

ARZUM ERDEM(ORCID 0000-0002-4375-8386)[a,b*],
SEYED MAJID HOSSEINI(ORCID 0000-0003-3952-084X)[a,b] AND
ESMA YILDIZ(ORCID 0000-0002-8094-8277)[a,b]

[a]Department of Analytical Chemistry, Faculty of Pharmacy, Ege University, Bornova, 35100 Izmir, Turkey; [b]The Institute of Natural and Applied Sciences, Biotechnology Department, Ege University, Bornova, 35100 Izmir, Turkey
*E-mail: arzum.erdem@ege.edu.tr

7.1 Introduction and History of CRISPR–Cas

7.1.1 Introduction to the CRISPR–Cas System

Over the past 40 years, there have been recurrent large-scale epidemics from emerging viruses such as human immunodeficiency virus (HIV), severe acute respiratory syndrome (SARS) and Middle East respiratory syndrome coronaviruses, 2009 pandemic influenza H1N1 virus, Ebola virus, Zika virus and most recently SARS coronavirus 2 (SARS-CoV-2).[1] All of these epidemics most

Detection Science Series No. 32
Advances in Virus Detection with Sensors
Edited by Adil Denizli
© The Royal Society of Chemistry 2026
Published by the Royal Society of Chemistry, www.rsc.org

probably resulted from an initial zoonotic animal-to-human transmission event, with either clinically apparent or occult spread into vulnerable human populations. Each time, a lack of rapid, accessible and accurate molecular diagnostic testing has hindered the public health response to the emerging viral threat.[1,2]

Clustered regularly interspaced short palindromic repeats (CRISPR)–CRISPR–associated (Cas) (CRISPR–Cas) systems are adaptive immune mechanisms naturally found in bacteria and archaea. These systems protect against invading viruses and plasmids by identifying and cutting specific sequences of foreign DNA or RNA. This ability to precisely target nucleic acids has made CRISPR a powerful tool in biotechnology. Over the past decade, CRISPR has become indispensable for genome editing, disease modelling and molecular diagnostics.[3–5]

Fundamentally, CRISPR technology consists of two essential parts. (I) CRISPR sequences: these brief repeating DNA sequences are present in bacterial genomes and originated from distinct DNA fragments that were derived from viruses. (II) Cas proteins: directed by RNA, these proteins function as molecular scissors that cleave particular nucleic acid sequences.

Utilizing these elements, CRISPR-based diagnostics produce extremely precise and programmable instruments for identifying infections, genetic alterations and even environmental pollutants. The system's programmability, combined with its high precision, has enabled groundbreaking innovations in healthcare, agriculture and environmental monitoring.[6,7]

7.1.2 History of Developments of CRISPR Technology

The origins of CRISPR research can be traced back to 1987, when repetitive DNA sequences were first identified in *Escherichia coli*.[8] At the time, the biological function of these sequences was unclear, but their discovery established the foundation for subsequent decades of research. By 2002, these repetitive elements were formally named clustered regularly ınterspaced short palindromic repeats (CRISPR). Concurrently, researchers identified CRISPR–associated (Cas) genes located adjacent to these sequences, which were suggested to play a role in an adaptive bacterial immune mechanism.[6–8] The first experimental proof of CRISPR's function in bacterial adaptive immunity was presented by researchers in 2007. It has been demonstrated that by identifying and cleaving foreign genetic material, CRISPR systems help bacteria protect themselves from bacteriophages.[9] The discovery of CRISPR RNA (crRNA) in 2008 broadened this understanding by demonstrating how these systems direct molecular machinery to certain target sequences. Additionally, the discovery of *trans*-activating CRISPR RNA[5] (tracrRNA) in 2011 revealed the mechanism of RNA-guided interference, in which CRISPR complexes precisely identify and cleave complementary

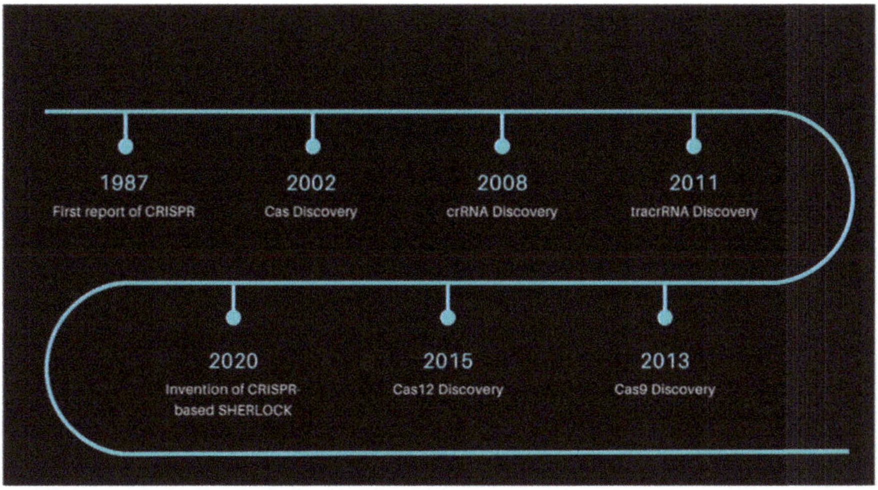

Figure 7.1 The timeline illustrates the historical development of CRISPR technology, beginning with the first report of CRISPR sequences in *Escherichia coli* in 1987. Major milestones include the discovery of Cas proteins in 2002, the identification of crRNA in 2008, and the discovery of tracrRNA in 2011. Subsequent breakthroughs include the discovery of Cas9 in 2013 and Cas12 in 2015. More recently, CRISPR-based diagnostic technologies have emerged, including the development of the SHERLOCK system in 2020. This timeline highlights the rapid evolution of CRISPR technology from fundamental biological discovery to powerful tools for genome editing and molecular diagnostics.

DNA or RNA sequences.[3,5,8] A pivotal breakthrough occurred in 2012, when Cas9 was identified as a programmable endonuclease.[7] This discovery revolutionized molecular biology by providing a tool for highly precise and targeted genome editing. Subsequently, in 2013, CRISPR–Cas9 technology was successfully applied to edit genes in eukaryotic cells, marking a transformative milestone in the fields of biotechnology and genomics. These advancements have since catalysed the development of numerous applications, including therapeutic genome editing, functional genomics and novel diagnostic platforms[5–8] (Figure 7.1).

7.1.3 Current Significance

CRISPR is currently at the forefront of scientific investigation. Beyond viral detection, it can be used for medical research, cancer screening and agricultural advancements. CRISPR-based techniques are extremely useful in responding to new diseases and global health emergencies because of their quick reprograming capabilities.

7.2 Fundamentals and Principles of the CRISPR–Cas System

7.2.1 Mechanism of Action of Cas Proteins

CRISPR–Cas systems function through a three-step process: adaptation, expression and interference. Each step is coordinated by Cas proteins and crRNAs, ensuring precise targeting and cleavage of nucleic acids:[8,9] (I) Adaptation: when a bacterium encounters a virus, it captures a fragment of the invader's DNA or RNA and incorporates it into its genome at the CRISPR loci. These fragments are known as protospacers. The protospacers are interspersed with repeat sequences, forming a memory bank of prior infections.[10,11] (II) Expression: upon subsequent infections, the CRISPR loci are transcribed into precursor CRISPR RNAs (pre-crRNAs), which are processed into mature crRNAs. These crRNAs, often combined with tracrRNA in some systems, guide Cas proteins to recognize complementary sequences in the invader's genome.[12,13] (III) Interference: guided by crRNAs, Cas proteins bind to the target sequence and cleave it, neutralizing the threat. This interference is precise, as it relies on sequence complementarity.[14,15]

7.2.2 Types of Cas Proteins

CRISPR–Cas systems are divided into two classes and multiple types, each with unique features.[28,29] The most widely studied Cas proteins include: (I) Class 1: which utilize multi-subunit effector complexes (*e.g.*, Cascade complex), (ii) Class 2: which rely on single, multi-domain proteins such as Cas9, Cas12 and Cas13.

Cas9 is a DNA endonuclease that introduces double-strand breaks at specific sites. Cas9 is widely used in genome editing and diagnostics. In viral detection, it is used to identify DNA viruses, such as human papillomavirus (HPV).[16] Its ability to target DNA with high specificity makes it ideal for detecting DNA-based pathogens. Cas12 targets DNA but also exhibits collateral cleavage activity, where it non-specifically cuts single-stranded DNA (ssDNA) near the target site. Collateral cleavage is harnessed in biosensors to amplify detection signals. Cas12 is commonly used in DNA endonuclease-targeted CRISPR trans reporter (DETECTR) systems for viruses like SARS-CoV-2.[17] Collateral cleavage activity enables signal amplification, improving sensitivity in diagnostics. Cas13 specifically targets RNA and exhibits collateral cleavage of single-stranded RNA (ssRNA). It is a cornerstone of specific high-sensitivity enzymatic reporter unlocking (SHERLOCK) systems, used to detect RNA viruses such as influenza, Zika and SARS-CoV-2.[18,19] Cas13's RNA-targeting capability makes it indispensable for detecting RNA-based viruses.[20]

7.2.3 Relationship with Virus Detection

CRISPR–Cas systems are uniquely suited for virus detection due to their precision and programmability. The mechanisms of action and versatility of Cas proteins allow for the development of diagnostics tailored to specific viral families. (I) Targeting RNA viruses: Cas13-based systems excel at detecting RNA viruses, which constitute the majority of human viral pathogens. For example, Cas13 has been used to detect SARS-CoV-2 and HCV with high sensitivity.[21] (II) Targeting DNA viruses: Cas12 and Cas9 are commonly used for DNA viruses such as HPV and hepatitis B virus (HBV), leveraging their ability to precisely cleave DNA targets.[22,23] (III) Multiplexing capabilities: CRISPR systems can be used to simultaneously detect multiple viruses by designing different crRNAs for each target, enabling comprehensive diagnostics. A CRISPR assay for influenza A and B detection used Cas13 to differentiate between strains, enabling co-infection diagnostics.[24]

CRISPR–Cas systems have revolutionized the field of molecular diagnostics, particularly in the detection of viral infections. These systems, which leverage programmable nucleases such as Cas9, Cas12 and Cas13, provide unparalleled sensitivity and specificity for nucleic acid detection. By harnessing the RNA or DNA recognition capabilities of CRISPR-associated proteins, these platforms have been adapted for use in a variety of biosensing methods, enabling rapid, accurate and field-deployable diagnostic solutions.[25,26]

The ability to combine signal amplification and transduction technologies with the molecular recognition and cleavage capabilities of Cas proteins supports the variety of CRISPR-based biosensing techniques. As a result of this versatility, a number of biosensor categories have been created, each suited to particular detection platforms and diagnostic requirements. These include fluorescence biosensors, electrochemical biosensors, colorimetric biosensors and other signal-based approaches (I) Fluorescence biosensors utilize fluorescent reporters cleaved by Cas enzymes, generating signals that can be quantified using simple optical devices. These are widely used for high-sensitivity detection of viral RNA or DNA.[27] (II) Electrochemical biosensors integrate CRISPR systems with electrodes to detect electrical signals corresponding to the presence of nucleic acids. These platforms are ideal for portable, point-of-care diagnostics.[28,29] (III) Colorimetric biosensors, by using enzymatic amplification or nanoparticles to cause a colour shift, allow for the visual identification of target nucleic acids. In environments with limited resources, they are especially beneficial. (IV) Other signal-based biosensors include surface-enhanced Raman scattering (SERS) and laser-based detection techniques, providing advanced sensitivity for challenging diagnostic scenarios[29,30] (Figure 7.2).

7.2.4 Signal Amplification in Diagnostics

Diagnosing illnesses with low virus loads requires the capacity to magnify detection signals. This is accomplished by CRISPR–Cas systems through the use of collateral cleavage and integration with additional methods,

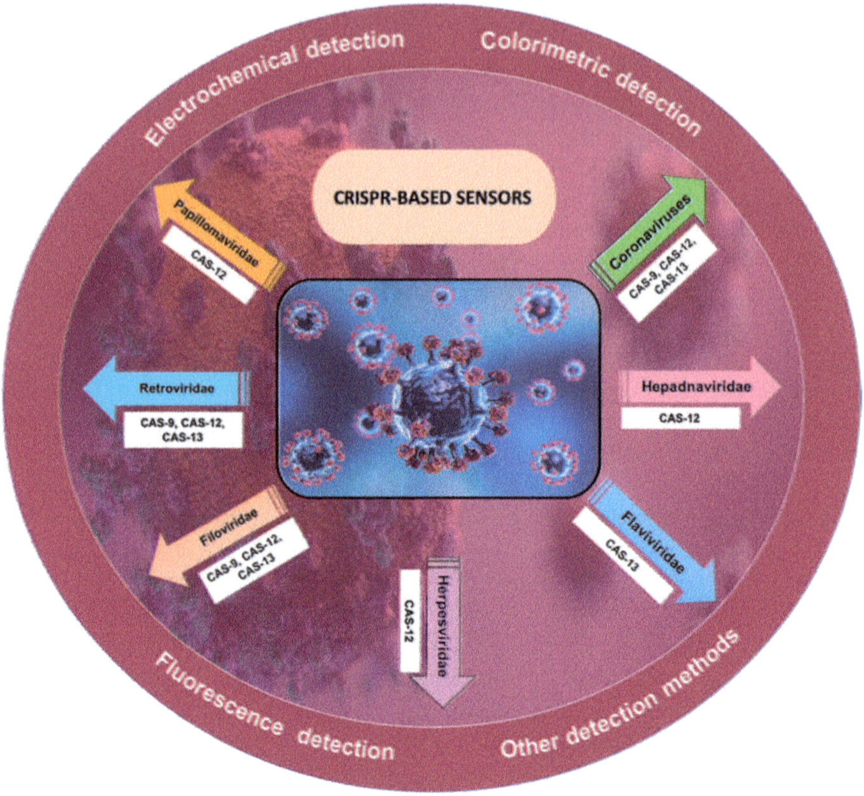

Figure 7.2 Applications of CRISPR-based biosensing for viral detection.

such as lateral flow assays (LFAs), electrochemical biosensors, isothermal amplification and fluorescence-based detection. Isothermal amplification techniques, including recombinase polymerase amplification (RPA) and loop-mediated isothermal amplification (LAMP), enhance the sensitivity of CRISPR-based diagnostics. These methods eliminate the need for thermal cycling, making them particularly suitable for point-of-care applications.[31] In fluorescence-based detection, CRISPR-mediated collateral cleavage releases fluorescent reporters, allowing real-time monitoring of viral nucleic acids. For example, the DETECTR system employs fluorescence signals to identify DNA targets.[26] Electrochemical biosensors incorporate CRISPR systems into electrochemical platforms, where cleavage activity alters electrical signals. These biosensors offer a highly sensitive and cost-effective diagnostic approach.[24,32] LFAs simplify diagnostics by providing visual readouts, often using gold nanoparticles. A notable example is the CRISPR–LFA system developed for Zika virus detection, which delivers accurate results in less than 30 minutes.[33]

7.2.5 Advantages of CRISPR–Cas Systems in Diagnostics

CRISPR–Cas systems offer several advantages over traditional diagnostic methods. The sequence-specific targeting of Cas proteins significantly reduces false positives, ensuring higher accuracy in detection. Additionally, the modular design of crRNAs allows for rapid adaptation to emerging pathogens, making these systems highly versatile. One of the key benefits of CRISPR-based diagnostics is their ability to deliver results in under an hour, making them particularly useful for urgent medical applications. These systems are also easily scalable, allowing for widespread deployment during pandemics or outbreaks. Furthermore, the low cost of CRISPR components makes this technology accessible even in resource-limited settings, increasing its potential for global use.[29,34–37]

CRISPR-based diagnostics, exemplified by platforms like DETECTR and SHERLOCK, represent highly efficient and innovative alternatives to traditional molecular methods such as reverse transcription–polymerase chain reaction (RT-PCR) for virus detection.[38,39] For detecting viral DNA or RNA, both methods show remarkable sensitivity and specificity. Nonetheless, CRISPR-based technologies are set apart with special benefits in terms of portability, speed and affordability. By using isothermal amplification methods like RPA or completely amplification-free workflows, these systems can frequently produce results in as little as one hour, while RT-PCR usually takes two to four hours or longer because of its reliance on heat cycling. This rapid turnaround time positions CRISPR diagnostics as particularly advantageous in settings where time is critical, such as during outbreak management or point-of-care testing.[40,41]

In addition to speed, CRISPR diagnostics offer substantial benefits in terms of accessibility and scalability. The relatively low cost of equipment and reagents makes CRISPR systems well-suited for deployment in resource-limited settings, contrasting with the high expense of thermal cyclers and specialized reagents required for RT-PCR. Furthermore, the portability of CRISPR platforms, combined with user-friendly readouts, such as lateral flow assays and colorimetric signals, facilitates their use by minimally trained personnel in decentralized environments.[41,42] The innovative use of collateral cleavage activity within CRISPR systems enhances detection signal amplification without the need for complex instrumentation, enabling the rapid and sensitive detection of pathogens like SARS-CoV-2, Zika virus and HIV.[29,43] While still in the process of gaining widespread regulatory approval and clinical adoption, CRISPR diagnostics represent a transformative advancement in molecular diagnostics, addressing the limitations of traditional RT-PCR methods and paving the way for broader applications in global health.[44,45]

Table 7.1 provides a structured comparison of CRISPR-based detection methods for various viral families, highlighting the pathogen type, CRISPR system utilized, detection platform, amplification techniques and key applications.

CRISPR–Cas biosensors have emerged as a revolutionary tool in molecular diagnostics, offering unparalleled specificity and sensitivity. In recent years, significant advancements have been made in integrating CRISPR systems with photoelectrochemical (PEC) biosensors, which utilize optical excitation and electronic detection to achieve rapid and precise diagnostics. PEC biosensors are particularly advantageous due to their high sensitivity, low background noise and ability to operate under ambient light conditions.[46,47] Additionally, PEC biosensors provide a label-free detection platform, reducing the complexity and cost of assays compared with traditional fluorescence or colorimetric methods. Their integration with CRISPR–Cas systems allows

Table 7.1 Overview of CRISPR-based biosensing for viral detection categorized by viral family.

Virus family	Pathogen	CRISPR system	Detection platform	Amplification	Applications
Coronaviridae	SARS-CoV-2	Cas12a, Cas13a	Fluorescence, electrochemical	PCR, RPA	Rapid diagnostics, multiplex detection
Flaviviridae	Zika virus (ZIKV), dengue virus (DENV)	Cas13a, Cas12a	Lateral flow assay, colorimetric	RT–RPA, nucleic acid sequence based amplification (NASBA)	Outbreak surveillance, point-of-care detection
Retroviridae	HIV	Cas12a, dead Cas9 (dCas9)	Nanopore sensor, fluorescence	RPA	Early HIV detection, drug resistance monitoring
Filoviridae	Ebola, Marburg	Cas13a	Fluorescence	PCR	Rapid diagnostics in outbreak scenarios
Papillomaviridae	HPV16, HPV18	Cas12a, *Alicyclobacillus acidiphilus* Cas12b (AaCas12b)	Lateral flow assay, electrochemical	RPA	Early screening for cervical cancer
Hepadnaviridae	HBV	Cas12a	Colorimetric, gel electrophoresis	LAMP, hybridization chain reaction (HCR)	HBV genotyping, liver cancer diagnostics
Herpesviridae	Epstein–Barr virus (EBV), HSV	Cas12a	Fluorescence, electrochemical	RPA	Detecting latent infections, clinical diagnostics

for programmable and highly specific recognition of viral DNA and RNA sequences, paving the way for rapid diagnostics at the point of care. This advancement not only accelerates the diagnostic process but also enhances the applicability of CRISPR biosensors in low-resource settings, demonstrating their transformative potential in global health security and outbreak management.[47,48]

7.3.1 CRISPR-based Detection of Coronaviridae Family Viruses

The CRISPR-based detection platforms for SARS-CoV-2 exhibit a range of strategies that improve sensitivity, specificity and usability by utilizing various CRISPR–Cas systems and biosensing techniques.

7.3.1.1 Cas12a Fluorescence-Based Detection

The fluorescence-based detection method powered by Cas12a has demonstrated a limit of detection (LOD) as low as one copy per microlitre. This approach employs Cas12a's *trans*-cleavage activity upon recognizing target sequences, breaking quencher-linked DNA strands, which results in a measurable fluorescent signal.[49] A notable example is the use of a smartphone-based readout system, allowing resource-limited settings to benefit from the technology's simplicity and portability. The Cas12a biosensors have proven similar to quantitative PCR (qPCR) in terms of accuracy, with 100% agreement in clinical trials.

7.3.1.2 Cas9 Lateral Flow Assay (Bio-SCAN)

The Bio-SCAN platform incorporates dead Cas9 (dCas9), a catalytically inactive variant of Cas9, with biotinylated detection complexes for visual readouts on LFA strips. This method achieved an LOD of four copies per microlitre, enabling detection of clinically relevant concentrations of SARS-CoV-2 RNA.[50] The technology is notable for its minimal equipment needs and compatibility with point-of-care settings, making it particularly suited for low-resource environments.[50] Unlike fluorescence-based systems, Bio-SCAN produces results visible to the naked eye, reducing reliance on advanced optical instruments.

7.3.1.3 Cas13a Electrochemical Detection

Cas13a-based biosensors focus on RNA targets and combine CRISPR technology with metal–carbon/nitrogen two dimensional compounds–gold nanoparticle (MXene–AuNP) composites, enabling ultrasensitive electrochemical detection with an LOD of 1 fM.[51] This method is especially advantageous for distinguishing viral variants such as omicron BA.2 and BA.5.

By eliminating the need for reverse transcription and amplification, Cas13a systems simplify the workflow and enhance detection accuracy. Furthermore, their integration with advanced nanomaterials ensures enhanced stability and conductivity, critical for point-of-care diagnostics.[51,52]

Different Cas proteins and biosensing methods are used by the different CRISPR-based detection technologies for SARS-CoV-2, each of which has unique benefits in terms of sensitivity, amplification requirements and detection speed. Table 7.2 provides a comparative summary of these methods, highlighting the CRISPR system used, target nucleic acid, amplification strategy, detection platform and reported LOD. This overview allows for a direct comparison of fluorescence, lateral flow and electrochemical detection techniques, demonstrating the adaptability of CRISPR diagnostics in both laboratory and point-of-care settings.

CRISPR–Cas-based biosensors have emerged as a rapidly advancing field, offering significant advantages in viral detection, including high sensitivity, specificity and portability. Cas13a-based electrochemical biosensors, which leverage *trans*-cleavage activity, have demonstrated the capability to detect SARS-CoV-2 RNA with high accuracy and speed without the need for nucleic acid amplification.[53] Furthermore, the integration of CRISPR–Cas systems with optical detection methods has facilitated practical applications in viral detection through signal amplification strategies, such as colorimetric and fluorescence-based approaches. These systems provide robust potential for the identification of biological molecules, particularly through their simplicity and high precision in nucleic acid detection.[54] Javalkote *et al.* explored the development of portable CRISPR-based devices designed for the rapid detection of SARS-CoV-2 and its variants. The authors presented a novel approach that employs CRISPR–Cas systems to detect viral RNA with high sensitivity and precision, while eliminating the need for complex laboratory equipment. The devices are not only portable but also user-friendly, making them applicable in non-clinical environments. The technology demonstrated high accuracy in identifying various SARS-CoV-2 variants, highlighting its

Table 7.2 CRISPR-based detection methods for SARS-CoV-2, categorized by CRISPR system and detection platform.

Pathogen	CRISPR system	Target	Amplification	Detection platform	LOD	References
SARS-CoV-2	Cas12a	RNA	PCR	AuNP fluorescence	1 copy per μl	49
SARS-CoV-2	Cas9	RNA	PCR	LFA	4 copies per μl	50
SARS-CoV-2	Cas13a	RNA	PCR	Electrochemical (AuNP composites)	1 fM	51
SARS-CoV-2	Cas9	RNA	PCR	AuNp	1.2 pM	52

potential as an effective tool for rapid, on-site diagnosis. This study underscores the transformative role of CRISPR-based biosensors in public health crises, such as pandemics, and their potential for broad-scale application in future diagnostic platforms.[55] Additionally, CRISPR–Cas-mediated electrochemical biosensors offer cost-effective and portable solutions for the diagnosis of bacterial and viral pathogens, establishing an innovative platform for clinical applications.[56] These advancements underscore the transformative impact of CRISPR–Cas-based biosensors on viral detection technologies.

7.3.2 CRISPR-based Detection of Flaviviridae Family Viruses

The CRISPR-based detection technologies for Flaviviridae family viruses involve diverse platforms targeting RNA and DNA with varying amplification and readout methods. Zika virus (ZIKV) diagnostics exemplify the flexibility of CRISPR systems, employing Cas9 in nucleic acid sequence based amplification (NASBA) assays for colorimetric detection and Cas13a for fluorescence and lateral flow detection. The NASBA-based approach demonstrates an impressive detection limit of 1 fM for RNA amplicons, providing a robust tool for sensitive ZIKV identification.[57] On the other hand, Cas13a-based fluorescence and LFA platforms achieve detection limits as low as 0.9 aM, especially when applied to saliva samples, making them suitable for field settings.[18]

For dengue virus (DENV), RT–RPA coupled with lateral flow assays leverages Cas13a's collateral cleavage activity to enable rapid and highly sensitive diagnostics. With detection limits of 0.9 aM in saliva, these methods are particularly advantageous for resource-limited environments prone to outbreaks.

In addition, recent studies on Japanese encephalitis virus (JEV) have demonstrated the use of Cas12a with RT–RPA, utilizing fluorescence and lateral flow assays for detection. Though specific detection limits have yet to be fully characterized, this approach represents a promising avenue for CRISPR-based diagnostics.[58] Similarly, Cas12a-based electrochemical detection for West Nile virus (WNV) utilizing graphene oxide sensors has achieved a notable detection limit of ten copies per ml, emphasizing the potential of nanotechnology-enhanced biosensors.[59]

Table 7.3 provides a structured comparison of CRISPR-based biosensing technologies for Flaviviridae family viruses, highlighting their respective CRISPR systems, amplification strategies and detection platforms.

7.3.3 CRISPR-based Detection of Filoviridae Family Viruses

CRISPR-based detection methods have been developed for Filoviridae family viruses, particularly Ebola virus (EBOV) and Marburg virus (MARV). These approaches primarily utilize the Cas13a system, which targets viral RNA sequences. The detection process typically involves PCR amplification followed by fluorescence-based readouts, achieving a LOD as low as 291

Table 7.3 CRISPR-based biosensing approaches for Flaviviridae family viruses, categorized by CRISPR system and detection platform.

Pathogen	CRISPR system	Target	Amplification	Detection platform	LOD	References
ZIKV	Cas9 [NASBA CRISPR cleavage (NASBACC)]	RNA	NASBA	Colorimetric	1 fM of RNA amplicon	57
ZIKV	Cas13a	DNA	RPA	Fluorescence	9 aM	60
ZIKV	Cas13a	RNA	RT–RPA	LFA	0.9 aM in saliva	18
ZIKV/DENV	Cas9 (NASBACC)	RNA	NASBA	NASBACC	1 fM of RNA amplicon	57
ZIKV/DENV	Cas13a	RNA	RT–RPA	LFA	0.9 aM in saliva	18
JEV	Cas12a	RNA	RT–RPA	Fluorescence and LFA	Not specified	58
WNV	Cas12a	RNA	RT-LAMP	Electrochemical (graphene oxide sensor)	10 copies per ml	59

attomolar (aM) for both EBOV and MARV In addition to the Cas13a-based methods, other CRISPR systems have been explored for viral diagnostics.[45,61] For instance, the Cas12a system has been employed in the detection of various viruses due to its collateral cleavage activity upon binding to target DNA sequences. This property has been harnessed in diagnostic platforms like DETECTR, which combines isothermal amplification with Cas12a-mediated detection, offering rapid and sensitive identification of viral DNA. Comparatively, while both Cas12a and Cas13a systems offer high specificity and sensitivity in viral detection, their applications differ based on the nature of the target nucleic acid.[61] Cas12a is more suited for DNA targets, whereas Cas13a is optimized for RNA targets, making it particularly useful for RNA viruses like EBOV and MARV. The choice between these systems depends on factors such as the type of viral genome, the required sensitivity and the available laboratory infrastructure[47] (Table 7.4).

7.3.4 CRISPR-based Detection of Retroviridae Family Viruses

CRISPR-based diagnostics for the Retroviridae family, particularly HIV, illustrate advanced methodologies tailored for clinical and research applications. The use of dCas9 in nanoelectrokinetic chips introduces a label-free, amplification-free approach for direct detection of target sequences. Such platforms provide rapid and precise measurements, although specific detection limits require further elucidation.[62]

Cas12a-based resistive pulse sensing harnesses nanopore technology to achieve a detection limit of 10 nM for DNA targets. This method offers high specificity and has the potential to detect even minor genetic variations, making it particularly useful for monitoring drug resistance in HIV treatment.[63,64]

Cas13a's versatility in RNA detection is evident in its application to fluorescence-based platforms coupled with RT–RPA. With a detection limit of 1 aM, these assays demonstrate extraordinary sensitivity, supporting early diagnosis and management of HIV infections. Additionally, microfluidic platforms combined with CRISPR–Cas12a enable efficient point-of-care testing, achieving a limit of detection of 43 copies in some assays.[65,66]

The integration of these CRISPR systems with advanced biosensing technologies highlights their transformative potential in combating Retroviridae family viruses, with implications for early detection, treatment monitoring and epidemiological surveillance.

Table 7.5 provides a structured comparison of CRISPR-based biosensing technologies for Retroviridae family viruses, particularly HIV. These approaches leverage diverse CRISPR systems, including dCas9, Cas12a and Cas13a, in combination with nanoelectrokinetic, microfluidic and fluorescence-based detection platforms. The integration of these biosensing technologies enables precise and rapid viral RNA or DNA detection, supporting early diagnosis and drug resistance monitoring. The table below summarizes

Table 7.4 CRISPR-based detection approaches for Filoviridae family viruses, categorized by CRISPR system and detection platform.

Pathogen	CRISPR system	Target	Amplification	Detection platform	LOD	References
EBOV	Cas13a	RNA	PCR	Fluorescence	291 aM	47
EBOV	Cas13	RNA	Off-chip RNA cleavage assays	Isothermal CRISPR	20 plaque forming units (pfu) per ml	61
MARV	Cas13a	RNA	PCR	Fluorescence	291 aM	47

Table 7.5 CRISPR-based detection approaches for Retroviridae family viruses, categorized by CRISPR system and detection platform.

Pathogen	CRISPR system	Target	Amplification	Detection platform	LOD	References
HIV	Cas12a	DNA	None	Resistive pulse (nanopore sensor)	10 nM	63
HIV	Cas13a	RNA	RT–RPA	Fluorescence	1 aM	65

key characteristics, including the target molecule, amplification strategy, detection platform and reported LOD.

7.3.5 CRISPR-based Detection of Hepadnaviridae Family Viruses

The diagnosis of Hepadnaviridae family viruses, especially hepatitis B virus (HBV), has greatly improved because of CRISPR-based diagnostic techniques. These novel techniques use the programmability and specificity of CRISPR systems to quickly and accurately identify viral nucleic acids.

HBV is a prototype member of the Hepadnaviridae family. It is an enveloped virus with a relaxed, circular, partially double-stranded DNA genome. Despite being a DNA virus, its replication occurs through reverse transcription of pre-genomic RNA (pgRNA).[67]

One notable method employs the CRISPR–Cas12a system in conjunction with RPA and a lateral flow biosensor. This combination enables the detection of HBV DNA or RNA at concentrations as low as one copy per microlitre, providing a user-friendly platform suitable for point-of-care testing.[68]

Another approach utilizes the CRISPR–Cas12a system paired with hybridization chain reaction (HCR) and gel electrophoresis. This method achieves a remarkable limit of detection of 1.5 femtomolar, highlighting its potential for identifying low-abundance viral targets in clinical samples.[69]

A summary of CRISPR-based HBV detection techniques is provided in Table 7.6, which also includes information on target types, amplification techniques, detection platforms, LODs and CRISPR systems utilized.

7.3.6 CRISPR-based Detection of Herpesviridae Family Viruses

Recent advancements in CRISPR-based diagnostics have significantly enhanced the detection of Herpesviridae family viruses, including Epstein–Barr virus (EBV) and herpes simplex virus type 1 (HSV-1).[71] These innovative approaches leverage the specificity and programmability of CRISPR

Table 7.6 CRISPR-based detection approaches for Hepadnaviridae family viruses, categorized by CRISPR system and detection platform.

Pathogen	CRISPR system	Target	Amplification	Detection platform	LOD	References
HBV	Cas12a	DNA/RNA	LAMP	Lateral flow fluorescent biosensor	1 copy per µl	68
HBV	Cas12a	DNA/RNA	HCR	Gel electrophoresis	1.5 fM	69
HBV	Cas12a	DNA/RNA	—	Electrochemical	30 pM	70

systems, particularly Cas12a, to achieve rapid and accurate identification of viral DNA.[71] One notable study utilized the CRISPR–Cas12a system to detect EBV DNA through fluorescence readouts. This method capitalizes on Cas12a's collateral cleavage activity, which, upon recognizing the target viral DNA, cleaves nearby fluorescently labelled reporter molecules. The result is a detectable fluorescence signal indicative of the presence of EBV. This assay demonstrated high sensitivity, making it a promising tool for early EBV detection.[72]

In the context of HSV-1 detection, researchers have developed quantum dot-labelled immunochromatographic strips incorporating the CRISPR–Cas12a system. This platform allows for visual detection of HSV-1 DNA without the need for sophisticated laboratory equipment. Although the specific limit of detection was not detailed, the assay's design emphasizes accessibility and ease of use, which are crucial for point-of-care diagnostics.[72–74]

Using CRISPR–Cas12a in conjunction with electrochemical sensors is another novel strategy. This technique uses electrodes coated with gold nanoparticles to detect HSV-1 DNA. The binding of the target DNA triggers a measurable electrochemical signal, providing a quantitative means of detection. While the exact sensitivity was not specified, this technique offers a promising avenue for developing portable and sensitive diagnostic tools.[75]

Collectively, these CRISPR-based diagnostic methods offer significant advantages over traditional techniques. They provide rapid, sensitive and specific detection of Herpesviridae viruses, with the added benefits of simplicity and potential for deployment in resource-limited settings. The versatility of CRISPR systems, particularly when combined with various detection platforms' such as fluorescence, lateral flow assays and electrochemical sensors, underscores their transformative potential in viral diagnostics.[76]

Huang *et al.* developed a highly sensitive and simple method for detecting HSV-1 using a combination of LAMP, CRISPR–Cas12 technology and gold nanoparticles. A diagnostic system that integrated these technologies was developed to enable rapid detection of HSV-1, which is crucial for timely treatment and preventing complications. The system allowed visual detection with the naked eye, using a colour change induced by gold nanoparticles, making it accessible and easy to use without complex equipment. The method demonstrated high sensitivity, achieving 93.9% positive detection consistency and 100% negative detection consistency when tested with clinical tear samples. The proposed approach provides a promising tool for point-of-care diagnostics, particularly for HSV-1, in clinical settings with limited resources.[77] (Table 7.7).

7.3.7 CRISPR-based Detection of Papillomaviridae Family Viruses

The application of CRISPR-based biosensors for detecting human papillomavirus (HPV), particularly high-risk types HPV16 and HPV18, has gained significant attention due to these viruses' association with cervical cancer

Table 7.7 CRISPR-based biosensing detection methods for Herpesviridae family viruses.

Pathogen	CRISPR system	Target	Amplification	Detection platform	LOD	References
EBV	Cas12a	DNA	—	Quantum dot-labelled immunochromatographic strips	High sensitivity	74
HSV-1	Cas12a	DNA	—	LFA (paper-based)	Not specified	76
	Cas12a	DNA	—	Electrochemical sensors (AuNP-coated electrodes)	Not specified	76

and other malignancies.[72] Traditional HPV diagnostic methods, such as PCR and hybridization techniques, offer high sensitivity but require laboratory-based infrastructure, limiting their accessibility in point-of-care settings.[73] CRISPR–Cas systems, especially Cas12a and Cas12b, provide a rapid, specific and cost-effective alternative for HPV detection.[74]

One of the most well-developed CRISPR-based detection methods for HPV16/18 is DETECTR, which utilizes Cas12a with RPA and fluorescence readout, achieving a LOD as low as two copies per microlitre.[72] This method enables fast and precise screening of HPV DNA, making it a valuable tool for clinical and field applications. Another variant of Cas12a-based HPV detection employs LFA combined with colorimetric readouts, improving visualization of test results, with an LOD reaching 0.28 fM.[73]

Recent advancements have also explored *Alicyclobacillus acidiphilus* Cas12b (AaCas12b)-based biosensors, offering even greater sensitivity, with fluorescence-based detection of 1×10^{-18} M for HPV DNA.[74] Additionally, electrochemical biosensors incorporating Cas12a have been developed, allowing for dual detection of DNA and protein markers associated with HPV infection. A LAMP-integrated Cas12a biosensor achieved an LOD of 10^{-4} M, facilitating point-of-care diagnosis.[75]

A promising alternative is CRISPR-activated PCR biosensors (CARP), which amplify target DNA sequences and integrate electrochemical readouts, achieving a detection threshold of 2 pg of DNA amplicon.[78] This hybrid method merges the high sensitivity of PCR with the specificity of CRISPR, further advancing the development of rapid molecular diagnostics for HPV detection.

Zhang *et al.* investigated the development of a novel electrochemical biosensor leveraging CRISPR–Cas12a technology for nucleic acid detection. The biosensor is designed to detect target nucleic acids through the Cas12a-mediated cleavage of hairpin DNA reporters at the sensor interface. This method combines the high specificity of CRISPR–Cas12a systems with the sensitivity of electrochemical sensing, enabling a rapid and accurate detection process. The proposed approach exhibits excellent sensitivity and a low detection limit, making it suitable for clinical applications. The study highlights the potential of integrating CRISPR-based molecular recognition with advanced sensing technologies to create efficient tools for nucleic acid diagnostics.[70] Dai *et al.* presented the *trans*-cleavage activity of CRISPR–Cas12a [CRISPR from *Prevotella* and *Francisella* 1 (Cpf1)] to develop a versatile electrochemical biosensor for nucleic acid detection. They designed a biosensing platform that utilizes the ability of Cas12a to cleave non-target single-stranded DNA upon recognition of a specific target DNA sequence. By integrating this molecular recognition mechanism with an electrochemical readout, the biosensor achieved high sensitivity and specificity. This universal platform demonstrated broad applicability across various targets, making it a promising tool for diagnostics in clinical and environmental settings. The work underscores the potential of CRISPR–Cas12a as a robust component in biosensor development.[78]

Zamani *et al.* introduced an electrochemical strategy for the affordable and sensitive detection of viral pathogens. The authors developed a low-cost biosensing platform that uses electrochemical techniques to detect viral nucleic acids with high precision. By leveraging simple materials and scalable methods, the platform minimizes costs while maintaining robust analytical performance. The biosensor demonstrated excellent sensitivity and specificity for viral detection, making it suitable for use in low-resource settings and point-of-care applications. The work highlights the potential of electrochemical approaches to democratize access to advanced diagnostic tools, particularly during global health emergencies.[79] Duan *et al.* presented a CRISPR–Cas12a-powered electrochemical sensor designed for highly sensitive nucleic acid detection. The sensor utilizes a hybrid composite of gold nanoparticles and MXene materials to enhance signal amplification and improve detection performance. By leveraging the *trans*-cleavage activity of Cas12a, the platform achieves highly specific and efficient target recognition. The integration of gold nanoparticles and MXene materials further increases sensitivity and provides a robust signal output. The biosensor demonstrated excellent detection performance, including a low detection limit and high reliability, making it suitable for clinical diagnostics and other nucleic acid detection applications. This work highlights the synergy between advanced materials and CRISPR technology in developing next-generation biosensors.[80]

Liu *et al.* reported the development of an ultrasensitive homogeneous electrochemical biosensor powered by CRISPR–Cas12a. The biosensor operates in a homogeneous reaction system, which simplifies the detection process and enhances performance. By utilizing the *trans*-cleavage activity of Cas12a, the biosensor is capable of highly specific and efficient nucleic acid detection. The design eliminates the need for complex electrode modifications,

Table 7.8 CRISPR-based biosensors detection methods for Papillomaviridae family viruses.

Pathogen	CRISPR system	Target	Amplification	Detection platform	LOD	References
HPV16/18	Cas12a (DETECTR)	DNA	RPA	Fluorescence	2 copies per µl	81
HPV16/18	Cas12a	DNA	RPA	Colorimetric (lateral flow assay)	0.28 fM	82
HPV16/18	AaCas12b	DNA	RPA	Fluorescence	1×10^{-18} M	83
HPV16/18	Cas12a	DNA, protein	LAMP	Electrochemical	10^{-4} M	79
HPV16	Cas12a	DNA	—	Electrochemical	3.22 pM	46
HPV16/18	Cas12a	DNA, protein	—	E-CRISPR electrochemical	10^{-12} M	84
HPV16/18	Cas12a	DNA, protein	PCR	CARP electrochemical	2 pg DNA amplicon	85

making it cost-effective and scalable for various applications. The platform demonstrates exceptional sensitivity and a low detection limit, showcasing its potential for clinical diagnostics and other nucleic acid-based assays. This work highlights the advantages of homogeneous systems in biosensor development for rapid and reliable detection.[46]

Table 7.8 summarizes the current CRISPR-based biosensor strategies for the detection of Papillomaviridae family viruses, outlining the target analytes, amplification techniques, detection platforms and limits of detection.

7.4 Conclusion

CRISPR–Cas systems have transformed the field of molecular diagnostics, offering unprecedented specificity, sensitivity and adaptability for viral detection. Rapid diagnoses in clinical and field settings are made possible by these technologies, which have demonstrated efficacy across a variety of viral families, including Coronaviridae, Flaviviridae, Retroviridae, Filoviridae and others. The integration of CRISPR-based platforms with advanced biosensing technologies, such as fluorescence, electrochemical and colorimetric methods, has further enhanced their utility for point-of-care diagnostics, particularly in resource-limited environments.

Despite these remarkable advancements, challenges remain. The scalability and affordability of CRISPR diagnostics must be further optimized to ensure global accessibility, particularly in underserved regions. Additionally, regulatory frameworks and clinical validation processes need to be streamlined to enable faster deployment of CRISPR-based diagnostics during health emergencies. Addressing these constraints may be possible with continued research into new Cas proteins, multiplexing properties and integration with nanotechnology.

In the future, the combination of CRISPR diagnostics with cutting-edge technologies like microfluidics and artificial intelligence may open the door to even more automated and flexible platforms. These innovations may not only improve the detection of viral pathogens but also expand the applications of CRISPR diagnostics to broader areas, including antimicrobial resistance monitoring and cancer screening. Ultimately, the continued development and deployment of CRISPR-based biosensing systems will play a crucial role in advancing global healthcare and preparedness for future pandemics.

7.5 Key Notes

Coronaviridae: Focuses on SARS-CoV-2 with highly sensitive detection platforms like fluorescence and electrochemical methods.

Flaviviridae: Includes viruses like ZIKV and DENV, with saliva-based LFA showing ultra-low detection limits.

Retroviridae: Highlights HIV detection using advanced nanopore sensors for drug resistance monitoring.

Filoviridae: Targets outbreak-prone viruses like Ebola and Marburg with Cas13a fluorescence-based platforms.

Papillomaviridae: Early HPV detection combines fluorescence and lateral flow assays with high precision.

Hepadnaviridae: HBV detection employs versatile colorimetric and electrophoresis methods for genotyping and diagnostics.

Herpesviridae: Focuses on EBV and HSV with RPA-enhanced fluorescence and electrochemical platforms.

Acknowledgements

A. E. would like to express her gratitude to the Turkish Academy of Sciences (TÜBA) as the Principal member of TÜBA for its partial support. E. Y. acknowledges to the PhD scholarship of YÖK-100/2000 Programme granded by Higher Education Council (YÖK) of Türkiye.

References

1. C. Wang, P. W. Horby, F. G. Hayden and G. F. Gao, A novel coronavirus outbreak of global health concern, *Lancet*, 2020, **395**(10223), 470–473.

2. N. Zhu, D. Zhang, W. Wang, X. Li, B. Yang and J. Song, *et al.*, A novel coronavirus from patients with pneumonia in China, 2019, *N. Engl. J. Med.*, 2020, **382**(8), 727–733.
3. M. Heidenreich and F. Zhang, Applications of CRISPR–Cas systems in neuroscience, *Nat. Rev. Neurosci.*, 2016, **17**(1), 36–44.
4. J. S. Godde and A. Bickerton, The repetitive DNA elements called CRISPRs and their associated genes: evidence of horizontal transfer among prokaryotes, *J. Mol. Evol.*, 2006, **62**, 718–729.
5. L. A. Marraffini and E. J. Sontheimer, CRISPR interference limits horizontal gene transfer in staphylococci by targeting DNA, *Science*, 2008, **322**(5909), 1843–1845.
6. P. Horvath, G. Gasiunas, V. Siksnys and R. Barrangou, Applications of the versatile CRISPR-cas systems, in *CRISPR-cas Systems: RNA-Mediated Adaptive Immunity in Bacteria and Archaea*, Springer, 2012, pp. 267–86.
7. J. A. Doudna and E. Charpentier, The new frontier of genome engineering with CRISPR-Cas9, *Science*, 2014, **346**(6213), 1258096.
8. Y. Ishino, M. Krupovic and P. Forterre, History of CRISPR-Cas from encounter with a mysterious repeated sequence to genome editing technology, *J. Bacteriol.*, 2018, **200**(7), e00580-17.
9. P. Horvath and R. Barrangou, CRISPR/Cas, the immune system of bacteria and archaea, *Science*, 2010, **327**(5962), 167–170.
10. D. C. Swarts, C. Mosterd, M. W. Van Passel and S. J. Brouns, CRISPR interference directs strand specific spacer acquisition, *PLoS One*, 2012, **7**(4), e35888.
11. H. Richter, L. Randau and A. Plagens, Exploiting CRISPR/Cas: interference mechanisms and applications, *Int. J. Mol. Sci.*, 2013, **14**(7), 14518–14531.
12. E. Charpentier, H. Richter, J. van der Oost and M. F. White, Biogenesis pathways of RNA guides in archaeal and bacterial CRISPR-Cas adaptive immunity, *FEMS Microbiol. Rev.*, 2015, **39**(3), 428–441.
13. F. Hille, H. Richter, S. P. Wong, M. Bratovič, S. Ressel and E. Charpentier, The biology of CRISPR-Cas: backward and forward, *Cell*, 2018, **172**(6), 1239–1259.
14. M. Jinek, A. East, A. Cheng, S. Lin, E. Ma and J. Doudna, RNA-programmed genome editing in human cells, *elife*, 2013, **2**, e00471.
15. J. S. Chen, E. Ma, L. B. Harrington, M. Da Costa, X. Tian and J. M. Palefsky, *et al.*, CRISPR-Cas12a target binding unleashes indiscriminate single-stranded DNase activity, *Science*, 2018, **360**(6387), 436–439.
16. X. Chen, Y. Tan, S. Wang, X. Wu, R. Liu and X. Yang, *et al.*, A CRISPR-Cas12b–Based platform for ultrasensitive, rapid, and highly specific detection of Hepatitis B Virus Genotypes B and C in Clinical Application, *Front. Bioeng. Biotechnol.*, 2021, **9**, 743322.
17. M. J. Kellner, J. J. Ross, J. Schnabl, M. P. Dekens, M. Matl and R. Heinen, *et al.*, A rapid, highly sensitive and open-access SARS-CoV-2 detection assay for laboratory and home testing, *Front. Mol. Biosci.*, 2022, **9**, 801309.

18. C. Myhrvold, C. A. Freije, J. S. Gootenberg, O. O. Abudayyeh, H. C. Metsky and A. F. Durbin, *et al.*, Field-deployable viral diagnostics using CRISPR-Cas13, *Science*, 2018, **360**(6387), 444–448.
19. Y. Xue, Z. Chen, W. Zhang and J. Zhang, Engineering CRISPR/Cas13 system against RNA viruses: from diagnostics to therapeutics, *Bioengineering*, 2022, **9**(7), 291.
20. Y. Shen, K. Hu, M. Yuan, G. Duan, Y. Guo and S. Chen, Progress and bioapplication of CRISPR-based one-step, quantitative and multiplexed infectious disease diagnostics, *J. Appl. Microbiol.*, 2023, **134**(3), lxad035.
21. X. Zhang, Y. Shi, G. Chen, D. Wu, Y. Wu and G. Li, CRISPR/Cas Systems-Inspired Nano/Biosensors for Detecting Infectious Viruses and Pathogenic Bacteria, *Small Methods*, 2022, **6**(10), 2200794.
22. M. M. Kaminski, O. O. Abudayyeh, J. S. Gootenberg, F. Zhang and J. J. Collins, CRISPR-based diagnostics, *Nat. Biomed. Eng.*, 2021, **5**(7), 643–656.
23. A. Erdem and M. Ozsoz, Chapter 2: Nucleic Acids as Biorecognition Element in Biosensor Development, in *Nucleic Acid Biosensors for Environmental Pollution Monitoring*, ed. I. Palchetti, and M. Mascini, Royal Society of Chemistry, UK, 2011, pp. 17–33.
24. Z. Weng, Z. You, J. Yang, N. Mohammad, M. Lin and Q. Wei, *et al.*, CRISPR-Cas biochemistry and CRISPR-based molecular diagnostics, *Angew. Chem., Int. Ed.*, 2023, **62**(17), e202214987.
25. A. Bonini, N. Poma, F. Vivaldi, A. Kirchhain, P. Salvo and D. Bottai, *et al.*, Advances in biosensing: The CRISPR/Cas system as a new powerful tool for the detection of nucleic acids, *J. Pharm. Biomed. Anal.*, 2021, **192**, 113645.
26. H. Li, Y. Xie, F. Chen, H. Bai, L. Xiu and X. Zhou, *et al.*, Amplification-free CRISPR/Cas detection technology: challenges, strategies, and perspectives, *Chem. Soc. Rev.*, 2023, **52**(1), 361–382.
27. Z. Huang, S. Liu, X. Pei, S. Li, Y. He and Y. Tong, *et al.*, Fluorescence Signal-Readout of CRISPR/Cas Biosensors for nucleic acid detection, *Biosensors*, 2022, **12**(10), 779.
28. D. W. Junior and L. T. Kubota, CRISPR-based electrochemical biosensors: an alternative for point-of-care diagnostics?, *Talanta*, 2024, 126467.
29. A. Kumaran, N. Jude Serpes, T. Gupta, A. James, A. Sharma and D. Kumar, *et al.*, Advancements in CRISPR-based biosensing for next-gen point of care diagnostic application, *Biosensors*, 2023, **13**(2), 202.
30. B. J. Yee, N. A. Ali, N. F. B. Mohd-Naim and M. U. Ahmed, Exploiting the Specificity of CRISPR/Cas System for Nucleic Acids Amplification-Free Disease Diagnostics in the Point-of-Care, *Chem. Bio. Eng.*, 2024, **1**(4), 330–339.
31. M. Kumar, S. Maiti and D. Chakraborty, Capturing nucleic acid variants with precision using CRISPR diagnostics, *Biosens. Bioelectron.*, 2022, **217**, 114712.
32. R. Nouri, Z. Tang, M. Dong, T. Liu, A. Kshirsagar and W. Guan, CRISPR-based detection of SARS-CoV-2: A review from sample to result, *Biosens. Bioelectron.*, 2021, **178**, 113012.

33. J. E. Garcia-Robledo, M. C. Barrera and G. J. Tobón, CRISPR/Cas: from adaptive immune system in prokaryotes to therapeutic weapon against immune-related diseases: CRISPR/Cas9 offers a simple and inexpensive method for disease modeling, genetic screening, and potentially for disease therapy, *Int. Rev. Immunol.*, 2020, **39**(1), 11–20.
34. A. Roberts and R. Barrangou, Applications of CRISPR-Cas systems in lactic acid bacteria, *FEMS Microbiol. Rev.*, 2020, **44**(5), 523–537.
35. A. Özcan, R. Krajeski, E. Ioannidi, B. Lee, A. Gardner and K. S. Makarova, *et al.*, Programmable RNA targeting with the single-protein CRISPR effector Cas7-11, *Nature*, 2021, **597**(7878), 720–725.
36. S. Kwon and H. Y. Shin, Advanced CRISPR-Cas effector enzyme-based diagnostics for infectious diseases, including COVID-19, *Life*, 2021, **11**(12), 1356.
37. J. Bond, CRISPR and COVID-19: Lessons Learned to Prepare for the Next Pathogen, *Georgetown Sci. Res J.*, 2022, **2**(1), 24–41.
38. T. Wang, Z. Wang, L. Bai, X. Zhang, J. Feng and C. Qian, *et al.*, Next-generation CRISPR-based diagnostic tools for human diseases, *TrAC, Trends Anal. Chem.*, 2023, **168**, 117328.
39. M. Bagi *et al.*, CRISPR–Cas based platforms for RNA detection: fundamentals and applications, *Chem. Commun.*, 2025, **61**(72), 13571–13600.
40. S. G. Jaybhaye, R. L. Chavhan, V. R. Hinge, A. S. Deshmukh and U. S. Kadam, CRISPR-assisted Diagnostics of Plant Viruses and Challenges, *Virology*, 2024, 110160.
41. J. de Dieu Habimana, R. Huang, B. Muhoza, Y. N. Kalisa, X. Han and W. Deng, *et al.*, Mechanistic insights of CRISPR/Cas nucleases for programmable targeting and early-stage diagnosis: A review, *Biosens. Bioelectron.*, 2022, **203**, 114033.
42. D. Huang, C. Xu, C. Jiang, Q. Chen, Z. Xu and X. Fang, Advances and challenges of signal readout systems in CRISPR-based biosensors for point-of-care testing of nucleic acid, *TrAC, Trends Anal. Chem.*, 2024, 117856.
43. S. Srivastava, D. J. Upadhyay and A. Srivastava, Next-generation molecular diagnostics development by CRISPR/Cas tool: rapid detection and surveillance of viral disease outbreaks, *Front. Mol. Biosci.*, 2020, **7**, 582499.
44. Z. Razavi, M. Soltani, M. Souri and H. Pazoki-Toroudi, CRISPR-driven biosensors: a new frontier in rapid and accurate disease detection, *Crit. Rev. Anal. Chem.*, 2024, 1–25.
45. Y. M. Hassan, A. S. Mohamed, Y. M. Hassan and W. M. El-Sayed, Recent developments and future directions in point-of-care next-generation CRISPR-based rapid diagnosis, *Clin. Exp. Med.*, 2025, **25**(1), 1–15.
46. J. Liu, Q. Wan, R. Zeng and D. Tang, An ultrasensitive homogeneous electrochemical biosensor based on CRISPR-Cas12a, *Anal. Methods*, 2021, **13**(29), 3227–3232.
47. Y. Li, R. Zeng, W. Wang, J. Xu, H. Gong and L. Li, *et al.*, Size-controlled engineering photoelectrochemical biosensor for human papillomavirus-16 based on CRISPR-Cas12a-induced disassembly of Z-scheme heterojunctions, *ACS Sens.*, 2022, **7**(5), 1593–1601.

48. H. Gong, Y. Wu, R. Zeng, Y. Zeng, X. Liu and D. Tang, CRISPR/Cas12a-mediated liposome-amplified strategy for the photoelectrochemical detection of nucleic acid, *Chem. Commun.*, 2021, **57**(71), 8977–8980.
49. L. Ma, L. Yin, X. Li, S. Chen, L. Peng and G. Liu, *et al.*, A smartphone-based visual biosensor for CRISPR-Cas powered SARS-CoV-2 diagnostics, *Biosens. Bioelectron.*, 2022, **195**, 113646.
50. Z. Ali, E. Sánchez, M. Tehseen, A. Mahas, T. Marsic and R. Aman, *et al.*, Bio-SCAN: A CRISPR/dCas9-based lateral flow assay for rapid, specific, and sensitive detection of SARS-CoV-2, *ACS Synth. Biol.*, 2021, **11**(1), 406–419.
51. Z. Chen, C. Wu, Y. Yuan, Z. Xie, T. Li and H. Huang, *et al.*, CRISPR-Cas13a-powered electrochemical biosensor for the detection of the L452R mutation in clinical samples of SARS-CoV-2 variants, *J. Nanobiotechnol.*, 2023, **21**(1), 141.
52. B. Yang, X. Zeng, J. Zhang, J. Kong and X. Fang, Accurate identification of SARS-CoV-2 variant delta using graphene/CRISPR-dCas9 electrochemical biosensor, *Talanta*, 2022, **249**, 123687.
53. X. Ge, T. Meng, X. Tan, Y. Wei, Z. Tao and Z. Yang, *et al.*, Cas14a1-mediated nucleic acid detectifon platform for pathogens, *Biosens. Bioelectron.*, 2021, **189**, 113350.
54. W. Su, J. Li, C. Ji, C. Chen, Y. Wang and H. Dai, *et al.*, CRISPR/Cas systems for the detection of nucleic acid and non-nucleic acid targets, *Nano Res.*, 2023, 1–14.
55. V. S. Javalkote, N. Kancharla, B. Bhadra, M. Shukla, B. Soni and A. Sapre, *et al.*, CRISPR-based assays for rapid detection of SARS-CoV-2, *Methods*, 2022, **203**, 594–603.
56. J. Ma, X. Li, C. Lou, X. Lin, Z. Zhang and D. Chen, *et al.*, Utility of CRISPR/Cas mediated electrochemical biosensors, *Anal. Methods*, 2023, **15**(31), 3785–3801.
57. K. Pardee, A. A. Green, M. K. Takahashi, D. Braff, G. Lambert and J. W. Lee, *et al.*, Rapid, low-cost detection of Zika virus using programmable biomolecular components, *Cell*, 2016, **165**(5), 1255–1266.
58. N. Kwak, B. J. Park and Y.-J. Song, A CRISPR-Cas12a-Based Diagnostic Method for Japanese Encephalitis Virus Genotypes I, III, and V, *Biosensors*, 2023, **13**(8), 769.
59. P. S. Tomar, J. S. Kumar, S. Patel and S. Sharma, Polymerase spiral reaction assay for rapid and real time detection of West Nile virus from clinical samples, *Front. Cell. Infect. Microbiol.*, 2020, **10**, 426.
60. C. M. Ackerman, C. Myhrvold, S. G. Thakku, C. A. Freije, H. C. Metsky and D. K. Yang, *et al.*, Massively multiplexed nucleic acid detection with Cas13, *Nature*, 2020, **582**(7811), 277–282.
61. P. Qin, M. Park, K. J. Alfson, M. Tamhankar, R. Carrion and J. L. Patterson, *et al.*, Rapid and fully microfluidic Ebola virus detection with CRISPR-Cas13a, *ACS Sens.*, 2019, **4**(4), 1048–1054.
62. M. Azimzadeh, M. Mousazadeh, A. Jahangiri-Manesh, P. Khashayar and P. Khashayar, CRISPR-powered microfluidics in diagnostics: a review of main applications, *Chemosensors*, 2021, **10**(1), 3.

63. J. Zhou, J. Hu, R. Liu, C. Wang and Y. Lv, Dual-amplified CRISPR-Cas12a bioassay for HIV-related nucleic acids, *Chem. Commun.*, 2022, **58**(26), 4247–4250.
64. X. Zhao, X. Tian, Y. Wang, L. Li, Y. Yu and S. Zhao, *et al.*, CRISPR-Cas12a-activated palindrome-catalytic hairpin assembly for ultrasensitive fluorescence detection of HIV-1 DNA, *Anal. Chim. Acta*, 2022, **1227**, 340303.
65. X. Li, B. Su, L. Yang, Z. Kou, H. Wu and T. Zhang, *et al.*, Highly sensitive and rapid point-of-care testing for HIV-1 infection based on CRISPR-Cas13a system, *BMC Infect. Dis.*, 2023, **23**(1), 627.
66. Y. Tian, Z. Fan, X. Zhang, L. Xu, Y. Cao and Z. Pan, *et al.*, CRISPR/Cas13a-Assisted accurate and portable hepatitis D virus RNA detection, *Emerging Microbes Infect.*, 2023, **12**(2), 2276337.
67. H. Bahrulolum, H. Tarrahimofrad, F. N. Rouzbahani, S. Nooraei, M. M. Sameh and A. Hajizade, *et al.*, Potential of CRISPR/Cas system as emerging tools in the detection of viral hepatitis infection, *Virol. J.*, 2023, **20**(1), 91.
68. R. Ding, J. Long, M. Yuan, X. Zheng, Y. Shen, Y. Jin and G. Duan, CRISPR/Cas12-based ultra-sensitive and specific point-of-care detection of HBV, *Int. J. Mol. Sci.*, 2021, **22**(9), 4842.
69. M. J. Kachwala, C. W. Smith, N. Nandu and M. V. Yigit, Reprogrammable gel electrophoresis detection assay using CRISPR-Cas12a and hybridization chain reaction, *Anal. Chem.*, 2021, **93**(4), 1934–1938.
70. D. Zhang, Y. Yan, H. Que, T. Yang, X. Cheng and S. Ding, *et al.*, CRISPR/Cas12a-mediated interfacial cleaving of hairpin DNA reporter for electrochemical nucleic acid sensing, *ACS Sens.*, 2020, **5**(2), 557–562.
71. G. Maini, G. Cianci, M. Ferraresi, V. Gentili and D. Bortolotti, DNA-based technology for herpesvirus detection, *DNA*, 2024, **4**(4), 553–581.
72. A. Binnie, E. Fernandes, H. Almeida-Lousada, R. A. De Mello and P. Castelo-Branco, CRISPR-based strategies in infectious disease diagnosis and therapy, *Infection*, 2021, **49**, 377–385.
73. S. Najafi, S. C. Tan, S. Aghamiri, P. Raee, Z. Ebrahimi and Z. K. Jahromi, *et al.*, Therapeutic potentials of CRISPR-Cas genome editing technology in human viral infections, *Biomed. Pharmacother.*, 2022, **148**, 112743.
74. H. Zhang, F. Tan, Q. Wang, P. Xu, M. Shen and X. Hu, *et al.*, CRISPR-Cas12a based HSV DNA detection method using quantum dot-labeled immunochromatographic strips, *Microchem. J.*, 2024, **207**, 112117.
75. S. S. Abolmaali, V. Alimardani, G. Farahavar, H. Najafi, M. Shafiee, N. Tanideh, *et al.*, Nanotechnology-based approaches against COVID-19, in *Emerging Nanomaterials and Nano-Based Drug Delivery Approaches to Combat Antimicrobial Resistance*, Elsevier, 2022, pp. 305–364.
76. H. Liang, L. Huang, P. Yan, F. Huang, H. Zhang and Q. Luo, Construction of Electrochemical Biosensors Based on the CRISPR/Cas12 System for Applications, *Mater. Sci. Appl.*, 2024, **15**(9), 285–304.
77. M. Huang, Y. Chen, L. Zheng and Y.-F. Yao, Highly sensitive and naked-eye detection of herpes simplex virus type 1 using LAMP-CRISPR/Cas12 diagnostic technology and gold nanoparticles, *Heliyon*, 2023, **9**(11), e22146.

78. Y. Dai, R. A. Somoza, L. Wang, J. F. Welter, Y. Li and A. I. Caplan, *et al.*, Exploring the trans-cleavage activity of CRISPR-Cas12a (cpf1) for the development of a universal electrochemical biosensor, *Angew. Chem.*, 2019, **131**(48), 17560–17566.
79. M. Zamani, J. M. Robson, A. Fan, Jr M. S. Bono, A. L. Furst and C. M. Klapperich, Electrochemical strategy for low-cost viral detection, *ACS Cent. Sci.*, 2021, 7(6), 963–972.
80. H. Duan, Y. Wang, S.-Y. Tang, T.-H. Xiao, K. Goda and M. Li, A CRISPR-Cas12a powered electrochemical sensor based on gold nanoparticles and MXene composite for enhanced nucleic acid detection, *Sens. Actuators, B*, 2023, **380**, 133342.
81. L. Yin, Z. Zhao, C. Wang, C. Zhou, X. Wu and B. Gao, *et al.*, Development and evaluation of a CRISPR/Cas12a-based diagnostic test for rapid detection and genotyping of HR-HPV in clinical specimens, *Microbiol. Spectrum*, 2025, **13**(1), e02253-24.
82. J. H. Tsou, Q. Leng and F. Jiang, A CRISPR Test for Detection of Circulating Nuclei Acids, *Transl. Oncol.*, 2019, **12**(12), 1566–1573.
83. Y. Zhao, D. Chen, Z. Xu, T. Li, J. Zhu and R. Hu, *et al.*, Integrating CRISPR-Cas12a into a microfluidic dual-droplet device enables simultaneous detection of HPV16 and HPV18, *Anal. Chem.*, 2023, **95**(6), 3476–3485.
84. F. Teng, L. Guo, T. Cui, X.-G. Wang, K. Xu and Q. Gao, *et al.*, CDetection: CRISPR-Cas12b-based DNA detection with sub-attomolar sensitivity and single-base specificity, *Genome Biol.*, 2019, **20**, 1–7.
85. B. Zhang, Q. Wang, X. Xu, Q. Xia, F. Long and W. Li, *et al.*, Detection of target DNA with a novel Cas9/sgRNAs-associated reverse PCR (CARP) technique, *Anal. Bioanal. Chem.*, 2018, **410**, 2889–2900.

CHAPTER 8

Quantum Dot Sensors

ÖZGE ALTINTAŞ[a], FATMA YILMAZ[b] AND ADIL DENIZLI[a*]

[a]Department of Chemistry, Hacettepe University, Ankara, Turkey;
[b]Department of Chemistry and Chemical Processing Technologies, Bolu Abant İzzet Baysal University, Bolu, Turkey
*E-mail: denizli@hacettepe.edu.tr

8.1 Introduction

In the modern era, the need for rapid, accurate and sensitive virus detection methods has become increasingly urgent, especially in the light of global pandemics and the growing prevalence of emerging infectious diseases.[1] Conventional diagnostic methods, while effective, often struggle to meet the demands of high-throughput testing, early-stage detection and cost-efficiency.[2] This has catalysed a paradigm shift toward advanced nanotechnology-driven approaches, among which quantum dot (QD) sensors stand out as transformative tools.[3] Quantum dots (QDs) are nanoscale semiconductor crystals that exhibit unique optical properties, such as size-dependent tuneable fluorescence, photostability and high signal-to-noise ratios.[4] These properties make them ideal candidates for developing high-precision biosensors. QD sensors, by leveraging these distinctive features, offer the potential to revolutionize virus detection through enhanced sensitivity, multiplexing capability and adaptability for point-of-care diagnostics.[5] Moreover, their ability to interface seamlessly with biological molecules, such as antibodies, nucleic acids and aptamers facilitates specific and targeted virus recognition.[6]

Detection Science Series No. 32
Advances in Virus Detection with Sensors
Edited by Adil Denizli
© The Royal Society of Chemistry 2026
Published by the Royal Society of Chemistry, www.rsc.org

This chapter explores the critical role of QD technology in addressing the challenges of virus detection. It begins with a detailed examination of quantum dot synthesis, functionalization and their core properties that underpin their use in sensor development. The discussion then transitions to the detection principles employed in QD sensors, highlighting their operational mechanisms, such as Förster resonance energy transfer (FRET) and fluorescence quenching. The chapter further delves into the diverse applications of QD sensors in detecting a range of viral pathogens, from well-known viruses, such as influenza and human immunodeficiency virus (HIV) to emerging threats like severe acute respiratory syndrome coronavirus 2 (SARS-CoV-2). By bridging the gap between nanotechnology and virology QD sensors are not just tools of innovation but are shaping the future of virus diagnostics.

8.2 Quantum Dot Technology

QDs are nanosized structures created from semiconductor materials, designed to confine electrons through quantum mechanical effects in all three spatial dimensions.[7] This unique confinement gives QDs characteristics that resemble those of atoms. They can absorb and emit light across specific wavelengths, depending on their size, making them highly versatile for optical and electronic applications.[8] Known as synthetic atoms, QDs display distinct energy states for charge carriers due to their nanoscale dimensions, similar to the discrete energy levels of atomic electrons.[9] When energized, QD electrons can jump from the valence band to the conduction band, creating a vacancy or hole in the valence band. This hole behaves like a positively charged particle capable of moving through the band.[10] Following excitation, the holes and electrons rapidly release energy and stabilize at the lowest energy states near the conduction and valence band edges.[11]

QDs were first discovered in 1981 by Alexey I. Ekimov within a glass matrix, marking the initial breakthrough in this field. Before the 1990s, conventional chemical techniques like microemulsions and co-precipitation were employed to synthesize colloidal nanoparticles.[12] Although these methods allowed for some control over nanocrystal size, the resulting materials lacked light-emitting properties and exhibited poor optical performance.[13] In the early 1990s, cadmium selenide (CdSe) nanocrystals were synthesized using dimethylcadmium ($CdMe_2$) in a metal–organic coordinated solvent at high temperatures. While effective, this approach was expensive, hazardous and highly toxic. By the early 2000s, $CdMe_2$ was replaced with the safer and less toxic cadmium oxide (CdO) for QD synthesis. By 2002, QDs had become commercially viable for use in life sciences. In 2014, a solution-based synthesis method enabled the creation of high-performance, cost-effective QD light-emitting diodes (QLEDs), combining the benefits of gallium nitride (GaN) LEDs and organic LEDs (OLEDs).[14]

The early QDs were primarily composed of group IV and III–V compounds, but substantial advancements were made with the introduction of other group elements. QDs are categorized on the basis of their chemical makeup. They may either consist of a single element, such as germanium (Ge) or silicon (Si), or be compound semiconductors, such as CdSe, CdTe, PbS or PbSe. Key QD types include core–shell structures, metal dichalcogenides and metal oxides.[15] Recently, carbon-based QDs (CQDs) have gained significant attention due to their many advantages, including straightforward synthesis, affordability of raw materials and excellent biocompatibility.[16]

8.2.1 Semiconductor-based Quantum Dots (SQDs)

SQDs, created from semiconductor materials, are widely recognized for their ability to emit light across a wide spectral range (from ultraviolet to near-infrared), coupled with their high quantum efficiency and outstanding photostability.[17] These QDs are classified according to the periodic table groups of the constituent elements. Group I–VI, examples include Cu_2S. Group I–VII includes AgBr. Group II–VI: common materials include CdTe, CdSe, CdS, ZnTe, ZnSe, ZnS and ZnO. Group III–V materials such as GaSb, GaP, GaAs, GaN, InSb, InAs, InP and InN. Group IV–VI includes PbTe, PbS and PbSe. Group IV includes Si.[18,19] The use of Pb, Cu and In in SQDs has been regulated due to their biotoxicity and poor environmental friendliness. Considering these constraints, the development of environmentally friendly and bio-friendly materials has become even more important.[20]

8.2.2 Carbon-based Quantum Dots (CQDs)

CQDs, sometimes referred to as group VI QDs, are a class of fluorescent nanomaterials. They encompass carbon-based nanomaterials with dimensions smaller than 10 nm, containing sp^3 and/or sp^2 carbon atoms. Nitrogen and oxygen atoms may also be integrated into their structures. These nanodots are synthesized through polymerization or pyrolysis processes involving small molecules or graphite-based precursors.[21] Their eco-friendly synthesis methods, coupled with their biocompatibility, make them highly desirable. Additionally, the tuneability of functional groups on CQD surfaces allows for diverse applications. For example, nitrogen-doped CQDs can be synthesized through hydrothermal treatment of citric acid mixed with ethylenediamine.[22]

8.2.3 Two-dimensional QDs (2D-QDs)

Graphene has been shown to exhibit luminescent properties when its size is reduced to below 20 nm. In this reduced form, graphene not only gains optical activity but also shows new attributes, such as a large surface area, ease of functionalization, and improved resolution. Graphene-derived QDs (GQDs) are part of the broader category of 2D-QDs, which also includes materials like hexagonal boron nitride, graphitic carbon nitride and transition metal dichalcogenides.[23]

8.2.4 Perovskite-based Quantum Dots (PQDs)

Perovskite, originally identified as a mineral with the formula $CaTiO_3$ in the Ural Mountains, has become a focus of research due to its exceptional photoelectronic properties. These materials follow the general formula ABX_3, where A is an alkali and X is oxygen or rare-earth metal ion.[24] There are two major perovskite types: hybrid organic–inorganic and entirely inorganic. However, their use in biomedical applications is restricted by their structural instability and the toxicity of certain components, such as lead.[25]

8.2.5 Applications

QD technology has a wide range of application areas. Some application areas of QDs are discussed in the following subsections.

8.2.5.1 Optoelectronic Devices

QDs are integral to LEDs, photovoltaic cells (solar panels), and display technologies due to their ability to emit and absorb light at specific wavelengths. QD-LED (QLED) screens, for instance, deliver vibrant colours and enhanced energy efficiency by producing precise red, green and blue hues with exceptional purity and contrast.[26]

8.2.5.2 Biomedical Applications

QDs play a critical role in biosensors and medical imaging, where their adjustable emission wavelengths allow for the labelling of cells and biomolecules. They offer more durable and highly sensitive fluorescent markers compared with traditional dyes.[27] QDs provide high-resolution imaging capabilities, enabling spatially resolved detection. QD fluorescence is used to map the location of targets in biological tissues or other environments. Imaging sensors use QD probes to identify specific areas where targets are concentrated. They are used to monitor cell surface interactions or intracellular processes and to image bacteria, viruses or fungi with functionalized QD probes.[28,29]

8.2.5.3 Laser Technology

QDs are employed to tune the wavelength of laser emissions, which is particularly advantageous in precision optical applications.[30] Near-range infrared (NIR) and short-wave infrared (SWIR) lasers and optical amplifiers in the 700–3000 nm range are of significant scientific and market importance because they cover many technologically important windows.[31]

8.2.5.4 Energy Storage and Solar Cells

QDs enhance solar energy capture by converting high-energy light into low-energy light more efficiently. Their size-tuneable energy levels make them suitable for developing advanced and highly efficient solar panels.[32]

8.2.6 Synthesis Methods

Scientists have focused on refining the synthesis methods of QDs to capitalize on their compact dimensions, particularly in biomedical applications. This involves modifying the surfaces of QDs by attaching various ligands, molecules or biomolecules through diverse synthesis techniques, both uniform and non-uniform. The methods of QD production can be grouped into two principal categories.[33]

8.2.6.1 Nanocrystal Synthesis

Nanocrystal synthesis is a key method for creating QDs with semiconductor properties. Unlike structures embedded in multilayered silicon wafers, this process results in free nanoscale particles suspended in a colloidal solution. The sharp contrast between the bandgap structure of the semiconductor core and the insulating nature of the solvent provides the conditions necessary for quantum confinement.[34] A notable benefit of this technique is its simplicity, as it avoids the need for advanced and expensive equipment like metal–organic chemical vapor deposition (MOCVD) or molecular beam epitaxy (MBE). This makes it feasible to produce highly controlled nanocrystals at an industrial level with minimal cost. The resulting QD nanocrystals are exceptionally adaptable due to their unique optical properties and suitability for biochemical applications. Recent advancements have explored their use in biomedical technologies, including immunoassays, DNA analysis and cancer diagnostics.[35]

8.2.6.2 QD Synthesis via Semiconductor Growth on Wafers

Another approach involves forming QDs during the synthesis of semiconductor devices on silicon wafers. Established techniques, such as MBE and MOCVD, have been employed for decades in the semiconductor industry. These methods allow precise control over layer thickness at the atomic level, enabling the creation of intricate multilayered semiconductor devices.[36] Since the 1970s, they have facilitated the development of quantum well (QW) structures, which are widely used in optoelectronic devices like lasers and photodetectors. However, the application of these structures in sensor technologies remains relatively restricted, with their primary contributions being in the fields of electronics and optoelectronics.[37] QDs are functionalized with specific ligands, antibodies or nucleic acids to achieve target-specific detection. Functionalization ensures selective binding to target molecules. Upon binding, the QD's properties (*e.g.*, fluorescence, charge state or aggregation behaviour) change, which is detected by optical, electrochemical or mechanical methods. This is used to detect biomarkers for diseases such as cancer or infectious agents and toxins or pollutants.[38,39]

8.3 Detection Principles with QD Sensors

QD-based sensors have garnered significant interest in the field of chemical and biological detection due to their unique optical and electronic characteristics. Their versatility in functionalization allows for easy modification with a range of biologically relevant and other essential molecules. Various strategies can be employed to leverage these distinctive properties of QDs for developing analytical techniques aimed at detecting chemicals and biomolecules.[40]

QD sensors leverage the unique optoelectronic properties of QDs to detect various biological, chemical and physical phenomena. The detection mechanisms are primarily based on the QDs' ability to absorb and emit light with high sensitivity and tuneability. The following sections give a detailed explanation of the principles and mechanisms behind detection with QD sensors.

8.3.1 Fluorescence-based Detection

QDs are well-known for their tuneable fluorescence, making them ideal for optical sensing applications. QDs fluoresce when excited by a light source (UV or visible). The emitted light's wavelength (colour) depends on the QDs' size. A target molecule (*e.g.*, a biomolecule, ion or gas) can interact with the QD's surface, altering its fluorescence properties, such as intensity, wavelength shift or lifetime.[41] When QDs are functionalized with antibodies or DNA strands, binding to specific antigens or complementary DNA changes the fluorescence, allowing highly specific sensing. Metal ions like cadmium or lead can quench or enhance fluorescence depending on the interaction mechanism.[42]

8.3.2 Förster Resonance Energy Transfer (FRET)

QDs are excellent energy donors in FRET-based detection systems. FRET occurs when energy from an excited QD (donor) is transferred non-radiatively to a nearby acceptor molecule (*e.g.*, dye or another QD) if the distance is within 1–10 nm.[43] Binding events (*e.g.*, between a target molecule and a receptor attached to the QD) bring the QD and acceptor molecule into proximity, causing a measurable energy transfer. As an application, FRET signals reveal the binding or conformational changes in proteins. In addition they are used in live-cell imaging to track dynamic biological processes.[44]

8.3.3 Photoelectrochemical (PEC) Detection

QDs can generate photocurrents under illumination, making them effective in PEC sensors. When exposed to light, QDs generate electron–hole pairs. These charge carriers can drive electrochemical reactions. The photocurrent changes upon interaction with specific target molecules, allowing detection. PEC is used for the detection of heavy metals, pesticides or organic pollutants in water and for the detection of glucose or cholesterol.[45,46]

8.3.4 Plasmon-enhanced Sensing

QDs can interact with plasmonic materials (*e.g.*, gold or silver nanoparticles) to enhance their sensing capabilities. The localized surface plasmon resonance (LSPR) of metal nanoparticles enhances the electric field near QDs. Changes in the local environment (*e.g.*, target binding or environmental shifts) modify the QD's optical properties, such as fluorescence enhancement or quenching. Applications include the detection of small molecules, DNA or proteins and monitoring of chemical changes in air or water.[47,48]

8.4 Applications in Virus Detection

The incorporation of functional materials into optoelectronic and photonic devices has significantly advanced the field of medical diagnostics. While traditional diagnostic materials and techniques have laid the groundwork for modern healthcare, they often fail to meet emerging clinical demands.[49] Organic dyes commonly used in imaging tend to undergo photobleaching, limiting their effectiveness in long-term monitoring. Likewise, silicon-based materials in photonic devices lack the necessary biocompatibility and flexibility for implantable and wearable applications.[50] To overcome these challenges, next-generation functional materials—such as perovskites and QDs—are being explored due to their exceptional photostability, tuneable enhanced compatibility and optical properties for modern diagnostic technologies.[51,52] Recent developments in wearable sensors have demonstrated the ability to accurately track heart rate variability and respiratory patterns even under dynamic conditions, making them invaluable for non-invasive health monitoring.[53] Furthermore, the integration of wireless data transmission with advanced biosensors is revolutionizing user-friendly healthcare electronics and medical diagnostics. These innovations enable real-time health tracking and remote patient monitoring, significantly improving accessibility and efficiency in healthcare.[54] In this context, a schematic representation illustrating the extraction of biometric data through integrated smart sensing wearable systems in human–machine interfaces is provided in Figure 8.1.

The coronavirus disease (COVID)-19 pandemic, caused by SARS-CoV-2, has resulted in an unprecedented number of transmissions, deaths worldwide

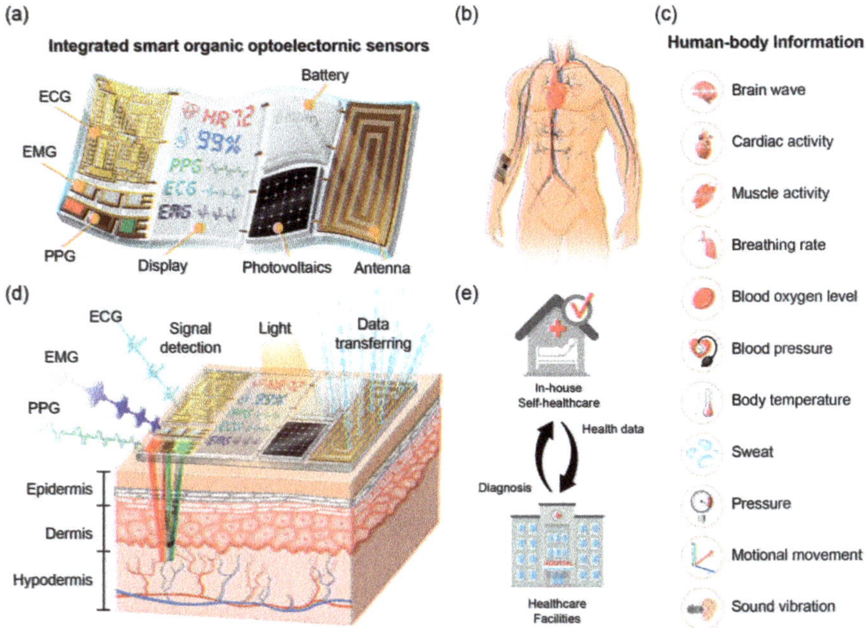

Figure 8.1 Perspective on next-generation stretchable organic optoelectronic devices. (a) Possible applications of integrated intelligent sensing systems. Abbreviations: electrocardiogram (ECG), electromyogram, (EMG), photoplethysmogram (PPG), heart rate (HR). (b) Utilization of smart sensing technology for capturing physiological signals from the human body. (c) Various types of biometric data that can be detected using smart sensors, (d) Real-time recording of multiple physiological signals with data transmission *via* wireless networks. (e) Seamless integration of wireless communication technologies for enhanced connectivity. Reproduced from ref. 55 with permission from Elsevier, Copyright 2021.

and hospitalizations. Despite significant efforts to control the outbreak, such as vaccine development and herd immunity strategies, the post-acute phase remains a major concern. However, it is still unclear how the virus has acquired so many advantageous mutations within such a short period. In such diseases, rapid and early detection of the virus is highly effective. Bardajee *et al.*, emphasized the importance and value of integrating various data sources for effective public health and research surveillance using advanced primer-based biosensors.[56] This system utilized cadmium telluride (CdTe) semiconductor nanocrystal QDs, which were modified with a thiol-labelled primer that served as a fluorescent tag. A complementary target DNA, designed to match a specific region of the COVID-19 genome, formed a sandwich hybrid on the QD surface. The primers selected for detection were adapted based on the most conserved and widespread strains of SARS-CoV-2, ensuring both specificity and accuracy in the test. This biosensor was capable of analysing nasopharyngeal swab samples through simple procedures, providing rapid results. The detection limit for this biosensor was determined to be 1.71×10^{-9} mol l^{-1}. The results of tests conducted on real patient samples demonstrated that FRET-based QD biosensor was well-aligned with real-time polymerase chain reaction (RT-PCR) results. Unlike traditional PCR methods used in virus processing facilities, this optical technique enables rapid and easy microorganism detection, facilitating critical decision-making in under 15 minutes to ensure product quality and safety. A schematic representation of this system is provided in Figure 8.2.

Tao *et al.* developed a clustered regularly interspaced short palindromic repeats (CRISPR)–CRISPR-associated 12a (Cas12a)-based colorimetric biosensor platform for detecting human monkeypox virus (MPXV) and human papillomavirus (HPV) DNA using molybdenum disulphide QDs (MoS_2 QDs).[57] In this study, a novel CRISPR–Cas12a-driven colorimetric biosensor was created by utilizing the enzyme-mimicking activity of MoS_2 QDs for virus detection (Figure 8.3). The system operates as follows. (I) Target DNA recognition: when MPXV or HPV DNA is present, CRISPR-Cas12a *trans*-cleavage activity is triggered, leading to probe DNA degradation. (II) Suppression of MoS_2 QDs activity; the degraded DNA reduces the peroxidase-like activity of MoS_2 QDs, causing a visible colour change. (III) Smartphone integration; the system enables real-time and smartphone-assisted detection of MPXV and

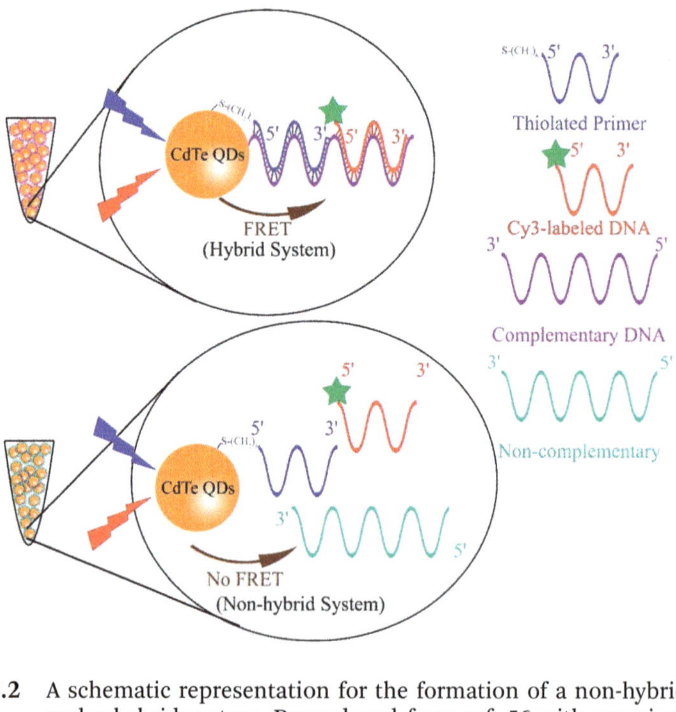

Figure 8.2 A schematic representation for the formation of a non-hybrid system and a hybrid system. Reproduced from ref. 56 with permission from Elsevier, Copyright 2025.

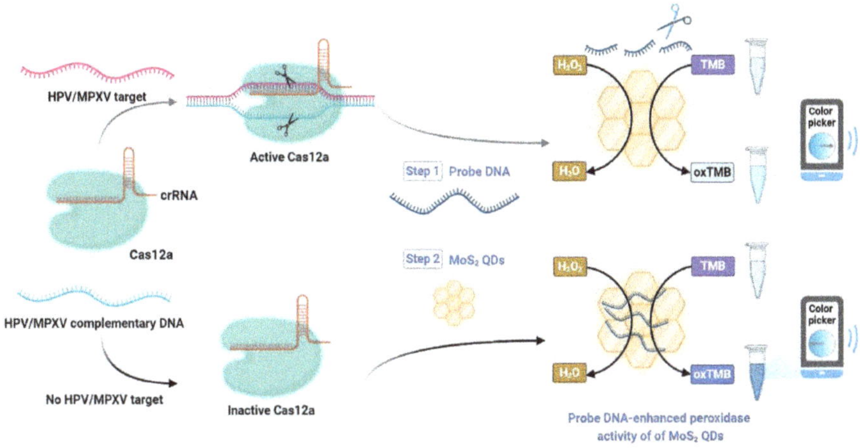

Figure 8.3 A conceptual depiction of CRISPR–Cas12-mediated modulation of peroxidase-like catalytic activity in MoS_2 QD–probe DNA nanohybrids for the colorimetric identification of HPV and MPXV. Reproduced from ref. 57 with permission from Elsevier, Copyright 2025.

HPV DNA, making it a portable and user-friendly tool. Transmission electron microscopy (TEM) confirmed the morphology of MoS_2 QDs. UV–visible spectroscopy was used for DNA detection. Fluorescence spectroscopy was used to assess Cas12a activity. The colorimetric response was proportional to DNA concentration, allowing quantitative detection. The biosensor demonstrated high specificity by successfully distinguishing target DNA from mismatched sequences. No cross-reactivity was observed with non-target viral DNA. Tests with human serum samples confirmed the sensor's high accuracy in biological samples. Recovery rates ranged from 98.5% to 100.5%, proving the system's high precision. The biosensor was successfully integrated with smartphones, making it a portable diagnostic system. This system holds great potential for diagnosing viral infections in low-resource settings, thanks to its high accuracy, portability and user-friendly design.

Influenza A virus (IAV) and Influenza B virus (IBV) share similar symptoms and can be easily confused with other respiratory infections such as the common cold, SARS-CoV-2, respiratory syncytial virus (RSV) and adenovirus. Zhang *et al.* investigated the ultrasensitive detection of the influenza virus using a QD microsphere (QDMS)-based lateral flow immunoassay (LFIA) system.[58] In this study, a new point-of-care (POC) biosensor (Figure 8.4) was developed for the rapid and simultaneous detection of IAV and IBV). Dual-layer QDMS were synthesized by immobilizing hydrophobic QDs onto silica microspheres *via* chemical bonding. QDMS exhibit high stability and strong fluorescence, ensuring highly sensitive detection. A handheld portable detector allows for visual detection. No cross-reactivity was observed with other pathogens, such as SARS-CoV-2, RSV, adenovirus or C-reactive protein (CRP). QDMS demonstrated excellent stability across various pH levels and

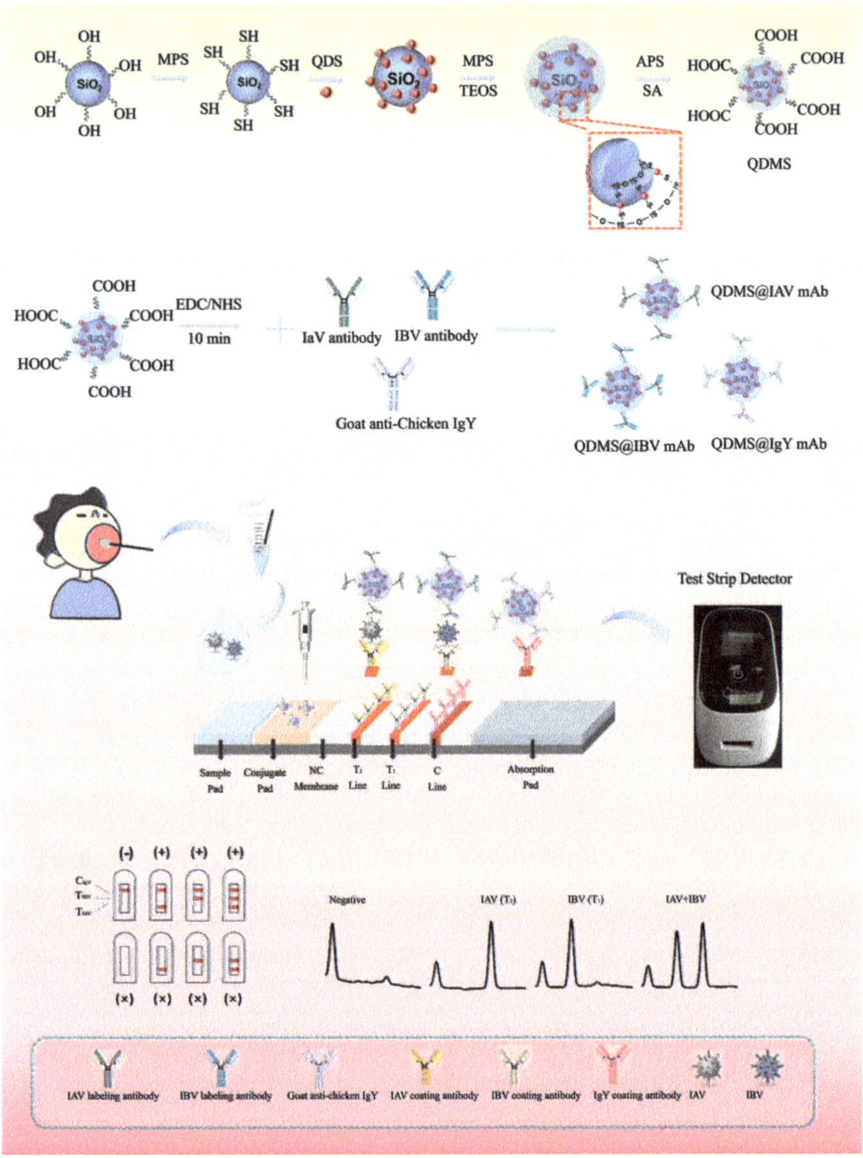

Figure. 8.4 Representation of QDMS-LFIA, QDMS fluorescent probes, and sandwich-structured SiO_2@(3-mercaptopropyl)trimethoxysilane(MPS)@QDs@SiO_2 spheres (QDMS). Abbreviations: tetraethyl orthosilicate (TEOS), 3-aminopropyltrimethoxysilane (APS), succinic anhydride (SA), 1-ethyl-3-(3-(dimethylamino)propyl)carbodiimide (EDC), *N*-hydroxysulfosuccinimide (NHS), nitrocellulose (NC). Reproduced from ref. 58, https://doi.org/10.1016/j.bios.2025.117187, under the terms of the CC BY 4.0 license, https://creativecommons.org/licenses/by/4.0/.

biological buffer solutions. The results of photostability and thermal stability tests confirmed that QDMS maintained strong stability. In real patient samples, the QDMS–LFIA system achieved an accuracy rate of 88%. A total of 28 IAV and 4 IBV samples from the Kaifeng Center for Disease Control and Prevention (Henan province, PR China) were tested. Compared with RT-PCR, this method exhibited a broad detection range and high sensitivity. This study validated the effectiveness of a high-sensitivity, low-cost LFIA biosensor platform using QDMS for influenza virus detection, providing a portable diagnostic system for home use and hospital applications and demonstrating high accuracy and sensitivity in clinical samples. The sensor represents a promising tool for epidemiological surveillance and early diagnosis.

Traditional fluorescence-linked immunosorbent assay (FLISA) platforms face limitations in clinical applications due to long response times and low sensitivity. Fan *et al.*, have developed a next-generation immunosensor with rapid response and dual signal amplification.[59] In this study, dual-functional nanobeads (MQBs) containing magnetic nanoparticles (MNPs) and QDs were developed. These MQBs enable the creation of a new immunosensor for the FLISA platform by providing a rapid magnetic response and strong fluorescence intensity. The immunosensor's ability to detect the SARS-CoV-2 antigen was tested. The results showed a 29-fold increase in fluorescence intensity compared with traditional FLISA-based QD nanobeads (SiO_2/QD/SiO_2, QBs). The sensitivity of the new system was enhanced 63-fold, reaching a detection limit of 7.50 pg ml^{-1} and the detection time was reduced by half compared with conventional FLISA methods, taking only 30 minutes. This dual signal amplification immunosensor demonstrated high specificity and accuracy, offering significant potential for clinical diagnosis and point-of-care testing. The magnetic-driven FLISA platform significantly improves sensitivity and reduces response time, marking a notable advancement in biomedical diagnostics. The development stages of this immunosensor are illustrated in Figure 8.5.

The white spot syndrome virus (WSSV) is a deadly pathogen that causes significant economic losses in shrimp. Due to global warming and increasing globalization, the emergence of new infectious diseases is expected. Takemura *et al.* developed a highly sensitive, disposable electrochemical sensor for detecting WSSV.[60] A highly conductive structure was obtained using nitrogen and sulphur-doped graphene QDs (N,S-GQDs) and a gold–polyaniline nanocomposite (AuNP–Pani). A specific anti-WSSV antibody was attached to the surface of this nanocomposite, enabling virus detection. Electrochemical impedance spectroscopy (EIS) was used to measure changes in the electrode in the presence of WSSV. A schematic representation of this electrode is shown in Figure 8.6. Immunoassay techniques were used to verify the electrode's effectiveness. Various microscopic analyses [transmission electron microscopy (TEM), scanning electron microscopy (SEM), atomic force microscopy (AFM)] were conducted to examine the structural properties of the nanomaterials. Gold nanoparticles incorporated into the polyaniline nanowire structure provided high conductivity. After antibody

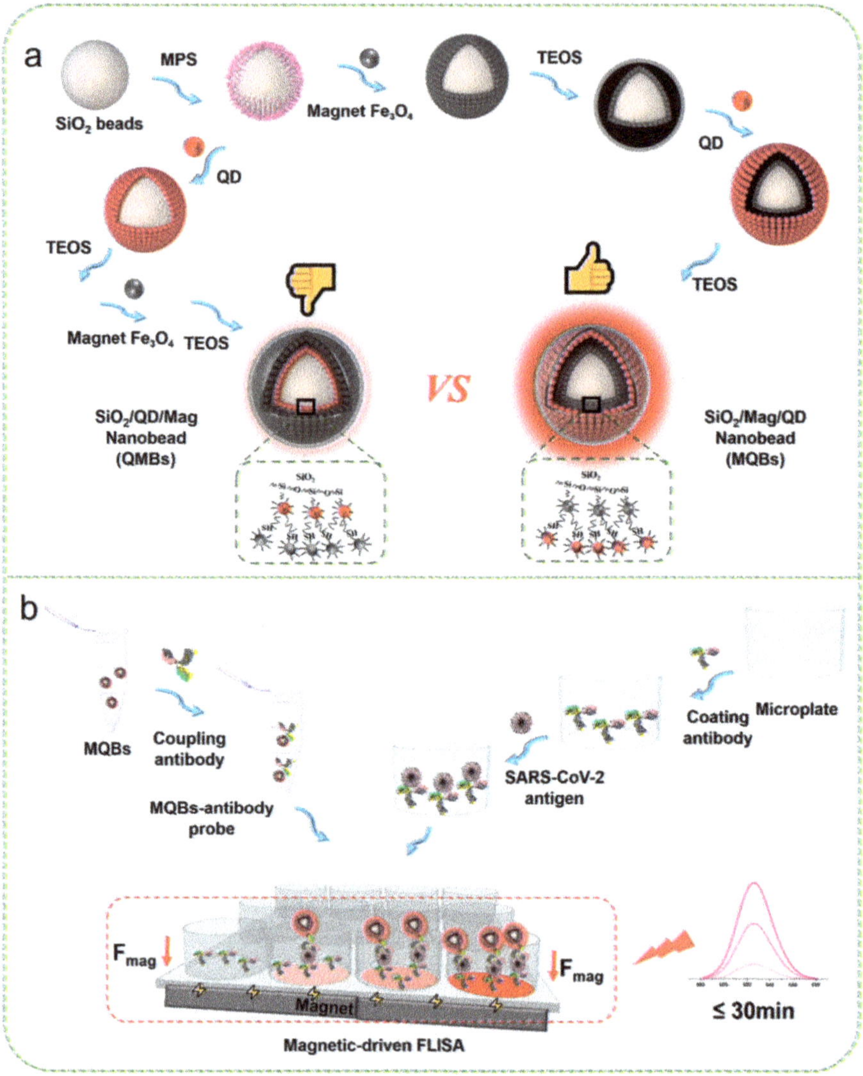

Figure 8.5 (a) Illustration of the process for creating photomagnetic dual-functional MQBs. (b) The development of dual-signal amplification sensors utilizing a magnetically driven FLISA platform with MQBs. Abbreviations: magnetic field (F_{mag}) Reproduced from ref. 59 with permission from Elsevier, Copyright 2025.

binding, the conductivity decreased and this decrease became more pronounced when the virus was detected. The developed sensor successfully detected WSSV in a concentration range of 1.45×10^2 to 1.45×10^5 DNA copies per ml. The LOD was determined to be 48.4 DNA copies per ml, which is highly sensitive compared with existing methods. The sensor demonstrated high selectivity against other viruses (such as Zika, hepatitis E and influenza

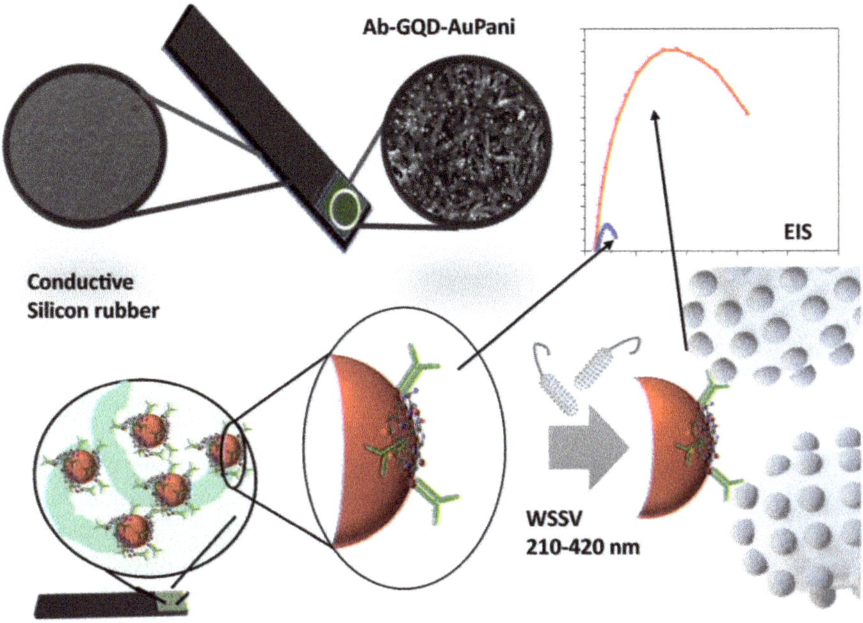

Figure 8.6 General view of the electrochemical electrode developed for WSSV detection. Reproduced from ref. 60, https://doi.org/10.1186/s12951-020-00712-4, under the terms of the CC BY 4.0 license, https://creativecommons.org/licenses/by/4.0/.

A) and various ions. Tests with real samples showed a 90% agreement with the RT-PCR method. A low-cost, rapid and highly sensitive WSSV detection method has been introduced. This system can also be adapted for the detection of other viral diseases, particularly for identifying foodborne and waterborne infections.

Yang *et al.* developed an immunochromatographic and homogeneous sensor based on quantum dot-functionalized polystyrene nanoprobes.[61] This sensor can detect respiratory viruses both qualitatively and quantitatively, offering results that are 51.2 to 1000 times more sensitive compared with commercial test kits. Two different detection methods were developed in the study. (I) LFIA; a portable, rapid and user-friendly test system that provides results within 10 minutes. (II) Homogeneous sensor; a system that ensures higher sensitivity and can deliver results within 30 minutes. Polystyrene (PS) nanosphere cores were coated with CdSe@ZnS–COOH QDs, forming dual-layer fluorescence structures that outperformed traditional fluorescent labels. Specific antibodies were attached to the surface of these nanocomposites to enable virus detection. The structural properties of the nanomaterials were confirmed using electrochemical and optical methods: EIS was used to measure changes in the electrode due to the presence of the virus. Fluorescence labelling was applied to enhance test sensitivity. Various microscopic analyses (SEM, TEM, AFM) were conducted to examine the structural characteristics of the nanomaterials. The LFIA system allowed successful visual detection of respiratory viruses, such as SARS-CoV-2, influenza A (IVA) and adenovirus (HAdV) within 10 minutes. The homogeneous sensor was capable of detecting viruses at much lower concentrations, demonstrating 291.7 to 480.5 times greater sensitivity compared with traditional methods like enzyme-linked immunosorbent assay (ELISA). The sensor exhibited high selectivity against other viruses (norovirus, rotavirus, dengue, *etc.*) and various ions. When tested with real clinical samples (throat swabs, saliva, *etc.*), the sensor achieved over 90% agreement with RT-PCR results. A rapid, low-cost and highly sensitive virus detection method was, thus, successfully developed. The dual-system sensor (Figure 8.7) is capable of both visual and quantitative detection, significantly outperforming traditional diagnostic methods. In the future it is planned to commercialize the sensor as a test kit and integrated with smartphones for broader usage.

Li *et al.* developed a new multiple aptamer recognition-based QD LFIA (MARQ–LFIA) diagnostic platform, which combines a multi-aptamer recognition strategy with QD-based fluorescence techniques.[62] This platform enhances diagnostic accuracy and sensitivity by allowing different aptamers to bind to the nucleocapsid (N) proteins of respiratory viruses. Components of the MARQ–LFIA are as follows. (I) Multiple high-affinity aptamers; used to recognize different binding regions of viruses. (II) QDs fluorescence technology; which enables highly sensitive signal detection. (III) Lateral flow test strips (LFIA); which provide a rapid and portable diagnostic solution. (IV) A portable fluorescence reader; which ensures accurate analysis of test results. In this study Scov1, Scov2 and Scov3 aptamers were used to detect the N protein of SARS-CoV-2, while FluB1, FluB2 and FluB3 aptamers were developed for the influenza B virus. These aptamers bind to different regions

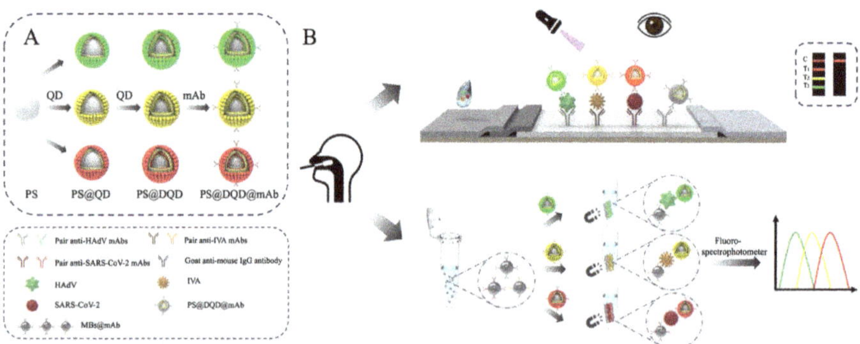

Figure 8.7 (a) The fabrication process of PS@ double QD (DQD) nanoparticles functionalized with antibodies and (b) a three-channel PS@DQD-based dual detection system designed for the simultaneous and fast identification of HAdV, IVA and SARS-CoV-2. Reproduced from ref. 61 with permission from Elsevier, Copyright 2025.

of the viruses, improving diagnostic precision. A patient sample (nasal swab or saliva) is applied to the test strip. The sample reacts with fluorescent quantum dot-functionalized nanoparticles [dendritic mesoporous silica nanoparticles/QDs/SiO_2 spheres (MQS)] modified with aptamers (Ap-MQS). If the target virus (*e.g.*, SARS-CoV-2 or influenza B) is present, the aptamers bind to its N protein. A fluorescent signal appears on the lateral flow strip. The portable fluorescence reader measures fluorescence intensity, identifying the presence and quantity of the virus. MARQ–LFIA did not show cross-reactivity with influenza A, other coronaviruses, adenoviruses or other respiratory viruses. The platform offers a low-cost, rapid and highly sensitive diagnostic solution. It can detect COVID-19, influenza B and other respiratory viruses, making it suitable for use in hospitals, airports, emergency rooms, rural areas and even home testing. By contributing to biosafety and public health, this technology can play a crucial role in pandemic control. Future plans include adapting the system for the detection of additional respiratory infections and integrating it with portable diagnostic devices for broader use.

Serological tests play a crucial role in detecting antibodies in the bloodstream, aiding in the diagnosis of infections, assessment of immune status and management of public health measures. The widespread prevalence of infectious diseases, such as dengue fever, influenza A and COVID-19, has further emphasized the importance of serological testing. Laser-induced graphene (LIG) has emerged as a promising material for electrochemical sensors due to its high conductivity, large surface area, low production cost and biocompatibility. Inlumphan *et al.* developed a new serological detection platform using LIG-based electrochemical immunosensors.[63] In this study, LIG electrodes were fabricated on polyimide (PI) surfaces using CO_2 laser induction. The laser power and scanning speed were optimized to achieve the best conductivity properties in the LIG electrodes. The electrode surfaces were functionalized with dengue virus antigen (DENV-Ag). All these steps are schematically summarized in Figure 8.8. The sensor performance was evaluated using differential pulse voltammetry (DPV) and EIS. Structural properties of the nanomaterials were analysed through X-ray diffraction (XRD), Raman spectroscopy, X-ray photoelectron spectroscopy (XPS), SEM and TEM. The results of XRD and Raman analyses confirmed the graphite-like structure of LIG. The results of electrochemical tests demonstrated that LIG possessed high conductivity and low internal resistance. Its porous structure provided a large surface area, enhancing sensitivity. The sensor exhibited high specificity and sensitivity for serological testing, successfully detecting specific interactions with the 4G2 antibody. Compared with ELISA, the LIG sensor achieved a lower LOD. It also performed well with various biological samples, including human blood serum. Tests conducted with human serum samples achieved over 90% accuracy. LIG-based electrochemical immunosensors were proven to be an effective platform for serological detection, particularly promising for the early diagnosis of viral diseases such as dengue fever and potentially adaptable for detecting other viral infections such as Zika, hepatitis E and influenza A. Further validation of LIG immunosensors in a broader range of clinical tests,

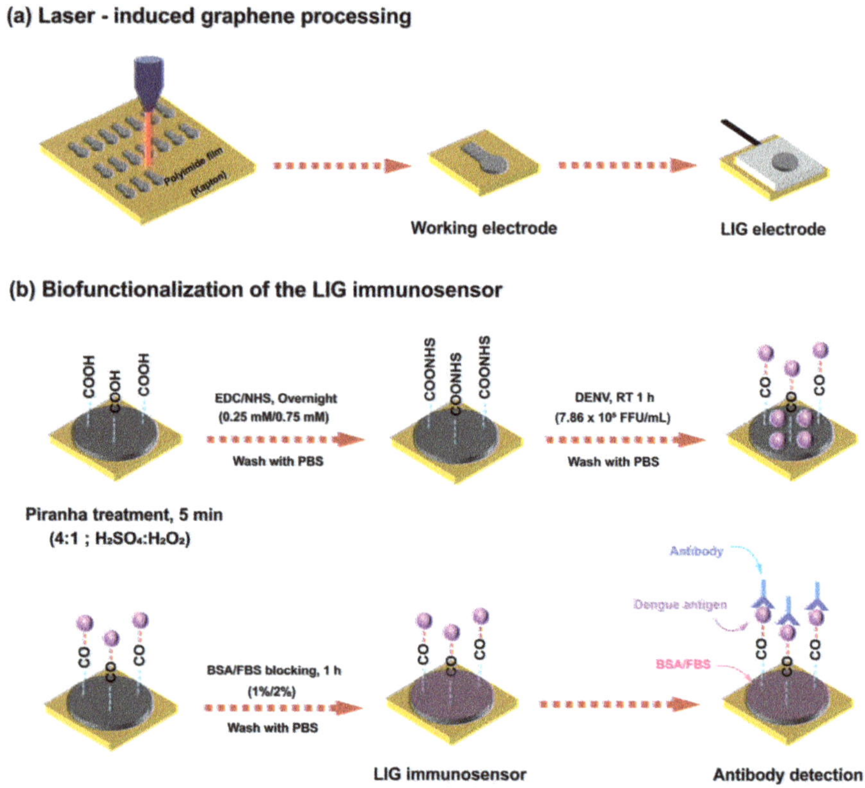

Figure 8.8 Fabrication of the laser-induced graphene (LIG) electrode and its biofunctionalization for immunosensor application. (a) Graphene formation on a polyimide (PI) sheet *via* laser irradiation, with optimized laser power and speed to achieve the desired electrode properties. (b) Functionalization of the LIG immunosensor, where the electrode underwent oxidation with Piranha solution for 5 minutes to enhance COOH-functional groups. The modified LIG surface was subsequently treated with EDC and NHS and functionalized with DENV antigen. To prevent non-specific binding, 1% bovine serum albumin (BSA) and 2% foetal bovine serum (FBS) were applied to prepare the immunosensor for antibody detection. Reproduced from ref. 63 with permission from Elsevier, Copyright 2025.

development of portable test kits for field and point-of-care applications is planned. This study highlights the potential of LIG-based sensors as a rapid, low-cost and highly sensitive diagnostic tool for viral infections, paving the way for widespread adoption in healthcare settings.

8.5 Conclusion

Respiratory infectious diseases spread rapidly and have far-reaching consequences, posing a significant threat to public health. The development of sensitive, accurate and rapid detection methods for respiratory viruses is crucial for effective disease prevention and control. However, existing diagnostic approaches often fail to meet the simultaneous demands for high accuracy and user convenience. Viral infections—such as hepatitis, HIV/AIDS, influenza, dengue, RSV and coronavirus—substantially impact global morbidity and mortality. Insights from the World Health Organization (WHO) and the Global Burden of Disease Study highlight the widespread consequences of these infections, reinforcing the urgent need for efficient and reliable diagnostic technologies. Early and precise detection is critical not only for initiating appropriate treatment but also for reducing transmission risks and implementing timely quarantine measures. The emergence of QD sensors represents a major breakthrough in virus detection, seamlessly integrating nanotechnology and biosensing to address critical diagnostic challenges. Unlike conventional methods, QD-based detection systems offer unparalleled sensitivity, precision and adaptability, enabling the identification of viral pathogens at ultra-low concentrations and during the early stages of infection. This capability is instrumental in enhancing clinical outcomes and curbing the spread of infectious diseases. Through their unique optical properties and compatibility with diverse detection platforms, QD sensors have demonstrated their applicability in detecting a wide range of viruses, including influenza, Zika and SARS-CoV-2. These advancements have paved the way for innovative diagnostic solutions, such as multiplexed detection, point-of-care testing and real-time monitoring. Furthermore, ongoing research continues to enhance their biocompatibility, scalability and cost-effectiveness, ensuring their accessibility in both developed and resource-limited settings. As the field advances, the potential applications of QD sensors extend beyond diagnostics to include therapeutic monitoring, vaccine evaluation and environmental surveillance of viruses. These developments highlight the transformative impact of QD technology in shaping the future of public health and global pandemic preparedness. Ultimately, QD sensors exemplify how cutting-edge nanotechnology can be leveraged to tackle one of humanity's most pressing challenges—combating viral diseases with precision, speed and efficiency.

References

1. N. Nova, T. S. Athni, M. L. Childs, L. Mandle and E. A. Mordecai, *Annu. Rev. Resour. Econ.*, 2022, **14**, 333–354.
2. S. Suleman, S. K. Shukla, N. Malhotra, S. D. Bukkitgar, N. P. Shetti, R. Pilloton, J. Narang, Y. N. Tan and T. M. Aminabhavi, *J. Chem. Eng.*, 2021, **414**, 128759.
3. Y. Zhang and T.-H. Wang, *Theranostics*, 2012, **2**, 631.
4. B. A. Kairdolf, A. M. Smith, T. H. Stokes, M. D. Wang, A. N. Young and S. Nie, *Annu. Rev. Anal. Chem.*, 2013, **6**, 143–162.
5. S. Das, H. Mazumdar, K. R. Khondakar, Y. K. Mishra and A. Kaushik, *ECS Sens. Plus*, 2024, **3**, 025001.

6. F. Sun, J. Zhang, Q. Yang and W. Wu, *Food Qual. Saf.*, 2021, **5**, fyab019.
7. A. P. Alivisatos, *Science*, 1996, **271**, 933–937.
8. U. Resch-Genger, M. Grabolle, S. Cavaliere-Jaricot, R. Nitschke and T. Nann, *Nat. Methods*, 2008, **5**, 763–775.
9. X. Wang, X. Ren, K. Kahen, M. A. Hahn, M. Rajeswaran, S. Maccagnano-Zacher, J. Silcox, G. E. Cragg, A. L. Efros and T. D. Krauss, *Nature*, 2009, **459**, 686–689.
10. L. Zhang, Y. Song, T. Fujita, Y. Zhang, M. Chen and T.-H. Wang, *Adv. Mater.*, 2014, **26**, 1289.
11. M. Dahan, S. Levi, C. Luccardini, P. Rostaing, B. Riveau and A. Triller, *Science*, 2003, **302**, 442–445.
12. A. Topcu, E. Bağda, T. Oymak and M. Durmuş, *Anal. Biochem.*, 2022, **654**, 114777.
13. A. L. Efros and L. E. Brus, *ACS Nano*, 2021, **15**, 6192–6210.
14. X. Dai, Z. Zhang, Y. Jin, Y. Niu, H. Cao, X. Liang, L. Chen, J. Wang and X. Peng, *Nature*, 2014, **515**, 96–99.
15. L. Wang, D. Xu, J. Gao, X. Chen, Y. Duo and H. Zhang, *Sci. China Mater.*, 2020, **63**, 1631–1650.
16. H. Park, D. J. Shin and J. Yu, *J. Chem. Educ.*, 2021, **98**, 703–709.
17. S. Chinnathambi and N. Shirahata, *Sci. Technol. Adv. Mater.*, 2019, **20**, 337–355.
18. P. Zhao, Q. Xu, J. Tao, Z. Jin, Y. Pan, C. Yu and Z. Yu, *Wiley Interdiscip. Rev.: Nanomed. Nanobiotechnol.*, 2018, **10**, e1483.
19. L.-W. Wang, C.-W. Peng, C. Chen and Y. Li, *Breast Cancer Res. Treat.*, 2015, **151**, 7–17.
20. K. Hagiwara, S. Horikoshi and N. Serpone, *Chem. – Eur. J.*, 2021, **27**, 9466–9481.
21. P. Namdari, B. Negahdari and A. Eatemadi, *Biomed. Pharmacother.*, 2017, **87**, 209–222.
22. S. Tao, T. Feng, C. Zheng, S. Zhu and B. Yang, *J. Phys. Chem. Lett.*, 2019, **10**, 5182–5188.
23. X. Wang, G. Sun, N. Li and P. Chen, *Chem. Soc. Rev.*, 2016, **45**, 2239–2262.
24. F. Bian, L. Sun, L. Cai, Y. Wang and Y. Zhao, *Wiley Interdiscip. Rev.: Nanomed. Nanobiotechnol.*, 2019, **11**, e1567.
25. S. Chandra, M. A. Mustafa, K. Ghadir, P. Bansal, M. Deorari, D. Y. Alhameedi, M. H. s. Alubiady, A. M. Al-Ani, S. O. Rab and S. S. Jumaa, *Naunyn-Schmiedebergs Arch.*, 2024, 1–42.
26. P. Bhattacharya, S. Ghosh and A. Stiff-Roberts, *Annu. Rev. Mater. Res.*, 2004, **34**, 1–40.
27. A. M. Wagner, J. M. Knipe, G. Orive and N. A. Peppas, *Acta Biomater.*, 2019, **94**, 44–63.
28. Y. R. Kumar, K. Deshmukh, K. K. Sadasivuni and S. K. Pasha, *RSC Adv.*, 2020, **10**, 23861–23898.
29. I. L. Medintz, H. T. Uyeda, E. R. Goldman and H. Mattoussi, *Nat. Mater.*, 2005, **4**, 435–446.
30. N. Ledentsov, *Semicond. Sci. Technol.*, 2010, **26**, 014001.

31. G. L. Whitworth, C. Roda, M. Dalmases, N. Taghipour, M. Dosil, K. Nikolaidou, H. Dehghanpour and G. Konstantatos, *Adv. Mater.*, 2025, **37**, 2410207.
32. Q. Xu, Y. Niu, J. Li, Z. Yang, J. Gao, L. Ding, H. Ni, P. Zhu, Y. Liu and Y. Tang, *Carbon Neutrality*, 2022, **1**, 13.
33. M. Fontanelli dos Santos, A. L. de Camargo Doimo, N. Aparecida Lima, J. de Jesus dos Santos, M. V. Palmeira-Mello and C. G. Marques Netto, *J. Chem. Educ.*, 2024, **101**, 5476–5483.
34. V. I. Klimov, *Nanocrystal Quantum Dots*, CRC press, 2017.
35. J. Joo, J. M. Pietryga, J. A. McGuire, S.-H. Jeon, D. J. Williams, H.-L. Wang and V. I. Klimov, *J. Am. Chem. Soc.*, 2009, **131**, 10620–10628.
36. S. Barth, F. Hernandez-Ramirez, J. D. Holmes and A. Romano-Rodriguez, *Prog. Mater. Sci.*, 2010, **55**, 563–627.
37. W. Zhou and J. J. Coleman, *Curr. Opin. Solid State Mater. Sci.*, 2016, **20**, 352–360.
38. A. S. Karakoti, R. Shukla, R. Shanker and S. Singh, *Adv. Colloid Interface Sci.*, 2015, **215**, 28–45.
39. F. Yan, Y. Jiang, X. Sun, Z. Bai, Y. Zhang and X. Zhou, *Mikrochim. Acta*, 2018, **185**, 1–34.
40. Z. Yue, F. Lisdat, W. J. Parak, S. G. Hickey, L. Tu, N. Sabir, D. Dorfs and N. C. Bigall, *ACS Appl. Mater. Interfaces*, 2013, **5**, 2800–2814.
41. S. Das, P. Somu, A. K. Yadav, P. K. Hopke and S. Paul, *Environ. Sci. Nano*, 2024, **11**, 739–765.
42. B. Li, Y. Wang, L. Huang, H. Qu, Z. Han, Y. Wang, M. J. Kipper, L. A. Belfiore and J. Tang, *Synth. Met.*, 2021, **276**, 116758.
43. M. C. Dos Santos, W. R. Algar, I. L. Medintz and N. Hildebrandt, *TrAC, Trends Anal. Chem.*, 2020, **125**, 115819.
44. E. Oh, M.-Y. Hong, D. Lee, S.-H. Nam, H. C. Yoon and H.-S. Kim, *J. Am. Chem. Soc.*, 2005, **127**, 3270–3271.
45. A. Qureshi, T. Shaikh and J. H. Niazi, *Analyst*, 2023, **148**, 1633–1652.
46. W.-W. Zhao, J. Wang, Y.-C. Zhu, J.-J. Xu and H.-Y. Chen, *Anal. Chem.*, 2015, **87**, 9520–9531.
47. M. Li, S. K. Cushing and N. Wu, *Analyst*, 2015, **140**, 386–406.
48. M. Bauch, K. Toma, M. Toma, Q. Zhang and J. Dostalek, *Plasmonics*, 2014, **9**, 781–799.
49. N. Thanjavur, L. Bugude and Y.-J. Kim, *Biosens*, 2025, **15**, 38.
50. K. Chen, J. Ren, C. Chen, W. Xu and S. Zhang, *Nano Today*, 2020, **35**, 100939.
51. N. Nazeer, V. Gurjar, P. Ratre, R. Dewangan, K. Zaidi, R. Tiwari, N. Soni, A. Bhargava and P. K. Mishra, *Microchim. Acta*, 2024, **191**, 1–11.
52. H.-R. Kim, J.-H. Bong, J.-H. Park, Z. Song, M.-J. Kang, D. H. Son and J.-C. Pyun, *ACS Appl. Mater. Interfaces*, 2021, **13**, 29392–29405.
53. Y. Cheng, K. Wang, H. Xu, T. Li, Q. Jin and D. Cui, *Anal. Bioanal. Chem.*, 2021, **413**, 6037–6057.
54. H. Wang, T. Wang, X. Yuan, Y. Wang, X. Yue, L. Wang, J. Zhang and J. Wang, *Sensors*, 2023, **23**, 8156.

55. H. Lee, Z. Jiang, T. Yokota, K. Fukuda, S. Park and T. Someya, *Mater. Sci. Eng.: R: Rep.*, 2021, **146**, 100631.
56. G. R. Bardajee, M. Sharifi and M. Khanahmadi, *Colloids Surf. A: Physicochem. Eng. Asp.*, 2025, **709**, 136152.
57. Y. Tao, H. Wang, E. Ju, Y.-H. Lao, Y. Zhang and M. Li, *Talanta*, 2025, **283**, 127153.
58. Y. Zhang, Y. Lv, X. Chen, Y. Ye, S. Yu, N. Li, H. Yao, Z. Wang, Q. Feng and R. Wu, *Biosens. Bioelectron.*, 2025, **273**, 117187.
59. J. Fan, Y. Shi, Y. Lv, N. Li, R. Wu, H. Shen and L. S. Li, *Sens. Actuators, B*, 2025, **423**, 136804.
60. K. Takemura, J. Satoh, J. Boonyakida, S. Park, A. D. Chowdhury and E. Y. Park, *J. Nanobiotechnol.*, 2020, **18**, 1–12.
61. S. Yang, W. Hu, S. Wang, X. Li, L. Lei, X. Wei and H. Lin, *Biosens. Bioelectron.*, 2025, **267**, 116716.
62. H. Li, X. Fu, Q. You, D. Shi, L. Su, M. Song, R. Peng, T. Fu, P. Wang and W. Tan, *J. Mater. Chem. B*, 2025, **13**, 1681–1689.
63. S. Inlumphan, W. Wongwiriyapan, N. Khemasiri, P. Rattanawarinchai, P. Leepheng, P. Luengrojanakul, T. Wuttikhun, M. Obata, M. Fujishige and K. Takeuchi, *Sens. Actuators Rep.*, 2025, **9**, 100276.

CHAPTER 9

Wearable and Portable Sensors

SEMRA AKGÖNÜLLÜ

Department of Chemistry, Hacettepe University, Ankara, Türkey
E-mail: semraakgonullu@hacettepe.edu.tr

9.1 Introduction

Biosensors are analytical devices that integrate a biological recognition element (receptor) with a physical component (transducer) to convert a biochemical interaction into a quantifiable signal.[1] The concept of the modern biosensor was first realized in 1962 by Clark and Lyons, who developed an amperometric sensor known as the "enzyme electrode" for glucose monitoring.[2,3] This innovation marked the beginning of practical biosensor applications and laid the foundation for the widespread use of electrochemical glucose sensors by individuals managing diabetes. These devices revolutionized diabetes care by empowering patients to monitor their glucose levels at home, without the need for laboratory visits. This level of convenience and autonomy has greatly enhanced disease control by enabling more frequent, cost-effective testing.[4,5] Glucose sensors serve as a model for point-of-care (POC) diagnostics—offering simple, reliable procedures that can be performed by users with no medical training. Typically, a compact device draws a small blood sample, and test strips inserted into a meter provide an immediate glucose reading. However, more complex biosensors, such as immunosensors used for detecting a broader range of biomolecules, often

Detection Science Series No. 32
Advances in Virus Detection with Sensors
Edited by Adil Denizli
© The Royal Society of Chemistry 2026
Published by the Royal Society of Chemistry, www.rsc.org

require additional reagents and procedural steps.[6] To be practical for POC applications, these systems must be further simplified, ideally automated, and packaged in a user-friendly format.[7]

Access to good quality healthcare is strongly influenced by both social and economic factors. Consequently, there is a growing demand for fast and accessible diagnostic tools. The incorporation of advanced technologies into healthcare offers significant opportunities to enhance patient engagement and improve the management of various health conditions. In this context, wearable biosensors have gained attention as effective alternatives to conventional diagnostic methods, providing mobile, continuous, and real-time health monitoring. These wearable devices can non-invasively analyze body fluids, such as sweat, saliva, and blood, to detect biochemical signals, allowing for early identification and ongoing assessment of medical issues. Their ability to be used to carry out on-site analysis at the molecular level, combined with high connectivity and minimal reliance on external lab processing, makes them a promising innovation. This chapter highlights fundamental principles and recent advancements in wearable and portable biosensor technologies, with a particular emphasis on tracking biomarkers associated with infectious diseases. It begins with an overview of key viruses, followed by a summary of the structural components of wearable and portable sensor systems. The specific functions and applications of each system are then discussed in detail. Recent advancements in wearable biosensors for use in various settings—including outdoor environments, clinical facilities, and home care—are also highlighted. The chapter concludes with a summary and a forward-looking discussion on future developments, potential challenges, and opportunities for growth in the field of wearable health monitoring technologies.

9.2 Wearable and Portable Sensors

Medical healthcare is crucial for all individuals and is related to social and economic growth.[8] Wearable biosensors represent a significant advancement in healthcare technology, offering a transformative alternative to traditional diagnostic and monitoring methods. Their potential for diverse clinical and physiological applications has garnered widespread interest. Recently, considerable progress has been made in developing wearable chemical sensors capable of detecting electrolytes, metabolites, toxic gases, and heavy metals in biofluids—such as sweat, saliva, and tears—directly on the skin. Some systems have also been tailored for monitoring wound healing. These sensors enable real-time, continuous tracking of biochemical markers, which is crucial for gaining deeper insights into an individual's health, stress levels, and physical performance at the molecular level. Despite their promise, several challenges remain before these devices can achieve widespread commercial use, including improving analyte selectivity, enhancing the stability of biorecognition elements, refining sample handling, minimizing invasiveness, and ensuring mechanical flexibility for user comfort.[9]

Wearable sensors have received much attention since the advent of smartphones and other mobile devices due to their ability to provide useful information about the performance and health of individuals.[10] In recent years, there has been growing interest in health monitoring technologies, leading to notable advancements in wearable biosensors. These devices offer non-invasive or minimally invasive ways to track an individual's physiological status.[11–17] When worn on the skin, wearable biosensors are particularly well-suited for the early detection of diseases by analyzing biochemical markers present in body fluids such as sweat, saliva, tears, and interstitial fluid.[18–21] Their flexibility and ability to conform to the skin make them ideal for continuous, real-time monitoring of health and physical performance.[22] In addition to widely used commercial devices like the Apple Watch, Fitbit, and WHOOP Strap—which monitor vital signs and physical activity—a new wave of wearable chemical and biochemical sensors is emerging. These next-generation systems provide detailed molecular-level data, supporting diverse applications in personalized healthcare, sports performance optimization, military readiness, nutrition tracking, and public safety.[9] By enabling dynamic monitoring of biochemical fluctuations in peripheral biofluids, wearable biosensors hold great potential for transforming how health and performance are assessed and managed.[9,23]

Initial developments in wearable health technology primarily targeted the measurement of physical parameters, such as skin temperature,[22] heart rate,[24] blood pressure,[25] respiration rate,[26] and bodily movement.[27] However, these metrics alone are insufficient for a comprehensive assessment of an individual's health. In response, recent advancements have focused on wearable chemical sensors and biosensors capable of detecting biochemical compounds directly from the skin.[28] These systems enable continuous, remote health and performance monitoring by integrating seamlessly with the human body.[29] A growing area of interest involves the detection of clinically relevant biomarkers through wearable platforms, broadening the scope of measurable physiological indicators.[30,31] Among various biofluids, sweat stands out due to its accessibility and the rich health-related information it carries, making it especially well-suited for non-invasive wearable biosensing applications.[32]

9.2.1 Important Biofluids and Biomarkers

Sweat carries a diverse array of important biomarkers—such as ions, metabolites, and proteins—that can provide valuable insights into a person's health and offer indications of disease development. Wearable sweat biosensors are capable of directly capturing and analyzing sweat on the skin, allowing for real-time, continuous, and non-invasive tracking of biochemical changes at the molecular level.[33] Accessible sweat comes from eccrine glands, consisting of the secretory coil where it is first produced and a dermal duct that can house the sweat in the skin.[34] During this process, ions, metabolites, acids, hormones, and analytes, including small proteins and peptides, are partitioned in sweat.[35] The most abundant ion species are Na^+ and Cl^-.[36] Sweat contains a variety of biologically significant substances—including electrolytes, like sodium and potassium, metabolites, such as glucose and lactate, as well as amino acids, proteins, and hormones—that reflect internal physiological processes. Because these biomarkers correlate with metabolic function, hydration status, hormonal activity, and toxin levels, analyzing sweat offers a non-invasive and practical approach to monitor metabolic diseases, evaluate overall health, and detect substance exposure or intoxication in real time.[35,37] Sweat contains a diverse range of analyte types, each with distinct diagnostic value depending on their physiological properties and how closely they correlate with blood levels, as summarized in Table 9.1.

9.2.2 Physiological Parameters

Basic physiological markers such as heart rate (HR), blood pressure and body temperature provide valuable information about an individual's overall health. Abnormal elevations in these parameters may indicate underlying health problems. For example, increased heart rate and blood pressure are often linked to cardiovascular disorders, while an increase in body

Table 9.1 An overview of the current understanding of the main analytical classes found in sweat and their respective roles or significance in sweat-based monitoring.[32]

Analyte type	Role in sweat samples
Electrolytes	Sodium (Na^+) and chloride (Cl^-), are considered active secretions with no consistent correlation to serum levels, whereas potassium (K^+) tends to reflect blood concentrations more accurately due to passive diffusion.
Metabolites	Lactate primarily indicate local sweat gland metabolism and do not typically mirror systemic levels, although certain metabolites may show correlation with serum concentrations under specific conditions.
Amino acids	These are consistently present in sweat, and their appearance is influenced by molecular weight and the rate of sweat production, which affects their filtration into the sweat.
Small molecules	Small molecules exhibit differing behaviors on the basis of their chemical properties: lipophilic molecules, such as cortisol, can demonstrate near 1:1 correspondence between sweat and serum (particularly the unbound fraction), while hydrophilic compounds are filtered to varying degrees depending on molecular size and sweat flow dynamics.
Proteins	Proteins, although significantly diluted in sweat compared with plasma, appear to maintain relative concentration ratios consistent with those in blood. Due to their generally low concentrations and limited need for rapid measurement, protein detection in sweat may benefit from pre-concentration methods or enhanced analytical techniques to overcome dilution-related challenges.

temperature often indicates the presence of infection or inflammation caused by pathogens. Next-generation wearable biosensors offer a cost-effective solution for continuous or frequent monitoring of physiological parameters. These devices support personalized health tracking by enabling routine measurement of vital signs and can potentially detect early changes in health status. Designed for ease of use, wearables can passively collect and transmit various health metrics in real time to users or healthcare providers, requiring little to no training. Beyond tracking indicators like heart rate and skin temperature, wearable technologies can also monitor daily physical activities—such as walking, cycling, or running—often integrated with the global positioning system (GPS) to provide precise data on both activity type and location.[38]

Body temperature is a key physiological marker that offers critical insight into a person's health, often increasing in response to infections or inflammatory conditions. As highlighted in the section on infectious diseases, fever is a common symptom, characterized by a controlled elevation of core body temperature above the normal range (36.5–37.5 °C) to enhance the body's immune response. While fever is often seen as a symptom of illness, it may also play a protective role during infections. It is important to note that not all increases in body temperature indicate fever—such changes may result from increased heat production or reduced heat loss. Depending on the duration, elevated body temperature can be categorized as acute (less than 7 days), subacute (up to 2 weeks), or chronic (more than 2 weeks).[39]

Effective management of infectious diseases relies on timely, accurate information at both the individual and population levels. For globally significant infections—ranging from emerging pandemics like corona virus disease (COVID)-19 to widespread endemic illnesses such as dengue—early identification of patients at risk of severe progression is often difficult, as most cases initially present with mild symptoms. Wearable technology offers a promising solution through continuous physiological monitoring in clinical and community settings. Integrating wearables into hospital care for infectious disease management, or deploying them for syndromic surveillance during outbreaks, has the potential to enhance healthcare delivery while offering a cost-effective, scalable approach. The low-cost healthcare wearable devices and rapid diagnostics could fulfill multiple roles for infectious disease management, as presented in Figure 9.1.[40]

9.3 Monitoring Vital Signs for Early Detection of Infectious Disease Symptoms

Portable and wearable devices equipped with multiple sensors are rapidly advancing technologies in the fields of health monitoring and disease detection. These systems hold great promise for enabling remote healthcare by providing continuous, real-time tracking of physiological data. Sensor-based health monitoring platforms serve as powerful tools for early diagnosis and

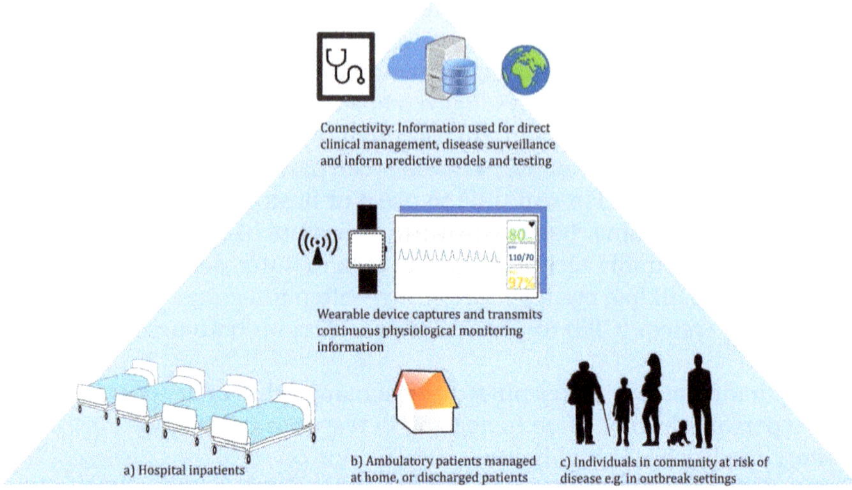

Figure 9.1 Healthcare wearables offer real-time physiological monitoring for various use cases: (a) detecting early signs of deterioration in hospitalized patients, (b) tracking recovery in patients at home, and (c) monitoring at-risk individuals during outbreaks for syndromic surveillance. Data is securely uploaded to cloud systems, combined with diagnostic and health service data, and used for both individual care and public health decision-making.[40] Reproduced from ref. 40, https://doi.org/10.1016/j.ijid.2020.05.086, under the terms of the CC BY 4.0 license, https://creativecommons.org/licenses/by/4.0/.

disease prediction, supporting timely treatment recommendations and proactive measures to prevent illness progression.[41]

The COVID-19 pandemic has underscored the critical importance of early detection in managing viral respiratory infections. Identifying infections before symptoms appear allows individuals to seek prompt medical attention and take preventive measures to reduce further transmission. Wearable sensors, widely available on the commercial market, have been employed to monitor users in their everyday environments, offering continuous insights into their physiological status and behavior. More recently, these devices have been adapted for detecting respiratory infections, such as COVID-19, influenza, and similar illnesses. Temple and co-workers[42] developed a method combining wearable electrocardiogram (ECG) sensors and machine learning to detect influenza infections before symptoms appear. They provide a noninvasive alternative that can supplement traditional diagnostic methods, including blood-based gene expression assays. In a controlled human challenge study, 20 healthy participants were monitored with wearable devices from 7 days before until 10 days after being exposed to the influenza A (H3N2) virus. A comprehensive methodology for data preprocessing, along with the extraction and standardization of heart rate (HR) and heart rate variability (HRV) features, was designed. These standardized physiological metrics were then used as input for semi-supervised machine learning models aimed at identifying viral respiratory infections, even in individuals who were presymptomatic or asymptomatic. The study demonstrated the potential of wearable sensors for early, non-invasive detection of respiratory infections, even in asymptomatic cases.

The COVID-19 pandemic prompted the rapid deployment of a wearable monitoring system within UK hospital isolation wards to support remote patient care and reduce healthcare worker exposure. Santos and co-workers[43] developed and implemented a virtual high-dependency unit (vHDU) wearable system using commercially available components to enable remote, real-time tracking of vital signs in ambulatory COVID-19 patients housed in individual rooms within hospital's isolation ward. The system combined a wearable chest patch and a pulse oximeter to continuously monitor heart rate, respiratory rate, and oxygen saturation in ambulatory COVID-19 patients. The system was tailored to fulfill six key requirements essential for the reliable, continuous monitoring of patients' cardio-respiratory function. Data were transmitted securely in real time to a clinician dashboard outside the isolation zone, allowing vital signs to be reviewed alongside nurse-recorded observations. Automated reconnection protocols minimized the need for staff to enter patient rooms. During the first wave of the pandemic (March–June 2020), wearable monitoring was used for nearly half of the patients in the isolation ward, with a median monitoring duration of 31.5 hours, covering over 88% of their registration time. Its usage notably increased alongside ward occupancy during the peak of the first COVID-19 wave in the UK, highlighting its practical clinical value. The system remained in use throughout subsequent waves, demonstrating sustained clinical utility. The data-flow diagram of the vHDU system is presented in Figure 9.2.

Sun and co-workers[44] investigated the potential of the recently introduced open-source mobile health platform, remote assessment of disease and relapse (RADAR)-base, as a flexible tool for rapidly evaluating the impact and response to non-pharmaceutical interventions (NPIs) designed to curb the spread of COVID-19. Their study utilized data collected *via* smartphones and wearable devices, processed through the RADAR-base platform, from 1062 participants across Italy, Spain, Denmark, the United Kingdom, and the Netherlands. They extracted nine daily behavioral and physiological indicators, including time spent at home, farthest distance traveled from home, number of nearby Bluetooth-enabled devices (used as an indicator of social proximity), step count, average heart rate, total sleep duration, bedtime, phone screen time, and time spent on social media applications. To examine variations in these metrics across different phases—baseline, pre-lockdown, and lockdown—they applied Kruskal–Wallis tests followed by Dunn's *post hoc* comparisons. Additionally, they analyzed behavioral trends based on demographic variables, such as age, gender, body mass index (BMI), and educational level.

Viral respiratory tract infections (VRTIs) are highly prevalent and, as underscored by the COVID-19 pandemic, present significant social and economic challenges. To mitigate the risk of future outbreaks, early detection

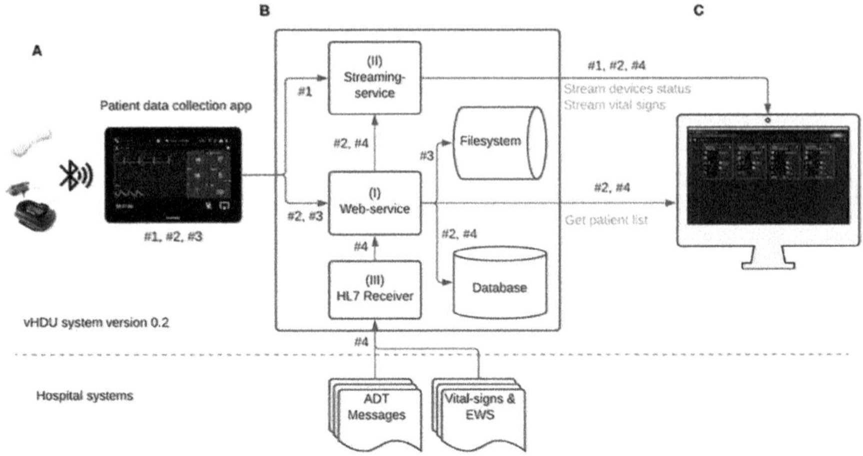

Figure 9.2 Virtual high-dependency unit (vHDU) System v0.2 – Data Flow Overview: The system comprises three main components. (a) A bedside tablet application that collects and forwards wearable sensor data. (b) A secure web-based back-end that stores and routes this information Abbreviations: health level seven standard (HL7), admission-discharge-transfers (ADT, early warning score (EWS). (c) A clinician dashboard accessible *via* any modern web browser, enabling remote viewing of real-time patient data.[43] Reproduced from ref. 43, https://doi.org/10.3389/fdgth.2021.630273, under the terms of the CC BY 4.0 license, https://creativecommons.org/licenses/by/4.0/.

and intervention strategies are critical. Wearable biosensors offer a promising approach for identifying asymptomatic infections, potentially reducing disease transmission and alleviating pressure on healthcare systems. Hadid and co-workers[45] have demonstrated that machine learning (ML) models can detect early immune activation solely on the basis of physiological data collected from wearable devices. Building on this, their study was intended to identify sensitive physiological and immunological signatures of VRTIs by applying ML techniques to continuously collected data from wearable vital sign sensors. In a controlled, prospective longitudinal study, 60 healthy adults aged 18–59 were exposed to a low-grade viral challenge *via* administration of the live attenuated influenza vaccine (LAIV). Participants were continuously monitored for 12 days—7 days before (baseline) and 5 days after LAIV administration—using wearable devices (a smart shirt, wristwatch, and ring) that track vital signs and physical activity. The Wearable Sensors for Early Detection and Tracking of Viral Respiratory Tract Infections (WE SENSE) study provides a framework for assessing the capability of wearable technology to detect asymptomatic VRTIs. It integrates multimodal data, including physiological measurements, PCR testing, inflammatory biomarker analysis, and app-based symptom tracking. ML algorithms will be developed to identify subtle physiological changes associated with immune activation, enabling early and non-invasive detection of viral infections. Three wearable biosensor platforms will be used for continuous monitoring of physiological and

Table 9.2 Physiological and activity parameters monitored by wearable vital signs systems in the WE SENSE study.[a]

	 Oura ring	 BioBeat watch	 Astroskin/ Hexoskin shirt
Heart rate	✔	✔	✔
Heart rate variability	✔	✔	✔
ECG	—	—	✔
Stroke volume, cardiac output, pulse pressure, systemic vascular resistance, mean arterial pressure	—	✔	—
Systolic blood pressure	—	✔	✔*
Peripheral capillary oxygen saturation	—	✔	✔*
Respiratory rate	✔	✔	✔
Tidal volume, minute ventilation	—	—	✔
Skin temperature	✔	✔	✔*
Activity levels (*e.g.*, steps, calories, cadence)	✔	—	✔
Sleep (NREM, REM, timing, quality)	✔	—	✔

[a]Parameters marked with $*$ are specific to the Astroskin device. Abbreviations: electrocardiogram (ECG); non-rapid eye movement sleep (NREM); rapid eye movement sleep (REM).[45] Reproduced from ref. 45 with permission from Elsevier, Copyright 2023.

activity parameters in the WE SENSE study: the Oura Ring (Oura Oy, Finland), the Biobeat Watch (Biobeat Technologies Ltd, Israel), and the Astroskin/Hexoskin Shirt (Hexoskin, Canada) (Table 9.2). The selected devices were chosen to provide a comprehensive and complementary assessment of physiological markers by covering different physiological states—such as sleep, rest, and physical activity—while also supporting high user adherence to ensure extended wear time. The Astroskin/Hexoskin shirts were included in the study for their distinct ability to deliver detailed, high-resolution recordings of ECG and respiratory activity.

A new generation of portable biosensors enables frequent tracking of key physiological metrics. Li and co-workers[38] assessed their ability to monitor physiological changes during daily activities and their potential in health management and disease detection. By collecting over 250 000 data points from 43 individuals, they observed personalized circadian patterns and confirmed known physiological trends. Notably, distinct changes occurred in specific environments—such as reduced oxygen levels and increased radiation exposure during air travel—linked to symptoms like fatigue. By integrating biosensor data with regular medical tests, they made two key findings: (1) wearable devices helped detect early signs of inflammation and Lyme disease using a personalized baseline model, and (2) they could differentiate between insulin-sensitive and insulin-resistant individuals. These findings highlight the value of wearable biosensors in tracking health, detecting disease, and

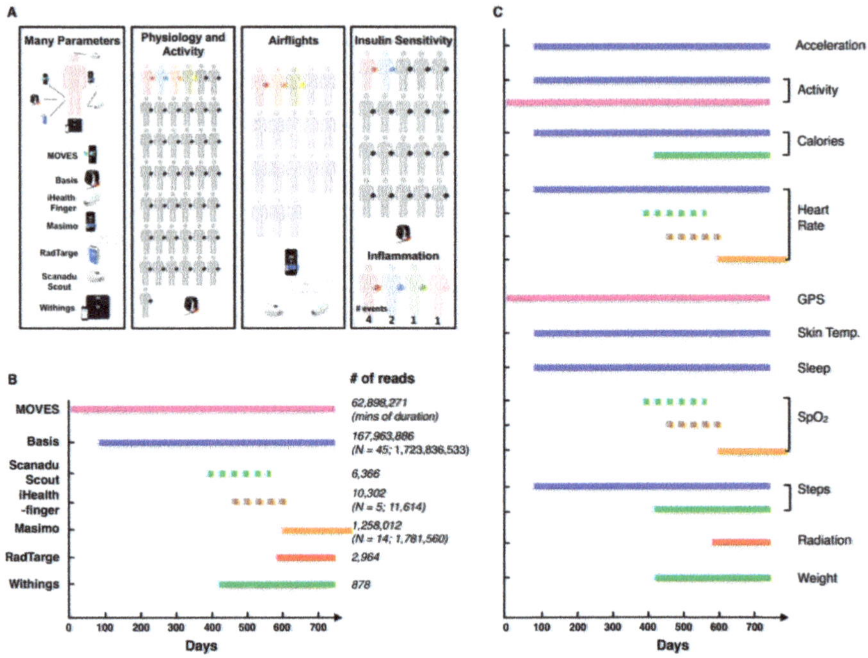

Figure 9.3 Project overview and device summary: (A) The wearable devices utilized in this study are illustrated alongside human figures colored to reflect participant involvement across different sub-studies. For example, red indicates participation in all five sub-studies, grey in two (physiology/activity and insulin sensitivity), blue in three (physiology/activity, insulin sensitivity, and inflammation), orange and yellow in two (physiology/activity and airflights), green and pink in one (Inflammation), and purple in one (airflights); (B) The timeline for device usage is shown, with the number of data points collected from participant #1 and others displayed on the right; (C) The parameters tracked by each device are detailed. Devices are color-coded: MOVES (magenta), Basis (dark blue), Scanadu Scout (light green), iHealth-finger (brown), Masimo (orange), RadTarge (red), and Withings (dark green). Continuous lines denote devices used for continuous monitoring, while dashed lines represent those employed intermittently for point-in-time measurements.[38] Reproduced from ref. 38, https://doi.org/10.1371/journal.pbio.2001402, under the terms of the CC0 license, https://creativecommons.org/publicdomain/zero/1.0/.

expanding access to care, particularly in underserved or remote populations. Figure 9.3 presents the list of wearable devices used in the study along with the physiological and activity-related metrics they capture. Together, these devices monitor: (a) three key physiological signals—heart rate (HR), blood oxygen saturation (SpO_2), and skin temperature; (b) six activity-related metrics—sleep, step count, walking, biking, running, caloric expenditure, and movement-related acceleration forces; (c) body weight; and (d) total exposure to gamma and X-ray radiation. In total, the integrated device system generates over 250 000 data points per day.

The Ebola outbreak in West Africa placed immense pressure on healthcare systems, surpassing their capacity to manage the high volume and severity of cases. Integrating digital health technologies into treatment centers may enhance both patient care and healthcare worker safety. A modular wireless patient monitoring system (MWPMS) was developed and tested in a proof-of-concept study at an Ebola treatment center (ETC) in Makeni, Sierra Leone by Steinhubl and co-workers.[46] The system used a wireless, multiparameter wearable patch for continuous monitoring and data transmission, supported by advanced analytics to enable early detection of changes in patient condition. Over three weeks, 26 eligible patients were enrolled, generating 1838 hours of continuous data (average 67.8 hours per patient). Parameters recorded included heart rate, heart rate variability, respiration rate, activity, posture, pulse transit time (an indirect indicator of blood pressure), uncalibrated skin temperature, and ECG waveforms. The wearable sensor was well tolerated, and automated measurements aligned closely with manual readings. Notably, changes in the multivariate change index, arrhythmia burden, and ECG features indicated early physiological shifts. The results from this study indicated that a portable, wireless monitoring system using wearable sensors and personalized analytics can deliver high-resolution, continuous monitoring in challenging healthcare environments. Such systems hold strong potential for improving clinical oversight and outcomes in low-resource or emergency settings.

9.4 Conclusions

This chapter aims to synthesize the research efforts on wearable sensors for health monitoring. The integration of multiple sensors into wearable devices for health monitoring and disease detection is a rapidly advancing area with strong potential to support remote healthcare solutions. These sensor-enabled systems play a crucial role in enabling continuous, real-time tracking of health metrics, allowing for early detection, diagnosis, and even prevention of various medical conditions. By facilitating timely interventions and personalized treatment recommendations, such technologies enhance the efficiency of modern healthcare delivery. Although the development of portable and wearable sensing technologies presents significant technical challenges, these devices offer the valuable advantage of enabling continuous collection of vital physiological data, allowing for long-term health monitoring of patients.

Emerging clinical research indicates that viral infections may trigger noticeable alterations in an individual's baseline physiological and behavioral patterns—such as changes in heart rate, respiratory rate, heart rate variability, body temperature, physical activity, and sleep—often before symptoms become apparent, and sometimes even in asymptomatic cases. Although the capability of wearable technologies to reliably detect viral infections in real-world environments is still under investigation, in numerous recent studies the feasibility of collecting continuous, individualized biometric data has been demonstrated. By applying machine learning algorithms to this data, it is possible to identify early physiological indicators and warning signs of infection.

Abbreviations

POC	Point-of-care
HR	Heart rate
EC	Gelectrocardiogram
HRV	Heart rate variability
vHDU	Virtual high-dependency unit
MWPMS	Modular wireless patient monitoring system
ETC	Ebola treatment center
VRTIs	Viral respiratory tract infections
ML	Machine learning
LAIV	Live attenuated influenza vaccine.

References

1. K. Conrad, D. Roggenbuck, D. Reinhold and U. Sack, Autoantibody diagnostics in clinical practice, *Autoimmun. Rev.*, 2012, **11**(3), 207–211.
2. L. C. Jr. Clark and C. Lyons, Electrode systems for continuous monitoring in cardiovascular surgery, *Ann. N. Y. Acad. Sci.*, 1962, **102**, 29–45.
3. S. Campuzano, M. Pedrero, A. González-Cortés, P. Yáñez-Sedeño and J. M. Pingarrón, Electrochemical biosensors for autoantibodies in autoimmune and cancer diseases, *Anal. Methods*, 2019, **11**, 871–887.
4. M. Thaler, A. Buhl, H. Welter, A. Schreiegg, M. Kehrel, B. Alber, J. Metzger and P. B. Luppa, Biosensor analyses of serum autoantibodies: application to antiphospholipid syndrome and systemic lupus erythematosus, *Anal. Bioanal. Chem.*, 2009, **393**, 1417–1429.
5. D. R. Thévenot, K. Toth, R. A. Durst, G. S. Wilson and D. R. Thévenot, Electrochemical biosensors: recommended definitions and classification, *Pure Appl. Chem.*, 2001, **16**, 121–131.
6. L. C. Clark and C. Lyons, Electrode systems for continuous monitoring in cardiovascular surgery, *Ann. N. Y. Acad. Sci.*, 1962, **102**, 29–45.
7. R. Golfinopoulou and S. Kintzios, Biosensing for autoimmune chronic disease—a review, *Chemosensors*, 2023, **11**, 366.
8. M. Bariya, H. Y. Y. Nyein and A. Javey, Wearable sweat sensors, *Nat. Electron.*, 2018, **1**, 160–171.
9. Y. Yang and W. Gao, Wearable and flexible electronics for continuous molecular monitoring, *Chem. Soc. Rev.*, 2019, **48**, 1465–1491.
10. A. J. Bandodkar, I. Jeerapan and J. Wang, Wearable chemical sensors: present challenges and future prospects, *ACS Sens.*, 2016, **1**, 464–482.
11. J. Kim, I. Jeerapan, J. R. Sempionatto, A. Barfidokht, R. K. Mishra, A. S. Campbell, L. J. Hubble and J. Wang, Wearable bioelectronics:

enzyme-based body-worn electronic devices, *Acc. Chem. Res.*, 2018, **51**, 2820–2828.
12. Y. Lin, M. Bariya and A. Javey, Wearable biosensors for body computing, *Adv. Funct. Mater.*, 2021, **31**, 1–21.
13. J. Min, J. R. Sempionatto, H. Teymourian, J. Wang and W. Gao, Wearable electrochemical biosensors in North America, *Biosens. Bioelectron.*, 2021, **172**, 112750.
14. S. Niu, N. Matsuhisa, L. Beker, J. Li, S. Wang, J. Wang, Y. Jiang, X. Yan, Y. Yun, W. Burnett, A. S. Y. Poon, J. B. H. Tok, X. Chen and Z. Bao, A wireless body area sensor network based on stretchable passive tags, *Nat. Electron.*, 2019, **2**, 361–368.
15. S. Shrivastava, T. Q. Trung and N. E. Lee, Recent progress, challenges, and prospects of fully integrated mobile and wearable point-of-care testing systems for self-testing, *Chem. Soc. Rev.*, 2020, **49**, 1812–1866.
16. J. Tu, R. M. Torrente-Rodríguez, M. Wang and W. Gao, The Era of digital health: a review of portable and wearable affinity biosensors, *Adv. Funct. Mater.*, 2020, **30**, 1–30.
17. X. Zeng, R. Peng, Z. Fan and Y. Lin, Self-powered and wearable biosensors for healthcare, *Mater Today Energy*, 2022, **23**, 100900.
18. M. C. Hartel, D. Lee, P. S. Weiss, J. Wang and J. Kim, Resettable sweat-powered wearable electrochromic biosensor, *Biosens. Bioelectron.*, 2022, **215**, 114565.
19. J. Madden, C. O'Mahony, M. Thompson, A. O'Riordan and P. Galvin, Biosensing in dermal interstitial fluid using microneedle based electrochemical devices, *Sens. Biosens. Res.*, 2020, **29**, 100348.
20. Y. Yang, Y. Song, X. Bo, J. Min, O. S. Pak, L. Zhu, M. Wang, J. Tu, A. Kogan, H. Zhang, T. K. Hsiai, Z. Li and W. Gao, A laser-engraved wearable sensor for sensitive detection of uric acid and tyrosine in sweat, *Nat. Biotechnol.*, 2020, **38**, 217–224.
21. Y. Zhu, M. C. Hartel, N. Yu, P. R. Garrido, S. Kim, J. Lee, P. Bandaru, S. Guan, H. Lin, S. Emaminejad, N. R. de Barros, S. Ahadian, H. J. Kim, W. Sun, V. Jucaud, M. R. Dokmeci, P. S. Weiss, R. Yan and A. Khademhosseini, Epidermis-inspired wearable piezoresistive pressure sensors using reduced graphene oxide self-wrapped copper nanowire networks, *Small Methods*, 2022, **6**, 1–13.
22. T. Q. Trung and N. E. Lee, Flexible and stretchable physical sensor integrated platforms for wearable human-activity monitoring and personal healthcare, *Adv. Mater.*, 2016, **28**, 4338–4372.
23. J. Kim, A. S. Campbell, B. E. F. de Ávila and J. Wang, Wearable biosensors for healthcare monitoring, *Nat. Biotechnol.*, 2019, **37**, 389–406.
24. K. Y. Shin, J. S. Lee and J. Jang, Highly sensitive, wearable and wireless pressure sensor using free-standing ZnO nanoneedle/PVDF hybrid thin film for heart rate monitoring, *Nano Energy*, 2016, **22**, 95–104.
25. N. Luo, W. Dai, C. Li, Z. Zhou, L. Lu, C. C. Y. Poon, S. C. Chen, Y. Zhang and N. Zhao, Flexible piezoresistive sensor patch enabling ultralow

power cuffless blood pressure measurement, *Adv. Funct. Mater.*, 2016, **26**, 1178–1187.
26. F. Q. Al-Khalidi, R. Saatchi, D. Burke, H. Elphick and S. Tan, Respiration rate monitoring methods: a review, *Pediatr. Pulmonol.*, 2011, **46**, 523–529.
27. C. S. Boland, U. Khan, C. Backes, A. O. Neill, J. Mccauley, S. Duane, R. Shanker, Y. Liu, I. Jurewicz, A. B. Dalton and J. N. Coleman, Terms of use bodily motion sensors based on graphene à rubber composites, *ACS Nano*, 2014, **8**, 8819–8830.
28. Q. Zhang, D. Jiang, C. Xu, Y. Ge, X. Liu, Q. Wei, L. Huang, X. Ren, C. Wang and Y. Wang, Wearable electrochemical biosensor based on molecularly imprinted Ag nanowires for noninvasive monitoring lactate in human sweat, *Sens. Actuators, B*, 2020, **320**, 128325.
29. L. Yin, M. Cao, K. N. Kim, M. Lin, J. Moon, J. R. Sempionatto, J. Yu, R. Liu, C. Wicker, A. Trifonov, F. Zhang, H. Hu, J. R. Moreto, J. Go, S. Xu and J. Wang, A stretchable epidermal sweat sensing platform with an integrated printed battery and electrochromic display, *Nat. Electron.*, 2023, **33**, 2303270.
30. A. J. Bandodkar and J. Wang, Non-invasive wearable electrochemical sensors: a review, *Trends Biotechnol.*, 2014, **32**, 363–371.
31. H. Lee, T. K. Choi, Y. B. Lee, H. R. Cho, R. Ghaffari, L. Wang, H. J. Choi, T. D. Chung, N. Lu, T. Hyeon, S. H. Choi and D. H. Kim, A graphene-based electrochemical device with thermoresponsive microneedles for diabetes monitoring and therapy, *Nat. Nanotechnol.*, 2016, **11**, 566–572.
32. J. Heikenfeld, Non-invasive analyte access and sensing through eccrine sweat: challenges and outlook circa 2016, *Electroanalysis*, 2016, **28**, 1242–1249.
33. W. Ji, J. Zhu, W. Wu, N. Wang, J. Wang, J. Wu, Q. Wu, X. Wang, C. Yu, G. Wei, L. Li and F. Huo, Wearable sweat biosensors refresh personalized health/medical diagnostics, *Research*, 2021, **2021**, 9757126.
34. J. Heikenfeld, Bioanalytical devices: technological leap for sweat sensing, *Nature*, 2016, **529**, 475–476.
35. W. Gao, S. Emaminejad, H. Y. Y. Nyein, S. Challa, K. Chen, A. Peck, H. M. Fahad, H. Ota, H. Shiraki, D. Kiriya, D.-H. Lien, G. A. Brooks, R. W. Davis and A. Javey, Fully integrated wearable sensor arrays for multiplexed in situ perspiration analysis, *Nature*, 2016, **529**, 509–514.
36. M. Bariya, H. Y. Y. Nyein and A. Javey, Wearable sweat sensors, *Nat. Electron.*, 2018, **1**, 160–171.
37. S. Emaminejad, W. Gao, E. Wu, Z. A. Davies, H. Y. Y. Nyein, S. Challa, S. P. Ryan, H. M. Fahad, K. Chen, Z. Shahpar, S. Talebi, C. Milla, A. Javey and R. W. Davis, Autonomous sweat extraction and analysis applied to cystic fibrosis and glucose monitoring using a fully integrated wearable platform, *Proc. Natl. Acad. Sci. U. S. A.*, 2017, **114**, 4625–4630.
38. X. Li, J. Dunn, D. Salins, G. Zhou, W. Zhou, S. M. Schüssler-Fiorenza Rose, D. Perelman, E. Colbert, R. Runge, S. Rego, R. Sonecha, S. Datta, T. McLaughlin and M. P. Snyder, Digital health: tracking physiomes and activity using wearable biosensors reveals useful health-related information, *PLoS Biol.*, 2017, **15**, e2001402.

39. A. Al-Halhouli, A. Albagdady, J. Alawadi and M. A. Abeeleh, Monitoring symptoms of infectious diseases: Perspectives for printed wearable sensors, *Micromachines*, 2021, **12**(6), 620.
40. D. K. Ming, S. Sangkaew, H. Q. Chanh, P. T. H. Nhat, S. Yacoub, P. Georgiou and A. H. Holmes, Continuous physiological monitoring using wearable technology to inform individual management of infectious diseases, public health and outbreak responses, *Int. J. Infect. Dis.*, 2020, 648–654.
41. A. Sharma, A. Singh, V. Gupta and S. Arya, Advancements and future prospects of wearable sensing technology for healthcare applications, *Sens. Diagn.*, 2022, **1**(3), 387–404.
42. D. S. Temple, M. Hegarty-Craver, R. D. Furberg, E. A. Preble, E. Bergstrom, Z. Gardener, P. Dayananda, L. Taylor, N. M. Lemm, L. Papargyris, M. T. McClain, B. P. Nicholson, A. Bowie, M. Miggs, E. Petzold, C. W. Woods, C. Chiu and K. H. Gilchrist, Wearable sensor-based detection of influenza in presymptomatic and asymptomatic individuals, *J. Infect. Dis.*, 2023, **227**, 864–872.
43. M. D. Santos, C. Roman, M. A. F. Pimentel, S. Vollam, C. Areia, L. Young, P. Watkinson and L. Tarassenko, A real-time wearable system for monitoring vital signs of COVID-19 patients in a hospital setting, *Front. Digital Health*, 2021, **3**, 630273.
44. S. Sun, A. A. Folarin, Y. Ranjan, Z. Rashid, P. Conde, C. Stewart, N. Cummins, F. Matcham, G. Dalla Costa, S. Simblett, L. Leocani, F. Lamers, P. S. Sørensen, M. Buron, A. Zabalza, A. I. Guerrero Pérez, B. W. Penninx, S. Siddi, J. M. Haro, I. Myin-Germeys, A. Rintala, T. Wykes, V. A. Narayan, G. Comi, M. Hotopf and R. J. Dobson, Using smartphones and wearable devices to monitor behavioral changes during COVID-19, *J. Med. Internet Res.*, 2020, **22**, e19992.
45. A. Hadid, E. G. McDonald, M. P. Cheng, J. Papenburg, M. Libman, P. C. Dixon and D. Jensen, The WE SENSE study protocol: a controlled, longitudinal clinical trial on the use of wearable sensors for early detection and tracking of viral respiratory tract infections, *Contemp. Clin. Trials*, 2023, **128**, 107103.
46. S. R. Steinhubl, D. Feye, A. C. Levine, C. Conkright, S. W. Wegerich and G. Conkright, Validation of a portable, deployable system for continuous vital sign monitoring using a multiparametric wearable sensor and personalised analytics in an Ebola treatment centre, *BMJ Global Health*, 2016, **1**, e000070.

CHAPTER 10

Artificial Intelligence and Machine Learning in Sensor-based Virus Detection

GARBIS ATAM AKCEOGLU

Institute of Innovation for Future Society, Nagoya University, Furo-cho, Chikusa-ku, Nagoya 464-8603, Japan
E-mail: akceoglu.garbis.atam.h8@f.mail.nagoya-u.ac.jp

Rapid and reliable virus detection requires very low limits of detection (LoD), short time-to-results (TTR), and stable performance across different samples and devices. Hand-tuned thresholds and *ad-hoc* analytics struggle to meet these demands. Artificial intelligence (AI) and machine learning (ML) close this gap by learning patterns directly from biosensor signals—fluorescence kinetics in clustered regularly interspaced short palindromic repeat (CRISPR) assays, impedance spectra from electrochemical immunosensors, lateral-flow line intensities, field-effect transistor (FET) transfer curves, nanopore ionic-current traces, and even smartphone audio from coughs—turning raw measurements into calibrated decisions about infection status, viral load, and variants for pathogens such as severe acute respiratory syndrome coronavirus 2 (SARS-CoV-2), influenza, dengue, respiratory syncytial virus (RSV), and hepatitis B virus (HBV).[1–7]

AI refers to computational systems that recognize patterns, generalize from examples, and adapt to new conditions—capabilities once associated with human judgment. Early "expert systems" relied on fixed rules; modern AI learns from data and improves with experience. This shift is especially

Detection Science Series No. 32
Advances in Virus Detection with Sensors
Edited by Adil Denizli
© The Royal Society of Chemistry 2026
Published by the Royal Society of Chemistry, www.rsc.org

important in biosensing, where signals are noisy, sample matrices vary, and device behavior drifts. In these settings, rigid thresholds miss weak positives or misclassify cross-reactive backgrounds—for example, separating true SARS-CoV-2 fluorescence curves from late-cycle artifacts in multiplex isothermal assays.[8–10]

ML provides several learning modes that map naturally to diagnostic tasks. In supervised learning, labeled data are used to train models to predict infection status or copies per ml. In unsupervised learning, clustering and anomaly detection reveal batch effects, drift, or unexpected coinfections. Semi-supervised learning uses a small labeled set to guide a much larger unlabeled set—common in point-of-care programs that upload thousands of influenza lateral flow assay (LFA) images while only a subset have real-time polymerase chain reaction (RT-PCR) confirmation.[11–14] Beyond these, reinforcement learning and Bayesian optimization tune assay conditions (*e.g.*, temperature, Mg^{2+}, guide concentration) to maximize early signal separation, cutting TTR and improving separability in CRISPR–CRISPR-associated 12 (Cas12) readouts for SARS-CoV-2.[15,16] Deep learning—one-dimensional convolutional neural networks (1D-CNNs), long short-term memories (LSTMs), and transformers—ingests raw time series and learns features end-to-end, enabling models that distinguish dengue serotypes in electrochemical or nanopore signals without manual feature engineering.[5,17–19]

A practical way to organize this space is as an end-to-end sensor → AI pipeline. First, a nanomaterial-enabled transducer produces raw signals [impedance spectra, fluorescence/surface enhanced Raman scattering (SERS) intensity, LFA line images, nanopore events]. Second, rigorous preprocessing corrects baselines, drift, and illumination. Third, supervised ML outputs calibrated infection probabilities, viral-load surrogates, or variant labels. Nanostructured electrodes and plasmonic/graphene optics raise signal-to-noise ratios and push LoDs down—often by orders of magnitude—while the ML layer converts these richer signals into robust decisions in the presence of real-world noise. For example, screen-printed gold electrodes (SPGE) achieved a detection limit for RSV by EIS that was roughly three orders of magnitude lower than reusable glassy-carbon electrodes; ML on the resulting spectra further stabilizes calls at low titres.[20]

Model choice follows the signal type. For images (LFA strips, fluorescence micrographs), CNNs quantify line intensities and spatial artifacts. For spectra and time series [electrochemical impedance spectroscopy (EIS) Nyquist or Bode plots, fluorescence kinetics, FET gate sweeps], 1D-CNNs or hybrid CNN–LSTM models capture frequency- and time-localized motifs. For nanopore event streams, temporal convolutional and attention models excel. For multi-sensor arrays (*e.g.*, electrochemical + temperature + turbidity), late fusion ensembles integration of heterogeneous evidence to stabilize decisions. In head-to-head tests, these approaches have improved sensitivity at fixed specificity for influenza A/B rapid tests compared with static thresholds.[4,18,21–23] Getting the pipeline right matters at every step: baseline wander in impedimetric chips, uneven illumination in LFA photos, or barcode bleed-through in fluorescent assays can dominate errors unless corrected, which otherwise show up as false positives in low-prevalence RSV screening.[24]

Evaluation must match clinical use. Alongside accuracy and area under the curve (AUC), virus diagnostics emphasize sensitivity/specificity, positive predictive value/negative predictive value (PPV/NPV) at intended prevalence, LoD, TTR, lot-to-lot robustness, and device generalization. Proper validation holds out a site-wise or device-wise external test set, uses RT-PCR or digital PCR as an orthogonal ground truth, and reports calibration (*e.g.*, Brier score, reliability curves) so probabilities can support triage. This is crucial when converting LFA intensities into actionable SARS-CoV-2 viral-load bands for return-to-work decisions.[21,25–29]

Real-world data bring practical challenges: label noise (*e.g.*, imperfect swabs), class imbalance (few positives in surveillance), and distribution shift (new lots, ambient lighting, variants).[30,31] Effective remedies include noise-robust losses, focal or cost-sensitive learning, strong data augmentation (color constancy for LFA; physics-aware simulation for EIS), self-training with confident pseudo-labels, and domain adaptation. Federated learning further enables multi-site training without centralizing patient data, helping preserve sensitivity as omicron shifted viral kinetics and threshold cycle (Ct) distributions.[32–35]

Case studies show how these pieces work together. CRISPR diagnostics [DNA endonuclease-targeted CRISPR trans reporter (DETECTR)/specific high-sensitivity enzymatic reporter unlocking (SHERLOCK)] pair Cas12/13 collateral cleavage with fluorescence; ML on early-time kinetics calls positives minutes sooner and at lower copy numbers than fixed thresholds for SARS-CoV-2 and Zika.[36] Electrochemical immunosensors turn antigen binding into impedance or voltammetric signals; 1D-CNNs on EIS spectra or learned embeddings from differential pulse voltammetry (DPV) peaks improve LoD and matrix tolerance for dengue nonstructural protein 1 (NS1) and hepatitis B surface antigen (HBsAg) compared with single-frequency readouts.[37,38] Smartphone analysis of LFA images standardizes line quantification across cameras and lighting, enabling semi-quantitative influenza viral-load estimates correlated with RT-PCR Ct, while on-device inference protects privacy and reduces latency.[39–41] Nanopore sensing goes further: deep models learn subtle dwell-time and blockade signatures characteristic of specific SARS-CoV-2 lineages, supporting variant detection from raw event streams.[42–44]

Moving from devices to systems, sensor outputs should not remain siloed. An AI broker can ingest data from readers and monitoring networks, map results to a regulated disease taxonomy, and drive actions in a central control center (risk alerts, prevention/containment, post-response). This integration supports both static analytics (rapid matching to known patterns) and dynamic analytics (anomaly detection for emerging variants), turning individual test results—such as SARS-CoV-2 LFA photos or EIS reads—into early warnings and coordinated responses across sites and regions.[45,46]

Finally, ethics and reproducibility are essential. Models trained in one clinic or region may underperform elsewhere; equitable datasets, clear data governance, and transparent reporting (*e.g.*, model cards) are required to avoid systematically worse sensitivity for under-represented groups—issues seen early in corona virus disease (COVID)-19 deployments.[47–49] Reproducibility depends on open protocols, reference datasets, and simulated stress tests (lighting, temperature, lot changes); without them, strong AUCs rarely translate to field performance for dengue and influenza screening.[50–52]

In summary, AI and ML convert sensor signals into clinical insights by replacing brittle thresholds with learned, uncertainty-aware predictions that generalize across devices and settings. When engineered as a full pipeline—with robust preprocessing, signal-appropriate models, calibrated evaluation, and careful deployment—AI enables earlier and more reliable virus detection, more informative reporting (viral load, variant), and faster public-health action for SARS-CoV-2, influenza, dengue, RSV, and other viruses.

References

1. M. J. Kellner, J. G. Koob and J. S. Gootenberg, *et al.*, SHERLOCK: nucleic acid detection with CRISPR nucleases, *Nat. Protoc.*, 2019, **14**, 2986–3012.
2. J. S. Gootenberg, O. O. Abudayyeh, J. W. Lee, P. Essletzbichler, A. J. Dy, J. Joung, V. Verdine, N. Donghia, N. M. Daringer, C. A. Freije, C. Myhrvold, R. P. Bhattacharyya, J. Livny, A. Regev, E. V. Koonin, D. T. Hung, P. C. Sabeti, J. J. Collins and F. Zhang, Nucleic acid detection with CRISPR-Cas13a/C2c2, *Science*, 2017, **356**(6336), 438–442.
3. J. S. Chen, E. Ma, L. B. Harrington, M. Da Costa, X. Tian, J. M. Palefsky and J. A. Doudna, CRISPR-Cas12a target binding unleashes indiscriminate single-stranded DNase activity, *Science*, 2018, **360**(6387), 436–439.
4. A. Yakoh, U. Pimpitak, S. Rengpipat, N. Hirankarn, O. Chailapakul and S. Chaiyo, Paper-based electrochemical biosensor for diagnosing COVID-19: detection of SARS-CoV-2 antibodies and antigen, *Biosens. Bioelectron.*, 2021, **176**, 112912.
5. A. Esteva, A. Robicquet, B. Ramsundar, V. Kuleshov, M. DePristo, K. Chou, C. Cui, G. Corrado, S. Thrun and J. Dean, A guide to deep learning in healthcare, *Nat. Med.*, 2019, **25**(1), 24–29.
6. D. M. El-Sherif, M. Abouzid, M. S. Gaballah, A. A. Ahmed, M. Adeel and S. M. Sheta, New approach in SARS-CoV-2 surveillance using biosensor technology: a review, *Environ. Sci. Pollut. Res. Int.*, 2022, **29**(2), 1677–1695.
7. Y. Wang, Y. Zhao, A. Bollas, Y. Wang and K. F. Au, Nanopore sequencing technology, bioinformatics and applications, *Nat. Biotechnol.*, 2021, **39**(11), 1348–1365.
8. E. J. Topol, High-performance medicine: the convergence of human and artificial intelligence, *Nat. Med.*, 2019, **25**(1), 44–56.
9. Y. LeCun, Y. Bengio and G. Hinton, Deep learning, *Nature*, 2015, **521**(7553), 436–444.
10. T. Fawcett, An introduction to ROC analysis, *Pattern Recognit. Lett.*, 2006, **27**, 861–874.

11. B. A. Zayed, A. N. Ali, A. A. Elgebaly, N. M. Talaia, M. Hamed and F. R. Mansour, Smartphone-based point-of-care testing of the SARS-CoV-2: a systematic review, *Sci. African*, 2023, **21**, e01757.
12. E. H. Shortliffe, R. Davis, S. G. Axline, B. G. Buchanan, C. C. Green and S. N. Cohen, Computer-based consultations in clinical therapeutics: explanation and rule acquisition capabilities of the MYCIN system, *Comput. Biomed. Res.*, 1975, **8**(4), 303–320.
13. C. M. Bishop and N. M. Nasrabadi, *Pattern recognition and machine learning*, Springer, 2006, vol. 4.
14. Z. Wang, P. Wang, K. Liu, P. Wang, Y. Fu, C.-T. Lu, C. C. Aggarwal, J. Pei and Y. Zhou, *A Comprehensive Survey on Data Augmentation*, 2024, *arXiv2405.09591*.
15. S. Kiranyaz, O. Avci, O. Abdeljaber, T. Ince, M. Gabbouj and D. J. Inman, 1D convolutional neural networks and applications: a survey, *Mech. Syst. Signal Process*, 2021, **151**, 107398.
16. A. G. Barto, Reinforcement learning: connections, surprises, and challenge, *AI Mag.*, 2019, **40**(1), 3–15.
17. S. Verma, R. P. Shukla and G. Dutta, Machine learning-enabled biosensors in clinical decision making, in *Next-Generation Nanobiosensor Devices for Point-Of-Care Diagnostics*, Springer, 2022, pp. 163–194.
18. Y. K. Wan, C. Hendra, P. N. Pratanwanich and J. Göke, Beyond sequencing: machine learning algorithms extract biology hidden in nanopore signal data, *Trends Genet.*, 2022, **38**(3), 246–257.
19. S. Lee, Y. K. Yoo, S. Il Han, D. Lee, S.-Y. Cho, C. Park, D. Lee, D. S. Yoon and J. H. Lee, Advancing diagnostic efficacy using a computer vision-assisted lateral flow assay for influenza and SARS-CoV-2 detection, *Analyst*, 2023, **148**(23), 6001–6010.
20. M. Ramalingam, A. Jaisankar, L. Cheng, S. Krishnan, L. Lan, A. Hassan, H. T. Sasmazel, H. Kaji, H.-P. Deigner, J. L. Pedraz, H.-W. Kim, Z. Shi and G. Marrazza, Impact of nanotechnology on conventional and artificial intelligence-based biosensing strategies for the detection of viruses, *Discover Nano*, 2023, **18**(1), 58.
21. J. Park, Smartphone based lateral flow immunoassay quantifications, *J. Immunol. Methods*, 2024, **533**, 113745.
22. K. Kitta, M. Sakamoto, K. Hayakawa, A. Nukazuka, K. Kano and T. Yamamoto, Nanopore impedance spectroscopy reveals electrical properties of single nanoparticles for detecting and identifying pathogenic viruses, *ACS Omega*, 2023, **8**(16), 14684–14693.
23. H. Cho, S. Shim, W. W. Cho, S. Cho, H. Baek, S.-M. Lee and D.-S. Shin, Electrochemical impedance-based biosensors for the label-free detection of the nucleocapsid protein from SARS-CoV-2, *ACS Sens.*, 2022, **7**(6), 1676–1684.
24. D. M. Kinyua, D. M. Memeu, C. N. Mugo Mwenda, B. D. Ventura and R. Velotta, Advancements and applications of lateral flow assays (LFAs): a comprehensive review, *Sensors*, 2025, DOI: 10.3390/s25175414.

25. W. Vargas, *FDA perspectives: SaMD, AI/ML and innovation in medical devices*, https://www.amed.go.jp/content/000131101.pdf.
26. M. K. Akhtar, R. K.-K. Leung and G. Khan, Rethinking COVID-19 test sensitivity-a strategy for improving the detection limit, *Pan Afr. Med. J.*, 2021, **39**, 244.
27. UNICEF, *TARGET PRODUCT PROFILE Acute Respiratory Infection Diagnostic Aid (ARIDA)*, 2014. https://www.unicef.org/supply/media/1356/file/Target product profile-Acute Respiratory Infection Diagnostic Aid-ARIDA.pdf.
28. G. S. Collins, K. G. M. Moons, P. Dhiman, R. D. Riley, A. L. Beam, B. Van Calster, M. Ghassemi, X. Liu, J. B. Reitsma, M. van Smeden, A.-L. Boulesteix, J. C. Camaradou, L. A. Celi, S. Denaxas, A. K. Denniston, B. Glocker, R. M. Golub, H. Harvey, G. Heinze, M. M. Hoffman, A. P. Kengne, E. Lam, N. Lee, E. W. Loder, L. Maier-Hein, B. A. Mateen, M. D. McCradden, L. Oakden-Rayner, J. Ordish, R. Parnell, S. Rose, K. Singh, L. Wynants and P. Logullo, TRIPOD+AI statement: updated guidance for reporting clinical prediction models that use regression or machine learning methods, *BMJ*, 2024, **385**, e078378.
29. G. L. Salvagno, B. M. Henry, S. De Nitto, L. Pighi and G. Lippi, Correlation between time to positive result of SARS-CoV-2 rapid antigen self-test and viral antigen concentration, *EJIFCC*, 2022, **33**(4), 309–316.
30. T. Roland, C. Böck, T. Tschoellitsch, A. Maletzky, S. Hochreiter, J. Meier and G. Klambauer, Domain shifts in machine learning based Covid-19 diagnosis from blood tests, *J. Med. Syst.*, 2022, **46**(5), 23.
31. I. Araf, A. Idri and I. Chairi, Cost-sensitive learning for imbalanced medical data: a review, *Artif. Intell. Rev.*, 2024, **57**(4), 80.
32. Z. Zhang and M. R. Sabuncu, Generalized cross entropy loss for training deep neural networks with noisy labels, *Adv. Neural Inf. Process Syst.*, 2018, **32**, 8792–8802.
33. R. Khattab, I. R. Abdelmaksoud and S. Abdelrazek, Automated detection of COVID-19 and pneumonia diseases using data mining and transfer learning algorithms with focal loss from chest X-ray images, *Appl. Soft Comput.*, 2024, **162**, 111806.
34. D. Doonyapisut, B. Kim, J. K. Kim, E. Lee and C.-H. Chung, Deep generative learning for exploration in large electrochemical impedance dataset, *Eng. Appl. Artif. Intell.*, 2023, **126**, 107027.
35. Q. Xie, M.-T. Luong, E. Hovy and Q. V. Le, Self-training with noisy student improves imagenet classification, in *Proceedings of the IEEE/CVF conference on computer vision and pattern recognition*, 2020, pp. 10687–10698.
36. M. M. Kaminski, O. O. Abudayyeh, J. S. Gootenberg, F. Zhang and J. J. Collins, CRISPR-based diagnostics, *Nat. Biomed. Eng.*, 2021, **5**(7), 643–656.
37. J. Upan, P. Banet, P.-H. Aubert, K. Ounnunkad and J. Jakmunee, Sequential injection-differential pulse voltammetric immunosensor for hepatitis b surface antigen using the modified screen-printed carbon electrode, *Electrochim. Acta*, 2020, **349**, 136335.

38. G. Han, B. Maranzano, C. Welch, N. Lu and Y. Feng, Deep-learning-guided electrochemical impedance spectroscopy for calibration-free pharmaceutical moisture content monitoring, *ACS Sens.*, 2024, **9**(8), 4186–4195.
39. S. Lee, S. Kim, D. S. Yoon, J. S. Park, H. Woo, D. Lee, S.-Y. Cho, C. Park, Y. K. Yoo, K.-B. Lee and J. H. Lee, Sample-to-answer platform for the clinical evaluation of COVID-19 using a deep learning-assisted smartphone-based assay, *Nat. Commun.*, 2023, **14**(1), 2361.
40. T. T. Le, P. Chang, D. J. Benton, J. W. McCauley, M. Iqbal and A. E. G. Cass, Dual recognition element lateral flow assay toward multiplex strain specific influenza virus detection, *Anal. Chem.*, 2017, **89**(12), 6781–6786.
41. S. Heydari and Q. H. Mahmoud, Tiny machine learning and on-device inference: a survey of applications, challenges, and future directions, *Sensors*, 2025, **25**(10), 3191.
42. A. Senanayake, H. Gamaarachchi, D. Herath and R. Ragel, DeepSelectNet: deep neural network based selective sequencing for oxford nanopore sequencing, *BMC Bioinf.*, 2023, **24**(1), 31.
43. A. M. Fleming, N. J. Mathewson, S. A. Howpay Manage and C. J. Burrows, Nanopore dwell time analysis permits sequencing and conformational assignment of pseudouridine in SARS-CoV-2, *ACS Cent. Sci.*, 2021, **7**(10), 1707–1717.
44. Y. Bao, J. Wadden, J. R. Erb-Downward, P. Ranjan, W. Zhou, T. L. McDonald, R. E. Mills, A. P. Boyle, R. P. Dickson, D. Blaauw and J. D. Welch, SquiggleNet: real-time, direct classification of nanopore signals, *Genome Biol.*, 2021, **22**(1), 298.
45. H. Burkom, W. Loschen, R. Wojcik, R. Holtry, M. Punjabi, M. Siwek and S. Lewis, Electronic surveillance system for the early notification of community-based epidemics (ESSENCE): overview, components, and public health applications, *JMIR Pub. Heal. Surveill.*, 2021, **7**(6), e26303.
46. A. Chatterjee, N. Pahari and A. Prinz, HL7 FHIR with SNOMED-CT to achieve semantic and structural interoperability in personal health data: a proof-of-concept study, *Sensors*, 2022, **22**(10), 3756.
47. L. Wynants, B. Van Calster, G. S. Collins, R. D. Riley, G. Heinze, E. Schuit, M. M. J. Bonten, D. L. Dahly, J. A. A. Damen, T. P. A. Debray, V. M. T. de Jong, M. De Vos, P. Dhiman, M. C. Haller, M. O. Harhay, L. Henckaerts, P. Heus, M. Kammer, N. Kreuzberger, A. Lohmann, K. Luijken, J. Ma, G. P. Martin, D. J. McLernon, C. L. Andaur Navarro, J. B. Reitsma, J. C. Sergeant, C. Shi, N. Skoetz, L. J. M. Smits, K. I. E. Snell, M. Sperrin, R. Spijker, E. W. Steyerberg, T. Takada, I. Tzoulaki, S. M. J. van Kuijk, B. van Bussel, I. C. C. van der Horst, F. S. van Royen, J. Y. Verbakel, C. Wallisch, J. Wilkinson, R. Wolff, L. Hooft, K. G. M. Moons and M. van Smeden, Prediction models for diagnosis and prognosis of Covid-19: systematic review and critical appraisal, *BMJ*, 2020, **369**, m1328.
48. M. Mitchell, S. Wu, A. Zaldivar, P. Barnes, L. Vasserman, B. Hutchinson, E. Spitzer, I. D. Raji and T. Gebru, Model cards for model reporting, *Proceedings of the Conference on Fairness, Accountability, and Transparency*, 2018.

49. J. R. Zech, M. A. Badgeley, M. Liu, A. B. Costa, J. J. Titano and E. K. Oermann, Variable generalization performance of a deep learning model to detect pneumonia in chest radiographs: a cross-sectional study, *PLoS Med*, 2018, **15**(11), e1002683.
50. J. Budd, B. S. Miller, N. E. Weckman, D. Cherkaoui, D. Huang, A. T. Decruz, N. Fongwen, G.-R. Han, M. Broto, C. S. Estcourt, J. Gibbs, D. Pillay, P. Sonnenberg, R. Meurant, M. R. Thomas, N. Keegan, M. M. Stevens, E. Nastouli, E. J. Topol, A. M. Johnson, M. Shahmanesh, A. Ozcan, J. J. Collins, M. Fernandez Suarez, B. Rodriguez, R. W. Peeling and R. A. McKendry, Lateral flow test engineering and lessons learned from COVID-19, *Nat. Rev. Bioeng.*, 2023, **1**(1), 13–31.
51. J.-H. Park, E.-K. Park, Y. K. Cho, I.-S. Shin and H. Lee, Normalizing the optical signal enables robust assays with lateral flow biosensors, *ACS Omega*, 2022, **7**(21), 17723–17731.
52. O. Ndiaye, K. Woolston, A. Gaye, C. Loucoubar, M. Cocozza, C. Fall, F. Dia, E. R. Adams, M. Samb and D. Camara, *et al.,* Laboratory evaluation and field testing of dengue NS1 and IgM/IgG rapid diagnostic tests in an epidemic context in Senegal, *Viruses*, 2023, **15**(4), 904.

CHAPTER 11

Introduction to Computational Studies in Virus Detection

MARYAM ATABAY[a,b], HOMA FARAJI[c] AND YEŞEREN SAYLAN[a*]

[a]Department of Chemistry, Hacettepe University, Ankara, Turkey; [b]UNAM-National Nanotechnology Research Center, Bilkent University, Ankara, Turkey; [c]Biosensor Research Center, Endocrinology and Metabolism Molecular-Cellular Sciences Institute, Tehran University of Medical Sciences, 10 Jalal-Al-Ahmad Street, Tehran, 1411713137, Iran
*E-mail: yeseren@hacettepe.edu.tr

11.1 Introduction

Viruses are found everywhere and depend on host organisms to reproduce. They live in all environments and can infect many kinds of life, from bacteria to plants and animals. Viruses cause many common illnesses in humans, like the common cold, cold sores, chickenpox, and the flu. They can also cause more serious and even life-threatening diseases, such as Ebola, hepatitis, rabies, acquired immune deficiency syndrome (AIDS), severe acute respiratory syndrome (SARS), Middle East respiratory syndrome (MERS), bird flu, and the most recently discovered corona virus disease (COVID)-19.[1] In terms of structure, viruses have at least two core components. One part is the DNA or RNA genome (double-stranded or single-stranded). The other part is the capsid, a symmetrical structure made from many copies of small proteins, which are encoded by the viral genome.[2,3] Viruses are obligate parasites that rely on host cells to replicate and initiate infection. They use inherent cellular

Detection Science Series No. 32
Advances in Virus Detection with Sensors
Edited by Adil Denizli
© The Royal Society of Chemistry 2026
Published by the Royal Society of Chemistry, www.rsc.org

mechanisms to initiate infection. While most viruses do not harm humans due to immune defenses, infection dynamics are complex, and not all individuals or cells are equally affected. In the early days, virology relied heavily on culture-based methods and electron microscopy to identify and study viruses. While foundational, these techniques were often time-consuming and limited in their ability to detect novel or fastidious viruses.[4] The advent of molecular biology introduced polymerase chain reaction (PCR) and sequencing technologies, enhancing sensitivity and specificity in virus detection.[5] However, these methods still required prior knowledge of viral genomes, posing challenges in identifying unknown pathogens.[6] To understand the ways viruses can cause diseases, researchers have developed quantitative methods that use genetic interference, the expression of dominant-negative proteins, and drugs. These methods help with studying infection processes and cellular functions.[7]

In recent years, biosensor technology has made great progress, especially with the use of smart healthcare. Biosensors are sensitive devices that transfer the concentration of biological analytes into an electrical signal for detection purposes. In their structure and detection processes, a receptor recognizes the biological substance and a transducer converts this recognition into a readable signal.[8–10] Biomarkers are biological indicators that reflecting physiological or pathological activities (like disease, aging, *etc.*) in an organism.[11] Biomarkers serve a vital function in the early diagnosis and assessment of treatment efficacy. They offer a wealth of information on various health conditions.[12,13] Early detection of infectious diseases is crucial for providing timely treatment and significantly lowering mortality rates. Employing specific and sensitive biomarkers for identifying infection sources is crucial to preventing epidemics and ensuring accurate diagnosis of infectious diseases. Even though experiments have helped in reconstructing cellular processes and structures utilizing a minimal set of components derived from cell extracts; how they interact during infection is still not fully understood. Computational modeling and simulation can help provide an experimental entity and reconstitute the interaction process in a controlled way.[14,15]

After discovery of the first virus in 1898; at the beginning of its second century, the research in the field of virology was marked with a new field called viroinformatics which comes from combining virology with bioinformatics. There are now more than 100 online databases and servers that focus on different viruses like influenza, dengue, human immunodeficiency virus (HIV), and human papilloma virus (HPV). In addition, advances in molecular biology and bioinformatics have also led to large amounts of genomic data. There are 112 viroinformatics resources available for studying RNA, protein interactions, and viral recombination that can lead to new drugs and vaccines.[16] Several computational approaches have been developed to detect virus structures. Depending on the abilities of the methods they have different applications in the study of the structure of viruses and related interactions.[17] The integration of computational biology into virology marked a paradigm shift. Bioinformatics tools enabled the analysis of vast amounts of genetic data, facilitating the discovery and characterization of viruses without the need for culturing.[18] Notably, the development of digital transcriptome subtraction techniques allowed researchers to computationally remove host sequences from datasets, unveiling hidden viral transcripts and leading to the discovery of new viruses implicated in diseases such as Merkel cell carcinoma.[19] High-throughput sequencing technologies, combined with metagenomic approaches, have further expanded the capacity to detect and study viruses directly from environmental and clinical samples. These methods have uncovered a vast "viral dark matter," revealing the immense diversity of the virosphere and identifying numerous previously uncharacterized viruses.[20,21] In recent years, computational methods have become indispensable in virology, especially during global health crises. For instance, during the COVID-19 pandemic, bioinformatics tools were pivotal in sequencing the severe acute respiratory syndrome coronavirus-2 (SARS-CoV-2) genome, analyzing its origin, and tracking its mutations.[22]

Among the computational techniques used in virus studies, methods such as molecular docking, molecular dynamics (MD) simulation, density functional theory (DFT), and quantum mechanics/molecular mechanics (QM/MM) can be mentioned. Molecular docking aims to evaluate the binding affinity and orientation between ligands and target receptors, like proteins, and it has become a powerful tool in the development of drugs.[23–26] MD simulation is another approach that enables the exploration of system behavior over time at the atomic level. This method can address the conformational landscape, binding free energy of the interacting structures, interatomic interactions, *etc.*[27,28] DFT is a popular method for the description of structural, energetic, and electronic proprieties of intermolecular interactions with good precision.[29,30] QM/MM simulations combine the accuracy of quantum mechanics with the efficiency of molecular mechanics and provide a more detailed and accurate representation of the region of the system where a chemical reaction takes place.[31,32] These models employ deep learning frameworks to anticipate potential mutations affecting antigenicity and viral infectivity, allowing for proactive adaptation of vaccines and treatment strategies.[33–35]

This chapter will present a brief explanation of each method and recent studies and applications in virus detection will be reviewed. By highlighting recent studies and advancements, we aim to explore both the current state and potential future of computational techniques in virology.

11.2 Computational Studies in Virus Detection

11.2.1 Molecular Docking

Molecular docking is a computational method employed to forecast how a molecule, typically a small compound such as a drug, aligns when it interacts with another molecule, often a receptor like a viral protein. The goal is to understand how well a ligand fits into a receptor's binding site.[36] Docking simulates the interaction between a ligand and its receptor through energy minimization, adjusting the ligand's conformation until the most energetically favorable binding pose is achieved.[37]

The energy of binding (E_{bind}) between a ligand and a receptor, representing the total energy of their interaction, is a key metric in docking to evaluate how well a ligand binds to a receptor. E_{bind} can be modeled by the following equation:

$$E_{\text{bind}} = E_{\text{ligand}} + E_{\text{receptor}} + E_{\text{interactions}} \tag{11.1}$$

where E_{ligand} is the internal energy of the ligand, including contributions from its conformation, bond lengths, angles, and torsions; E_{receptor} is the internal energy of the receptor, similarly including contributions from its conformation and structural stability; $E_{\text{interactions}}$ represents the non-covalent forces between the ligand and the receptor, encompassing van der Waals interactions, electrostatic forces, hydrogen bonding, and solvation energies.[38,39] This equation is a simplified representation of the binding energy calculation used in molecular docking. In practice, docking algorithms often incorporate additional terms, such as entropic contributions and desolvation penalties, to improve the accuracy of binding energy predictions.[36]

The docking process involves three major components: the docking algorithm, scoring function, and molecular flexibility. The docking algorithm determines the method for searching the space of possible ligand conformations and orientations.[40] The scoring function is a mathematical model that evaluates the quality of the interaction between the ligand and receptor.[39] The flexibility of molecules refers to the extent to which the receptor or ligand can change shape to accommodate the other molecule.[41] On the basis of molecular flexibility, docking approaches can be categorized into rigid docking, flexible docking, and hybrid docking.

In rigid docking, both the ligand and receptor are held fixed, and only the ligand's orientation is varied.[36] This method is computationally fast and simple[40] and is useful for preliminary screening when computational resources are limited,[38] but it cannot account for conformational changes in the ligand or receptor, which may lead to inaccurate predictions.[39] In flexible docking, the receptor or ligand is allowed to change shape during the docking process to better mimic real-life binding events.[40] This approach is more accurate than rigid docking as it accounts for molecular flexibility[39] and is suitable for detailed studies of ligand–receptor interactions, especially when conformational changes are significant[38] but it is computationally expensive and time-consuming.[36] Hybrid docking combines rigid and flexible docking approaches, balancing accuracy and computational cost.[39] This method is widely used in drug discovery and biosensor design, where both speed and accuracy are important,[38] but it requires careful tuning of parameters to achieve optimal results.[36]

11.2.1.1 Advantages and Disadvantages of Molecular Docking

Molecular docking is a high-throughput computational technique that can be applied to various targets, such as viral proteins and biosensors.[38,41] It can be used to screen thousands of ligands in a short time,[40] making it a cost-effective approach that reduces the need for expensive experimental screening.[36] Additionally, it provides detailed information about ligand–receptor interactions at the atomic level[39] and is widely used in drug discovery and biosensor applications.[38,41] However, docking has limitations that must be considered. Scoring functions may not always accurately predict binding affinities,[39] and precision of the method is highly dependent on the initial configurations of the ligands and receptors—errors in their preparation can lead to incorrect predictions.[36] Furthermore, explicit solvent molecules are often omitted, which can influence the accuracy of the binding energy estimations.[38]

11.2.1.2 Recent Research on Molecular Docking in Virus Detection

In 2024, Hegde *et al.* used protein–protein docking to validate the selective binding of dengue virus non-structural protein 1 (NS1) to its antibody on a porous carbon-based electrochemical biosensor with a high surface area. Docking confirmed a strong affinity with multiple hydrogen bonds, ensuring high specificity and minimal interference, thereby reinforcing the biosensor's ultrasensitive detection of dengue NS1 protein. The improved sensitivity of the biosensor facilitated the detection of even minute quantities of NS1, making it possible to diagnose dengue infection in its early stages.[42] A study by Wu *et al.* utilized structure-based virtual screening to identify potential antiviral compounds by docking ligands to target proteins and optimizing their binding using computational modeling. Various viral and host proteins were screened against large compound libraries, ranking candidates on the basis of their binding affinities. The results highlighted molecular docking as a powerful approach for detecting viral targets and aiding antiviral drug discovery.[43] Rabiei *et al.* used molecular docking to validate the binding of the G-quadruplex aptamer (Apt31) to the E2 glycoprotein of bovine viral diarrhea virus genotype 1 (BVDV-1), ensuring its specificity. HDOCK and PatchDock servers ranked aptamers on the basis of binding affinity, and the results of analysis confirmed nine hydrogen bonds and multiple hydrophobic interactions with conserved, non-glycosylated E2 regions. Control docking with bovine serum albumin showed minimal binding, further confirming Apt31's specificity. These results supported the aptamer's use in a colorimetric biosensor for rapid and selective BVDV-1 detection.[44] Molecular docking also helped Huang *et al.* identify an inflammatory sensor for Seneca valley virus 3C (SVV 3C) protease (Figure 11.1).[45]

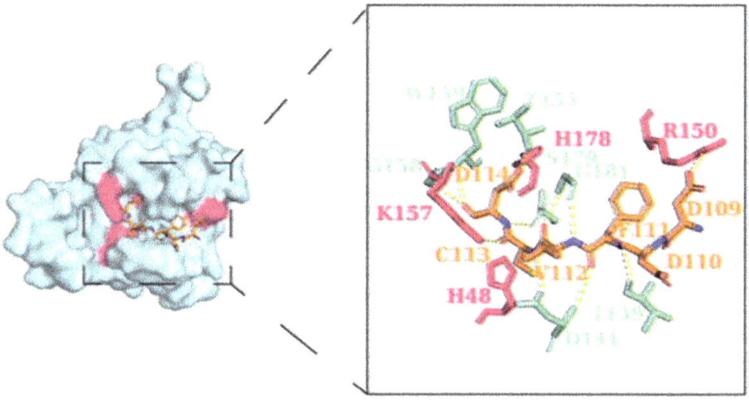

Figure 11.1 Detailed interaction interface of the biosensor and virus using molecular docking. The positively charged region of SVV 3C (H48, R150, K157, H178) is highlighted in bright pink, while the interacting domain of the discovered sensor is marked in orange. Reproduced from ref. 45, https://doi.org/10.1371/journal.ppat.1012398, under the terms of the CC BY 4.0 license, https://creativecommons.org/licenses/by/4.0/.

11.2.2 Molecular Dynamics Simulation

Molecular dynamics (MD) simulation is a computational approach that simulates the movement of atoms and molecules over time, achieved by solving Newton's equations of motion for a system of interacting particles, MD simulations provide a detailed view of molecular interactions, conformational changes, and dynamic behavior.[46] This makes MD an invaluable tool for studying biological processes, including the mechanisms of virus detection and infection.[47]

At its core, MD simulation is grounded in Newton's second law of motion:

$$F = m \times a \tag{11.2}$$

where F represents the force applied to an atom, m is its mass, a is its acceleration.

The interactions between atoms are calculated using interatomic potentials or force fields, which provide an approximation of the system's potential energy surface. These force fields capture the various atomic interactions, including the stretching of bonds, bending of angles, rotational movements around bonds, and non-covalent interactions like van der Waals forces and electrostatic attractions.[46] Commonly used force fields include Assisted Model Building with Energy Refinement (AMBER),[48] Chemistry at HARvard Macromolecular Mechanics (CHARMM),[49] and GROningen MOlecular Simulation (GROMOS).[50] By numerically integrating Newton's equations of motion, MD simulations predict atomic trajectories over time, offering a dynamic view of molecular systems.[46]

11.2.2.1 Advantages and Disadvantages of MD Simulation

MD simulations provide a high-resolution view of molecular behavior, enabling researchers to observe conformational changes, binding processes, and dynamic interactions that are typically challenging to capture through experimental methods. They can bridge gaps in experimental data by offering atomistic insights that help in interpreting results from techniques such as X-ray crystallography or cryo-electron microscopy. Additionally, MD simulations enable precise control over parameters such as temperature, pressure, and pH, allowing the study of specific variables in isolation.[47,51–53] If researchers conduct long simulations or repeat their runs multiple times, they can apply statistical analysis to assess the significance of their observations.[52] However, MD simulations are computationally intensive, limiting the size of systems (typically thousands to millions of atoms) and the timescales that can be simulated (nanoseconds to microseconds).[54] The accuracy of MD simulations relies on the quality of the force fields employed, and improper parameterization can result in inaccurate outcomes.[48] Furthermore, MD often involves approximations, such as treating electrons implicitly, which may not fully capture quantum mechanical effects.[47]

11.2.2.2 Recent Research on MD Simulation in Virus Detection

MD simulations are particularly useful for understanding interactions between viral components, such as proteins and nucleic acids, and host molecules.[55]

Lidya *et al.* investigated why commercial rapid tests exhibit low sensitivity to the Indonesian chikungunya virus (Ind-CHIKV). Using MD simulations, they found that mutations G194S and V318R in the E1–E2 antigenic proteins weaken antibody binding, reducing diagnostic accuracy. Their findings indicated that modifying antibodies to accommodate these mutations could improve test sensitivity, offering insights for developing region-specific biosensors and diagnostics.[56] Another study focused on developing a simple and rapid colorimetric detection method for SARS-CoV-2 utilizing peptide-functionalized gold nanoparticles (GNPs). The results of MD simulations indicated that peptides derived from the angiotensin converting enzyme-2 (ACE2) receptor stably functionalized GNPs and strongly bound to the spike protein's receptor binding domain (RBD), leading to GNP aggregation in the presence of the virus. These findings indicate that a peptide-based approach could serve as an alternative to antibody-based biosensors, offering a scalable and effective diagnostic tool.[57] Mehranfar *et al.* evaluated the potential of phosphorene (Figure 11.2) and graphene nanosheets as biosensors for SARS-CoV-2 detection using MD simulations. Their findings revealed that while both materials interact strongly with the RBD of the virus, phosphorene exhibited higher affinity and sensitivity owing to enhanced electrostatic and van der Waals interactions. This indicates that phosphorene is a superior candidate for designing SARS-CoV-2 biosensors.[58] Another study

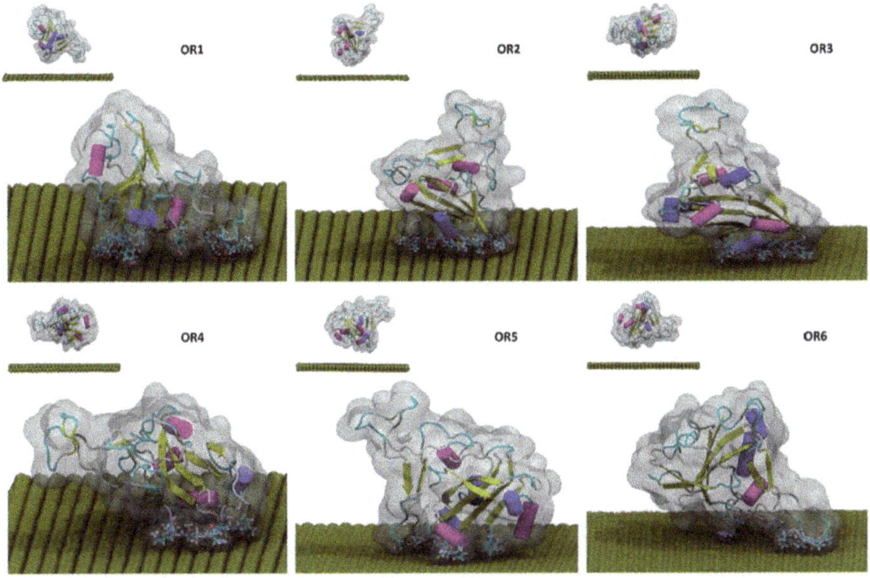

Figure 11.2 The obtained structures of the RBD complexes with phosphorene nanosheet after the simulation time with different initial orientations relative to the surface. The starting configurations are shown as the small images in the top corner. Reproduced from ref. 58 with permission from Elsevier, Copyright 2022.

employed *in silico* peptide design to identify candidates that interact with the RBD of SARS-CoV-2 and potentially block viral entry. MD simulations screened 12 peptides on the basis of binding affinity and stability in aqueous conditions, narrowing them down to two strong candidates. These peptides were functionalized on graphene- and carbon nanotube-based biosensors, demonstrating stable interactions with RBD, making them promising for SARS-CoV-2 detection.[59]

11.2.3 Density Functional Theory

Density functional theory (DFT) is based on quantum mechanics applied in chemistry, and material science to describe the electronic structures of many-body systems.[60,61] There are two main numerical approaches: the wave function theory and the DFT. The first one deals with the many-body wave function and tries to find the exact wave function with a good approximation. However, finding the exact ground-state wave function is extremely difficult when the number of N electrons is more than a few dozen. DFT uses the electron density [$\rho(r)$] as the fundamental variable. This approach aims to solve the Schrödinger equation, treating the total energy of the system as a function of the electron density [$\rho(r)$] based on the Hohenberg–Kohn theorem and Kohn–Sham equation. In the full non-relativistic Hamiltonian, atomic nuclei and electrons interact with each other. Because the nuclei are much heavier than electrons, the kinetic energy of electrons is much bigger than that of nuclei. Therefore, consideration of the electron and nuclear dynamics separately can be a good approximation; and DFT is a theory that uses this approximation.[62] Eqn (11.3) presents the electronic part of Hamiltonian ($\hat{H}_e$) describing the electron–nucleus interaction:

$$\hat{H}_e = -\sum_i \left(-\frac{\hbar^2}{2m}\frac{\partial^2}{\partial r_i^2} - \sum_I \frac{Z_I e^2}{|r_i - R_I|} \right) + \sum_{i<j} \frac{e^2}{|r_i - r_j|} \tag{11.3}$$

where i and j label electrons; I refers to atomic nuclei; r and R represent the electron and nucleus coordinates; $\hbar$ indicates reduced Planck constant; and Z represents atomic number of nucleus I respectively.

Specifically, both the ground-state kinetic energy $E_{\text{kin}}[\rho]$ and the electron–electron interaction energy $E_{\text{ee}}[\rho]$ can be expressed in terms of the electron density:

$$E_v[\rho] = E_{\text{kin}}[\rho] + \int \rho(r)\upsilon(r)\mathrm{d}r + E_{\text{ee}}[\rho] \tag{11.4}$$

$$E_v[\rho] \geq E_v[\rho_0] = E_0 \tag{11.5}$$

where $E_v[\rho]$ indicates total energy functional of the electron density and $\rho_0(E_0)$ is the electron density of the ground state (energy) under the potential $v(r)$. Although eqn (11.5) looks simple, the complexity is that there is no exact form of the function $E_{\text{kin}}[\rho]$ and $E_{\text{ee}}[\rho]$. To solve this problem Kohn and Sham introduced an auxiliary non-interacting electron system that reproduces the same ground-state electron density as the interacting system. In this approach, the electron density is expressed in terms of single-particle orbitals obtained from a one-electron Schrödinger-like equation:

$$\left[-\frac{\hbar^2}{2m}\frac{\partial^2}{\partial r^2} + v_{\text{eff}}(r) \right] \varnothing_i(r) = E_i \varnothing_i(r) \tag{11.6}$$

where E_i is the Kohn–Sham energy and $\varnothing_i$ is the Kohn–Sham (KS) state. $v_{\text{eff}}(r)$ is an effective potential and is defined in eqn (11.9).

The electron density is given by summation over occupied Khon–Sham states:

$$\rho(r) = \sum_{i=1}^{\text{occ}} |\varnothing_i(r)|^2 \tag{11.7}$$

This theorem introduces exchange–correlation function summation $E_{\text{XC}}[\rho]$ of the exact $E_{\text{kin}}[\rho]$ and $E_{\text{ee}}[\rho]$ energies of the original interacting system:

$$E_{\text{XC}}[\rho] = \left(E_{\text{kin}}[\rho] - E_{\text{kin}}^{\text{KS}}[\rho]\right) + \left(E_{\text{ee}}[\rho] - E_{\text{H}}[\rho]\right) \tag{11.8}$$

where $E_{\text{H}}[\rho]$ is Hartree energy function and:

$$v_{\text{eff}}(r) = v(r) + e^2 \int \frac{\rho(r')}{|r - r'|}\mathrm{d}r' + v_{\text{XC}}(r);\ v_{\text{XC}}(r) = \frac{\partial E_{\text{XC}}[\rho]}{\partial \rho(r)} \tag{11.9}$$

The theorem explains that once the electron density in the ground state is known, it is possible to determine the external potential. Therefore, all the physical properties related to the ground-state wave function can be calculated just from the electron density.

11.2.3.1 Advantages and Disadvantages of DFT

Besides the advantages of DFT including finding the adsorption energy calculations, band gap calculation in semiconductors, predicting the surface energy of a material, identification of defect bands, prediction of electron–phonon coupling in superconductors, and charge transfer predictions;[61] a major limitation of DFT is complexity. In fact, DFT loses its simplicity when the function becomes complicated. In addition, if the exchange–correlation function of DFT was exact, the quantum nature of matter would be described correctly. While the approximate nature of this function can be the reason for the success or failure of DFT applications.[63] In the following section, some applications of DFT computation in the study of virus detection will be discussed.

11.2.3.2 Recent Research on DFT in Virus Detection

Viruses are genetic entities composed of either DNA or RNA, enclosed within a protein capsid, and they replicate exclusively within a host organism. The viral proteins play critical roles in facilitating infection, promoting viral spread, and modulating the host immune response. Their structure is composed of 20 different amino acids and the most important amino acids are glycine, proline, and aspartic acid which are present in the structure of proteins of viruses like SARS-CoV-2, MERS-CoV, influenza virus, and HIV.

Panigrahi *et al.* used DFT computations and a green phosphorene monolayer (GPM) as a biosensor to detect glycine, proline, and aspartic acid. They applied vacancy defects like Ag, Cu, and Zn to increase the sensitivity of GPM toward the mentioned amino acids and employed thermodynamic analysis to determine the sensing mechanism under varying pressure and temperature conditions.[64] Liv and Özerdem developed a DFT-supported point-of-care electrochemical biosensing platform to detect the NS1 antibody of yellow fever virus in human plasma. Their biosensor is based on cysteine, yellow fever antigen protein (YFAG), and bovine serum albumin (BSA)-functionalized screen-printed gold electrodes [(BSA/YFAG/Cys/Au)]. In this study, DFT serves as a vital tool in understanding the adsorption behavior of cysteine (Cys) on the gold surface, which forms the foundation of this biosensor. By optimizing the geometry and calculating the energy changes of these interactions, DFT contributes significantly to understanding of the principles underlying biosensor design and improves the sensitivity of the biosensor for detecting YF antibodies.[65] Figure 11.3 illustrates the total energy changes calculated by DFT in designing the biosensor structure.

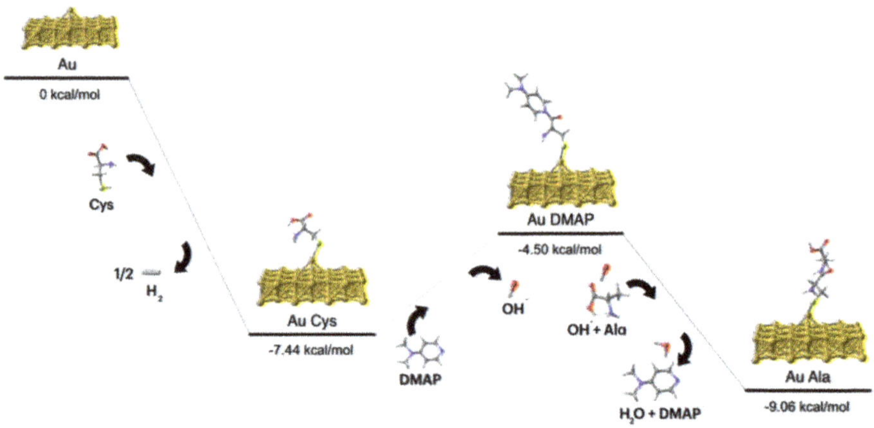

Figure 11.3 Total energy scheme of the gold surface. All structures are DFT-optimized structures. Abbreviations: 4-(dimethylamino)pyridine (DMAP), alanine (Ala) Reproduced from ref. 65 with permission from Elsevier, Copyright 2024.

Rahimi and Solimannejad introduced a B_3O_3 nanosheet as a biosensor to detect coronaviruses' biomarkers. Through DFT computations, the sensory characteristics of this nanosheet were studied to detect ethyl butyrate, a biomarker associated with coronavirus infection in the breath of patients. It was shown that the nanostructure can be an excellent substrate with adsorption energy in the range of −2.421 to −15.243 kcal mol^{-1}.[66] For biosensor development, Saha *et al.* investigated the interaction between boron arsenide, gallium arsenide, and aluminum arsenide monolayers and SARS-CoV-2 biomarkers using DFT computations. In that study, they tried to modify the optical properties, band structure, and electron density difference of monolayers and studied their abilities in the detection of biomarkers along with the interfering gases CO_2 and H_2O. Their findings indicated that gallium arsenide has a significant sensitivity towards the SARS-CoV-2 biomarkers.[67]

11.2.4 Quantum Mechanics/Molecular Mechanics

The concept of quantum mechanics/molecular mechanics (QM/MM) was first introduced in biological systems by Warshel and Levitt in 1976.[68] This method has become increasingly important in recent decades for studying chemical reactions in biological systems.[68] This approach can balance accuracy and computational speed, which are two major challenges in biomolecular modeling.[69] Molecular mechanics (MM) simulations are widely used because they rely on well-tested force fields that provide accurate results with reasonable computational cost. Additionally, many biomolecular systems are too large to be studied using electronic-structure methods.[70,71] However, quantum mechanics can be helpful in some cases, including studying biomolecules with metal atoms, electronic excitations, charge transfer processes, and chemical reactions where bond breaking and formation happens.[72] Traditional force fields cannot describe such reactions because they assume a fixed molecular structure. QM-based methods can accurately describe these processes but are computationally expensive, often scaling exponentially with system size.[73] To address this challenge, QM/MM approaches offer a solution by combining QM and MM methods. The portion of the system that includes the active site is treated using quantum mechanics to accurately model the chemical reactions. The rest of the system, which does not directly take part in the reaction but contains many more atoms, is handled using molecular mechanics, typically with a biomolecular force field.

11.2.4.1 Advantages and Disadvantages of QM/MM

The wide success of QM/MM methods is due to their ability to be used to describe quantum phenomena and compare different reaction mechanisms, they help eliminate unlikely paths and indicate new ones that match experimental results. In addition to enhancing accuracy and efficiency, comparing different reaction pathways, and reducing computational cost while maintaining accuracy can be other advantages of this method.[68,74] The main disadvantage of this method arises from the difficulty in finding the coupling term accurately between the two parts, QM and MM. In addition, the higher computational costs associated with the process can be another limitation of this approach.[75] The following section will describe some studies carried out on applications of the QM/MM approach in virus studies.

11.2.4.2 Recent Research on QM/MM in Virus Detection

Palacio *et al.* created highly sensitive biosensors to detect the hepatitis C virus core protein. Their sensors are very specific and reliable, able to detect the viral protein at very low levels in human blood plasma. In their study, DFT and QM/MM simulations were used to understand the covalent functionalization of graphene, explore the process of attaching molecules to it, and enhance the sensitivity of the biosensor.[76] In 2011 Makatini *et al.* studied HIV protease inhibitors, specifically targeting the resistance-prone wild-type C-South African HIV protease. Experimental findings indicate that most of the HIV-1 PR inhibitors have drug resistance because of the site-specific mutations occurring in the enzyme structure. Using NMR investigations and QM/MM/MD simulations and docking studies, this group identified a unique binding mode at the active site region within the HIV protease enzyme. Their results led to a more comprehensive understanding of the functional mechanisms of novel inhibitors that can enhance the transport of drugs across cell membranes.[77] In similar work Alves *et al.*, under the QM/MM paradigm, investigated the interaction energy of a protein and ligand for the wild-type and a N155S HIV-1 integrase mutant, both complexed with a β-diketo acid motif (DKA). The objective of their study was to investigate the molecular mechanism of drug resistance in HIV-1 integrase that can be helpful in the design of anti-HIV inhibitors.[78] Water plays an important role in biomolecular recognition. Bridging water molecules within the binding site can affect the interaction strength of a ligand with a protein. Woods *et al.* applied QM/MM Monte Carlo simulations to study the binding affinity of water molecules to influenza neuraminidase. They assessed the free-energy change when transitioning from a QM to a MM representation of the water molecule bound to the system. By comparing this change with the same transition in bulk water, they found that this shift plays a role in the binding process.[79] Currently, rare research focuses on using QM/MM computations for virus detection with biosensors. This presents a potential direction for future work and offers more accurate and sensitive diagnostic tools. To aid better understanding of the computational methods discussed, some of their key features are outlined in Table 11.1.

11.3 Conclusion

In conclusion, viruses are complex obligate parasites that rely on host cells for replication and infection. Despite significant advances in understanding of viral infections through experimental techniques, a comprehensive understanding of how viruses interact with host cells during infection is still lacking. Various computational approaches have been developed to investigate virus structures, and each method has different applications on the basis of its capabilities for studying viral structures and their related interactions. In this chapter, computational methods, such as molecular docking, MD simulations, DFT, and QM/MM have been briefly discussed, and their critical roles in virus detection summarized across several studies. Among these approaches, there is limited research on using QM/MM computations for virus detection. This represents a promising avenue for future research and could lead to more accurate and sensitive diagnostic tools. Overall, as computational methods continue to evolve, they will play an increasingly vital role in virology, providing new insights and shaping the future of virus detection, prevention, and therapeutic development.

Acknowledgements

Maryam Atabay and Yeşeren Saylan gratefully acknowledge the support from Scientific and Technological Research Council of Turkey (TÜBİTAK) Directorate of Science Fellowships and Grant (BİDEB) 2247-D National Young Researchers (Project No: 121C226). Yeşeren Saylan also thanks the Young Scientist Awards Program (BAGEP) award from the Science Academy.

Table 11.1 Summarized information on simulation and computational methods.

Molecular docking	Definition	Molecular docking is a computational technique used to predict the preferred orientation of a ligand when bound to a receptor. It is used to help in evaluating ligand–receptor interactions and binding affinity.[36]
	Software	AutoDock & AutoDock Vina, DOCK, MOE (Molecular Operating Environment), Schrödinger Glide, SwissDock, GOLD (Genetic Optimization for Ligand Docking)
	Applications	a. Drug discovery and development. b. Analysis of ligand–receptor interactions. c. Screening antiviral compounds against viral proteins. d. Biosensor design for virus detection. e. Identification of potential biomarkers for disease diagnostics.[36,38–41,43]
	Limitations	a. Limited accuracy of scoring functions in predicting true binding affinity.[39] b. Challenges in modeling receptor and ligand flexibility due to computational costs.[36] c. Dependence on high-quality ligand and receptor structures.[36] d. Omission of solvent effects, which can impact accuracy.[38]
MD simulation	Definition	MD simulation models the physical movements of atoms and molecules over time by solving Newton's equations of motion to address the conformational changes and energy of interacting system.[46,47]
	Software	AMBER, CHARMM, GROMOS, GROMACS (Groningen Machine for Chemical Simulations), NAMD (Nanoscale Molecular Dynamics), OpenMM, YASARA (Yet Another Scientific Artificial Reality Application)
	Applications	a. Biomolecular dynamics. b. Drug discovery. c. Protein folding. d. Viral protein interaction studies.[47,51–53]
	Limitations	a. High computational cost. b. Dependence on force field accuracy. c. Limited timescales. d. System size constraints. e. Simplifications of quantum effects.[47,48,54]

Table 11.1 (*Continued*)

DFT	Definition	DFT is a computational quantum mechanical modeling method used in surface science and catalysis for computation of electronic configurations of solids and atoms.[80]
	Software	ABINIT, SIESTA (Spanish Initiative for Electronic Simulations with Thousands of Atoms), Quantum Espresso, GAUSSIAN, ORCA, VASP (Vienna *Ab initio* Simulation Package), OpenMx, GAMESS (General Atomic and Molecular Electronic Structure System), MOPAC (Molecular Orbital Package), Qbox.
	Applications	Is used to solve problems in the following areas. a. Catalysts (charge transfer, surface energy, and adsorption energy). b. Molecular electronics (current–voltage characteristics and conductance). c. Semiconductors (band structure, band gap, defect bands). d. Superconductors (critical transition temperature and electron–phonon coupling).[61,80,81]
	Limitations	a. Underestimation of key quantities such as barriers to chemical reactions, charge transfer excitation energies, and energies of dissociation of molecular ions. b. DFT struggles with degenerate or near-degenerate states, which are critical for systems like strongly correlated materials, transition metals, and bond-breaking scenarios.[82]
QM/MM	Definition	An approach in which the part of the system where the chemical reaction happens is treated using QM, while the rest is described using a MM force field.[31]
	Software	Besides CHARMM codes that have included a QM module, AMBER and GROMACS support QM/MM calculations. NAMD has an interface with ORCA and MOPAC to carry out QM/MM computations.
	Applications	a. Study of enzymatic reaction mechanisms. b. Understanding and predicting drug resistance and drug metabolism. c. Studying the binding of ligands to biomolecules.[68,69,83]
	Limitations	a. QM/MM methods are not yet capable of providing quantitatively accurate results for more complex systems, especially in terms of free energy profiles or exact transition state structures. b. Deciding which atoms to include in the QM region is a significant challenge. c. Accurate free energy profiles are still difficult to achieve with QM/MM methods. d. Higher accuracy generally comes with a significant increase in computational cost.[84–86]

References

1. E. Vermisoglou, D. Panáček, K. Jayaramulu, M. Pykal, I. Frébort, M. Kolář, M. Hajdúch, R. Zbořil and M. Otyepka, *J. Biosens. Bioelectron.*, 2020, **166**, 112436.
2. E. Domingo, *Virus as Populations*, 2nd edn, 2019, pp. 1–33.
3. A. Cassedy, A. Parle-McDermott and R. O'Kennedy, *Front. Mol. Biosci.*, 2021, **8**, 637559.
4. P. Roingeard, *Biol. Cell*, 2008, **100**, 491–501.
5. K. B. Mullis and F. A. Faloona, *Methods Enzymol.*, 1987, **155**, 335–350.
6. J. Parker and J. Chen, *J. Clin. Virol.*, 2017, **86**, 20–26.
7. A. P. Liu and D. A. Fletcher, *Nat. Rev. Mol. Cell Biol.*, 2009, **10**, 644–650.
8. K. Zhang, J. Wang, T. Liu, Y. Luo, X. J. Loh and X. Chen, *Adv. Healthcare Mater.*, 2021, **10**, 2100734.
9. Y. Yang, X. Yang, Y. Yang and Q. Yuan, *Carbon*, 2018, **129**, 380–395.
10. M. Vestergaard, K. Kerman and E. Tamiya, *Sensors*, 2007, **7**, 3442–3458.
11. B. V. Chikkaveeraiah, A. A. Bhirde, N. Y. Morgan, H. S. Eden and X. Chen, *ACS Nano*, 2012, **6**, 6546–6561.
12. S. G. Meirinho, L. G. Dias, A. M. Peres and L. R. Rodrigues, *Biotechnol. Adv.*, 2016, **34**, 941–953.
13. Y. Aslan, M. Atabay, H. K. Chowdhury, I. Göktürk, Y. Saylan and F. Inci, *Biosensors*, 2023, **13**, 569.
14. I. F. Sbalzarini and U. F. Greber, *Influenza Virus*, 2018, 609–631.
15. M. Atabay, F. Inci and Y. Saylan, *J. Biosens. Bioelectron.*, 2025, **277**, 117275.
16. D. Sharma, P. Priyadarshini and S. Vrati, *J. Virol.*, 2015, **89**, 1489–1501.
17. A. K. Halder, P. Dutta, M. Kundu, S. Basu and M. Nasipuri, *Briefings Funct. Genomics*, 2018, **17**, 381–391.
18. W. I. Lipkin, *Nat. Rev. Microbiol.*, 2013, **11**, 133–141.
19. H. Feng, M. Shuda, Y. Chang and P. S. Moore, *Science*, 2008, **319**, 1096–1100.
20. Y. Z. Zhang, M. Shi and E. C. Holmes, *Cell*, 2018, **172**, 1168–1172.
21. G. A. Pavlopoulos, F. A. Baltoumas, S. Liu, O. Selvitopi, A. P. Camargo, S. Nayfach, A. Azad, S. Roux, L. Call, N. N. Ivanova, I. M. Chen, D. Paez-Espino, E. Karatzas, I. Iliopoulos, K. Konstantinidis, J. M. Tiedje, J. Pett-Ridge, D. Baker, A. Visel, C. A. Ouzounis, S. Ovchinnikov, A. Buluç and N. C. Kyrpides, *Nature*, 2023, **622**, 594–602.
22. Z. Shen, Y. Xiao, L. Kang, W. Ma, L. Shi, L. Zhang, Z. Zhou, J. Yang, J. Zhong, D. Yang, L. Guo, G. Zhang, H. Li, Y. Xu, M. Chen, Z. Gao, J. Wang, L. Ren and M. Li, *Clin. Infect. Dis.*, 2020, **71**, 713–720.
23. K. Fouda and R. S. Mohamed, *Food Funct.*, 2024, **15**, 3566–3582.
24. Ö. Erdem, I. Eş, Y. Saylan, M. Atabay, M. A. Gungen, K. Ölmez, A. Denizli and F. Inci, *Nat. Commun.*, 2023, **14**, 4840.
25. M. Atabay and M. R. Ejtehadi, *J. Mol. Struct.*, 2023, **1284**, 135355.
26. P. C. Agu, C. A. Afiukwa, O. Orji, E. Ezeh, I. Ofoke, C. Ogbu, E. I. Ugwuja and P. Aja, *Sci. Rep.*, 2023, **13**, 13398.

27. V. T. Sabe, T. Ntombela, L. A. Jhamba, G. E. Maguire, T. Govender, T. Naicker and H. G. Kruger, *Eur. J. Med. Chem.*, 2021, **224**, 113705.
28. S. O. Oselusi, P. Dube, A. I. Odugbemi, K. A. Akinyede, T. L. Ilori, E. Egieyeh, N. R. Sibuyi, M. Meyer, A. M. Madiehe, G. J. Wyckoff and S. A. Egieyeh, *Comput. Biol. Med.*, 2024, **169**, 107927.
29. H. Zhao, Y. Yang, X. Shu, Y. Wang and Q. Ran, *Adv. Colloid Interface Sci.*, 2018, **256**, 230–241.
30. M. Benafqir, A. Hsini, M. Laabd, T. Laktif, A. A. Addi, A. Albourine and N. El Alem, *Sep. Purif. Technol.*, 2020, **236**, 116286.
31. G. Groenhof, *Methods Mol. Biol.*, 2012, 43–66.
32. L. Böselt, M. Thürlemann and S. Riniker, *J. Chem. Theory Comput.*, 2021, **17**, 2641–2658.
33. A. P. Hederman and M. E. Ackerman, *Trends Immunol*, 2023, **44**, 333–344.
34. W. Han, N. Chen, X. Xu, A. Sahil, J. Zhou, Z. Li, H. Zhong, E. Gao, R. Zhang, Y. Wang, S. Sun, P. P. Cheung and X. Gao, *Nat. Commun.*, 2023, **14**, 3478.
35. B. Bravi, *NPJ Vacc.*, 2024, **9**, 15.
36. D. B. Kitchen, H. Decornez, J. R. Furr and J. Bajorath, *Nat. Rev. Drug Discovery*, 2004, **3**, 935–949.
37. X. Y. Meng, H. X. Zhang, M. Mezei and M. Cui, *Curr. Comput.-Aided Drug Des.*, 2011, **7**, 146–157.
38. S. Y. Huang and X. Zou, *J. Comput. Chem.*, 2006, **27**, 1876–1882.
39. I. A. Guedes, F. S. S. Pereira and L. E. Dardenne, *Front. Pharmacol.*, 2018, **9**, 1089.
40. O. Trott and A. J. Olson, *J. Comput. Chem.*, 2010, **31**, 455–461.
41. S. Y. Huang and X. Zou, *Proteins*, 2007, **66**, 399–421.
42. S. S. Hegde, Shivakumar, B. R. Bhat, P. Mishra, U. Dalimba, M. U. Ahmed and G. N. Santos, *Biosens. Bioelectron.: X*, 2024, **20**, 100525.
43. C. Wu, Y. Liu, Y. Yang, P. Zhang, W. Zhong, Y. Wang, Q. Wang, Y. Xu, M. Li, X. Li, M. Zheng, L. Chen and H. Li, *Acta Pharm. Sin. B*, 2020, **10**, 766–788.
44. P. Rabiei, H. Mohabatkar and M. Behbahani, *PloS one*, 2024, **19**, e0293561.
45. X. Huang, Z. Zhao, C. Zhu, L. Chai, Y. Yan, Y. Yuan, L. Wu, M. Li, X. Jiang, H. Wang, Z. Liu and P. Li, *PLOS Pathog.*, 2024, **20**, e1012398.
46. D. Frenkel and B. Smit, *Understanding molecular simulation: from algorithms to applications*, 2023, vol. 1, p. 658.
47. M. Karplus and J. A. McCammon, *Nat. Struct. Biol.*, 2002, **9**, 646–652.
48. D. A. Case, T. E. Cheatham III, T. Darden, H. Gohlke, R. Luo, K. M. Merz Jr, A. Onufriev, C. Simmerling, B. Wang and R. J. Woods, *J. Comput. Chem.*, 2005, **26**, 1668–1688.
49. B. R. Brooks, R. E. Bruccoleri, B. D. Olafson, D. J. States, S. a. Swaminathan and M. Karplus, *J. Comput. Chem.*, 1983, **4**, 187–217.
50. W. R. Scott, P. H. Hünenberger, I. G. Tironi, A. E. Mark, S. R. Billeter, J. Fennen, A. E. Torda, T. Huber, P. Krüger and W. F. Van Gunsteren, *J. Phys. Chem. A*, 1999, **103**, 3596–3607.
51. J. C. Phillips, R. Braun, W. Wang, J. Gumbart, E. Tajkhorshid, E. Villa, C. Chipot, R. D. Skeel, L. Kale and K. Schulten, *J. Comput. Chem.*, 2005, **26**, 1781–1802.

52. H. Faraji and A. Ebrahim-Habibi, *J. Biol. Phys.*, 2024, **50**, 89–118.
53. Y. Duan, C. Wu, S. Chowdhury, M. C. Lee, G. Xiong, W. Zhang, R. Yang, P. Cieplak, R. Luo, T. Lee, J. Caldwell, J. Wang and P. Kollman, *J. Comput. Chem.*, 2003, **24**, 1999–2012.
54. A. Warshel, P. K. Sharma, M. Kato, Y. Xiang, H. Liu and M. H. Olsson, *Chem. Rev.*, 2006, **106**, 3210–3235.
55. H. Ode, M. Nakashima, S. Kitamura, W. Sugiura and H. Sato, *Front. Microbiol.*, 2012, **3**, 258.
56. B. Lidya, M. Yusuf, U. Baroroh, K. Novitriani, B. Alisjahbana, I. Rahayu and T. Subroto, *Indones J. Chem.*, 2023, **23**, 1236–1247.
57. N. Verma, Y. Badhe, R. Gupta, A. K. Maparu and B. Rai, *J. Mol. Model.*, 2022, **28**, 202.
58. A. Mehranfar, M. Khavani and M. Izadyar, *J. Mol. Liq.*, 2022, **345**, 117852.
59. Y. Badhe, R. Gupta and B. Rai, *RSC Adv.*, 2021, **11**, 3816–3826.
60. H.-P. Zhang and Y. Tang, *Hybrid Polymer Composite Materials*, 2017, pp. 267–287.
61. D. Sibanda, S. T. Oyinbo and T. C. Jen, *Nanotechnol. Rev.*, 2022, **11**, 1332–1363.
62. Y. Nomura and R. Akashi, *Encyclopedia of Condensed Matter Physics*, 2022, pp. 867–878.
63. A. J. Cohen, P. Mori-Sánchez and W. Yang, *Chem. Rev.*, 2012, **112**, 289–320.
64. P. Panigrahi, Y. Pal, A. Panigrahi, H. Bae, H. Lee, R. Ahuja and T. Hussain, *Adv. Theory Simul.*, 2022, **5**, 2200357.
65. L. Liv and Z. Özerdem, *Int. J. Biol. Macromol.*, 2024, **269**, 132169.
66. R. Rahimi and M. Solimannejad, *J. Mol. Liq.*, 2022, **354**, 118855.
67. S. Saha, D. I. Sajib and M. K. Alam, *Phys. Chem. Chem. Phys.*, 2024, **26**, 6242–6255.
68. S. F. Sousa, A. J. Ribeiro, R. P. Neves, N. F. Brás, N. M. Cerqueira, P. A. Fernandes and M. J. Ramos, *WIREs Comput. Mol. Sci.*, 2017, **7**, e1281.
69. J. Mieres-Perez and E. Sanchez-Garcia, *Adv. Phys. Org. Chem.*, 2020, **54**, 143–183.
70. A. D. Mackerell, Jr., *J. Comput. Chem.*, 2004, **25**, 1584–1604.
71. B. Abhinav, K. Sameer, M. Chao, J. C. Phillips, Z. Gengbin and L. V. Kale, *IEEE International Symposium on Parallel and Distributed Processing*, 2008, pp. 1–12.
72. D. C. Young, *Computational Drug Design: A Guide for Computational and Medicinal Chemists*, John Wiley & Sons, 2009.
73. L. Ratcliff, S. Mohr, G. Huhs, T. Deutsch, M. Masella and L. Genovese, *Wiley Interdiscip. Rev.: Comput. Mol. Sci.*, 2016, 7, 1–24.
74. E. Brunk and U. Rothlisberger, *Chem. Rev.*, 2015, **115**, 6217–6263.
75. H. M. Senn and W. Thiel, *Angew. Chem.*, 2009, **48**, 1198–1229.
76. I. Palacio, M. Moreno, A. Náñez, A. Purwidyantri, T. Domingues, P. D. Cabral, J. Borme, M. Marciello, J. I. Mendieta-Moreno, B. Torres-Vázquez, J. I. Martínez, M. F. López, M. García-Hernández, L. Vázquez, P. Jelínek, P. Alpuim, C. Briones and J. Martín-Gago, *Biosens. Bioelectron.*, 2023, **222**, 115006.

77. M. M. Makatini, K. Petzold, S. N. Sriharsha, N. Ndlovu, M. E. Soliman, B. Honarparvar, R. Parboosing, A. Naidoo, P. I. Arvidsson, Y. Sayed, P. Govender, G. E. Maguire, H. G. Kruger and T. Govender, *Eur. J. Med. Chem.*, 2011, **46**, 3976–3985.
78. C. N. Alves, S. Martí, R. Castillo, J. Andrés, V. Moliner, I. Tuñón and E. Silla, *Biophys. J.*, 2008, **94**, 2443–2451.
79. C. J. Woods, K. E. Shaw and A. J. Mulholland, *J. Phys. Chem. B*, 2015, **119**, 997–1001.
80. T. van Mourik, M. Bühl and M. P. Gaigeot, *Philos. Trans. R. Soc., A*, 2014, **372**, 20120488.
81. J. Hafner, C. Wolverton and G. Ceder, *MRS Bull.*, 2006, **31**, 659–668.
82. A. J. Cohen, P. Mori-Sánchez and W. Yang, *Science*, 2008, **321**, 792–794.
83. M. W. van der Kamp and A. J. Mulholland, *Biochemistry*, 2013, **52**, 2708–2728.
84. M. Garcia-Viloca, J. Gao, M. Karplus and D. G. Truhlar, *Science*, 2004, **303**, 186–195.
85. F. Claeyssens, J. N. Harvey, F. R. Manby, R. A. Mata, A. J. Mulholland, K. E. Ranaghan, M. Schütz, S. Thiel, W. Thiel and H. J. Werner, *Angew. Chem.*, 2006, **45**, 6856–6859.
86. A. L. Bowman, L. Ridder, I. M. Rietjens, J. Vervoort and A. J. Mulholland, *Biochemistry*, 2007, **46**, 6353–6363.

Section IV: Applications of Sensor Technologies in Virus Detection

CHAPTER 12

Point of Care Diagnostics for Virus Detection

EYLUL GULSEN YILMAZ[a,b], BEYZA NUR KÜÇÜK[a,b], SELCAN GENC[a,b], YUSUF ASLAN[a,b], ÖZGECAN ERDEM[a] AND FATIH INCI[a,b*]

[a]UNAM—National Nanotechnology Research Center, Bilkent University, 06800 Ankara, Turkey; [b]Institute of Materials Science and Nanotechnology, Bilkent University, Ankara, Turkey
*E-mail: finci@bilkent.edu.tr

12.1 Introduction

Infections caused by viruses are responsible for millions of deaths that occur every year around the globe.[1] In situations when there is no vaccination that has been proven to be successful, the public health approach is to make an early diagnosis in order to either place those who may be sick under quarantine or recommend that they make changes to their lifestyle in order to avoid the spread of infection.[2] Centralized laboratory procedures, such as the enzyme-linked immunosorbent assay (ELISA) and the polymerase chain reaction (PCR), are currently the diagnostic methods that are considered to be the standards for the diagnosis of viral infections.[3] Due to the fact that these methods require a considerable amount of time, they are not suitable for widespread implementation in nations where the health care systems are not sufficient to give adequate services to all of the population.[4] Point-of-care (PoC)

Detection Science Series No. 32
Advances in Virus Detection with Sensors
Edited by Adil Denizli
© The Royal Society of Chemistry 2026
Published by the Royal Society of Chemistry, www.rsc.org

diagnostics have been presented as a means of overcoming these disadvantages. If such diagnostics are successful, they have the potential to significantly enhance public health worldwide.[5]

Among the many benefits of PoC testing (PoCT) is adaptability to different stages of patient treatment. A hospital's operating room, intensive care unit, or emergency room can all make use of it. Importantly, it can also be used in areas with low resources, in the community, as an outpatient environment, and in the home. The development of PoC devices has great potential for improving health monitoring and early disease identification, facilitating the transition to personalized treatment, and enhancing preventative care[6] (Figure 12.1a). For PoC tests, a number of different physiological fluids (urine, tears, blood, saliva, or sweat) can be utilized.[7–9] Electrolytes, hormones, and early indicators of infection are just a few of the many biomarkers that may be examined in these samples to track an individual's health.

Affordable, sensitive, specific, user-friendly, rapid-robust, equipment-free, and deliverable to end-users are the ASSURED characteristics that the World Health Organization (WHO) uses to define the ideal PoC test[10] (Figure 12.1b). Although numerous PoC technologies have been created for the diagnosis of viral infections in the last decades, the majority of them fall short of even near-PoC standards and do not meet the ASSURED requirements. Crucially, a test's market competitiveness increases in direct proportion to how closely it matches the ASSURED standards. For example, out of all the PoC tests, approximately 20% are fully ASSURED-compliant lateral flow immunoassays.[11] However, regulatory considerations also play a role in determining a

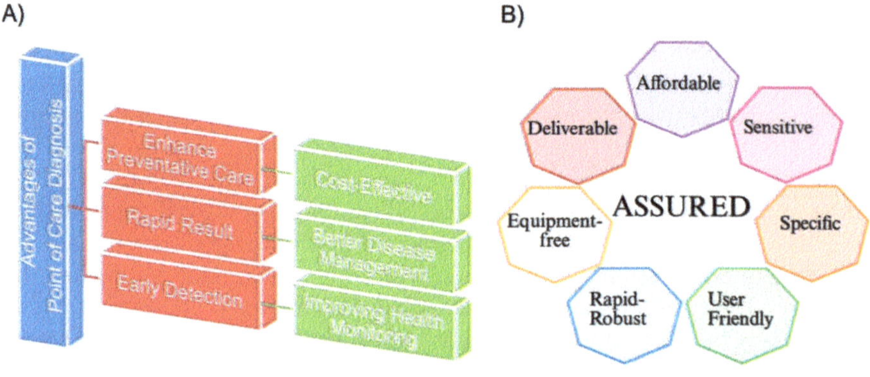

Figure 12.1 (a) Key advantages PoC diagnostics, including early detection, rapid results, enhanced preventative care, cost-effectiveness, better disease management, and improved health monitoring. (b) The ASSURED criteria defined by the WHO for ideal PoC diagnostics: affordable, sensitive, specific, user-friendly, rapid–robust, equipment-free, and deliverable. Created in BioRender. E. Yılmaz, 2025. https://BioRender.com/xu5d72u.

test's efficacy; for example, Australia's prohibition on severe acute respiratory syndrome coronavirus 2 (SARS-CoV-2) PoC tests.[12]

This chapter offers a thorough summary of the most recent developments in PoC diagnostics for virus detection. It begins by exploring the sensor technologies that form the foundation of PoC systems, including optical, electrochemical, and microfluidic-based biosensors. Then, with a focus on their function in identifying respiratory, sexually transmitted, blood-borne, vector-borne, and foodborne viruses, PoC diagnostic clinical uses are addressed. Finally, the chapter discusses the technical and legal issues confronting PoC technologies as well as future trends—such as the integration of artificial intelligence (AI), internet of things (IoT), and new materials—which are set to transform the sector. PoC technologies are poised to be crucial in fighting viral infections and enhancing worldwide health results by closing the gap between centralized laboratory testing and real-time, on-site diagnostics.

12.2 Sensor Technologies for PoC Virus Detection

Contagious pandemics, including H1N1, the Zika Virus, the SARS and Middle East respiratory syndrome (MERS) outbreaks, the Ebola virus outbreak, and most recently, coronavirus disease (COVID)-19, have significantly affected millions of people. Early detection of viral infections is essential to control the spread of the disease worldwide.[13] The common conventional methods for virus detection are the real-time PCR (RT-PCR) assay, ELISA and the hemagglutination assay.[14] Even though they are highly sensitive, they typically require a prolonged assay time and an expensive setup. The urgent need for a rapid, precise, and cost-effective diagnostic test for severe viral diseases to diagnose at the earlier stages and decrease the mortality rate has led to the development of PoC biosensor applications.

Biosensors are analytical tools utilized to measure biological components in diverse samples quantitatively or semi-quantitatively. Typically, there are three main segments in a biosensor that permit the detection of target elements: (i) a biorecognition molecule (antibody, molecularly imprinted polymer (MIP), aptamer, *etc.*), (ii) a transducer (optical, electrochemical, *etc.*), and (iii) signal processing. The initial biorecognition molecule attachment can be provided by a variety of techniques, such as physical methods (*e.g.*, entrapment, encapsulation, and adsorption) as well as chemical methods (*e.g.*, covalent bonding and cross-linking).[15] The overall performance of the biosensor is significantly affected by the bioreceptor immobilization process and needs to be controlled to enhance the sensitivity and specificity of the system. Upon the adsorption of the target on the sensing substrate, binding events on the surface can be converted to readable output signals *via* transducers and can be measured. PoC sensing systems can be classified depending on the transducer type used, which is selected by considering the physicochemical changes during sensing.[16]

12.2.1 Optical PoC Biosensors

When light illuminates the material, there is an interaction, and different phenomena may be observed, such as reflection, absorption, and scattering.[17] Fluorescence, luminescence, or plasmonic properties of the optic materials may be enhanced by altering certain features of the utilized material (size, aspect ratio, structure, *etc.*) and the light (wavelength, polarization, *etc.*). Optical biosensors quantify the alterations in the aforementioned features resulting from the binding of a target analyte to a detecting molecule immobilized on the surface of the optical sensor.

Plasmonic optical biosensor technology is a powerful diagnostic instrument that has become more widespread in the last decade.[18,19] The virus detection process can be altered since plasmonic biosensors are highly versatile. For instance, the immobilized recognition element could be a monoclonal antibody [immunoglobulin M (IgM)],[20] PCR-amplified viral bodies,[21] or DNA probes.[22] One of the most common plasmonic optical biosensors is surface plasmon resonance (SPR), which quantitatively assesses binding events in real time without the need to tag the interacting molecules.[23] It sensitively quantifies the alterations in the refractive index resulting from molecular interactions at the sensor surface. It has been reported that Ebola viruses (EBOV) have been detected in real-time and without labels *via* SPR with a limit of detection (LOD) of 0.5 pg ml^{-1}.[24] RNA, receptor binding domain (RBD) glycoprotein, and immunoglobulin G (IgG) protein associated with the COVID-19 virus has also been screened with SPR.[25,26] The SPR techniques can be tuned and confined around nanostructures fabricated on the optical surface, depending on their size and morphology; this phenomenon is termed localized surface plasmon resonance (LSPR). A nanoantenna-based biosensor has been developed that effectively identifies EBOV soluble glycoprotein (sGP) in human plasma at concentrations as low as 220 fg ml^{-1}, representing a remarkable 240 000-fold enhancement in sensitivity relative to the 53 ng ml^{-1} detection threshold of current EBOV immunoassays.[27] Another report described the development of an LSPR biosensor that utilizes silver nanostructures, produced through thermal annealing of thin silver films, to detect the dengue non-structural protein 1 (NS1) antigen, which can be identified early in the infection process. The biosensor incorporates blood-plasma separation to create a compact PoC lab-on-chip device that enables quick diagnosis in a total of 30 minutes of dengue NS1 antigen from 10 μl of whole blood.[28] Different human immunodeficiency virus (HIV) subtypes (A, B, C, D, E, G, and subtype panel) were captured and detected in another nanoplasmonic integrated platform with 98 ± 39 copies per ml sensitivity in approximately 70 minutes from HIV subtype spiked whole blood samples (Figure 12.2a and b).[29] The second plasmonic optical biosensor type is Raman spectroscopy, which can be described as a method employed to determine molecular structure by examining the vibrational frequencies of bonds through the analysis of light scattered from the sample.[30] The low inelastic scattering of biological materials constrains the efficacy of conventional

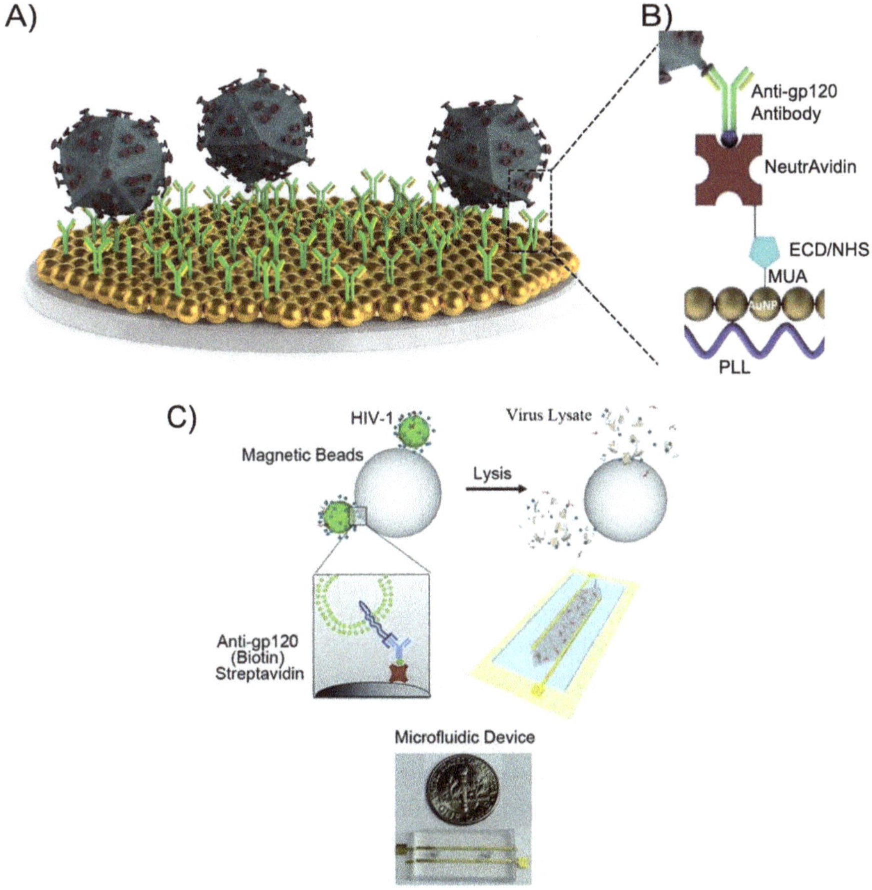

Figure 12.2 (a) Antibody immobilized nanoplasmonic surface and (b) the surface chemistry applications illustrated for HIV detection. Abbreviations: *N*-ethyl-*N'*-(3-dimethylaminopropyl)carbodiimide hydrochloride (EDC), 11-mercaptoundecanoic acid (MUA), *N*-hydroxysulfosuccinimide (NHS), gold nanoparticle (AuNP), poly-L-lysine (PLL). Reproduced from ref. 29 with permission from American Chemical Society Copyright, 2013. (c) Impedance-based HIV detection *via* a microfluidic PoC device is illustrated. Reproduced from ref. 50 with permission from John Wiley & Sons, Copyright 2013 WILEY-VCH Verlag GmbH & Co. KGaA, Weinheim.

Raman spectroscopy for sensitive virus detection due to insufficient signal strength. Further methods are required for signal amplification for more sensitive Raman signal detection. The plasmonic nanostructure-modified version of the system, known as surface-enhanced Raman spectroscopy (SERS), can magnify the signal intensity by over one million times and enable up to single-molecule detection.[31] Additionally, SERS offers numerous advantages over conventional Raman spectroscopy for virus detection, including (i) non-destructive analysis, (ii) reduced sample requirements, (iii) compatibility with liquid-phase sample measurements, (iv) multi-channel detection capabilities, and (v) the feasibility of on-site analysis using portable instruments.[32] In the literature regarding the detection of various virus types, including SARS-CoV-2,[33] influenza[34] and respiratory infection virus (RIV),[35] Raman-based approaches, particularly the SERS method, have demonstrated significant potential for reliable viral identification with satisfactory LOD values.

Photonic crystals (PC) are also emerging technologies for biosensing applications, which can be defined as periodic configurations of uniformly formed materials with varying dielectric constants. Periodicity might vary as one-dimensional (1D), two-dimensional (2D), or three-dimensional (3D). PCs typically have specific colours owing to their unique wavelengths of reflection, which can be defined by the distance between layers. Alteration of the crystal's periodicity, such as through biochemical stimulation, will result in a corresponding shift in the wavelength of the reflected light.[36] Therefore, the PC effect provides the basis for practical instruments for biosensing applications. PC platforms can be fabricated by inexpensive methods and be used with complex media (such as urine, blood, sweat, *etc.*).[37] PC results are also easily accessible for visual interpretation by untrained operators. Furthermore, the procedures are expected to be cost-effective, and the handling process is rather straightforward. A label-free optical detection approach employing nanostructured PC for capturing and quantifying HIV-1 viruses from physiologically relevant samples was demonstrated in a study.[38] It was noted that surface-adsorbed biotargets caused a shift in the resonant peak wavelength value (PWV), which may be detected with a resolution of less than 10 pm, facilitating the identification of both biomolecular layers and a limited number of viruses that are poorly distributed on the transducer surface. The successful collection and detection of HIV-1 in serum and phosphate-buffered saline (PBS) samples with virus loads between 10^4 and 10^8 copies per ml were noted.

Fluorescent and colourimetric optical methods serve as alternative techniques for the detection and visualization of biomolecules owing to their high sensitivity, selectivity, rapid response, practical fabrication, ease of operation, and stability.[39] In a study, the sensing surfaces of zeolitic imidazolate frameworks (ZIFs) were modified with single-stranded DNA (ssDNA) to detect HIV with LOD as low as 1.2 nmol ml^{-1}. When probe ssDNA was bound to the target ssDNA on HIV, double-stranded DNA (dsDNA) formed, and the fluorescent signal increased.[40] In another study, the H5N1 avian influenza virus was detected by using a polydiacetylene-based colorimetric biosensor, which is a polymer that may switch colour from blue to red in response to external stimuli, such as mechanical stress. In the study, the H5 influenza virus was sensitively detected after binding to the recognition molecule with a detection limit of 0.53 copies per μl, exhibiting a pronounced blue-to-red colour shift that is observable with the unaided eye or using a UV-visible spectrophotometer.[41]

12.2.2 Electrochemical PoC Biosensors

Electrochemistry is a fundamental analytical technique for observing reactions related to electron transfer. Electrochemical biosensors, which quantitatively measure the electron movements after the target and recognition element are bound, represent a prominent detection approach for numerous illnesses and demonstrate significant potential for PoC diagnostic applications owing to certain advantages. Rapid detection and data collection using electrochemical biosensors decreases the diagnosis time compared with conventional methods.[42] They can reach high sensitivity and selectivity with miniaturized, portable, cost-effective and user-friendly structures.[43] When a viral target binds to the detection site, the reaction under investigation may produce measurable biochemical changes in: (i) the current (amperometric), (ii) the potential or charge accumulation (potentiometric), (iii) the voltage properties (voltammetric), or (iv) the impedance (resistance and capacitance) (impedimetric).[44]

Amperometric measurement is an analytical method in which an applied voltage drives electrocatalytic redox reactions, producing measurable currents proportional to analyte concentration. A controlled-potential system is used, typically with a two- or three-electrode electrochemical cell, where one electrode serves as a reference. The working electrode facilitates the reaction, while the reference electrode (*e.g.*, Ag–AgCl or Hg–Hg_2Cl_2) maintains a constant potential. An auxiliary electrode (typically platinum or graphite) is also used. Erdem *et al.* utilized an amperometric disposable screen-printed carbon electrode (SPCE) for SARS-CoV-2 spike protein (S1) detection with an LOD of 0.19 ng ml^{-1} in buffer and 0.13 ng ml^{-1} in a saliva sample (10 µl volumes).[45] An anti-SARS-CoV-2 S1 monoclonal antibody was presented as the immobilized recognition molecule. The sandwich-type structure was established in the presence of the target S1 protein through the application of a secondary antibody and horseradish peroxidase (HRP)–IgG, respectively. 3,3′,5,5′-Tetramethylbenzidine (TMB) was used as a mediator for oxidation/reduction, and quantitative amperometric measurements were recorded.

Potentiometric biosensors primarily measure voltage changes between electrodes and depend on a biological reaction that produces less complex ion compounds, which are then detected electrochemically (such as NH_4OH, CO_2, pH, H_2O_2, *etc.*).[46] The voltage fluctuation between a reference electrode and an indicator (measurement) electrode is detected using a high-impedance voltmeter, under conditions of nearly zero current flow in potentiometric biosensors. In a recent study, cyclic voltammetry (CV) and electrochemical impedance spectroscopy (EIS) measurements were conducted to detect potentiometric alterations of the Zika virus on a molecularly imprinted (MI) gold electrode.[47] Since antibody-based detection methods may result in false positive results in other types of flaviviruses (in particular dengue virus), it was assessed that the MI method is a more selective detection method. The system detected 0.1 plaque forming units (PFU) ml^{-1} in buffer and 10 PFU ml^{-1} in human saliva samples that contained diverse biomolecules.

Voltammetric biosensors are another type that detect current in relation to the applied voltage, producing unique oxidation and reduction patterns related to the analyte. Voltammetry commonly employs techniques such as cyclic voltammetry, differential pulse voltammetry (DPV), and square wave voltammetry. They typically offer high selectivity with reduced cost and rapid analysis time. Abd Muain *et al.* have established a voltammetric immunosensor for the purpose of identification of Newcastle disease virus (NDV) utilizing polyclonal antibodies as the bioreceptor.[48] Antibodies were attached to a polyethylene glycol (PEG) self-assembled monolayer (SAM) surface. NDV was investigated electrochemically *via* DPV analysis, which presents an appealing alternative to standard detection methods owing to its simplicity, speed, and reduced labor requirements.

Finally, an impedimetric biosensor quantifies the impedance, defined as the resistance applied against the electrical current, of a system to indicate the presence or level of a target substance. Impedance is a combined measure of resistance, which opposes the flow of direct current, and reactance, which opposes the flow of alternating current. It can be affected by factors such as changes in the conductivity of the medium or the formation of biomolecular interactions on the surface of the sensor. In contrast to voltammetry and amperometry methods, which rely on direct current for measurements, EIS uses an alternating current potential or a sinusoidal signal applied to the electrochemical cell.[49] In an impedimetric sensor application study, a novel label-free approach, capable of detecting lysed viruses by impedance analysis was demonstrated (Figure 12.2c).[50] This method could be a feasible alternative to ELISA. It was evaluated with HIV-1 throughout a dynamic range of viral concentrations from 10^6 to 10^9 copies per ml; however, it was suggested that the system can be modified for other infectious disorders, including herpes, influenza, hepatitis, and poxes.

12.2.3 Microfluidics-based PoC Biosensors

To provide a comprehensive point-of-care system for viral detection, the previously described biosensors have typically been integrated with microfluidic devices in recent years. The most significant advantage of a microfluidic system is its small size, which permits reduced reagent volumes and portable devices with enhanced sensitivity. Recent advancements in portable microfluidic devices have increased their utilization, which has provided faster diagnostic studies and shortened analysis times.[51] Considering these advantages, microfluidic on-chip sensors can be integrated with the aforementioned optical and electrochemical methods.[45,50] Moreover, loop-mediated isothermal amplification (LAMP) and reverse-transcription LAMP (RT-LAMP) techniques can be integrated into microfluidic systems for more efficient and compact testing of DNA and RNA viruses. LAMP and RT-LAMP are isothermal nucleic acid amplification methods that are alternatives to conventional PCR methods without the need for thermal cycling.[52] In the literature, Zika, dengue (types 1 and 3) and chikungunya viruses have been reported to be detected with microfluidic PoC LAMP systems.[53,54]

Paper-based microfluidics are a more cost-effective and easier alternative to conventional lithography-based microfluidic devices. Theillet and coworkers utilized a laser-cut microfluidic paper-based analytical device (PAD) to detect chikungunya virus from 30 µl of patient serum.[55] The viral disease was detected in less than 10 minutes by the PAD with 70.6% sensitivity and 98% specificity, and the system showed reduced cross-reaction with other tropical viruses. Lateral flow assays (LFAs) are also practical tools characterized by their disposable design for determining the presence of an analyte. They are commonly used for PoC diagnostics, providing quick results with minimal equipment and technical expertise. During the COVID-19 pandemic, LFA strip applications provided significant help in diagnosis at the earlier stages.[56,57]

12.3 Clinical Applications of PoC Diagnostics in Virus Detection

By allowing fast, precise, and easily accessible testing at or near the site of patient treatment, PoC diagnostics have changed the scene of virus detection. In clinical environments, these technologies are very important since they provide quick results that guide public health activities, infection control policies, and treatment decisions.[58] New PoC technologies have completely changed the way infectious disease viruses, particularly those preventable by vaccinations, are detected. Among the disadvantages of traditional laboratory procedures are slow turnaround times and inadequate sensitivity and specificity in response to new viral epidemics.[59,60] These advancements in diagnostic tools seek to solve these problems. Examples of technological innovations include biosensors, clustered regularly interspaced short palindromic repeats (CRISPR)–CRISPR-associated (Cas), and smartphone-enabled diagnostic tools.[61] These technologies have changed significantly over time. Thanks to technological developments that have greatly improved patient outcomes, healthcare providers may now give patients accurate and fast results just when they need them. Molecular PoC testing now detects many viral infections, including respiratory, bloodborne, sexually transmitted diseases, vector-borne, and emerging viruses, as well as gastrointestinal and foodborne viruses. Every category has particular diagnostic difficulties; PoC technologies have been essential in removing these obstacles, therefore improving patient outcomes and facilitating disease management in several healthcare environments.[62]

12.3.1 Respiratory Viruses

Respiratory infections are very common and can spread quickly, as seen with the COVID-19 pandemic. This can lead to major public health problems. In the last 20 years, respiratory viruses (RVs) such as SARS in 2003,[63] the influenza A virus (H1N1) in 2009,[64] and the MERS coronavirus (MERS-CoV)[65] in 2015 have caused major outbreaks that have had effects on society and the economy. The COVID-19 pandemic was the worst. It killed over 5 million people around the world by early 2022 and made it clear that the world requires better diagnosis tools right away.[66]

The pandemic has made it clear how important PoC tests are for quickly treating respiratory virus infections. RVs like influenza, coronaviruses, respiratory syncytial virus, parainfluenza virus (PIV), adenovirus, rhinovirus, and other secondary viruses often have symptoms that are similar to each other, which makes it hard to make a correct diagnosis[60,67] PoC technologies have improved, allowing for faster infection diagnosis, more personalized treatment, and fewer complications.[68] They also help stop people from using too many antibiotics by identifying the difference between viral and bacterial illnesses.

In the absence of vaccines or antiviral medications, rapid and accurate diagnoses are of the highest priority in containing newly developing viruses. For the nucleocapsid (N) gene of MERS-CoV, a one-pot RT-LAMP approach was devised.[65] Within 60 minutes, this approach could identify as few as four viral genome copies using six primers and optimal enzyme conditions. For improved DNA polymerase compatibility, the EvaGreen dye outperformed SYBR Green I in signal read-out. Even more sensitive detection down to 0.4 viral genome copies was achieved using field-deployable microchamber devices, that did not exhibit cross-reactivity with other respiratory viruses, such as influenza A (H1N1, H3N2), influenza B, or human coronavirus 229E. Integrated into PoC equipment for quick, on-site detection, this sensitive and specific technology permits large-scale testing during emergencies (Figure 12.3a).

The current gold standard real-time quantitative PCR (RT-qPCR) method is one of the few SARS-CoV-2 detection methods that can identify two genes simultaneously, which is a critical step in improving detection accuracy. However, very few methods can achieve this. Xiong *et al.* presented a novel method for the quick and simultaneous detection of SARS-CoV-2 in a single strip test. It utilized a CRISPR–Cas9-mediated triple-line lateral flow assay (TL-LFA) in conjunction with multiplex reverse transcription–recombinase polymerase amplification (RT–RPA)[69] (Figure 12.3b). This test can identify 100 RNA copies per response ($25\ \mu l^{-1}$) of SARS-CoV-2 viral RNA standards or cell-cultured SARS-CoV-2 using the envelope (E) and open reading frame 1ab (Orf1ab) genes. Also, analysis of 64 nasopharyngeal swab samples using dual-gene technology resulted in a near perfect score of 97.14 for positive predictive agreement and a perfect score of 100 for negative predictive agreement.

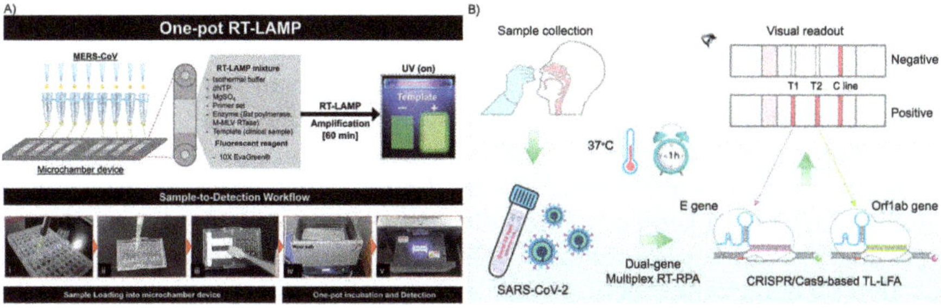

Figure 12.3 (a) The "sample-to-detection" scheme depicts the whole one-pot RT–loop-mediated isothermal amplification (RT-LAMP) process. (i)–(iii): Sample preparation and loading methods, (iv) and (v): One-pot incubation in water bath and LAMP signal read-out analysis. Reproduced from ref. 65, https://doi.org/10.3389/fmicb.2016.02166, under the terms of the CC BY 4.0 license, https://creativecommons.org/licenses/by/4.0/ (B) A CRISPR–Cas9-mediated triple-line lateral flow assay combined with multiplex RT–RPA for simultaneous dual-gene diagnosis of SARS-CoV-2. Reproduced from ref. 69 with permission from John Wiley & Sons, Copyright 2020 Wiley-VCH GmbH.

Using this platform, people in areas with limited resources will have an easier and more precise way to diagnose infectious disorders like COVID-19.

12.3.2 Sexually Transmitted and Blood-borne Viruses

HIV, hepatitis B (HBV) and hepatitis C (HCV), human papillomavirus (HPV),[70] *Chlamydia trachomatis* (CT), *Neisseria gonorrhoeae* (NG), syphilis, and *Trichomonas vaginalis*[71] are among the sexually transmitted and blood-borne illnesses (STBBI) that remain significant public health issues. Because of the stigma and discrimination surrounding these diseases, which may deter people from getting tested and receiving treatment, STBBI are especially difficult to eradicate. Because of this, the general public have a low risk perception and a low acceptance rate of STBBI testing, and healthcare providers are reluctant to regularly give STBBI testing.[72]

In order to eradicate HIV, HCV, and other STBBI as risks to public health by 2030, diagnostic testing is essential. In fact, prompt and precise STBBI diagnosis is crucial for connecting patients with the right care and treatment, averting negative consequences, and reducing the spread of infection.[73,74] For many populations, PoCT has been demonstrated to be usually acceptable, and frequently even better than traditional testing techniques. Reaching first-time testers, increasing testing frequency among current testers, increasing the percentage of testers receiving their results, accelerating linkage to treatment and care, and lowering loss to follow-up can all be facilitated by this high acceptability and some of the unique features of PoCT (such as the availability of rapid results and flexibility in the settings in which it can be administered). PoCT also helps reduce the danger of further transmission in outbreak situations by giving early access to test results and treatment.

Using an expressed HCV core protein as the recognition molecule, Aronoff-Spencer and colleagues developed a quantitative diagnostic for anti-HCV that is more relevant to proof-of-concept testing.[75] As the anti-HCV receptor, yeast cells expressing the HCV core protein on the surface were attached to a gold screen-printed electrode (SPE) (Figure 12.4a). Using a secondary alkaline phosphatase (ALP) tagged anti-human antibody, which catalyses a detectable reaction when exposed to its substrate utilizing cyclic

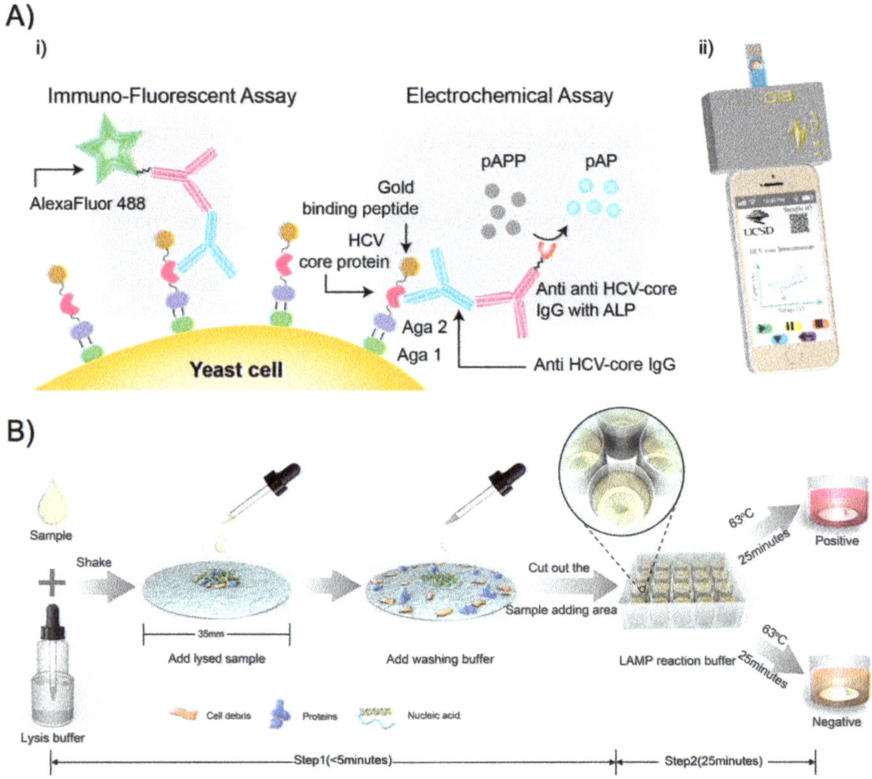

Figure 12.4 (a) Representation of a yeast biobrick chimera facilitating the detection of anti-HCV core antibodies using a PoC smartphone potentiostat platform. (i) Detection of anti-HCV core antibody *via* fluorescence and electrochemical techniques. Abbreviations: A-agglutinin (Aga) *p*-aminophenol (pAP), *p*-aminophenyl phosphate (pAPP) (ii) Depiction using a smartphone-integrated potentiostat that interfaces with a host device *via* the audio jack. Reproduced from ref. 75 with permission from Elsevier, Copyright 2016. (b) Schematic representation of the paper-based LAMP technique for the extraction, amplification, and visual interpretation of data for rotavirus A diagnosis at the point of care. Reproduced from ref. 91 with permission from Elsevier, Copyright 2018.

CV, it was possible to make the binding of the anti-HCV to the yeast protein detectable. A potential sweep technique, CV helps by finding the electrochemical signal strength (current) *vs.* concentration. In a buffer solution with quantified readout on a smartphone, this approach was able to reach a LOD of 2 nM and a detectable range of 4–250 nM for anti-HCV.

The GeneXpert platform (Cepheid) can perform a qualitative, *in vitro* real-time PCR Xpert CT/NG assay to identify and differentiate genomic DNA from *Chlamydia trachomatis* and *Neisseria gonorrhoeae*.[76] The modular platform allows the assay to be performed immediately on patient specimens without manual or off-board preparation. A sample adequacy control ensures that patient samples and testing are adequate in the assay. Within 90 minutes of loading the specimen, results are accessible. The assay and platform are Conformité Européenne (CE)-marked and Food and Drug Administration (FDA)-approved for urine, endocervical, vaginal, rectal, and pharyngeal samples. The Xpert CT/NG is one of the best PoC tests for genitourinary specimens, with sensitivities over 97% and specificities over 95%.[77] The FDA has also cleared the assay for testing in pharyngeal and rectal specimens with sensitivity and specificity of 94% and 98%, respectively.[78]

12.3.3 Vector-borne and Emerging Viruses

Rapidly spreading, highly lethal, and capable of causing epidemics in areas with limited resources, vector-borne and emerging viruses frequently present formidable public health concerns; here, PoC diagnostics play an essential role in their diagnosis and control. Dengue, malaria, Zika, chikungunya, West Nile, and yellow fever are just a few of the viruses that can be spread by arthropods known as vectors. Because of their high mortality rates and unpredictable behaviour, emerging viruses like Ebola, Marburg, and Nipah further hamper worldwide health initiatives. PoC diagnostics provide for efficient disease management and outbreak control by providing testing solutions that are accessible, accurate, and can be done quickly.

Mauk *et al.* used RT-LAMP to produce affordable molecular detection of Zika virus (ZIKV) without complicated instrumentation.[79] They constructed RT-LAMP primers for the Americas-dominant Zika lineage using highly conserved ZIKV genomic regions. They created a disposable microfluidic cassette for viral nucleic acid capture, concentration, isothermal amplification, and detection for PoC molecular diagnostics. Their disposable microfluidic cassette has numerous amplification reactors with silica-based nucleic acid isolation membranes at their inlets. By decoupling the sample volume from the reaction volume, this design allows great sensitivity with relatively large sample volumes. The isolation membrane captures nucleic acids that can be used as templates in RT-LAMP amplification without elution, simplifying flow management. Their microfluidic cassette can be incubated with or without battery-power. A simple, thermally insulated portable cup heated by an exothermic process was employed for chip-based isothermal amplification without power. Raw saliva samples spiked with various ZIKV concentrations were used to test the PoC diagnostic system. The results indicated that electricity-free PoC diagnostic device could detect 5 PFU of ZIKV per saliva sample in 40 min. Without laboratory facilities, expensive equipment, or well-trained staff, this PoC diagnostic technique is similar to the benchtop assay.

Benzine and colleagues detail a molecular diagnostic approach that can reliably detect EBOV in blood samples taken at the PoC.[80] This test relies on RT-LAMP, which is a method for amplifying the EBOV glycoprotein gene. PCR tubes with a volume of 0.2 ml were used to lyophilize the final reaction formulations. A battery-operated isothermal apparatus was used to conduct the RT-LAMP reactions. The RT-LAMP assay had a limit of detection of 2.8×10^2 PFU per test for EBOV-Kikwit and 1×10^3 PFU per test for EBOV-Makona within 40 minutes, respectively. Since no nonspecific amplification was observed in blood samples spiked with closely similar viruses and other infections, it was determined that this assay was specific for the detection of EBOV. On the basis of these findings, this diagnostic test is suitable for application in the PoC setting, where it can detect EBOV in blood samples with a high degree of specificity within 40 minutes.

12.3.4 Gastrointestinal and Foodborne Viruses

Around the world, gastrointestinal and foodborne viruses still create many intestinal disorders in humans, hence generating a great health and economic burden.[81] Estimates from the WHO indicate that over two billion people die each year from diarrhoea or sickness brought on by tainted food.[82] Interactions among foodborne bacteria, hosts, food, and the environment determine the occurrence and relevance of various foodborne disorders.[83] Viruses including norovirus, rotavirus, hepatitis A virus (HAV), and enteric adenoviruses have been reported to cause viral foodborne diseases.[84] Common signs of foodborne disease are stomach discomfort, diarrhoea, vomiting, nausea, fever, respiratory problems, and even death in extreme situations.[85] Most of the outbreaks have been attributed to vegetables, fruits, dairy products, shellfish, ground meat, and poultry.[86] Detecting these viruses is thus crucial since they endanger public health significantly. Traditional techniques for identifying food pathogens consist mostly of plate separation, chemical analysis, and immunoassay.[87] These techniques may have drawbacks, including laborious procedures, protracted detection cycle, high expense, and high standards for a professional level of operators.[88] Thus, it is absolutely necessary to create low-cost, quick, simple, sensitive, and easy ways to identify harmful bacteria in complicated food samples. In recent years, PoC technology has become a fast evolving foodborne pathogen detection technique with benefits including easy operation, quick operation, portability, and automation.[89]

The most common viral cause of acute gastroenteritis is human norovirus (HuNoV). Rapid diagnosis of HuNoVs has been significantly aided by PoC based on lateral flow immunochromatography (LIFC). Traditional LIFC, on the other hand, has low sensitivity and a lack of quantification, which limits its usefulness. Gao *et al.* therefore created a quick and precise method combining LFIC with immunomagnetic enrichment (IM) to find genogroup II (GII) HuNoVs in faecal samples.[90] Amino-functionalized magnetic beads (MBs) and carboxylated quantum dots (QDs) were connected at a mass ratio of 4:10 before production of immunofluorescent nanomagnetic microspheres and to obtain HuNoV enrichment in IM and fluorescent signals in LFIC. QDs–MBs were then coupled with anti-HuNoV monoclonal antibody. Under ideal conditions, the limit of detection was 1.56×10^4 copies per ml and the quantitative detection range was 1.56×10^4–1×10^6 copies per ml. Though the prevalent HuNoV genotypes GII.2, GII.3, GII.4, and GII.17 could be found, there was no cross-reaction with other enteric viruses, including rotavirus, astrovirus, enterovirus, and sapovirus. Among 87 faecal samples, IM–LFIC and RT-qPCR produced a good degree of concordance (kappa = 0.799). This implied that the approach is quick and sensitive, so it may be useful for point-of-care testing in the future.

Ye *et al.* described a very straightforward paper-based test for the PoC identification of rotavirus A, one of the most prevalent viruses responsible for paediatric gastroenteritis[91] (Figure 12.4b). This paper-based test could be used to perform nucleic acid extraction in under 5 minutes, then to amplify the target sequence in 25 minutes; the outcome was apparent to the naked eye right after or quantifiable by the UV-visible absorbance. This inexpensive technique is simple to employ in a lab or at the PoC and does not call for additional equipment. Rotavirus A's detection limit was 1×10^3 copies per ml. Testing of 48 clinical stool samples yielded 100% sensitivity and specificity. Ultimately, the paper-based test met the key criteria for a PoC diagnostic tool and could be used for precision diagnosis, control, and disease prevention.

12.4 Challenges and Future Perspectives

12.4.1 Technical and Regulatory Challenges

PoC diagnostics have the potential to transform the way infectious diseases, particularly viral infections, are detected and managed. As demand for quick, on-site diagnostic solutions increases, PoC technologies provide a number of advantages, including reduced time to diagnosis, decentralization of healthcare, and the ability to deliver real-time results. However, several technical and regulatory barriers need to be overcome before PoC diagnostics can reach their full potential, especially for viral detection.

Most PoCT platforms have limited capabilities for multiplex detection, which restricts their use in diagnosing co-infections or analysing biomarker panels. Moreover, these platforms often lack advanced data interpretation tools, making it difficult to extract clinically relevant insights from complex biological signals. The interpretation of results by untrained users—especially when dealing with faint or ambiguous test lines—introduces variability and reduces diagnostic confidence.[92] This variability can significantly affect the accuracy of diagnoses, particularly in non-laboratory settings.

PoCT devices are classified as medical devices, which means they must adhere to regulatory standards. Obtaining approval and ensuring compliance with these regulations can be a complicated and time-consuming process.[93] Currently, clear regulatory standards for PoCT are still lacking. PoCT is primarily guided by laboratory standards, but there is an urgent need for a regulatory framework that facilitates access to and reimbursement for these technologies.[94] The exclusion of certain tests from Essential Diagnostic Lists creates ambiguity regarding who is responsible for covering the costs, which negatively affects access to testing. Furthermore, policies for PoCT should address all stages, from test procurement and approval to the registration of results for surveillance purposes. In the post-pandemic era, funding for PoCT may be limited, requiring strategic deployment of testing strategies to ensure access for those in need. To establish whether differential diagnosis should be a public health priority, it is essential to demonstrate the value and cost-effectiveness of these testing strategies, highlighting the potential savings and reduction in future disease burden on the healthcare system and society.[95]

In addition, there is a growing need for real-time data processing and automated error detection systems to improve the accuracy and reliability of results. Regulatory frameworks must also evolve rapidly to accommodate the fast-paced advancements in PoCT technology, ensuring that quality, reliability, and safety are maintained. Addressing these technical and regulatory issues is essential to enable broader and more effective integration of PoCT into global healthcare systems.[96]

12.4.2 Future Trends in PoC Diagnostics

Laboratory-based diagnostic methods are well-established and continue to serve as the gold standard for virus detection. However, commercial PoCT technologies are rapidly advancing and are expected to emerge as a promising alternative for viral detection, offering convenience and accessibility outside traditional laboratory settings. The ability to detect and quantify whole viruses or virus-specific and host-derived biomarkers plays a critical role in differential diagnosis and immune monitoring, ultimately supporting more effective disease management. Thanks to recent advances in synthetic biology, signal amplification strategies, and smart materials, biosensing technologies are rapidly evolving into powerful tools for decentralized point-of-care testing. These innovations offer strong analytical performance and hold particular promise for improving viral diagnostics in low-resource settings, where access to conventional laboratory infrastructure is limited.

Advancements in materials science and nanotechnology have paved the way for the development of highly effective biosensors for pandemic control.

Among these, conjugated polymers have emerged as highly promising candidates due to their unique molecular structures, flexibility, and high sensitivity for virus detection.[97] Colourimetric biosensors, in particular, stand out for their ability to detect viruses through simple, visible colour changes, making them accessible, low-cost, and easy to use with tools like smartphones. These sensors offer excellent sensitivity and specificity, making them invaluable for both medical professionals and researchers.[98] In the future, new materials like colourimetric paper will further enhance the performance of these biosensors, opening up new possibilities for rapid and accurate virus detection.[99] On the other hand, various electrochemical biosensing techniques have been developed to achieve high detection and quantification limits for viruses, suitable for real-world applications.[100] Many of these biosensors are built using compact potentiostats, which are portable, battery-operated, and capable of wirelessly connecting to smartphones *via* apps that can interpret electrochemical results. Micro-nanofabrication and 3D printing techniques have played a key role in miniaturizing and enhancing the portability of these devices. For example, compact discs have been repurposed into biopotential, electrochemical, stretchable, and flexible wearable sensors, which can communicate *via* Bluetooth® with smartphones, offering a practical solution for virus detection on the go.[15,18] Finally, the fusion of the Internet of Things (IoT) and artificial intelligence (AI) technologies will redefine the future of PoC diagnostics. IoT connectivity will enable devices to transmit real-time data wirelessly, while AI algorithms will process and interpret this data, facilitating predictive diagnostics, personalized treatment, and remote patient monitoring.[101] Together, IoT and AI will transform PoC devices into fully integrated, intelligent systems capable of continuous health monitoring and supporting advanced, real-time clinical decision-making.

As PoC diagnostics continue to evolve, they will become more accessible, personalized, and efficient, enhancing healthcare delivery and disease management. The potential for a patient-centred, proactive, data-driven healthcare revolution is increasingly evident, particularly through the integration of optical biosensors with emerging technologies like IoT and AI. Portable optical biosensors will play a pivotal role in shaping the future of healthcare, leading to significant advancements in diagnosis and continuous monitoring, and ushering in a new era of medical care.

12.4.3 Impact of PoC Diagnostics on Global Healthcare

The global expansion of molecular PoCT is rapidly reshaping healthcare delivery, particularly in settings with limited access to centralized laboratories. As the potential of PoCT to deliver rapid, accurate diagnostics at or near the site of patient care becomes increasingly recognized, international efforts have intensified to integrate these technologies into both high-resource and resource-constrained healthcare systems.[102] Collaboration between researchers, engineers, and healthcare professionals is essential to further enhance the selectivity, sensitivity, and specificity of these devices. It is expected that these biosensors will become indispensable tools for proactive health monitoring and on-the-spot disease diagnosis, both in clinical settings and for individual use. Continuous innovation in materials and technologies will further propel this transformation, fostering broader accessibility and enhanced patient outcomes globally.

Future PoC diagnostics will focus on making these technologies more affordable, easier to use, and compliant with necessary regulations, ensuring that they can be deployed widely and fairly. While pandemics and epidemics continue to pose significant challenges, particularly in underserved areas, ensuring the equitable distribution of PoC diagnostics in regions with limited healthcare access will be pivotal in bridging healthcare disparities. As PoC technologies evolve, they have the potential to significantly strengthen healthcare systems worldwide. By providing rapid diagnostics, improving global surveillance, and enhancing the management of viral infections both at the individual and community levels, these advancements could have a profound effect on public health worldwide.

12.5 Conclusion

This chapter has reviewed the evolving field of sensor technologies for viral detection, highlighting their working principles, clinical relevance, and possible impact on diagnostic procedures. The technological viability of portable, rapid virus testing devices and their wider consequences for clinical practice and public health have been discussed. From numerous platforms providing quick and reliable diagnosis, PoC biosensor technology has evolved. Optical PoC biosensors, employing fluorescence, colorimetric, or plasmonic detection, offer exceptional sensitivity and are optimal for real-time viral load monitoring. Moreover, portable and low-cost electrochemical PoC biosensors provide precise detection by means of measurable electrical signals produced by biological interactions. Microfluidic-based biosensors, on the other hand, combine sample processing and analysis into miniature chips, hence enabling multiplexed detection with low sample volumes. The combination of biosensors, microfluidic systems, and miniaturized electronics indicates that user-friendly, selective, and sensitive diagnostic tools are available. Especially in circumstances when time, infrastructure, and access are constrained, these technologies offer a viable alternative for conventional laboratory-based assays. In the clinical detection of many viral infections, PoC diagnostic tools are increasingly important. Rapid testing for respiratory viruses, including influenza and SARS-CoV-2, facilitates prompt isolation and treatment. Particularly in resource-limited settings, these devices enable early diagnosis and continuous monitoring for sexually transmitted and blood-borne viruses like HIV and hepatitis B/C. During outbreak scenarios, vector-borne and emerging viruses such as Zika, dengue, and chikungunya can be detected by exploiting the portability of PoC platforms.

Furthermore, foodborne and gastrointestinal viruses, like norovirus and rotavirus can be rapidly identified to prevent further spread, particularly in healthcare and community settings. Such systems, deployed in PoC surveillance, exhibit potential in providing timely information that can assist therapeutic decisions. Furthermore, the emergence of new viruses and the mutation of existing ones emphasize the need for flexibility and modularity in sensor design. Preparedness for future epidemics will require systems that can quickly change to find new viral targets. At the same time, real-time disease monitoring and customized treatment will be made possible by the combination of sensor data with AI-driven analytics, mobile health platforms, and cloud-based systems. Several elements influence the successful transformation of technological developments from laboratory to practical applications: (i) clinical validation, (ii) production scalability, (iii) user education, and (iv) data integration. Complicated and often behind the pace of innovation, the regulatory approval procedures highlight the need for cooperative systems that can accelerate assessment without compromising safety or dependability. The field is located at a crossroads of technical possibility and worldwide need. The results for PoC systems indicate the possibilities and challenges involved in building next-generation diagnostic tools. In summary, new approaches in responsive viral diagnostics will be produced by coordinated efforts among scientific, clinical, and regulatory domains thus enabling continuous development and efficient use of PoC biosensor technologies.

Acknowledgements

Fatih Inci gratefully acknowledges the support from TÜBİTAK 2232 International Fellowship for Outstanding Researchers (Project No. 118C254) and the Turkish Academy of Sciences Outstanding Young Scientists Award Program (TÜBA GEBİP). This work was also supported by the Young Scientist Awards Program (BAGEP) award from the Science Academy.

References

1. L. Singh, H. G. Kruger, G. E. M. Maguire, T. Govender and R. Parboosing, The role of nanotechnology in the treatment of viral infections, *Ther. Adv. Infect. Dis.*, 2017, **4**, 105–131.
2. F. S. R. R. Teles, Biosensors and rapid diagnostic tests on the frontier between analytical and clinical chemistry for biomolecular diagnosis of dengue disease: a review, *Anal. Chim. Acta*, 2011, **687**, 28–42.
3. H. Zhu, Z. Fohlerová, J. Pekárek, E. Basova and P. Neužil, Recent advances in lab-on-a-chip technologies for viral diagnosis, *Biosens. Bioelectron.*, 2020, **153**, 112041.
4. R. McNerney, Diagnostics for developing countries, *Diagnostics*, 2015, **5**, 200–209.

5. S. Choi, Powering point-of-care diagnostic devices, *Biotechnol. Adv.*, 2016, **34**, 321–330.
6. A. M. Vicente, W. Ballensiefen and J.-I. Jönsson, How personalised medicine will transform healthcare by 2030: the ICPerMed vision, *J. Transl. Med.*, 2020, **18**, 180.
7. J. Choi, R. Ghaffari, L. B. Baker and J. A. Rogers, Skin-interfaced systems for sweat collection and analytics, *Sci. Adv.*, 2025, **4**, eaar3921.
8. A. J. Steckl and P. Ray, Stress biomarkers in biological fluids and their point-of-use detection, *ACS Sens.*, 2018, **3**, 2025–2044.
9. K. E. Kaczor-Urbanowicz, F. Wei, S. L. Rao, J. Kim, H. Shin, J. Cheng, M. Tu, D. T. W. Wong and Y. Kim, Clinical validity of saliva and novel technology for cancer detection, *Biochim. Biophys. Acta, Rev. Cancer*, 2019, **1872**, 49–59.
10. M. Naseri, Z. M. Ziora, G. P. Simon and W. Batchelor, ASSURED-compliant point-of-care diagnostics for the detection of human viral infections, *Rev. Med. Virol.*, 2022, **32**, e2263.
11. R. Reid, B. Chatterjee, S. J. Das, S. Ghosh and T. K. Sharma, Application of aptamers as molecular recognition elements in lateral flow assays, *Anal. Biochem.*, 2020, **593**, 113574.
12. F.-T.-Z. Khanam, L. A. Chahl, J. S. Chahl, A. Al-Naji, A. G. Perera, D. Wang, Y. H. Lee, T. T. Ogunwa, S. Teague and T. X. B. Nguyen, Noncontact sensing of contagion, *J. Imaging*, 2021, **7**, 28.
13. S. Dorlikar, M. Gautam, N. Vats, M. Shakir, A. K. Kaushik and P. K. Sharma, State-of-the-art biosensor technologies for point-of-care applications, in *Next-Generation Smart Biosensing*, ed. K. Reza Khondakar, and A. K. B. T Kaushik, Academic Press, 2024, pp. 209–227.
14. J. Wang, Z. Xu and D. G. Kotsifaki, Plasmonic and metamaterial biosensors: a game-changer for virus detection, *Sens. Diagn.*, 2023, **2**, 600–619.
15. G. Marrazza, M. Ramalingam, A. Jaisankar, L. Cheng, G. Selvolini and I. A. Vitale, Advancements and emerging technologies in biosensors for rapid and accurate virus detection, *TrAC, Trends Anal. Chem.*, 2024, **172**, 117609.
16. S. P. Mohanty and E. Kougianos, Biosensors: a tutorial review, *IEEE Potentials*, 2006, **25**, 35–40.
17. R. W. Raut, H. S. Naik, P. M. Sah, P. Golińska and A. Gade, A comparative analysis of optical biosensors for rapid detection of SARS-CoV-2 and Influenza, *Biotechnol. Bioeng.*, DOI: 10.1002/bit.28956
18. O. Tokel, F. Inci and U. Demirci, Advances in plasmonic technologies for point of care applications, *Chem. Rev.*, 2014, **114**, 5728–5752.
19. Z. Li, L. Leustean, F. Inci, M. Zheng, U. Demirci and S. Wang, Plasmonic-based platforms for diagnosis of infectious diseases at the point-of-care, *Biotechnol. Adv.*, 2019, **37**, 107440.
20. Y.-F. Chang, W.-H. Wang, Y.-W. Hong, R.-Y. Yuan, K.-H. Chen, Y.-W. Huang, P.-L. Lu, Y.-H. Chen, Y.-M. A. Chen and L.-C. Su, Simple strategy for rapid and sensitive detection of avian influenza A H7N9 virus based on intensity-modulated SPR biosensor and new generated antibody, *Anal. Chem.*, 2018, **90**, 1861–1869.

21. L. Shi, Q. Sun, J. He, H. Xu, C. Liu, C. Zhao, Y. Xu, C. Wu, J. Xiang and D. Gu, Development of SPR biosensor for simultaneous detection of multiplex respiratory viruses, *Biomed. Mater. Eng.*, 2015, **26**, S2207–S2216.
22. W. Diao, M. Tang, S. Ding, X. Li, W. Cheng, F. Mo, X. Yan, H. Ma and Y. Yan, Highly sensitive surface plasmon resonance biosensor for the detection of HIV-related DNA based on dynamic and structural DNA nanodevices, *Biosens. Bioelectron.*, 2018, **100**, 228–234.
23. D. J. Schofield and N. J. Dimmock, Determination of affinities of a panel of IgGs and Fabs for whole enveloped (influenza A) virions using surface plasmon resonance, *J. Virol. Methods*, 1996, **62**, 33–42.
24. P. K. Sharma, J. S. Kumar, V. V Singh, U. Biswas, S. S. Sarkar, S. I. Alam, P. K. Dash, M. Boopathi, K. Ganesan and R. Jain, Surface plasmon resonance sensing of Ebola virus: a biological threat, *Anal. Bioanal. Chem.*, 2020, **412**, 4101–4112.
25. R. Sun, Y. Zhou, Y. Fang, Y. Qin, Y. Zheng and L. Jiang, DNA aptamer-linked sandwich structure enhanced SPRi sensor for rapid, sensitive, and quantitative detection of SARS-CoV-2 spike protein, *Anal. Bioanal. Chem.*, 2024, **416**, 1667–1677.
26. R. Kumar, S. Singh, H. Bouandas and J. Alam, Detection of COVID-19 using surface plasmon resonance sensor for sensitivity enhancement: theoretical analysis, *Plasmonics*, 2025, **20**(9), 7049–7059.
27. F. Zang, Z. Su, L. Zhou, K. Konduru, G. Kaplan and S. Y. Chou, Ultrasensitive Ebola virus antigen sensing via 3D nanoantenna arrays, *Adv. Mater.*, 2019, **31**, 1902331.
28. P. P. Austin Suthanthiraraj and A. K. Sen, Localized surface plasmon resonance (LSPR) biosensor based on thermally annealed silver nanostructures with on-chip blood-plasma separation for the detection of dengue non-structural protein NS1 antigen, *Biosens. Bioelectron.*, 2019, **132**, 38–46.
29. F. Inci, O. Tokel, S. Wang, U. A. Gurkan, S. Tasoglu, D. R. Kuritzkes and U. Demirci, Nanoplasmonic quantitative detection of intact viruses from unprocessed whole blood, *ACS Nano*, 2013, **7**, 4733–4745.
30. J. Lukose, A. K. Barik, N. Mithun, S. Pavithran, S. D. George, V. M. Murukeshan and S. Chidangil, Raman spectroscopy for viral diagnostics, *Biophys. Rev.*, 2023, **15**, 199–221.
31. J. Langer, D. Jimenez de Aberasturi, J. Aizpurua, R. A. Alvarez-Puebla, B. Auguié, J. J. Baumberg, G. C. Bazan, S. E. J. Bell, A. Boisen and A. G. Brolo, Present and future of surface-enhanced Raman scattering, *ACS Nano*, 2019, **14**, 28–117.
32. B. Cao, P. Lin, Y. Wang, W. Yang, L. Ren, Z. Ge, H. Sui, Y. Gao, M. Liu, G. Bei, S. Zhou, Q. Zhou and F. Cui, Intelligent and ligand-based surface-enhanced Raman spectroscopy biosensors for the instant screening of viral respiratory infections, *Biosens. Bioelectron. X*, 2024, **20**, 100527.
33. Y. Yang, Y. Peng, C. Lin, L. Long, J. Hu, J. He, H. Zeng, Z. Huang, Z.-Y. Li and M. Tanemura, Human ACE2-functionalized gold "virus-trap" nanostructures for accurate capture of SARS-CoV-2 and single-virus SERS detection, *Nano-micro Lett.*, 2021, **13**, 1–13.

34. Y. Peng, J. Liu, X. Zhao, H. Sun, Z. Liu, X. Jing and L. Meng, A nanozyme-catalyzed nasopharyngeal swab sensor for ultra-fast colorimetric detection and highly sensitive Raman spectroscopic qualitative analysis of influenza virus, *Biosens. Bioelectron.*, 2025, **277**, 117280.
35. S. Shanmukh, L. Jones, J. Driskell, Y. Zhao, R. Dluhy and R. A. Tripp, Rapid and sensitive detection of respiratory virus molecular signatures using a silver nanorod array SERS substrate, *Nano Lett.*, 2006, **6**, 2630–2636.
36. C. Fenzl, T. Hirsch and O. S. Wolfbeis, Photonic crystals for chemical sensing and biosensing, *Angew. Chem., Int. Ed.*, 2014, **53**, 3318–3335.
37. H. Inan, M. Poyraz, F. Inci, M. A. Lifson, M. Baday, B. T. Cunningham and U. Demirci, Photonic crystals: emerging biosensors and their promise for point-of-care applications, *Chem. Soc. Rev.*, 2017, **46**, 366–388.
38. H. Shafiee, E. A. Lidstone, M. Jahangir, F. Inci, E. Hanhauser, T. J. Henrich, D. R. Kuritzkes, B. T. Cunningham and U. Demirci, Nanostructured optical photonic crystal biosensor for HIV viral load measurement, *Sci. Rep.*, 2014, **4**, 4116.
39. E. Grueso, R. M. Giráldez-Pérez and R. Prado-Gotor, Chapter 15 – colorimetric and fluorescence, in *Handbook of Biomolecules*, ed. C. Verma, and D. K. Verma, Elsevier, 2023, pp. 299–332.
40. Y. Pan, S. Zhan and F. Xia, Zeolitic imidazolate framework-based biosensor for detection of HIV-1 DNA, *Anal. Biochem.*, 2018, **546**, 5–9.
41. L. Jiang, J. Luo, W. Dong, C. Wang, W. Jin, Y. Xia, H. Wang, H. Ding, L. Jiang and H. He, Development and evaluation of a polydiacetylene based biosensor for the detection of H5 influenza virus, *J. Virol. Methods*, 2015, **219**, 38–45.
42. J. Baranwal, B. Barse, G. Gatto, G. Broncova and A. Kumar, Electrochemical Sensors and Their Applications: A Review, *Chemosensors*, 2022, **10**(9), 363.
43. C. Wang, M. Liu, Z. Wang, S. Li, Y. Deng and N. He, Point-of-care diagnostics for infectious diseases: from methods to devices, *Nano Today*, 2021, **37**, 101092.
44. D. Grieshaber, R. MacKenzie, J. Vörös and E. Reimhult, Electrochemical biosensors - sensor principles and architectures, *Sensors*, 2008, **8**, 1400–1458.
45. A. Erdem, H. Senturk, E. Yildiz and M. Maral, Amperometric immunosensor developed for sensitive detection of SARS-CoV-2 spike S1 protein in combined with portable device, *Talanta*, 2022, **244**, 123422.
46. S. Yunus, A. M. Jonas and B. Lakard, Potentiometric biosensors, in *Encyclopedia of Biophysics*, ed. G. C. K. Roberts, Springer Berlin Heidelberg, Berlin, Heidelberg, 2013, pp. 1941–1946.
47. V. Ricotta, Y. Yu, N. Clayton, Y.-C. Chuang, Y. Wang, S. Mueller, K. Levon, M. Simon and M. Rafailovich, A chip-based potentiometric sensor for a Zika virus diagnostic using 3D surface molecular imprinting, *Analyst*, 2019, **144**, 4266–4280.

48. M. F. Abd Muain, A. S. Amir Hamzah, S. L. Chia, K. Yusoff, H. N. Lim, I. Shinya and A. Ahmad Tajudin, Voltammetric-based immunosensing of Newcastle disease virus on polyethylene glycol-containing self-assembled monolayer modified gold electrode, *Anal. Biochem.*, 2025, **697**, 115700.
49. N. Manring, M. M. N. Ahmed, N. Tenhoff, J. L. Smeltz and P. Pathirathna, Recent advances in electrochemical tools for virus detection, *Anal. Chem.*, 2022, **94**, 7149–7157.
50. H. Shafiee, M. Jahangir, F. Inci, S. Wang, R. B. M. Willenbrecht, F. F. Giguel, A. M. N. Tsibris, D. R. Kuritzkes and U. Demirci, Acute on-chip HIV detection through label-free electrical sensing of viral nano-lysate, *Small*, 2013, **9**, 2553–2563.
51. M. A. Abdul Ghani, A. N. Nordin, M. Zulhairee, A. Che Mohamad Nor, M. Shihabuddin Ahmad Noorden, M. K. Muhamad Atan, R. Ab Rahim and Z. Mohd Zain, Portable electrochemical biosensors based on microcontrollers for detection of viruses: a review, *Biosensors*, 2022, **12**(8), 666.
52. G. Alhamid, H. Tombuloglu and E. Al-Suhaimi, Development of loop-mediated isothermal amplification (LAMP) assays using five primers reduces the false-positive rate in COVID-19 diagnosis, *Sci. Rep.*, 2023, **13**, 5066.
53. A. Ganguli, A. Ornob, H. Yu, G. L. Damhorst, W. Chen, F. Sun, A. Bhuiya, B. T. Cunningham and R. Bashir, Hands-free smartphone-based diagnostics for simultaneous detection of Zika, Chikungunya, and Dengue at point-of-care, *Biomed. Microdevices*, 2017, **19**, 73.
54. B. S. Batule, Y. Seok and M.-G. Kim, Paper-based nucleic acid testing system for simple and early diagnosis of mosquito-borne RNA viruses from human serum, *Biosens. Bioelectron.*, 2020, **151**, 111998.
55. G. Theillet, G. Grard, M. Galla, C. Maisse, M. Enguehard, M. Cresson, P. Dalbon, I. L. Leparc-Goffart and F. Bedin, Detection of chikungunya virus-specific IgM on laser-cut paper-based device using pseudo-particles as capture antigen, *J. Med. Virol.*, 2019, **91**, 899–910.
56. C. Wang, X. Yang, B. Gu, H. Liu, Z. Zhou, L. Shi, X. Cheng and S. Wang, Sensitive and simultaneous detection of SARS-CoV-2-specific IgM/IgG using lateral flow immunoassay based on dual-mode quantum dot nanobeads, *Anal. Chem.*, 2020, **92**, 15542–15549.
57. M. A. Black, G. Shen, X. Feng, W. F. Garcia Beltran, Y. Feng, V. Vasudevaraja, D. Allison, L. H. Lin, T. Gindin, M. Astudillo, D. Yang, M. Murali, A. J. Iafrate, G. Jour, P. Cotzia and M. Snuderl, Analytical performance of lateral flow immunoassay for SARS-CoV-2 exposure screening on venous and capillary blood samples, *J. Immunol. Methods*, 2021, **489**, 112909.
58. K. Lakshmanan and B. M. Liu, Impact of point-of-care testing on diagnosis, treatment, and surveillance of vaccine-preventable viral infections, *Diagnostics*, 2025, **15**, 123.

59. N. J. Maclachlan and E. J. Dubovi, *Fenner's Veterinary Virology*, Academic press, 2010.
60. Z. Zhang, P. Ma, R. Ahmed, J. Wang, D. Akin, F. Soto, B. F. Liu, P. Li and U. Demirci, Advanced point-of-care testing technologies for human acute respiratory virus detection, *Adv. Mater.*, 2022, **34**, 2103646.
61. M. Xiao, F. Tian, X. Liu, Q. Zhou, J. Pan, Z. Luo, M. Yang and C. Yi, Virus detection: from state-of-the-art laboratories to smartphone-based point-of-care testing, *Adv. Sci.*, 2022, **9**, 2105904.
62. S. Basharat and J. Horton, An overview of emerging point-of-care tests for differentiating bacterial and viral infections, *Can. J. Heal. Technol*, 2021, **1**(12), DOI: 10.51731/cjht.2021.222.
63. World Health Organization, in *WHO guidelines for the global surveillance of severe acute respiratory syndrome (SARS): Updated recommendations, October 2004*, 2004.
64. World Health Organization, Human infection with new influenza A (H1N1) virus: clinical observations from Mexico and other affected countries, May 2009, *Weekly Epidemiological Record*, 2009, **84**(21), 185–189.
65. S. H. Lee, Y. H. Baek, Y.-H. Kim, Y.-K. Choi, M.-S. Song and J.-Y. Ahn, One-pot reverse transcriptional loop-mediated isothermal amplification (RT-LAMP) for detecting MERS-CoV, *Front. Microbiol.*, 2017, **7**, 2166.
66. C. Huang, Y. Wang, X. Li, L. Ren, J. Zhao, Y. Hu, L. Zhang, G. Fan, J. Xu and X. Gu, Clinical features of patients infected with 2019 novel coronavirus in Wuhan, China, *Lancet*, 2020, **395**, 497–506.
67. N. J. Brendish, H. F. Schiff and T. W. Clark, Point-of-care testing for respiratory viruses in adults: the current landscape and future potential, *J. Infect.*, 2015, **71**, 501–510.
68. Y. Seok, M. G. Mauk, R. Li and C. Qian, Trends of respiratory virus detection in point-of-care testing: a review, *Anal. Chim. Acta*, 2023, **1264**, 341283.
69. E. Xiong, L. Jiang, T. Tian, M. Hu, H. Yue, M. Huang, W. Lin, Y. Jiang, D. Zhu and X. Zhou, Simultaneous dual-gene diagnosis of SARS-CoV-2 based on CRISPR/Cas9-mediated lateral flow assay, *Angew. Chem., Int. Ed.*, 2021, **60**, 5307–5315.
70. M. D. Pauly and L. Ganova-Raeva, Point-of-care testing for hepatitis viruses: a growing need, *Life*, 2023, **13**(12), 2271.
71. J. D. Tucker, C. H. Bien and R. W. Peeling, Point-of-care testing for sexually transmitted infections: recent advances and implications for disease control, *Curr. Opin. Infect. Dis.*
72. S. Vargas, G. Calvo, J. Qquellon, F. Vasquez, K. Blondeel, R. Ballard and I. Toskin, Point-of-care testing for sexually transmitted infections in low-resource settings, *Clin. Microbiol. Infect.*, 2022, **28**, 946–951.
73. A. S. Baber, B. Suganthan and R. P. Ramasamy, Current advances in hepatitis C diagnostics, *J. Biol. Eng.*, 2024, **18**, 48.
74. R. Simmons, J. Plunkett, L. Cieply, S. Ijaz, M. Desai and S. Mandal, Blood-borne virus testing in emergency departments – a systematic review of seroprevalence, feasibility, acceptability and linkage to care, *HIV Med*, 2023, **24**, 6–26.

75. E. Aronoff-Spencer, A. G. Venkatesh, A. Sun, H. Brickner, D. Looney and D. A. Hall, Detection of Hepatitis C core antibody by dual-affinity yeast chimera and smartphone-based electrochemical sensing, *Biosens. Bioelectron.*, 2016, **86**, 690–696.
76. P. C. Adamson, M. J. Loeffelholz and J. D. Klausner, Point-of-care testing for sexually transmitted infections: a review of recent developments, *Arch. Pathol. Lab. Med.*, 2020, **144**, 1344–1351.
77. S. H. de Cortina, C. C. Bristow, D. J. Davey and J. D. Klausner, A systematic review of point of care testing for Chlamydia trachomatis, Neisseria gonorrhoeae, and Trichomonas vaginalis, *Infect. Dis. Obstet. Gynecol.*, 2016, **2016**(1), 4386127.
78. S. B. Doernberg, L. Komarow, T. T. T. Tran, Z. Sund, M. W. Pandori, D. Jensen, E. L. Tsalik, C. D. Deal, H. F. Chambers and V. G. Fowler Jr., Simultaneous evaluation of diagnostic assays for pharyngeal and rectal Neisseria gonorrhoeae and Chlamydia trachomatis using a master protocol, *Clin. Infect. Dis.*, 2020, **71**, 2314–2322.
79. M. G. Mauk, J. Song, H. H. Bau and C. Liu, Point-of-care molecular test for Zika infection, *Clin. Lab. Int.*, 2017, **41**, 25.
80. J. W. Benzine, K. M. Brown, K. N. Agans, R. Godiska, C. E. Mire, K. Gowda, B. Converse, T. W. Geisbert, D. A. Mead and Y. Chander, Molecular diagnostic field test for point-of-care detection of Ebola virus directly from blood, *J. Infect. Dis.*, 2016, **214**, S234–S242.
81. R. Akter, M. H. Rahman, T. Bhattacharya, D. Kaushik, V. Mittal, J. Parashar, K. Kumar, M. T. Kabir and P. Tagde, Novel coronavirus pathogen in humans and animals: an overview on its social impact, economic impact, and potential treatments, *Environ. Sci. Pollut. Res.*, 2021, 1–19.
82. D. Belina, Y. Hailu, T. Gobena, T. Hald and P. M. K. Njage, Prevalence and epidemiological distribution of selected foodborne pathogens in human and different environmental samples in Ethiopia: a systematic review and meta-analysis, *One Heal. outlook*, 2021, **3**, 19.
83. A. Saravanan, P. S. Kumar, R. V Hemavathy, S. Jeevanantham, R. Kamalesh, S. Sneha and P. R. Yaashikaa, Methods of detection of food-borne pathogens: a review, *Environ. Chem. Lett.*, 2021, **19**, 189–207.
84. H. Bozkurt, K.-Y. Phan-Thien, F. van Ogtrop, T. Bell and R. McConchie, Outbreaks, occurrence, and control of norovirus and hepatitis a virus contamination in berries: a review, *Crit. Rev. Food Sci. Nutr.*, 2021, **61**, 116–138.
85. P. Myintzaw, A. K. Jaiswal and S. Jaiswal, A review on campylobacteriosis associated with poultry meat consumption, *Food Rev. Int.*, 2023, **39**, 2107–2121.
86. A. Leon Madrazo and M. R. Segura Campos, Review of antimicrobial peptides as promoters of food safety: Limitations and possibilities within the food industry, *J. Food Saf.*, 2020, **40**, e12854.

87. X. Han, Y. Liu, J. Yin, M. Yue and Y. Mu, Microfluidic devices for multiplexed detection of foodborne pathogens, *Food Res. Int.*, 2021, **143**, 110246.
88. G. Vidyadharani, H. K. Vijaya Bhavadharani, P. Sathishnath, S. Ramanathan, P. Sariga, A. Sandhya, S. Subikshaa and S. Sugumar, Present and pioneer methods of early detection of food borne pathogens, *J. Food Sci. Technol.*, 2022, **59**, 2087–2107.
89. Y. Xu, T. Wang, Z. Chen, L. Jin, Z. Wu, J. Yan, X. Zhao, L. Cai, Y. Deng and Y. Guo, The point-of-care-testing of nucleic acids by chip, cartridge and paper sensors, *Chin. Chem. Lett.*, 2021, **32**, 3675–3686.
90. J. Gao, L. Xue, Y. Li, W. Cai, S. Miao, L. Meng, S. Ren, J. Zhang, J. Wang, S. Wu, Q. Ye, L. Chen, Q. Gu and Q. Wu, Rapid and sensitive lateral flow biosensor for the detection of GII human norovirus based on immunofluorescent nanomagnetic microspheres, *J. Med. Virol.*, 2024, **96**, e29487.
91. X. Ye, J. Xu, L. Lu, X. Li, X. Fang and J. Kong, Equipment-free nucleic acid extraction and amplification on a simple paper disc for point-of-care diagnosis of rotavirus A, *Anal. Chim. Acta*, 2018, **1018**, 78–85.
92. P. Mukadi, P. Gillet, B. Barbé, J. Luamba, A. Lukuka, J. Likwela, D. Mumba, J.-J. Muyembe, P. Lutumba and J. Jacobs, SMS photograph-based external quality assessment of reading and interpretation of malaria rapid diagnostic tests in the Democratic Republic of the Congo, *Malar. J.*, 2015, **14**, 1–12.
93. A. R. Khan, W. L. Hussain, H. C. Shum and S. U. Hassan, Point-of-care testing: a critical analysis of the market and future trends, *Front. Lab a Chip Technol*, 2024, **3**, 1394752.
94. P. Mitra and P. Sharma, POCT in developing countries, *Ejifcc*, 2021, **32**, 195.
95. C. A. Alvarez-Moreno, E. S. A. de Araújo, E. Baumeister, K. A. Nogales Crespo, A. M. Kalergis, J. E. Muñoz Medina, P. Tsukayama and C. Ugarte-Gil, Differential diagnosis in the management of acute respiratory infections through point-of-care rapid testing in a post-pandemic scenario in latin america: special focus on COVID-19, influenza, and respiratory syncytial virus, *COVID*, 2024, **4**, 221–260.
96. G.-R. Han, A. Goncharov, M. Eryilmaz, S. Ye, B. Palanisamy, R. Ghosh, F. Lisi, E. Rogers, D. Guzman and D. Yigci, Machine learning in point-of-care testing: innovations, challenges, and opportunities, *Nat. Commun.*, 2025, **16**, 3165.
97. S. Malik, J. Singh, R. Goyat, Y. Saharan, V. Chaudhry, A. Umar, A. A. Ibrahim, S. Akbar, S. Ameen and S. Baskoutas, Nanomaterials-based biosensor and their applications: a review, *Heliyon*, 2023, **9**, e19929.
98. R. Liu, J. Li, B. J. Salena and Y. Li, Aptamer and DNAzyme based colorimetric biosensors for pathogen detection, *Angew. Chem., Int. Ed.*, 2025, **64**, e202418725.
99. Z. Geng, Y. Miao, G. Zhang and X. Liang, Colorimetric biosensor based on smartphone: State-of-art, *Sens. Actuators, A*, 2023, **349**, 114056.

100. M. S. Brown, L. Somma, M. Mendoza, Y. Noh, G. J. Mahler and A. Koh, Upcycling compact discs for flexible and stretchable bioelectronic applications, *Nat. Commun.*, 2022, **13**, 3727.
101. T. Moeinfard, E. Ghafar-Zadeh and S. Magierowski, CMOS point-of-care diagnostics technologies: recent advances and future prospects, *Micromachines*, 2024, **15**, 1320.
102. Y. Zu, H. Chang and Z. Cui, Molecular point-of-care testing technologies: current status and challenges, *Nexus*, 2025, **2**(2), 100059.

CHAPTER 13

Remote Patient Monitoring

MERVE ASENA ÖZBEK, MERVE ÇALIŞIR AND ERDOĞAN ÖZGÜR*

Department of Chemistry, Hacettepe University, Ankara, Turkey
*E-mails: erdoganozgur@hacettepe.edu.tr, mervecalisir@hacettepe.edu.tr, asenaozbek@hacettepe.edu.tr

13.1 Introduction

Remote patient monitoring (RPM) has grown more important over time, mainly because people are living longer and facing more health issues as a result. As technology keeps advancing, it's become clearer that traditional healthcare methods have some shortcomings, making RPM a more appealing option.[1] One key reason why RPM is preferred is its ability to manage diseases that limit a patient's ability to travel or involve contagious conditions. These systems, which aim to make patients more comfortable, also help by managing illnesses that require isolation and by easing the workload of healthcare staff and facilities.[2,3]

Although RPM systems were first developed to improve the quality of life for patients, their potential for virus detection has become more noticeable over time. The recent pandemic highlighted the importance of RPM once again. It made monitoring patients in quarantine much easier and helped slow down the spread of diseases, showing how crucial these systems are. RPM is also useful for tracking long-term symptoms that may appear after viral infections. For example, it helps with managing recovery for patients dealing with post-corona virus disease (COVID)-19 conditions, like long

Detection Science Series No. 32
Advances in Virus Detection with Sensors
Edited by Adil Denizli
© The Royal Society of Chemistry 2026
Published by the Royal Society of Chemistry, www.rsc.org

COVID symptoms.[4] In addition, the use of RPM has expanded not only for detecting viruses but also for monitoring other health issues like diabetes and heart diseases that might develop after viral infections.[5]

Before RPM technologies came into play, patient monitoring mostly relied on wired sensors. These sensors were usually large, limited in number in hospitals, and made it hard for patients to move around freely. As the population grew and the need for home healthcare services increased with longer lifespans, RPM systems started becoming more common. RPM works by using sensors to collect data on patients' physical health and sending that information wirelessly to healthcare professionals. The data gathered includes things like blood values, oxygen saturation, and blood pressure levels.[6–8] Because of this, development of better sensor technologies is essential for improving RPM systems. On top of that, communication technology must work smoothly to ensure the data from these wireless sensors is transmitted accurately.

This chapter will look at how sensor technology can be used in RPM for detecting viral infections. It highlights the role RPM systems play in spotting early symptoms in infected people and controlling the spread of infectious diseases. The use of RPM in diseases like COVID-19, influenza, Zika, and Ebola is discussed, as well as how it could be useful in future pandemics. Moreover, new technologies like nanotechnology, biosensors, artificial intelligence, and machine learning, which could make RPM systems more accurate, are examined. The goal here is to show how RPM contributes to public health improvements and helps in creating sustainable healthcare strategies.

13.2 Sensor Technologies in Remote Patient Monitoring

Sensor technology has significantly enhanced RPM systems by introducing a wide range of advantages that improve both efficiency and effectiveness. The unique characteristics of sensors align perfectly with the requirements of RPM, making them indispensable components in remote healthcare. One of their most notable benefits is their ability to collect data continuously and in real-time, surpassing many traditional diagnostic methods.[9] By recording various physiological parameters, these devices facilitate remote monitoring, helping to prevent potential health complications. In diseases where early intervention is crucial, the combination of sensors and RPM represents a highly promising approach.

A key priority in RPM is patient comfort, and sensor systems play a pivotal role in ensuring that patients receive maximum benefits from treatment without experiencing excessive fatigue or discomfort.[10] These user-friendly solutions make remote healthcare more accessible and manageable. Additionally, as sensors have long been recognized in the medical field for their precision and reliability, their integration into RPM contributes to reducing the economic burden on healthcare facilities.[11,12] They minimize the need for prolonged hospital stays while maintaining high-quality care.

With the integration of wireless connectivity, sensor systems emerge as an efficient and high-quality option for modern healthcare. Their application in RPM encompasses wearable, implantable, and contactless sensors, all of which enhance remote monitoring and improve patient outcomes.

13.2.1 Wearable Sensors

Wearable sensors have gained significant attention in recent years for their ability to provide real-time information about an individual's health status. Initially, these sensors were developed to track basic physiological parameters, such as calorie expenditure and heart rate. However, advancements in the field have led to their evolution into more sophisticated devices capable of detecting specific disease markers.[13–17] Wearable sensors have become particularly prominent in the continuous monitoring of cardiovascular diseases and blood glucose levels in diabetic patients, where frequent assessments are essential.[18–20]

The integration of wearable sensors into medical applications is closely linked to the development of biosensors. Biosensors are defined as sensor systems that incorporate biological recognition elements, such as enzymes, antibodies, cell receptors, or organelles. With their ability to detect a wide range of biomarkers in biological fluids, biosensors offer high adaptability and biocompatibility, making them ideal for use in wearable sensor technologies.[21] Their potential to replace painful and time-consuming laboratory analyses with real-time, non-invasive alternatives is a driving force behind their adoption in healthcare.

In RPM systems, the incorporation of wearable sensors provides substantial advantages, particularly in real-time tracking and early disease diagnosis, where timely intervention is critical. Data collected by these sensors are transmitted *via* wireless body sensor networks (BSNs) using short-range wireless communication protocols to smartphones, computers, and other processing units.[22,23] These data are then relayed to healthcare professionals *via* the internet, enabling continuous RPM and facilitating timely medical intervention.

Several key attributes are sought in wearable sensors, including ease of use, ergonomic design, and, most importantly, biocompatibility. Additionally, as biological fluids—particularly blood–are complex matrices, wearable sensors must feature highly sensitive and selective detection technologies to ensure precise analysis of specific biomarkers. It has been demonstrated that biosensors utilizing electrochemical, piezoelectric, and optical measurement principles can detect trace amounts of biomarkers with high precision.[24] The integration of smart designs into wearable sensor technology is expected to bring forth numerous solutions, particularly in the healthcare sector.

Despite their promising applications in RPM, wearable sensors also face several challenges. The commercialization of non-invasive biosensors for glucose monitoring, for instance, holds tremendous market potential but is hindered by challenges related to the sensitivity and reliability of biological recognition elements.[25] Another major issue is the lifespan of these sensors, as they are exposed to significantly higher levels of biofouling compared with traditional assays. Since highly sensitive measurements require well-maintained sensor surfaces, prolonged use often leads to surface degradation and decreased accuracy. Therefore, for wearable sensors to be effectively integrated into RPM, their robustness and reusability must be prioritized in the design phase.

13.2.2 Implantable Sensors

Implantable sensors can be highly beneficial in managing chronic diseases and enabling early diagnosis.[26] With advancements in sensor technologies, they can detect changes in a wide range of physiological parameters, making them very useful devices for RPM. Implantable sensors are basically microelectronic devices that are placed inside the body and can continuously measure a target parameter within the organism.[27] Because these sensors are implanted into the body, using biocompatible materials is one of the primary requirements.[28] In addition, compatibility with body temperature, pH levels, and tissues is also crucial. For RPM, it is important that these devices are both physically and chemically durable to ensure continuous data transmission.[29]

An implantable sensor generally consists of four main components, including a sensor unit that measures biochemical or physical parameters, a data processing unit that analyses the collected information, an energy source to power the device, and a transmission module that enables data communication with external systems.[30] When it comes to implantation methods, options include subcutaneous (under the skin), intravascular (inside blood vessels), organ-specific implants, digestive system implants, and muscle–joint implants.

In terms of sensor modules, biochemical sensors are often preferred for RPM. There are many types of biosensors that can measure specific biomarkers using various techniques, such as glucose sensors, pH sensors, and electrolyte sensors.[31–36] With biosensors that use optical, electrochemical, or piezoelectric detection methods, it is possible to monitor a wide range of changes.

Implantable sensors used in RPM can also be highly effective for measuring electrical activities, not just chemical changes. Examples include electrocardiography (ECG) sensors, electroencephalography (EEG) sensors, and electromyography (EMG) sensors.[37–39] Besides electrical activity, implantable sensors can also be designed to monitor other physiological changes. For example, pressure sensors can be used for hypertension monitoring, accelerometer-based sensors can detect movement and balance, while temperature sensors and respiration sensors are also types of implantable sensors that can be used for RPM.[40]

13.2.3 Contactless Sensors

Contactless sensors are innovative technologies developed to monitor individuals' vital signs without the need for physical contact. They play a significant role, especially in the healthcare sector, within RPM applications.[41] Using various methods, such as electromagnetic waves, radar systems, optical technologies, and even sound waves, these sensors can accurately track heart rate, respiratory rate, and other vital indicators. These technologies not only enhance patient comfort but also reduce the risk of infection for healthcare workers, while providing continuous monitoring that facilitates early diagnosis and intervention.[42]

In recent years, it has become possible to monitor vital signs without attaching any devices to patients' bodies. Contactless monitoring systems stand out for their accuracy and ease of use, improving patient comfort while simultaneously reducing costs.[43]

Traditional remote patient monitoring applications generally involve placing various sensors on the patient's body to track heart rate and respiratory rate during sleep. Masks or nasal cannulas are used for respiratory monitoring, while electrodes are applied for heart monitoring. Healthcare professionals regularly check patients' vital signs using these devices. However, these methods are often intrusive and uncomfortable, potentially affecting sleep quality negatively. Moreover, prolonged use may cause skin irritation and allergic reactions in individuals with sensitive skin.[44]

On the other hand, contactless sensors attract attention due to their less intrusive nature and comfortable usage. They provide significant advantages, especially for the elderly or individuals with physical limitations. Additionally, since they minimize direct physical contact between healthcare workers and patients, they considerably reduce the risk of infection.

13.2.4 Key Parameters Monitored by Sensors

The human body undergoes continuous metabolism, with certain signals indicating vital functions and contributing significantly to evaluating health conditions.[45] Keeping track of these signals is essential for diagnosing illnesses and monitoring health. In medical environments, instruments like blood pressure cuffs, pulse oximeters, ECGs, EMGs, and thermometers are commonly employed. Yet, these tools are often bulky and inconvenient to transport. Wearable tech, such as smartwatches, lacks the precision required for medical use. These modern sensors bring benefits like compact design, lightweight structure, multifunctionality, and compatibility with biological systems, making them practical for telemedicine, health assessments, and athletic performance monitoring.[46] A major focus in sensor development is enhancing the materials utilized. Choosing flexible and skin-friendly materials that can endure prolonged body contact is vital.[47] Materials capable of self-repair and biodegradable sensors improve longevity and reduce environmental impact.[48]

The body emits physiological signals, like heartbeats and breathing, even when stationary. These signals fall into physical and chemical groups. Physical signals cover electrical activity, blood pressure, body heat, breathing, and movement acceleration.[49–51] Regular blood pressure tracking is critical and can be done using wearable devices.[52] Body temperature monitoring aids in identifying infections or hypothermia.[53] Analysis of breathing patterns contributes to health evaluations. Acceleration sensors are indispensable for detecting movement changes and falls.[54]

Chemical signals enable cell-to-cell communication and are present in bodily fluids like blood, saliva, tears, sweat, and urine. These biochemical markers include pH levels, lactate, glucose, electrolytes, and oxygen levels. Non-invasive collection of fluids like saliva, tears, sweat, and urine allows for hassle-free health monitoring.[55] Electrochemical sensors are effective in fluid detection due to their affordability and high sensitivity.[56] Such innovations significantly enhance non-invasive health monitoring techniques.

13.3 Sensor Detection of Viruses in Remote Patient Monitoring Applications

The literature states that typical RPMs are designed to continually collect a large amount of clinical data from patients and enable doctors to be continuously observed by a variety of internal and exterior sensors.[57] Sareen *et al.*[58] offered a method to manage and control the Zika virus. In order to reduce the latency and expense of communication in such high-end cloud-based systems, they employed fog computing as an intermediary layer between users and the cloud server. With a similar technique, the COVID-19 patient monitoring system developed by El-Rashidy *et al.*[59] relies on a combination of cloud and fog computing to enhance monitoring performance.

In recent years, point-of-care testing (POCT) using smartphones has quickly attracted great interest as a viable substitute for clinical diagnosis conducted in a laboratory. This emerging diagnostic platform offers several advantages, including simpler fabrication, faster analysis times, and reduced reagent consumption, significantly enhancing the early detection, monitoring, and management of viral diseases.[60] The results of studies published in the literature have indicated that colorimetric analyses may detect and differentiate colour tone variations using smartphone cameras.[61] Hsu *et al.*[62] successfully developed a highly sensitive immunosensor (1 pg ml^{-1}) for point-of-care testing for Zika virus. In the accessory-free approach, a smartphone was utilized to photograph the colorimetric signal change, measure the test findings, and create a graphical display of a surveillance map through a tailored application. Because of their affordability and convenience of use, lateral flow immunoassays (LFIAS), the sample-to-answer POCT platforms, have garnered a lot of interest and are frequently employed for the surveillance of viral disease epidemics. In order to semi-quantify Ebola-specific antibodies in 15 minutes, Brangel *et al.*[63] created a serological platform that included an in-house colloidal gold LFIAS and a smartphone reader that was compatible. On the test strip, there were three test lines plotted with three recombinant Ebola virus proteins in addition to a control line. Using a specially designed smartphone application, a semi-quantitative readout of the colour intensity of the strip's test zone was converted. The enhanced analytical performance of microfluidic chips, which have greater surface-to-volume ratios and faster mass and heat transfer rates, has made them a viable alternative to centralized laboratories. Also, higher reproducibility results can be obtained in pathogen analyses performed with systems integrated with microfluidic chips. Ma *et al.*[64] designed a portable platform integrated with a self-driven loop-mediated amplification (LAMP) microfluidic device to enable rapid detection of influenza A (H1N1) virus *via* a smartphone interface. In the system, all necessary reagents were preloaded into the microfluidic chip, which included a sample pretreatment module and a LAMP reaction module. In 40 minutes, the microfluidic chip integrated system was able to measure the outcomes of colorimetric tests, purify, lyse viral pathogens, and perform isothermal nucleic acid amplification.

Nowadays, quantitative detection of virus-induced diseases can be realized through smartphone-based applications of fluorescent LFIAS sensors. By creating quantum dot-based lateral flow immunochromatographic strips (QD-LFICS), Xiao *et al.*[65] aimed to detect high-titre avian influenza virus (AIV) H7N9. In the study, it was observed that the results achieved in the analyses performed on real samples were consistent with the results obtained using the conventional technique, real-time PCR. Similarly, Yeo *et al.*[66] used an effective reflector to produce a fluorescent strip reader for smartphones that could identify three distinct AIV subtypes (H5N3, H7N1, and H9N2). An application was used to determine the ratio between the test and control lines after the fluorescence intensities of LFIAS were photographed with a smartphone camera. The developed smartphone-based diagnostic device presented quite sensitive measurement results for H5N1 determination within 15 minutes.

Smartphone-based electrochemical detection platforms for POCT have emerged as a result of the convenience and benefits of electrochemical techniques in portable devices. In their study, Aronoff-Spencer *et al.*[67] described a potentiostat that uses cyclic voltammetry (CV) to detect the hepatitis C virus (HCV). Functionalized gold screen-printed electrodes (SPE) were employed to detect HCV antibodies using an economical potentiostat, which facilitated the direct conversion of immune complex production into a signal that is sensed, processed, and communicated by a smartphone. A smartphone-controlled amperometric biosensor for the quantitative assessment of hepatitis B Virus (HBV) was introduced by Teengam *et al.*[68] The entire system included a card-sized electrochemical near field communication (NFC) tag sensor combined with a smartphone, and an antibody-modified electrode sensor. Gold nanoparticles were used to enhance the sensitivity of screen-printed graphene electrodes (SPGE), and β-cyclodextrin (β-CD) was electropolymerized on the surface to collect antibodies. The results demonstrated that the limit of detection for hepatitis B surface antigen (HBsAg) was 0.17 $\mu g\ ml^{-1}$ and the linear calibration curve was 10–200 $\mu g\ ml^{-1}$.

Essential criteria for wearable sensors encompass comfort during wear, affordability, and the capability for non-invasive application on the body. Durability, accuracy, sensitivity, and rapid responsiveness are also crucial.[69] The feasibility of identifying coronavirus from patients' exhaled breath was investigated by Grassin-Delyle *et al.*[70] This investigation involved the invasive collection of 303 samples from 40 adult patients, 28 of whom were verified to have COVID-19, in intensive care units utilizing breathing systems by a heated transfer line directly linked to the endotracheal tube. In the exhaled breath of the COVID-19 patients, they found a particular breathprint of volatile organic chemicals. Similarly, Giovannini *et al.*[71] suggested that non-volatile substances, such as viruses, DNA, RNA, and microorganisms, may be detected using standard polymerase chain reaction (PCR) based techniques by turning the breath into a liquid and examining the condensate or aerosol. Wang *et al.*[72] reported a study on a wearable bioelectronic mask system for the detection of various respiratory infectious diseases (COVID-19, H5N1, and H1N1). Stretchable polyvinyl alcohol (PVA)-based ionic gels were synthesized as dielectric layers for flexible electronics in a wearable bioelectronic mask system that is integrated with ion-gated transistors (Figure 13.1). It has been observed that with the developed device, low concentration detections of 0.1–10 $fg\ ml^{-1}$ for COVID-19, H5N1, and H1N1 viral proteins can be achieved in as little as 10 minutes.

13.4 Future Directions and Conclusions

Advancements in sensor technology are bringing noticeable changes to remote patient monitoring (RPM) and healthcare services. As sensor abilities improve, they open up more possibilities to enhance patient health, manage medical conditions better, and change how healthcare is delivered.

Patients' preferences are pushing for smaller, more comfortable sensors, making devices easier to use without being obvious. With mini sensors now added to wearables, like smartwatches, fitness trackers, and smart clothes, people can wear these devices all the time without discomfort. This allows health conditions to be tracked in real-time while patients continue with their daily activities without interruption.

Bringing multiple sensors together in one device makes it possible to monitor different health indicators at the same time. This gives a fuller picture of a patient's health and removes the need to handle many devices or apps. Combining optical, acoustic, and biochemical sensors helps with collecting different types of health data efficiently.

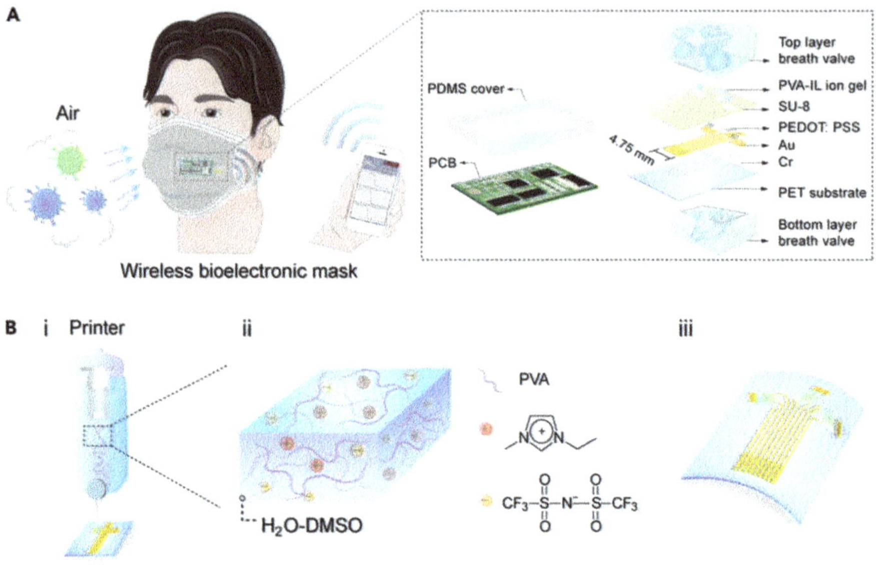

Figure 13.1 (A) Schematic representation of a wireless bioelectronic mask, including its key components such as the PIL-IGT, breath valve, and printed circuit board (PCB). (B) Schematic illustration of the fabrication process of the ion gel-based dielectric layer used in the PIL-IGT, including the 3D printing of the PVA-IL ion gel and the structure of the flexible device. Abbreviations: polydimethylsiloxane (PDMS); printed circuit board (PCB); polyvinyl alcohol (PVA); ionic liquid (IL); structured by UV (SU); poly(2,3-dihydrothieno-1,4-dioxin)-poly(styrenesulphonate) (PEDOT:PSS); polyethylene terephthalate (PET); dimethyl sulfoxide (DMSO). Reproduced from ref. 72 with permission from Elsevier, Copyright 2022.

Developing sensors that give real-time feedback and alerts increases patient safety and makes timely interventions easier. With advanced algorithms and machine learning, sensors can spot critical health issues and permit fast reactions. These alert systems need to be simple to use and customizable for both patients and healthcare workers.

Personalized medicine is another major area growing in RPM. When sensors are combined with artificial intelligence (AI)-driven analysis, it becomes possible to create treatment plans tailored to each patient's unique data. This leads to more precise healthcare responses and helps with adjusting treatments to better fit individual needs.

Saving and analysing patient data over long periods is important for spotting health trends and doing predictive analysis. It is necessary to create data storage systems that can safely handle large amounts of sensor data over time or work with cloud-based providers. This makes it easier to do retrospective studies, supports medical research, and helps improve treatment methods.

While non-invasive sensors and wearables are usually preferred for everyday health tracking and short-term care, implantable sensors offer big benefits for people with chronic diseases. Development of biocompatible, implantable sensors ensures constant and accurate health monitoring while reducing the risk of negative reactions.

To make sensor technologies more sustainable, energy-harvesting methods can be used. Self-powered sensors that gather energy from the environment or the human body reduce the need for external power. This innovation cuts down the need for frequent battery changes, extends the lifespan of devices, and increases patient comfort.

Predictive analysis and early intervention algorithms based on sensor data allow for proactive healthcare management. Advanced algorithms that process real-time sensor data can spot early signs of health issues and trigger necessary medical actions. This proactive method improves patient outcomes and lowers the chances of complications.

To stay competitive in this fast-changing market, it's important to keep up with new trends in sensor technology. Companies that include innovations like AI integration, cloud storage, and personalized medicine in their products will lead the way in RPM. This helps improve patient outcomes, advance healthcare delivery, and shape the future of healthcare.

In summary, the future of sensors in RPM looks bright and full of potential. Investments should focus on areas like miniaturization, multi-sensor integration, real-time feedback, personalized medicine, remote diagnostics, long-term data management, biocompatible materials, energy harvesting, and predictive analysis. The combination of advanced sensors with connected devices and AI-supported analysis will keep transforming healthcare, offering more personalized, efficient, and accessible care worldwide.

References

1. K. Boikanyo, A. M. Zungeru, B. Sigweni, A. Yahya and C. Lebekwe, *Sci. Afr.*, 2023, **20**, e01638.
2. F. A. C. D. De Farias, C. M. Dagostini, Y. D. A. Bicca, V. F. Falavigna and A. Falavigna, *Telemed. e-Health*, 2020, **26**, 576–583.
3. L. P. Malasinghe, N. Ramzan and K. Dahal, *J. Ambient Intell. Humaniz. Comput.*, 2019, **10**, 57–76.
4. A. Zito, A. Restivo, G. Ciliberti, R. Laborante, G. Princi, G. F. Romiti, M. Galli, D. Rodolico, E. Bianchini, L. Cappannoli, M. D'Oria, C. Trani, F. Burzotta, A. Cesario, G. Savarese, F. Crea and D. D'Amario, *Int. J. Cardiol.*, 2023, **388**, 131163.
5. Y. Sahraoui, A. Korichi, C. A. Kerrache, M. Bilal and M. Amadeo, *Trans. Emerg. Telecommun. Technol.*, 2022, **33**, 1–17.
6. S. K. Prabhakar and H. Rajaguru, *IFMBE Proc.*, 2017, **61**, 78–85.
7. V. A. J. Block, E. Pitsch, P. Tahir, B. A. C. Cree, D. D. Allen and J. M. Gelfand, *PLoS One*, 2016, **11**, 1–41.
8. D. A. Askew, L. Crossland, R. S. Ware, S. Begg, P. Cranstoun, P. Mitchell and C. L. Jackson, *Contemp. Clin. Trials*, 2012, **33**, 969–975.

9. P. Kakria, N. K. Tripathi and P. Kitipawang, *Int. J. Telemed. Appl.*, 2015, **2015**, 373474.
10. S. Y. Tan, J. Sumner, Y. Wang and A. Wenjun Yip, *npj Digit. Med.*, 2024, 7, 192.
11. K. R. De Guzman, C. L. Snoswell, M. L. Taylor, L. C. Gray and L. J. Caffery, *Value Heal.*, 2022, **25**, 897–913.
12. B. Devine, *Value Heal.*, 2022, **25**, 887–889.
13. A. J. Bandodkar, I. Jeerapan and J. Wang, *ACS Sens.*, 2016, **1**, 464–482.
14. J. Kim, A. S. Campbell, B. E. F. de Ávila and J. Wang, *Nat. Biotechnol.*, 2019, **37**, 389–406.
15. A. Pantelopoulos and N. G. Bourbakis, *IEEE Trans. Syst. Man Cybern. Part C Appl. Rev.*, 2010, **40**, 1–12.
16. S. U. Singh, S. Chatterjee, S. A. Lone, H. H. Ho, K. Kaswan, K. Peringeth, A. Khan, Y. W. Chiang, S. Lee and Z. H. Lin, Advanced wearable biosensors for the detection of body fluids and exhaled breath by graphene, *Microchim. Acta*, 2022, **189**, 236.
17. Y. Liu, M. Pharr and G. A. Salvatore, *ACS Nano*, 2017, **11**, 9614–9635.
18. J. Kim, A. S. Campbell and J. Wang, *Talanta*, 2018, **177**, 163–170.
19. M. J. Tierney, J. A. Tamada, R. O. Potts, L. Jovanovic and S. Garg, *Biosens. Bioelectron.*, 2001, **16**, 621–629.
20. K. C. Tseng, B. S. Lin, L. De Liao, Y. Te Wang and Y. L. Wang, *IEEE Syst. J.*, 2014, **8**, 900–906.
21. G. S. Wilson and R. Gifford, *Biosens. Bioelectron.*, 2005, **20**, 2388–2403.
22. S. Ullah, H. Higgins, B. Braem, B. Latre, C. Blondia, I. Moerman, S. Saleem, Z. Rahman and K. S. Kwak, *J. Med. Syst.*, 2012, **36**, 1065–1094.
23. M. Al Ameen, J. Liu and K. Kwak, *J. Med. Syst.*, 2012, **36**, 93–101.
24. J. S. Daniels and N. Pourmand, *Electroanalysis*, 2007, **19**, 1239–1257.
25. M. Chan, D. Estève, J. Y. Fourniols, C. Escriba and E. Campo, *Artif. Intell. Med.*, 2012, **56**, 137–156.
26. K. Bazaka and M. V. Jacob, Implantable devices: Issues and challenges, *Electronics*, 2012, **2**, 1–34.
27. A. Demosthenous, *Adv. Electron.*, 2014, **2014**, 1–21.
28. A. Ben Amar, A. B. Kouki and H. Cao, *Sensors*, 2015, **15**, 28889–28914.
29. E. S. Hosseini, S. Dervin, P. Ganguly and R. Dahiya, *ACS Appl. Bio Mater.*, 2021, **4**, 163–194.
30. S. Y. Yang, V. Sencadas, S. S. You, N. Z. X. Jia, S. S. Srinivasan, H. W. Huang, A. E. Ahmed, J. Y. Liang and G. Traverso, *Adv. Funct. Mater.*, 2021, **31**, 2009289.
31. M. Çalışır, M. Bakhshpour, H. Yavuz and A. Denizli, *Sens. Actuators, B*, 2020, **306**, 127561.
32. Ç. Çiçek, F. Yilmaz, E. Özgür, H. Yavuz and A. Denizli, *Chemosensors*, 2016, **4**, 21.
33. M. Ovezova, F. Yılmaz, I. Göktürk and A. Denizli, *Microchem. J.*, 2024, **206**, 111545.
34. M. Bakhshpour, E. Chiodi, I. Celebi, Y. Saylan, N. L. Ünlü, M. S. Ünlü and A. Denizli, *Biosens. Bioelectron.*, 2022, **201**, 113961.

35. G. Sener, E. Ozgur, A. Y. Rad, L. Uzun, R. Say and A. Denizli, *Analyst*, 2013, **138**, 6422–6428.
36. M. Erkek, S. Akgonullu, A. Denizli and N. Bereli, *IEEE Sens. J.*, 2024, **24**, 17410–17417.
37. J. H. Lee and D. W. Seo, *Micromachines*, 2019, **10**, 38.
38. M. Rehman, L. M. Higdon and M. R. Sperling, *J. Clin. Neurophysiol.*, 2024, **41**, 200–206.
39. M. M. Lowery, R. F. F. Weir and T. A. Kuiken, *IEEE Trans. Biomed. Eng.*, 2006, **53**, 1926–1933.
40. V. T. N. Linh, S. Han, E. Koh, S. Kim, H. S. Jung and J. Koo, *Biomaterials*, 2025, **314**, 122865.
41. S. Suresh Kumar, K. Dashtipour, Q. H. Abbasi, M. A. Imran and W. Ahmad, *Front. Commun. Networks*, 2021, **2**, 1–10.
42. M. S. Raheel, F. Tubbal, R. Raad, P. Ogunbona, J. Coyte, C. Patterson, D. Perlman, S. Iranmanesh, N. Odeh and J. Foroughi, *Sens. Diagn.*, 2024, **3**, 1085–1118.
43. H. Gjoreski and M. Gams, *Proc. IASTED Int. Conf. Artif. Intell. Soft Comput. ASC*, 2011, **2011**, 340–347.
44. L. J. Delaney, M. J. Currie, H. C. C. Huang, V. Lopez and F. Van Haren, *BMC Health Serv. Res.*, 2018, **18**, 1–9.
45. L. Zhao, H. Li, J. Meng and Z. Li, *InfoMat*, 2020, **2**, 212–234.
46. Q. Zheng, Q. Tang, Z. L. Wang and Z. Li, *Nat. Rev. Cardiol.*, 2021, **18**, 7–21.
47. Z. Lin, G. Li, K. Jiang, Z. Li and T. Liu, *Mol. Cancer*, 2024, **23**, 191.
48. T. Karasu, F. Z. Erkoc-Biradli, M. Ö. Öztürk-Öncel, C. Armutcu, L. Uzun, B. Garipcan and M. E. Çorman, *Biomed. Phys. Eng. Express*, 2022, **8**, 055004.
49. J. Park, M. Kim, Y. Lee, H. S. Lee and H. Ko, 2015, *Sci. Adv.*, **1**, e1500661.
50. G. Ge, Y. Cai, Q. Dong, Y. Zhang, J. Shao, W. Huang and X. Dong, *Nanoscale*, 2018, **10**, 10033–10040.
51. C. S. Boland, U. Khan, C. Backes, A. O. Neill, J. Mccauley, S. Duane, R. Shanker, Y. Liu, I. Jurewicz, A. B. Dalton and J. N. Coleman, *ACS Nano*, 2014, **8**, 8819–8830.
52. N. Luo, W. Dai, C. Li, Z. Zhou, L. Lu, C. C. Y. Poon, S. C. Chen, Y. Zhang and N. Zhao, *Adv. Funct. Mater.*, 2016, **26**, 1178–1187.
53. P. Mair, E. Kornberger, W. Furtwaengler, D. Balogh and H. Antretter, *Resuscitation*, 1994, **27**, 47–54.
54. M. Ferrari, B. Harrison, O. Rawashdeh, R. Hammond, Y. Avery, M. Rawashdeh, W. Sa'deh and M. Maddens, *Geriatr. Nurs.*, 2012, **33**, 177–183.
55. A. J. Bandodkar and J. Wang, *Trends Biotechnol.*, 2014, **32**, 363–371.
56. A. Deniz, T. Karasu, E. Özgür and L. Uzun, *Bioelectrochemistry*, 2025, **161**, 1–7.
57. N. El-Rashidy, S. El-Sappagh, S. M. Riazul Islam, H. M. El-Bakry and S. Abdelrazek, *Diagnostics*, 2021, **11**, 1–32.
58. S. Sareen, S. K. Gupta and S. K. Sood, *Enterp. Inf. Syst.*, 2017, **11**, 1436–1456.
59. N. El-Rashidy, S. El-Sappagh, S. M. R. Islam, H. M. El-Bakry and S. Abdelrazek, *Electron*, 2020, **9**, 1–25.

60. M. Xiao, F. Tian, X. Liu, Q. Zhou, J. Pan, Z. Luo, M. Yang and C. Yi, *Adv. Sci.*, 2022, **9**, 2105904.
61. A. Priye, C. S. Ball and R. J. Meagher, *Anal. Chem.*, 2018, **90**, 12385–12389.
62. Y. P. Hsu, N. S. Li, Y. T. Chen, H. H. Pang, K. C. Wei and H. W. Yang, *Biosens. Bioelectron.*, 2020, **151**, 111960.
63. P. Brangel, A. Sobarzo, C. Parolo, B. S. Miller, P. D. Howes, S. Gelkop, J. J. Lutwama, J. M. Dye, R. A. McKendry, L. Lobel and M. M. Stevens, *ACS Nano*, 2018, **12**, 63–73.
64. Y. D. Ma, K. H. Li, Y. H. Chen, Y. M. Lee, S. T. Chou, Y. Y. Lai, P. C. Huang, H. P. Ma and G. Bin Lee, *Lab Chip*, 2019, **19**, 3804–3814.
65. M. Xiao, L. Huang, X. Dong, K. Xie, H. Shen, C. Huang, W. Xiao, M. Jin and Y. Tang, *Analyst*, 2019, **144**, 2594–2603.
66. S. J. Yeo, K. Choi, B. T. Cuc, N. N. Hong, D. T. Bao, N. M. Ngoc, M. Q. Le, N. L. K. Hang, N. C. Thach, S. K. Mallik, H. S. Kim, C. K. Chong, H. S. Choi, H. W. Sung, K. Yu and H. Park, *Theranostics*, 2016, **6**, 231–242.
67. E. Aronoff-Spencer, A. G. Venkatesh, A. Sun, H. Brickner, D. Looney and D. A. Hall, *Biosens. Bioelectron.*, 2016, **86**, 690–696.
68. P. Teengam, W. Siangproh, S. Tontisirin, A. Jiraseree-amornkun, N. Chuaypen, P. Tangkijvanich, C. S. Henry, N. Ngamrojanavanich and O. Chailapakul, *Sens. Actuators, B*, 2021, **326**, 128825.
69. Q. Gao, J. Zhang, Z. Xie, O. Omisore, J. Zhang, L. Wang and H. Li, *J. Mater. Sci.*, 2019, **54**, 5187–5223.
70. S. Grassin-Delyle, C. Roquencourt, P. Moine, G. Saffroy, S. Carn, N. Heming, J. Fleuriet, H. Salvator, E. Naline, L. J. Couderc, P. Devillier, E. A. Thévenot and D. Annane, *EBioMedicine*, 2021, **63**, 103154.
71. G. Giovannini, H. Haick and D. Garoli, *ACS Sens.*, 2021, **6**, 1408–1417.
72. B. Wang, D. Yang, Z. Chang, R. Zhang, J. Dai and Y. Fang, *Matter*, 2022, **5**, 4347–4362.

CHAPTER 14

Screening and Early Detection

ZEHRA TUĞÇE KURT[a], HANDAN YAVUZ ALAGÖZ[b] AND NILAY BERELI[b*]

[a]Department of Bioengineering, Hacettepe University, Ankara, Turkey;
[b]Department of Chemistry, Hacettepe University, Ankara, Turkey
*E-mails: zehratugcekurt@hacettepe.edu.tr, handany@hacettepe.edu.tr, bereli@hacettepe.edu.tr

14.1 Introduction

Viruses are among the most infectious pathogens that can cause high mortality rates worldwide. Since effective clinical drugs are not available for all viral infections, rapid and accurate diagnosis in the early stages of infection is essential to prevent the rapid spread of pathogens.[1] Although viruses have different pathogenic mechanisms, all viruses require a host for replication and survival within a biological system.[2] Easy and rapid detection of viruses plays an important role in diagnosis and treatment. The importance of these methods becomes even more significant during virus-based outbreaks. Viruses are genetically classified as deoxyribonucleic acid (DNA)- or ribonucleic acid (RNA)-based and are detected using different diagnostic methods. The detection of viruses is generally based on antigen detection, nucleic acid amplification through polymerase chain reaction (PCR), and serological tests that identify antibodies after infection.[3] Thanks to sensor technologies in medical applications, negative health consequences can be prevented through the rapid detection of viruses.[4,5] Thus, rapid diagnosis ensures

Detection Science Series No. 32
Advances in Virus Detection with Sensors
Edited by Adil Denizli
© The Royal Society of Chemistry 2026
Published by the Royal Society of Chemistry, www.rsc.org

quick virus detection and effective control. Moreover, some sensor-based applications not only detect viruses but also enable their rapid identification through integration with imaging systems.[6,7]

The detection of viruses can be classified into two main types: indirect and direct.[8] Indirect methods involve detecting the immune response (*e.g.*, antibodies, cytokines) or secondary effects of the virus rather than the virus itself. Virus isolation *via* culturing is generally considered a direct method since it directly detects viable viruses.[9] The classical form of viral culturing usually requires an incubation period of 1–4 weeks, but faster alternatives, such as the shell vial assay, allow viral detection within 1–2 days. After 1–2 days of incubation, viral proteins can be detected in cells by immunofluorescence or similar procedures.[10] Direct methods include microscopic methods,[11–13] PCR,[14] reverse transcription PCR (RT-PCR),[15,16] loop-mediated isothermal amplification (LAMP),[17] clustered regularly interspaced short palindromic repeat (CRISPR)–CRISPR-associated (Cas),[18] enzyme-linked immunosorbent assay (ELISA),[19] and sensors[20]-based approaches. Direct methods offer significant advantages over indirect methods in terms of speed, convenience, and safety. Advances in sensor technology have facilitated the development of faster, more efficient, and cost-effective virus detection techniques.[21]

In recent years, biomedical sensors have become important for early diagnosis, continuous monitoring, and pre-disease detection in health and medical applications.[22] Advances in microfabrication and nanofabrication techniques and advanced materials have made sensor technologies more efficient, cost-effective, and multifunctional, opening up new opportunities for health monitoring and disease diagnosis. These sensors have great potential in areas such as biomarker detection, physiological signal analysis, and environmental factor assessment.[23] Viral infections are caused by pathogenic agents that pose serious health risks, and treatments for certain types remain challenging. In the event of an outbreak, rapid detection and isolation of the virus are crucial. Mutations in viral genomes can reduce the effectiveness of existing vaccines. Additionally, the development and implementation of treatment strategies and antiviral drugs require time. For this reason, efficient detection methods are essential, along with real-time imaging capabilities to support timely clinical decisions. The corona virus disease (COVID)-19 pandemic has accelerated research on rapid virus detection and screening through sensor technologies.[24–28] Sensor technologies in this field are integrated into existing analytical systems, as well as smartphones,[29,30] and point-of-care (POC)[31,32] systems, enabling colorimetric analysis for detecting the presence or absence of the virus. Today, sensors offer a significant time advantage over traditional diagnostic methods in virus detection.

This chapter focuses on the applications of virus sensors. Sensor applications for screening and easy detection of viruses in medical applications and the mechanisms of these sensors are explained. In addition, research on virus sensors, integrated systems, and future perspectives will be discussed.

14.2 Fundamentals of Sensor Technologies in Virus Detection

Sensor systems generally consist of two main components: the recognition region and the transducer, which converts the detected signal into a meaningful output.[33] Sensor mechanisms operate by generating a signal when the analyte interacts with the recognition region through chemical or physical interactions. The produced signal is then processed and converted into a meaningful form by the transducer. The interpretation of the signal is a critical aspect of sensor functionality.

A viral biosensor consists of a bioreceptor unit to bind a specific virus of interest and a transducer mechanism to convert the information about the bound target analyte (virus or biomolecule) into a measurable value through optical, electrical, magnetic, or thermal signals. On the basis of the type of transducer used and the physical principles involved, biosensors can be classified into several categories, such as optical electrochemical, piezoelectric, and thermal.[34–36] Optical and electrochemical biosensors are the most widely used among the common biosensors.[37] The functionality of electrochemical biosensors relies on the direct generation of voltage or current proportional to the analyte concentration, whereas optical or photonic biosensors utilize series of steps to produce an electronic or optical signal indicative of analyte concentration.[38]

14.2.1 Screening and Easy Detection Mechanisms

In virus detection, biological recognition within the sensing region of sensor systems requires precise and specific immobilization. Proper immobilization enhances sensitivity, selectivity, speed, and ease of detection. Various strategies, including DNA or RNA hybridization, viral protein interactions, and antigen–antibody binding, are employed for identification within the recognition region. Different material technologies are utilized to immobilize biologically relevant viral particles. With advancements in material technology, nanoparticles, graphene, and molecularly imprinted polymers have been increasingly used. Quantum dots are known for their superior optical properties that enhance sensitivity. Silver particles,[39] valued for their antimicrobial properties, are integrated into sensor surfaces of sensor systems to facilitate rapid detection and easy screening.[40] The enhancing effects of surface technology on both immobilization and sensor detection capacity are further integrated with optical, electrochemical, microfluidic, and chip-based systems. As a result of technological advances, sensors created with different methods have provided easily obtained screening results (Figure 14.1).

14.2.2 Material Technologies and Virus Detection

Nanoparticles, characterised by their distinctive optical and electronic properties, are used for the development of highly sensitive and easy to use diagnostic analyses.[41] Gold nanoparticles (AuNPs),[42] silver nanoparticles (AgNPs),[43]

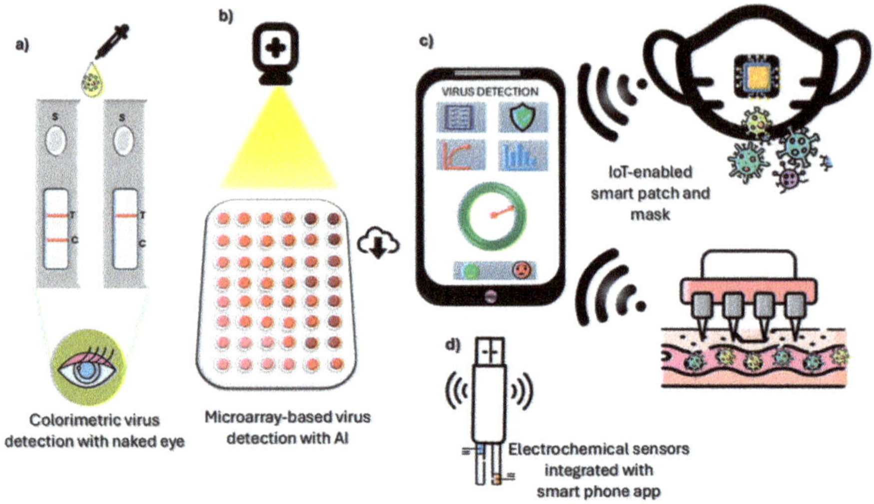

Figure 14.1 Sensors with different methodologies for screening and early detection. (a) Colorimetric virus detection with the naked eye. (b) Microarray-based virus detection, these systems generally include artificial intelligence (AI)-assisted analyses and are available on smartphones or specialized portable devices. (c) Internet of things (IoT)-enabled virus detection of wearable biosensors. (d) Electrochemical biosensors integrated in a universal serial bus (USB) system, results can be obtained using a smartphone or computer.

magnetic nanoparticles (MNPs),[44] quantum dots (QDs),[45] and nano-molecularly imprinted polymers (NMIPs)[46] significantly contribute to improved virus detection, offering higher sensitivity and multiple diagnostic possibilities. These nanomaterials can be easily integrated into biosensors and diagnostic platforms, which is highly advantageous for improving viral detection and contributing to more effective disease control and prevention strategies.

Molecularly imprinted polymers (MIPs) exhibit strong resistance to chemical and biological inactivation or damage. These artificial receptor ligands possess the unique ability to detect and bind to molecular targets.[47] Target-specific MIP nanoparticles (NPs) can be used as biosensors for sensitively recognizing pathogenic viruses that cause various diseases or death due to environmental contamination.[48] Typically, two-dimensional surface imprinting is performed for the detection of very small amounts of pathogenic viruses, while three-dimensional imprinting is used to bind and extract larger amounts of viruses from solutions. However, MIPs usually consist of two-dimensional surface imprinting patterns of viruses bound to a sensor on a cross-linked surface of the polymer. Following the imprinting and removal of the viral templates, the sensors can rebind a very small amount of the target virus to the surface of the polymer.[49] MIPs are durable and highly selective materials, described as plastic antibodies. Given the recurrent outbreaks of pathogenic viruses and their profound effects on public health and economies, the need for more effective detection methods has increased. Therefore, MIP technology is attracting more and more attention in the field of virus detection.[50]

AuNPs range in size from 1 nm to 8 μm and exhibit various morphologies, including spherical, rod-shaped, shell-like, cubic, cage-like, and branched structures. Size and shape play a crucial role in biological sensing; smaller AuNPs often demonstrate higher sensitivity, thereby enhancing biosensors' performance.[51] Due to their tuneable optical properties over a broad wavelength range, AuNPs are widely utilized in virus sensors.[52,53] In virus sensors, surface-modified AuNPs facilitate specific interactions with target viral particles, enabling highly sensitive detection mechanisms.[54] AuNPs in aqueous solutions exhibit different colours, ranging from red to blue, as their size varies between 1 and 100 nm. Additionally, depending on their size, they exhibit a characteristic absorption peak within the 500–550 nm range. This absorption is associated with the surface plasmon resonance (SPR) effect, which results from the collective oscillation of conduction electrons induced by incident photons.[55]

Graphene is a single-layer planar film with a hexagonal honeycomb lattice of carbon atoms. The size, number of layers, shape, and chemical groups of graphene significantly affect the sensor's performance.[56] Graphene-based composites enhance the sensitivity and flexibility of conventional sensors and shorten the reaction time. The high electrical conductivity of graphene is primarily attributed to the presence of delocalized free electrons. The results of studies on biosensors have demonstrated that graphene materials, due to their strong adsorption capacities and excellent conductivity, provide high sensitivity and improve sensor performance. Graphene quantum dots (GQDs) enable the simultaneous detection of multiple biomolecules owing to their biocompatibility and catalytic properties. GQDs can serve as electrocatalysts and nanozymes for label-free target analyte detection. Furthermore, graphene exhibits excellent performance when integrated with MIP-based SPR sensors.

With advancements in technology and scientific research, the design and efficiency of biosensors have improved considerably. Biosensors with different designs and mechanisms of action are used to detect viral particles. Biosensors can significantly reduce analysis time due to the relative simplicity of the procedures and, as published data show, are highly sensitive and require minimal pre-treatment of the test material. Biosensors typically comprise a sensitive biological component and a detection system. The detection system operates through various mechanisms, including optical, colorimetric, acoustic, and electrical *etc.* Additionally, the integration of multiple detection elements is possible, highlighting the potential for the development of highly versatile and efficient biosensor systems.[57]

14.3 Virus Sensors for Screening and Early Detection: Types and Applications

14.3.1 Electrochemical Sensors

Electrochemical biosensors mainly consist of three basic components: biorecognition elements, transducers, and electronic systems. The biorecognition elements selectively identify the target virus, which is a critical step in virus detection. The differences in transducers in electrochemical biosensors lie in the electrode interface, which mainly includes immobilization polymer matrices and nanomaterials. Detection signals are processed and analysed using various electrochemical techniques *via* electronic systems.[58] The currents and voltages generated by the immunocomplex are measured using electrochemical techniques such as amperometry, voltammetry, potentiometry, electrochemical impedance spectroscopy, and conductometry. Genosensors utilize immobilized DNA probes as recognition elements and measure the formation of DNA hybrids and interactions between DNA and proteins on the surface. In electrochemical genosensors, the probe consists of a small oligonucleotide sequence that recognizes the target analyte (nucleic acids) by forming duplexes on the transducer surface, which are then electrochemically detected. These sensors are often combined with nanocomposites or NPs to improve the immobilization and sensitivity of the probe, thus offering a rapid and cost-effective detection method, especially in POC facilities.[59]

Another type of sensor demonstrating rapid virus detection is the microelectromechanical systems (MEMS) sensor. Muhsin *et al.* developed a MEMS-based impedance biosensor for the rapid and sensitive detection of severe acute respiratory syndrome corona virus 2 (SARS-CoV-2) using clinical samples. The sensor comprises three regions, for virus concentration, trapping, and sensing using an electrode coated with a SARS-CoV-2-specific antibody. The impedance change upon antigen–antibody binding indicates the test result. The biosensor demonstrated a detection limit of 50 50% tissue culture infectious dose (TCID50) per ml in phosphate buffered saline (PBS) and a sensitivity of 26 TCID50 per ml in clinical swab samples. Specificity was confirmed using the influenza virus, while selectivity was verified with influenza polyclonal sera. The results indicated that the biosensor enables SARS-CoV-2 detection in 40 minutes with high sensitivity and specificity.[60]

Zhou *et al.* developed a novel paper-printed electrochemical biosensor based on a nickel single-atom electrocatalyst for pre-screening of COVID-19 *via* fractional exhaled nitric oxide (FeNO) detection. The electrocatalyst, consisting of single nickel atoms with N_2O_2 coordination embedded in porous acetylene black, showed an ultrasensitive response to NO in the range of 0.3–180 parts per billion. The results of clinical studies confirmed a correlation between FeNO levels and COVID-19, and a machine-learning model was established to discriminate between healthy subjects and COVID-19 patients. As seen in Figure 14.2, this sensor offers a rapid, cost-effective, and non-invasive approach for COVID-19 screening in daily life.[61]

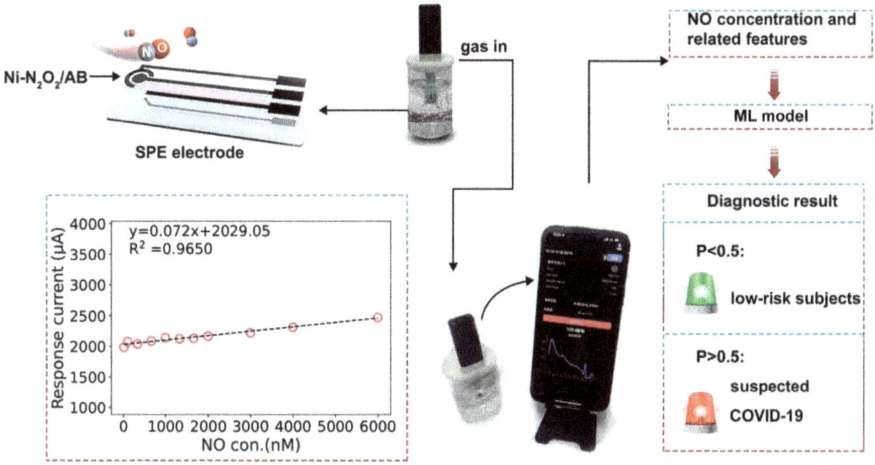

Figure 14.2 The FeNO detection analyser includes a portable mini-detector and a machine learning (ML) model integrated into a mobile app. The mini-detector measures FeNO levels and transmits the data to the app, where the ML model analyses it for preliminary screening. Abbreviations: acetylene black (AB), screen printed electrode (SPE). Reproduced from ref. 61 with permission from American Chemical Society, Copyright 2022.

A novel sponge-based electrochemical sensor designed for the rapid *in situ* detection and analysis of infectious viruses on solid surfaces with a different design by utilizing material technology has been developed. The sensor comprises a biofunctionalized porous sponge modified with graphene, graphene oxide, and specific antibodies, enabling both sample collection and electrochemical measurement. The hydrophilic sponge efficiently captures viral particles through process steps including surface fogging, sponge sliding, finger pressure electrolyte discharge, and *in situ* electrochemical measurement. The instrument demonstrated a detection limit of 0.4 TCID50 per ml and an effective concentration range of 1–10^6 TCID50 per ml for the H1N1 virus. The innovative push-release electrolyte dispensing mechanism improves portability and provides 15% collection efficiency with a single swipe across surfaces. This versatile system offers a promising and practical approach for rapid, *in situ* virus detection without the need for extensive sample preparation.[62]

PCR analyses are a time-consuming method for virus detection that require expertise. An electrochemical paper-based analytical device (ePAD) for multiplexed, label-free detection of SARS-CoV-2, H1N1, and respiratory syncytial virus (RSV) infections has been described. The device uses pyrrolidinyl peptide nucleic acid probes to bind viral nucleic acids, resulting in electrochemical signal changes. It shows high sensitivity and no cross-reactivity with detection limits of 0.12, 0.35, and 0.36 pM for SARS-CoV-2, H1N1, and RSV, respectively. The platform has been validated with nasopharyngeal swab samples and compared with RT-PCR, showing 100% clinical sensitivity and ≥97% specificity, making it highly suitable for screening and diagnosis of respiratory diseases.[63]

In another interesting study, Szydlowska *et al.* demonstrated the potential of additive manufacturing for biosensing applications by developing a screen-printed graphene electrode platform for electrochemical detection of SARS-CoV-2. The electrodes were thermally annealed to increase surface area and electrical conductivity, enabling sensitive biosensing *via* electrochemical impedance spectroscopy. Thermally annealed graphene electrodes provided oxygen moieties for robust antibody binding, resulting in immunosensors with superior performance metrics. The sensors achieved rapid detection with a 55-second readout after a 15-minute incubation and a detection limit approaching 500 ag ml^{-1} for the omicron variant. The results of clinical validation demonstrated that SARS-CoV-2 omicron was successfully detected in saliva samples in as few as 10^3 copies per ml, adhering to Centers for Disease Control and Prevention (CDC) protocols.[64]

A genomagnetic assay paired with a disposable electrochemical device for gene detection and quantification for the selective capture of viral RNA from influenza A and SARS-CoV-2 viruses in saliva samples has been described. Figure 14.3 shows how RNA-based viral detection was performed. The method showed excellent performance with limits of detection and limits of quantification of 5.0 fmol l^{-1} and 8.6 fmol l^{-1} for SARS-CoV-2 and 1.0 fmol l^{-1} and 108.9 fmol l^{-1} for influenza A, respectively. Results from testing 36 saliva samples were similar to those for real-time RT-PCR, demonstrating the reliability and potential of this approach for rapid, POC diagnosis of influenza and COVID-19.[65]

Non-structural protein 1 (NS1), an important biomarker present in the bloodstream during the acute infection stage, has been widely used for the diagnosis of dengue. The development of an electrochemical immunosensor for NS1 detection using a nanobio-based platform consisting of reduced graphene oxide (rGO), polydopamine (pDA), and AuNPs immobilized with anti-NS1 antibodies has been described. The nanocomposite structure and antibody immobilization were characterized using microscopic, spectroscopic, and electrochemical techniques. Using differential pulse voltammetry (DPV), the electrochemical properties and detection performance of the immunosensor were confirmed. The sensor exhibited a linear detection response from 1 ng ml^{-1} to 100 μg ml^{-1}, a sensitivity of 1.78 ng ml^{-1} mA^{-1}, and good stability. With its extremely low detection limit, this immunosensor has strong potential for early diagnosis of dengue virus infections.[66]

Another study presented a portable electrochemical biosensor for rapid rabies virus (RABV) detection in nasopharyngeal swabs. The sensor electrode, made of rGO with immobilized cDNA, increases specificity through π–π bonds. The sensor's performance was evaluated using RNA and swab samples collected from bats. The RNA detection demonstrated high selectivity, while the quantification exhibited a strong linear correlation ($R^2 = 0.990$) across a concentration range of 0.145–25.39 ng μl^{-1}. The sensor achieved a

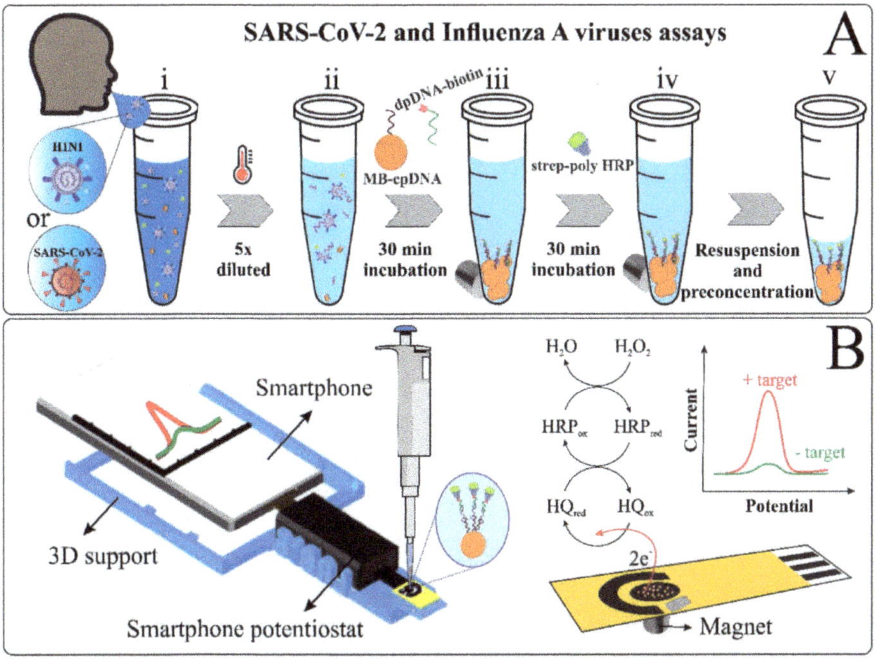

Figure 14.3 (a) The RNA viral detection procedure for SARS-CoV-2 or influenza A involves: (i) thermal inactivation of the virus, (ii) one-step hybridization with magnetic bead-capture probe DNA (MB-cpDNA) and detection probe DNA (dpDNA)–biotin, (iii) magnetic capture of viral RNA and separation from unreacted dpDNAs, followed by streptavidin (strep)-poly-horseradish peroxide (HRP) conjugation, (iv) removal of unbound strep-poly-HRP, and (v) resuspension and preconcentration of the magneto-bioconjugate. Abbreviations: hydroquinone (HQ), reduced (red) oxidised (ox). Finally, (b) electrochemical detection is performed to analyse the viral RNA. Reproduced from ref. 65 with permission from Elsevier, Copyright 2024.

limit of detection (LOD) of 0.104 ng μl^{-1} with a sensitivity of 0.321 μA (ng $\mu l^{-1})^{-1}$. In nasopharyngeal swab samples, RABV detection effectively distinguished positive samples from negative ones, delivering results within seconds, which was significantly faster than conventional techniques. This device offers a rapid, portable solution for RABV detection and can be adapted for other RNA viruses.[67] The portable sensor, capable of detecting RNA-based viruses, also allows for easy screening in the presence of viruses, as shown in Figure 14.4.

14.3.2 Optical Sensors

The optical transducer is closely integrated with the biorecognition element in the optical biosensor. Optical biosensors have provided an additional advantage in disease diagnosis and medical services due to their selective,

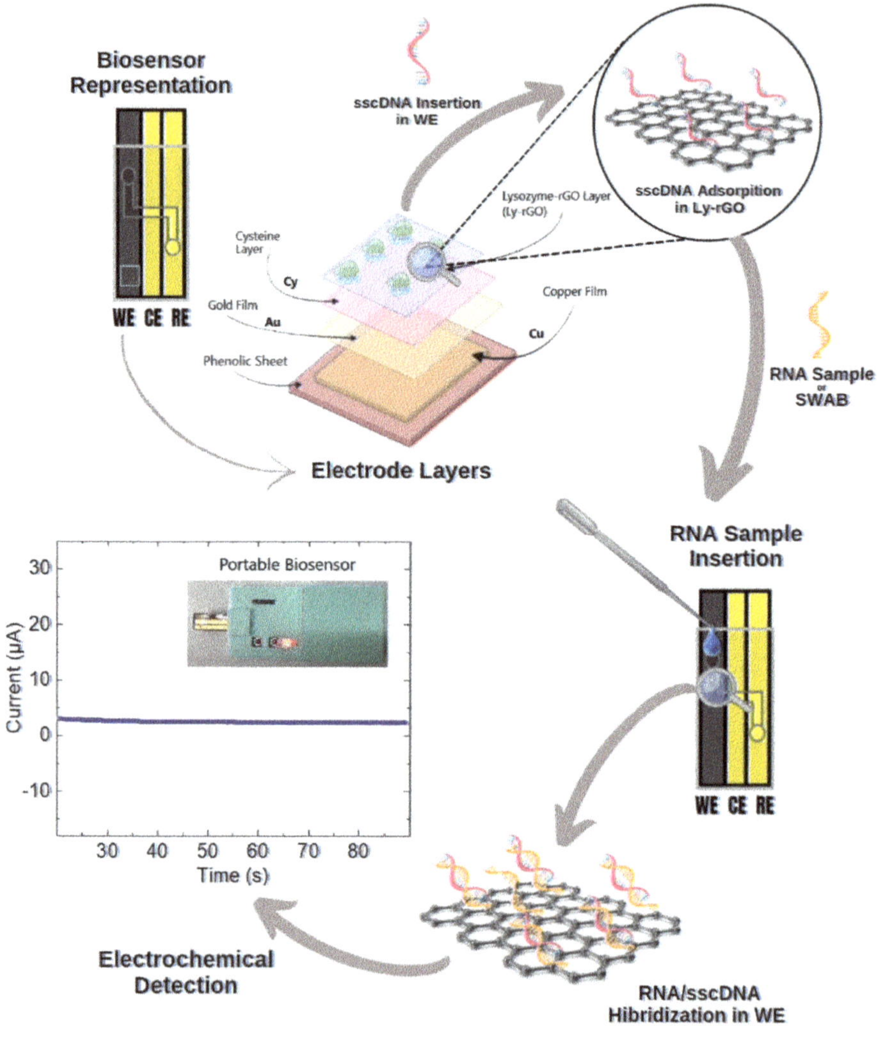

Figure 14.4 Schematic illustration of the biosensor: working electrode (WE) layers and detection of RNA concentration by chronoamperometry technique. Abbreviations: counter electrode (CE), reference electrode (RE), single stranded complementary DNA (sscDNA), lysozyme-reduced graphene oxide (Ly-rGO) Reproduced from ref. 67 with permission from Elsevier, Copyright 2023.

fast, and highly sensitive measurements. The analyte is detected using light with label-based or label-free techniques. In label-based biosensors, a fluorescent or colorimetric label is used and the intensity of fluorescence or change in colour is observed to detect an analyte. On the basis of design aspects and transducers, optical biosensors include surface plasmon resonance, optical waveguides, optical fibres, ring resonators, interferometers, photonic crystals, and fluorescence or luminescence sensors.[68]

Plasmonic biosensors are mainly divided into two types, using thin metal films and inorganic plasmon resonance nanostructures. One of the most common types is surface plasmon resonance sensors, which are used to study biomolecular interactions with metal films, usually made of gold. These sensors can be used independently or integrated with nanostructures and have broad detection capabilities. In the detection of infectious viruses, the need for cost-effective and rapid diagnosis has stimulated the development of biosensor platforms. Recently, plasmonic biosensors have been developed for on-site diagnostic approaches in addition to traditional methods such as PCR and ELISA. Plasmonic applications are important in viral infectious disease detection due to their label-free monitoring method, short response times, and versatile use.[69]

A portable and selective fibre-optic SPR-based sensor for the early detection of dengue virus by targeting NS1 protein as a biomarker has been developed. The sensor utilizes antibody–antigen binding, facilitated by a self-assembled monolayer of alkanethiols on a silver-coated optical fibre. A wavelength interrogation based SPR technique was employed for the detection of NS1 antigen within a dynamic concentration range of 0.2–2.0 μg ml^{-1}. The optical fibre, coated with a 40 nm thick silver layer, demonstrated a resonance wavelength centred around 500 nm. The shift in resonance wavelength was systematically recorded at each stage of molecular attachment to the fibre surface. The sensitivity of the sensor at the lowest NS1 antigen concentration was determined to be 54.7 nM (μg ml^{-1}). Furthermore, the LOD was calculated as 0.06 μg ml^{-1}, which falls within the physiological concentration range of NS1 protein in infected blood. This indicates that the proposed approach has the potential to enable the early detection of NS1 antigen.[70]

Molecularly imprinted polymers are widely utilized in the functionalization of SPR sensor chip surfaces, and a noteworthy study has demonstrated their potential in virus detection. Denizli and colleagues presented a SPR based sensor for the detection of hepatitis B surface antibody (HBsAb) in human serum using a molecularly imprinted poly(hydroxyethyl methacrylate-*N*-methacryloyl-L-tyrosine methyl ester) (PHEMAT) film. The gold SPR chip was first modified with allyl mercaptan, followed by the formation of a HBsAb-imprinted PHEMAT layer. The results of kinetic studies confirmed that the sensor followed the Langmuir adsorption model, with a maximum detection limit of 208.2 mIU ml^{-1}. The chip exhibited high selectivity, showing no significant response to HBsAb-negative serum, highlighting its potential for accurate HBsAb detection in clinical diagnostics.[71]

14.3.3 Microfluidics and Lab-on-a-Chip Devices

The lateral flow assay (LFA) is another sensor system commonly used for the easy detection and visualization of viruses. Due to its low cost and ease of production, it is a widely preferred diagnostic method for virus detection. This test typically consists of a test strip and, when necessary, a reader device. As a paper-based platform, LFA can detect specific virus biomolecules in complex biological samples within 5 to 30 minutes. LFA is categorized on the basis of the type of analyte used. The first category is the lateral flow immunoassay (LFIA), which relies on the binding of antigens and antibodies, while the second category is the nucleic acid-based LFA (NALFA), used to detect PCR products.

The test consists of four main components. First, there is the sample pad, where the sample is applied. Due to capillary action, the sample moves toward the conjugate pad. In the conjugate pad, the analytes in the sample interact with conjugate molecules and are transported to the reaction membrane. On the reaction membrane, test and control bands indicate the interaction with the target molecule. Finally, the sample reaches the absorbent pad, which is necessary to prevent false results by ensuring that the sample does not backflow. Membranes used in LFA tests may include nitrocellulose, nylon, composite, or paper. The results can be read either with special devices, depending on the type of label used, or directly with the naked eye.[72]

Analytical devices using smartphones as detectors and systems based on the colorimetric method are among the most promising technologies. Colorimetry is a widely used method in analytical and biochemical fields. The application of colorimetric methods in smartphone-based imaging devices has contributed to the development of POC testing (POCT) systems. Furthermore, quantification of colour-based reactions can be easily performed thanks to the advanced image processing and computational capabilities of smartphones. Colorimetric methods are widely preferred in smartphone-based biosensors due to ease of use.[73]

In addition to selecting the paper substrate, the functionalization of nanomaterials within the paper matrix plays a significant role in meeting the criteria of simplicity, efficiency, and low-cost production. Various strategies, such as direct deposition, wax printing, inkjet printing, and screen printing are employed to integrate nanomaterials into the paper substrate. Moreover, the structure of the nanomaterial and the type of modification required play a crucial role in the development of suitable colorimetric and fluorescent probes. Techniques such as colorimetry, fluorescence, chemiluminescence, and phosphorescence are precise and selective methods used to analyse various chemical components from different sample matrices. These approaches have gained significant prominence in paper-based platforms, which are widely used for detecting various analytes. Due to their advantages, such as speed, sensitivity, cost-effectiveness, accuracy, and specificity, these methods have attracted considerable interest in analytical and biosensor applications.[74]

Paper-based colorimetric sensor arrays have the advantages of rapid detection, visualized readout, high sensitivity and selectivity, and generally do not require large instruments.[75] Paper-based analytical devices (PADs) are used in LFAs, dipsticks, and micro PADs (μPADs) with 2D or 3D structures, depending on the complexity of the application. These devices change the cellulose regions from hydrophilic to hydrophobic and can be fabricated using techniques such as wax printing, inkjet printing, photolithography, laser processing, and screen printing.[76] Colorimetric sensors are classified under optical sensors, which show single, double, or multicolour change in the presence of detected target molecules.[77]

Gong *et al.* developed a low-cost, paper-based molecularly printed sensor for rapid visual detection of influenza H5N1 virus. H5N1 aptamer (Apt) was covalently attached to the surface of zeolite imidazolate framework (ZIF-8)-NH_2, which was coated with rhodamine B (RhB), *via* an amide bond, forming the fluorescent and colorimetric probe ZIF-8-NH2@RhB~Apt. The sensor demonstrated ultra-high sensitivity, with a LOD of 0.58 fmol l^{-1} when measured using fluorescence spectroscopy. Furthermore, semi-quantitative signals could be observed with the naked eye within 10 minutes.[78] In this study, the presence or absence of H5N1 virus could be analysed with the naked eye. Figure 14.5 illustrates this portable sensor. It was also shown that the virus could be easily detected using this sensor.

COVID-19 and influenza A are respiratory infections with similar clinical symptoms, making their differentiation crucial for outbreak control. A tetrahedral DNA framework (TDF)-based rotational paper analytical device was developed for the simultaneous detection of SARS-CoV-2 and influenza A H1N1. The assay was based on colorimetric analysis using horseradish peroxidase (HRP) and 3,3′5,5′-tetramethylbenzidine (TMB)–H_2O_2, and grayscale signal quantification was performed *via* ImageJ software. The sensor showed a good linear response in the range 10^{-14} to 10^{-8} g ml^{-1} for both viruses, without cross-reactivity. This method provides a rapid, reliable and quantitative approach for the clinical differentiation and diagnosis of COVID-19 and influenza A viruses.[79]

Pixel-diversity interferometric reflectance imaging sensor (IRIS) (PD-IRIS) is a new label-free digital detection platform that improves on conventional IRIS technology. Unlike the time-consuming z-scan acquisition of the previous single particle IRIS (SP-IRIS), PD-IRIS generates an optical signature of target nanoparticles from a single image, making it more suitable for POC applications. The platform demonstrated superior sensitivity in detecting monkeypox virus (MPXV) with a LOD of 200 plaque forming units (PFU) per ml, outperforming laboratory-based ELISA (1800 PFU per ml LOD). MPXV detection specificity was confirmed against herpes simplex virus (HSV-1) and cowpox virus (CPXV). The modular design of PD-IRIS offers significant potential for clinical diagnostics in POC environments by allowing multiplex pathogen detection.[80]

The system proposed in a recent study utilizes displaceable probe loop-mediated isothermal amplification (DP-LAMP) with displaced probes in combination with sucrose-enhanced reagents, allowing infectious mosquitoes to feed and expel viral particles directly into the reaction mixture. Pathogenic arboviruses, including West Nile and dengue viruses, were successfully detected from mosquito saliva within 45 min without the need for RNA extraction. Sucrose demonstrated its effectiveness in stabilizing viral RNA for up to 48 h at field temperatures. Fluorescence-based detection enabled visual

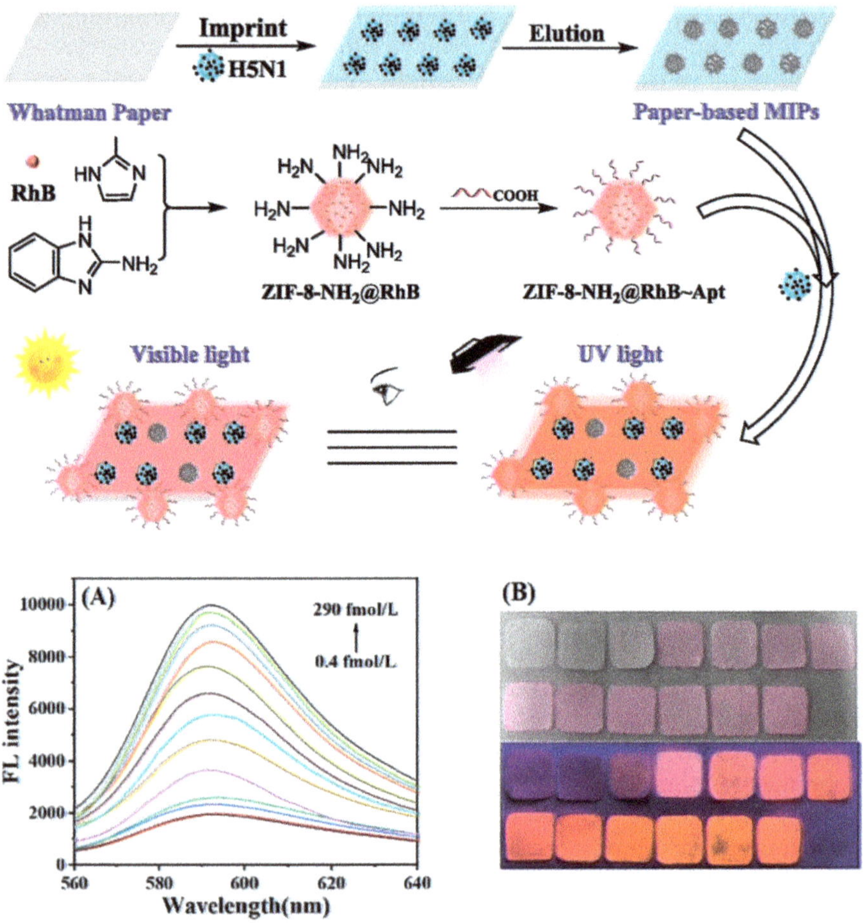

Figure 14.5 The construction and operation of the portable visual virus sensor involve: (a) molecularly imprinted polymer (MIP) detection of H5N1 at varying concentrations (0.4–290 fmol l^{-1}) with a linear response, and (b) visualization of detection results under both visible light (top) and UV illumination (bottom), enabling clear differentiation of target concentrations. Reproduced from ref. 78 with permission from Elsevier, Copyright 2023.

discrimination between positive and negative samples *via* a simple optical image sensor. This innovative approach has the potential to revolutionize arbovirus monitoring by simplifying sample handling, reducing detection time and enabling more comprehensive field surveillance.[81]

Wan *et al.* developed an innovative diagnostic test, human adenoviruses–multiple cross-displacement amplification–lateral flow biosensor (HAdV–MCDA–LFB), for the easy and rapid diagnosis of HAdV, especially HAdV-3 and HAdV-7. The test combines isothermal MCDA and a nanoparticle-based LFB. The assay demonstrated high sensitivity by targeting the conserved hexon gene and detected hexon-containing plasmid at 10 fg per reaction under optimized conditions. HAdV–MCDA–LFB displayed similar diagnostic accuracy to real-time PCR in clinical sample analysis, demonstrating its practical applicability. The method can be completed in less than one hour, including rapid template preparation (15 min), MCDA reaction (40 min at 67 °C) and result interpretation (less than 5 min). Due to its simplicity, high accuracy, and rapid result-generation capacity, this method has attracted attention as a promising POC diagnostic tool for the diagnosis of HAdV-3 and HAdV-7, especially in resource-limited areas.[82]

Li *et al.* developed a highly sensitive and cost-effective immunoresistive sensor on a polyethylene terephthalate (PET) film for POC COVID-19 screening. The sensor used single-walled carbon nanotubes (SWCNTs) functionalized with monoclonal antibodies targeting the SARS-CoV-2 spike protein. Screen-printed silver electrodes reduced contact resistance, while dual-sensor measurements increased detection sensitivity and specificity. The system achieved a lower limit of detection (LLD) of 350 genome equivalents per ml with 100% sensitivity and 90% specificity in clinical sample testing. Its compact design, low power requirement, and simple operation make it a promising tool for rapid COVID-19 diagnostics.[83]

A portable, multi-coloured, and colorimetric MIP-based sensor capable of detecting H5N1 virus with high specificity and sensitivity has been developed. The sensor works by MIPs formed on Fe_3O_4@QDs and H5N1-specific aptamer modification. This sensor was developed in three different modes: fluorescence quenching, visible change of solution colour, and absorption measurement by UV-visible spectrophotometry. This sensor, which has a LOD 11.6 fM, contributes to the effective management of H5N1 infections by providing a reliable method for POC diagnosis.[84]

Xue *et al.* introduced an innovative smart face mask that integrates a flexible immunosensor, a miniaturized impedance circuit, and wireless communication units for *in situ* COVID-19 detection. Figure 14.6 illustrates the appearance and design of the sensor mask. The sensor utilizes a high-density conductive nanowire array where nanoscale voids enhance the capture efficiency of virus particles. The system has demonstrated rapid detection of coronavirus spike protein and whole virus aerosol in simulated human breath, reaching a detection limit of 7 PFU per ml in just 5 minutes. This POC platform offers a promising solution for preliminary COVID-19 screening and monitoring of disease progression during treatment.[85]

Chen *et al.* presented a low-cost diagnostic platform for early and accessible detection of influenza A (FluA) virus. To overcome the limitations of smartphone-based systems, an optical collection unit (OPU) from commercial optical drives with fluorescence LFIA test strips was used. Integrated with 3D printed parts and a customized control system, the device rapidly detected FluA by scanning the test strip with the help of a stepper motor. The prototype offers an important innovation in influenza diagnostics with a detection limit of 2.91 ng ml^{-1}, analysis time under 15 minutes and compact dimensions.[86]

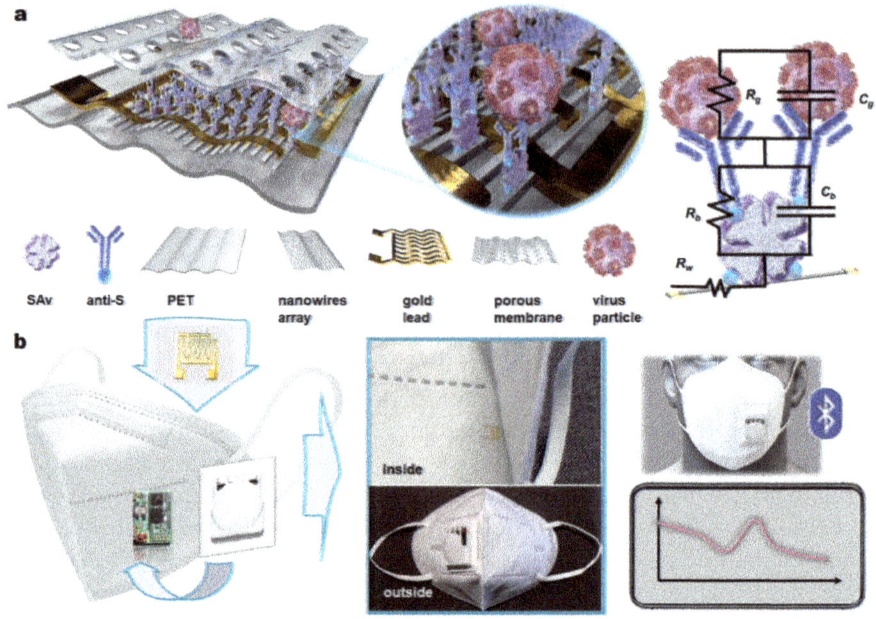

Figure 14.6 Nanoscale sensor design including key components such as polyethylene terephthalate (PET), a nanowire array, gold leads, streptavidin (SAv), anti-spike proteins (anti-S), a porous membrane, and virus particles. (a) The virus sensing mechanism is represented by equivalent resistance (R) and capacitance (C) values for surface-bound antibodies [(R_b) and (C_b)] and antigens [(R_g) and (C_g)], along with nanowire resistance (R_w). (b) The structural images illustrate the intelligent face mask integrated with the sensor for real-time detection. Reproduced from ref. 85 with permission from Elsevier, Copyright 2021.

Another study introduced a new molecular diagnostic method for hepatitis C virus (HCV), reverse transcription loop-mediated isothermal amplification integrated with gold nanoparticle-based lateral flow biosensor (RT-LAMP–AuNPs–LFB). This assay enables sensitive, rapid, and visual detection of common HCV genotypes in China by targeting the 5′ untranslated region (5′UTR). The optimization process involved approximately 5 minutes of nucleic acid extraction, 30 minutes of RT-LAMP reaction at 67 °C, and 2 minutes of visual assessment, for a total of 40 minutes. The assay demonstrated a detection limit of 20 copies per tube, high specificity, and strong anti-parasite properties. These findings demonstrate the potential of the RT-LAMP–AuNPs–LFB method as a rapid, reliable, and low-cost POC test for HCV diagnosis, especially in resource-limited areas.[87]

Wang *et al.* developed a highly sensitive nucleic acid lateral flow assay (NALFA) based on Janus Fe_3O_4–Au@Pt nanoparticles (Fe_3O_4–Au@Pt NPs) to solve the problem of low sensitivity of conventional colloidal gold based lateral flow assays. Due to their strong plasmonic properties and high peroxidase-like activity, these special nanoparticles significantly enhanced the colour change in the test strips, enabling qualitative and quantitative detection with the naked eye. By oxidising TMB to form a dark blue complex (oxTMB), the sensitivity of the assay was increased 20-fold and the LOD was reduced to 50 pM. This method improved the performance of conventional LFA and provided a powerful alternative for sensitive diagnostic systems.[88]

14.3.4 Virus Sensors from Other Perspectives

Sensors can be classified according to the transducer method used in virus detection, but it is also possible to classify them in different ways. A sensor that utilizes an aptamer as a biomolecule on its surface is called an aptasensor. Aptamers are single-stranded oligonucleotide sequences that interact with target molecules through steric and electrostatic interactions.[89,90] Aptasensor and DNA sensors are different from each other in terms of their scope of target recognition and interactions and can be called DNA sensors if virus DNA or RNA is used.

A low-cost, label-free surface acoustic wave (SAW) biosensor that can detect the target DNA sequence of the H1N1 virus quickly, *in situ* and with high sensitivity has been developed. The SAW biosensor was fabricated on a 128° YX–$LiNbO_3$ substrate and DNA probes with phosphate groups were immobilized on the sensor surface by chemical functionalisation with (3-aminopropyl) triethoxysilane (APTES). The specific binding of the target DNA sequence with the probes was detected by the shift in the resonance frequency of the SAW biosensor. The sensor showed a linear response between 50 pM and 10 nM, reaching a detection limit of 14.4 pM. It also exhibited high selectivity and stability in natural environments.[91]

Immunosensors are biosensors based on the high binding affinity between antibodies and target antigens. Antibodies are immunoglobulins produced by the immune system against pathogens and toxins and can recognize and bind to specific antigens.[92] An immunobiosensor was developed using AuNPs and camel nanocores [variable domain of a heavy chain-only antibodies (VHHs)] for rapid and sensitive diagnosis of hantavirus infection. Recombinant anti-hantavirus nucleocapsid nanocores (anti-prNΔ85 VHH) were purified and bioconjugated with AuNPs. The results of spectrophotometry and transmission electron microscopy (TEM) analyses confirmed stable and homogeneous bioconjugation conditions. The assays were able to detect up to 25 ng μl^{-1} of hantavirus antigen, showing specific interaction within 5–10 min. The results indicated that AuNPs–VHH bioconjugates offer an effective biosensor platform for hantavirus diagnostics.[93]

It is also possible to classify biosensors according to the size of the material used in the detection method. Electrochemical sensors have taken various forms, such as SPR sensors, fluorescence-based sensors, and lateral flow tests, and have a wide range of applications in the clinical diagnosis of viruses. A dual-mode aptasensor combining colorimetric and electrochemical methods has been developed for norovirus detection (Figure 14.7). The colorimetric

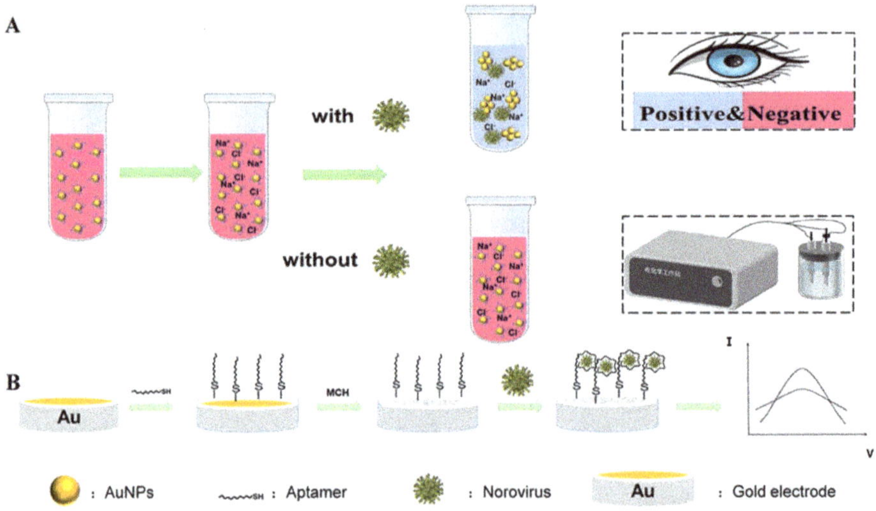

Figure 14.7 Norovirus detection by dual mode aptasensor. (a) Schematic diagram of AuNPs colorimetric mode. (b) Schematic diagram of electrochemical mode. Abbreviation: 6-mercapto-1-hexanol (MCH). Reproduced from ref. 94 with permission from Elsevier, Copyright 2024.

method provides naked-eye detection at 0.1 μg ml^{-1} using AuNPs. The electrochemical method achieved a detection limit of 1.365×10^{-10} μg ml^{-1} by measuring the interaction with norovirus *via* an aptamer-coated gold electrode. The aptasensor stands out as a sensitive and reliable diagnostic tool, showing high specificity in faecal samples.[94]

These sensors provide fast, sensitive, and specific detection of targets and offer effective solutions in important health and safety areas, such as disease diagnosis and detection.

14.4 Future Directions

Developments in artificial intelligence technology, the internet of things (IoT), real-time monitoring systems, materials, and device technologies are increasing day by day in medical detection and diagnosis. It is expected that multiplex sensors will be developed for the simultaneous detection of multiple viruses and that detection and diagnosis will be made in a short time by utilizing artificial intelligence technology in these sensor systems. Advances in transducer design, nanofabrication, and bio-interface innovation have increased the sensitivity, accuracy, and selectivity of multiplex virus biosensors. With scientific advances in the field of sensors, the amount of samples required for detection has been reduced to picolitres, and advances in artificial intelligence enable easy detection of viral species.[95] Especially in sensors, it will be possible to easily detect a virus in a short time by processing meaningful data after the transducer and there will be no need for an expert to evaluate the results. Personalized medicine approaches will be used in the early detection of viruses. Individuals will be detected and evaluated by taking into account their previous viral diseases or chronic diseases. All these virus detection techniques can be integrated with smartphone applications. IoT biosensor systems provide safe detection by preventing the spread of viral infection risk to the environment.

14.5 Conclusion

Screening and rapid virus detection are critical to controlling the spread of infectious diseases, particularly in the face of global viral outbreaks. Early and accurate identification of viruses helps prevent further transmission and allows for timely intervention. Advances in sensor technologies, such as biosensors based on DNA or RNA hybridization, antigen–antibody interactions, and the use of nanoparticles, have significantly improved the speed, sensitivity, and accessibility of virus detection methods. Technologies like electrochemical and optical biosensors are widely used due to their high precision and cost-effectiveness. With the continuous development of sensor materials, such as graphene and MIPs, virus detection has become faster and more efficient. These innovations have expanded the capabilities of sensors, enabling real-time and POC diagnostic systems. Moreover, the integration of biosensors with mobile devices and IoT platforms has enhanced the accessibility and usability of virus-screening tools, even in remote areas. In the future, the combination of advanced sensor technologies with artificial intelligence, machine learning, and personalized medicine will allow for even more rapid and accurate detection of viral infections. As sensor technologies continue to evolve, they promise to revolutionize virus screening and rapid detection, offering a more resilient and adaptive healthcare infrastructure.

References

1. P. Bhadoria, G. Gupta and A. Agarwal, *J. Fam. Med. Primary Care*, 2021, **10**, 2745–2750.
2. J. Kang, A. Tahir, H. Wang and J. Chang, *Wiley Interdiscip. Rev. Nanomed. Nanobiotechnol.*, 2021, **13**, e1700.
3. P. Drori, O. Mouhadeb, G. G. Moya Muñoz, Y. Razvag, R. Alcalay, P. Klocke, T. Cordes, E. Zahavy and E. Lerner, *iScience*, 2024, **27**, 110982.
4. E. Vermisoglou, D. Panáček, K. Jayaramulu, M. Pykal, I. Frebort, M. Kolar, M. Hajduch, R. Zbořil and M. Otyepka, *Biosens. Bioelectron.*, 2020, **166**, 112436.
5. Q. Wang, M. Su, M. Zhang and R. Li, *Int. J. Environ. Res. Pub. Health*, 2021, **18**, 6053.
6. F. Nasrollahi, R. Haghniaz, V. Hosseini, E. Davoodi, M. Mahmoodi, S. Karamikamkar, M. A. Darabi, Y. Zhu, J. Lee and S. E. Diltemiz, *Small*, 2021, **17**, 2100692.

7. Y. Saylan, Ö. Erdem, S. Ünal and A. Denizli, *Biosensors*, 2019, **9**, 65.
8. N. Zhang, L. Wang, X. Deng, R. Liang, M. Su, C. He, L. Hu, Y. Su, J. Ren and F. Yu, *J. Med. Virol.*, 2020, **92**, 408–417.
9. P. S. Morley and N. Pusterla, in *Equine Sports Medicine and Surgery*, ed. K. W. Hinchcliff, A. J. Kaneps, R. J. Geor, and W. B. Saunders, 2nd edn, 2014, pp. 649–664.
10. S. Modrow, D. Falke, U. Truyen and H. Schatzl, *Mol. Virol.*, 2013, **163**.
11. R. Witte, V. Andriasyan, F. Georgi, A. Yakimovich and U. F. Greber, *Viruses*, 2018, **10**, 202.
12. A. Pennesi, G. Ferrari, F. Giardina, S. Piselli, R. L. Savio, A. Carloni, S. Paolucci and F. Baldanti, *Fut. Virol.*, 2024, **19**, 47–57.
13. P. Roingeard, P. I. Raynal, S. Eymieux and E. Blanchard, *Rev. Med. Virol.*, 2019, **29**, e2019.
14. A. A. Kojabad, M. Farzanehpour, H. E. G. Galeh, R. Dorostkar, A. Jafarpour, M. Bolandian and M. M. Nodooshan, *J. Med. Virol.*, 2021, **93**, 4182–4197.
15. J. D. Pearson, D. Trcka, S. Lu, S. J. Hyduk, M. Jen, M. M. Aynaud, J. J. Hernández, P. Peidis, M. Barrios-Rodiles and K. Chan, *Virol. J.*, 2021, **18**, 99.
16. M. Teymouri, S. Mollazadeh, H. Mortazavi, Z. Naderi Ghale-noie, V. Keyvani, F. Aghababaei, M. R. Hamblin, G. Abbaszadeh-Goudarzi, H. Pourghadamyari, S. M. R. Hashemian and H. Mirzaei, *Pathol., Res. Pract.*, 2021, **221**, 153443.
17. N. Garg, F. J. Ahmad and S. Kar, *Curr. Res. Microb. Sci.*, 2022, **3**, 100120.
18. J. Dronina, U. Samukaite-Bubniene and A. Ramanavicius, *J. Nanobiotechnol.*, 2021, **19**, 348.
19. A. Cassedy, A. Parle-McDermott and R. O'Kennedy, *Front. Mol. Biosci.*, 2021, **8**, 637559.
20. S. Akgönüllü and A. Denizli, in *Nanotechnology-Based Smart Remote Sensing Networks for Disaster Prevention*, ed. A. Denizli, M. S. Alencar, T. A. Nguyen, and D. E. Motaung, Elsevier, 2022, pp. 43–57.
21. H. Khalil, F. Mogahed, E. Elkhawaga, O. Sleem, H. Habib and R. Emara, in *Handbook of Nanosensors: Materials and Technological Applications*, Springer, 2024, pp. 963–1012.
22. M. Padash, C. Enz and S. Carrara, *Sensors*, 2020, **20**, 4236.
23. W. Zheng, Y. Yang, C. Liu and W. Zhou, *Sensors*, 2023, **23**, 3218.
24. J. Lukose, S. Chidangil and S. D. George, *Biosens. Bioelectron.*, 2021, **178**, 113004.
25. A. Parihar, S. Yadav, M. A. Sadique, P. Ranjan, N. Kumar, A. Singhal, V. Khare, R. Khan, S. Natarajan and A. K. Srivastava, *Bioeng. Transl. Med.*, 2023, **8**, e10481.
26. E. Mauriz, *Sensors*, 2020, **20**, 4745.
27. R. De Fazio, A. Sponziello, D. Cafagna, R. Velázquez and P. Visconti, *Int. J. Smart Sens. Intell. Syst.*, 2021, **14**, 1–28.
28. T. S. Dhahi, A. K. Y. Dafhalla, S. A. Saad, D. M. I. Zayan, A. E. T. Ahmed, M. E. Elobaid, T. Adam and S. C. Gopinath, *Biotechnol. Appl. Biochem.*, 2024, **71**, 429–445.

29. J. A. Pandit, J. M. Radin, G. Quer and E. J. Topol, *Nat. Biotechnol.*, 2022, **40**, 1013–1022.
30. F. Beierle, J. Schobel, C. Vogel, J. Allgaier, L. Mulansky, F. Haug, J. Haug, W. Schlee, M. Holfelder, M. Stach, M. Schickler, H. Baumeister, C. Cohrdes, J. Deckert, L. Deserno, J.-S. Edler, F. A. Eichner, H. Greger, G. Hein, P. Heuschmann, D. John, H. A. Kestler, D. Krefting, B. Langguth, P. Meybohm, T. Probst, M. Reichert, M. Romanos, S. Störk, Y. Terhorst, M. Weiß and R. Pryss, *Int. J. Environ. Res. Pub. Health*, 2021, **18**, 7395.
31. S. G. Yedire, H. Khan, T. AbdelFatah, R. S. Moakhar and S. Mahshid, *Sens. Diagn.*, 2023, **2**, 763–780.
32. K. Zarean Mousaabadi, Z. Talebi Vandishi, M. Kermani, N. Arab and A. A. Ensafi, *TrAC, Trends Anal. Chem.*, 2023, **169**, 117361.
33. S. Shajari, K. Kuruvinashetti, A. Komeili and U. Sundararaj, *Sensors*, 2023, **23**, 9498.
34. P. Roy, in *Biosensors for Emerging and Re-Emerging Infectious Diseases*, Elsevier, 2022, pp. 137–168.
35. L. Castillo-Henríquez, M. Brenes-Acuña, A. Castro-Rojas, R. Cordero-Salmerón, M. Lopretti-Correa and J. R. Vega-Baudrit, *Sensors*, 2020, **20**, 6926.
36. J. B. Kaushal, P. Raut and S. Kumar, *Biosensors*, 2023, **13**, 976.
37. Y. Cao, L. Zhou, Z. Fang, Z. Zou, J. Zhao, X. Zuo and G. Li, *Chem. Commun.*, 2023, **59**, 3383–3398.
38. S. B. Nanjunda, V. N. Seshadri, C. Krishnan, S. Rath, S. Arunagiri, Q. Bao, K. Helmerson, H. Zhang, R. Jain, A. Sundarrajan and B. Srinivasan, *Nanophotonics*, 2022, **11**, 5041–5059.
39. Z. A. Ratan, F. R. Mashrur, A. P. Chhoan, S. M. Shahriar, M. F. Haidere, N. J. Runa, S. Kim, D.-H. Kweon, H. Hosseinzadeh and J. Y. Cho, *Pharmaceutics*, 2021, **13**, 2034.
40. C.-W. Yen, H. de Puig, J. O. Tam, J. Gómez-Márquez, I. Bosch, K. Hamad-Schifferli and L. Gehrke, *Lab Chip*, 2015, **15**, 1638–1641.
41. G. A. Naikoo, T. Awan, I. U. Hassan, H. Salim, F. Arshad, W. Ahmed, A. M. Asiri and A. Qurashi, *IEEE Sens. J.*, 2021, **21**, 17643–17656.
42. M. Pirzada and Z. Altintas, *Chem. Soc. Rev.*, 2022, **51**, 5805–5841.
43. P. Tan, H. Li, J. Wang and S. C. Gopinath, *Biotechnol. Appl. Biochem.*, 2021, **68**, 1236–1242.
44. S. Khizar, A. A. Al-Dossary, N. Zine, N. Jaffrezic-Renault, A. Errachid and A. Elaissari, *Talanta*, 2022, **241**, 123243.
45. U. Saurabh, D. Abhisek and W. Xing, in *Nano-Engineering at Functional Interfaces for Multi-Disciplinary Applications*, ed. S. S. Ramamurthy, S. Bhaskar, and N. Reddy, Elsevier, 2025, pp. 445–472.
46. A. Denizli, *Molecular Imprinting for Nanosensors and Other Sensing Applications*, Elsevier, 2021.
47. N. Bereli, S. Akgönüllü, S. Aslıyüce, D. Çimen, İ. Göktürk, D. Türkmen, H. Yavuz and A. Denizli, *Hacettepe J. Biol. Chem.*, 2020, **48**, 575–601.

48. N. Idil, M. Bakhshpour, I. Perçin, B. Mattiasson and A. Denizli, in *Molecular Imprinting for Nanosensors and Other Sensing Applications*, Elsevier, 2021, pp. 255–281.
49. G. Jamalipour Soufi, S. Iravani and R. S. Varma, *Analyst*, 2021, **146**, 3087–3100.
50. M. S. Amorim, M. G. F. Sales and M. F. Frasco, *Biosens. Bioelectron.: X*, 2022, **10**, 100131.
51. M. R. Kumalasari, R. Alfanaar and A. S. Andreani, *Talanta Open*, 2024, **9**, 100327.
52. E. Ferrari, *Biosensors*, 2023, **13**.
53. B. Gurung, A. Lama, T. Pokhrel, B. B. Sinjali, S. Thapa, M. Bhusal and A. Adhikari, *J. Nanomater.*, 2023, 8091118.
54. H. Teimouri, S. Taheri, F. E. Saidabad, G. Nakazato, Y. Maghsoud and A. Babaei, *Biomed. Pharmacother.*, 2025, **183**, 117844.
55. V. S. Parkhe and A. P. Tiwari, *World J. Microbiol. Biotechnol.*, 2024, **40**, 269.
56. J. Liu, S. Bao and X. Wang, *Micromachines*, 2022, **13**, 184.
57. O. I. Guliy, B. D. Zaitsev, O. S. Larionova and I. A. Borodina, *Biophysics*, 2019, **64**, 890–897.
58. X. Ma, Y. Shi, G. Gao, H. Zhang, Q. Zhao and J. Zhi, *J. Electroanal. Chem.*, 2023, **950**, 117867.
59. N. Manring, M. M. Ahmed, N. Tenhoff, J. L. Smeltz and P. Pathirathna, *Anal. Chem.*, 2022, **94**, 7149–7157.
60. S. A. Muhsin, Y. He, M. Al-Amidie, K. Sergovia, A. Abdullah, Y. Wang, O. Alkorjia, R. A. Hulsey, G. L. Hunter, Z. K. Erdal, R. J. Pletka, H. S. George, X. F. Wan and M. Almasri, *Anal. Chim. Acta*, 2023, **1275**, 341378.
61. W. Zhou, Y. Tan, J. Ma, X. Wang, L. Yang, Z. Li, C. Liu, H. Wu, L. Sun and W. Deng, *ACS Sens.*, 2022, 7, 3422–3429.
62. Y. Tian and L. Dong, *ACS Sens.*, 2024, **9**, 1978–1991.
63. A. Lomae, K. Teekayupak, P. Preechakasedkit, E. Pasomsub, T. Ozer, C. S. Henry, D. Citterio, T. Vilaivan, O. Chailapakul and N. Ruecha, *Talanta*, 2024, **279**, 126613.
64. B. M. Szydlowska, C. C. Pola, Z. Cai, L. E. Chaney, J. Hui, R. Sheets, J. Carpenter, D. Dean, J. C. Claussen and C. L. Gomes, *ACS Appl. Mater. Interfaces*, 2024, **16**, 25169–25180.
65. D. J. A. D. Santos, T. R. D. Oliveira, G. M. D. Araújo, H. Pott-Junior, M. E. Melendez, E. C. Sabino, O. D. Leite and R. C. Faria, *Biosens. Bioelectron.*, 2024, **255**, 116210.
66. M. Chaturvedi, M. Patel, D. P. Mondal, A. K. Srivastava, N. Dwivedi and C. Dhand, *Electrochim. Acta*, 2024, **475**, 143630.
67. R. Challhua, L. Akashi, J. Zuñiga, H. Beatriz de Carvalho Ruthner Batista, R. Moratelli and A. Champi, *Biosens. Bioelectron.*, 2023, **232**, 115291.
68. A. K. Singh, S. Mittal, M. Das, A. Saharia and M. Tiwari, *Alexandria Eng. J.*, 2023, **67**, 673–691.
69. J. Hasan and S. Bok, *Biosensors*, 2024, **14**, 130.
70. S. K. Gahlaut, A. Pathak, B. D. Gupta and J. P. Singh, *Biosens. Bioelectron.*, 2022, **196**, 113720.

71. L. Uzun, R. Say, S. Ünal and A. Denizli, *Biosens. Bioelectron.*, 2009, **24**, 2878–2884.
72. M. Abdolhosseini, F. Zandsalimi, F. S. Moghaddam and G. Tavoosidana, *J. Virol. Methods*, 2022, **301**, 114461.
73. T. Sahare, N. Singh, B. N. Sahoo and A. Joshi, *Biosens. Bioelectron.: X*, 2024, **21**, 100544.
74. S. Patel, R. Jamunkar, D. Sinha, Monisha, T. K. Patle, T. Kant, K. Dewangan and K. Shrivas, *Trends Environ. Anal. Chem.*, 2021, **31**, e00136.
75. Y.-Q. Li and L. Feng, *Chin. J. Anal. Chem.*, 2020, **48**, 1448–1457.
76. T. Ozer and C. S. Henry, *TrAC Trends Anal. Chem.*, 2021, **144**, 116424.
77. S. Krishnan and Z. u. Q. Syed, *Sens. Actuators Rep.*, 2022, **4**, 100078.
78. H. Gong, L. Tang, C. Chen, F. Chen and C. Cai, *Chem. Eng. J.*, 2023, **477**, 146990.
79. F. Li, J. Qi, Z. Ren, X. Hu, Y. Chen, B. Li and X. Fu, *Microchem. J.*, 2023, **185**, 108304.
80. M. Aslan, E. Seymour, H. Brickner, A. E. Clark, I. Celebi, M. B. Townsend, P. S. Satheshkumar, M. Riley, A. F. Carlin, M. S. Ünlü and P. Ray, *medRxiv*, 2024.
81. D. Kim, T. J. DeBriere, B. H. Eastmond, A. A. Alomar, O. Yaren, J. McCarter, K. M. Bradley, S. A. Benner, B. W. Alto and N. D. Burkett-Cadena, *PLOS ONE*, 2024, **19**, e0298805.
82. L. Wan, X. Huang, J. Fu, F. Xiao, N. Jia, Y. Zhang, M. Chen, X. Zhu, S. Li, J. Zhou and Y. Wang, *Sens. Actuators Rep.*, 2024, **8**, 100252.
83. T. Li, S. D. Soelberg, Z. Taylor, V. Sakthivelpathi, C. E. Furlong, J.-H. Kim, S.-g. Ahn, P. D. Han, L. M. Starita, J. Zhu and J.-H. Chung, *Biosensors*, 2022, **12**, 149.
84. H. Gong, L. Tang, F. Chen, C. Chen, Y. Cheng and C. Cai, *Chem. Eng. J.*, 2024, **483**, 148946.
85. Q. Xue, X. Kan, Z. Pan, Z. Li, W. Pan, F. Zhou and X. Duan, *Biosens. Bioelectron.*, 2021, **186**, 113286.
86. X. Chen, X. Huang, S. Kanwal, J. Wang, J. Wen and D. Zhang, *Biosensors*, 2024, **14**, 263.
87. Y. Shi, Q. Zhou, S. Dong, Q. Zhao, X. Wu, P. Yang, X. Zeng, X. Yang, Y. Tan, X. Luo, Z. Xiao and X. Chen, *BMC Microbiol.*, 2024, **24**, 68.
88. K. Wang, D. Yu, X. Liang, W. Lu, X. Jiang, K. Tashpulatov, J. Zeng and C.-Y. Wen, *Microchem. J.*, 2024, **204**, 111104.
89. B. Lou, Y. Liu, M. Shi, J. Chen, K. Li, Y. Tan, L. Chen, Y. Wu, T. Wang, X. Liu, T. Jiang, D. Peng and Z. Liu, *Trends Analyt Chem*, 2022, **157**, 116738.
90. D. S. D. Divya, R. Kumari, S. Mahapatra, R. Kumar and P. Chandra, *Biosensors*, 2022, **12**, 81.
91. H. Sun, Y. Tao, Y. Li, M. Wang and H. Wang, *IEEE Sens. J.*, 2024, **24**, 17570–17578.
92. D. Bhatia, S. Paul, T. Acharjee and S. S. Ramachairy, *Sens. Int.*, 2024, **5**, 100257.

93. E. A. Bastos-Soares, M. S. da Silva Morais, M. Funes-Huacca, R. M. O. Sousa, N. Brilhante-Da-Silva, S. A. Roberto, N. D. R. Prado, C. N. D. dos Santos, A. C. M. Marinho, A. M. Soares, R. G. Stabeli, S. D. S. Pereira and C. F. C. Fernandes, *Mol. Diagn. Therapy*, 2024, **28**, 479–494.
94. X. Li, Y. Zhao, W. Gu, Y. Qian, Q. Huang, X. Hu and H. Xing, *Anal. Biochem.*, 2024, **687**, 115444.
95. Y. Zheng, X. Song, Z. Fredj, S. Bian and M. Sawan, *Anal. Chim. Acta*, 2023, **1244**, 340860.

CHAPTER 15

Environmental Monitoring

MEHMET KARAGÖZLÜ[a,b], FULYA YONUCU[c], VALARIE ORU AGBOR[b,c], BEHNAZ SHIRGIR[c] AND SÜLEYMAN AŞIR[b,c*]

[a]Department of Food Engineering, Faculty of Agriculture, Near East University, Cyprus, Mersin 10, Nicosia 99138, Turkey; [b]Research Center for Science, Technology and Engineering (BILTEM), Near East University, Cyprus, Mersin 10, Nicosia 99138, Turkey; [c]Department of Biomedical Engineering, Faculty of Engineering, Near East University, Cyprus, Mersin 10, Nicosia 99138, Turkey
*E-mail: suleyman.asir@neu.edu.tr

15.1 Introduction

The detection of viruses in environmental samples is critical for safeguarding public health. Environmental monitoring acts as an early warning system for potential viral outbreaks, enabling timely interventions and preventing widespread transmission. Sensor technologies play a vital role in this effort by providing rapid, sensitive, and cost-effective methods for detecting viral pathogens in various environmental matrices. This chapter provides a comprehensive overview of the applications of sensor technologies in virus detection for both environmental monitoring and wildlife surveillance. The chapter is structured to cover water and air quality assessment, the surveillance of public spaces, and monitoring of virus transmission in wildlife.

Detection Science Series No. 32
Advances in Virus Detection with Sensors
Edited by Adil Denizli
© The Royal Society of Chemistry 2026
Published by the Royal Society of Chemistry, www.rsc.org

15.2 Water and Air Quality Assessment

Air pollution can be defined as the decrease in air quality due to the spread of particulate matter, volatile organic compounds, oxides and microorganisms into the environment and their suspension in the air. Sources of air pollution include dust from ventilation or solid combustibles, carbon monoxide (CO), nitrogen oxide (NO), sulphur dioxide (SO_2), cigarette smoke and airborne microorganism pollutants, and products from human activities such as fossil fuel combustion and exhaust from vehicles and industries.[1] Air pollution is an increasing problem and it brings with it different health problems. In recent years, with the increase in health problems, the importance of air quality has been emphasized in scientific research.[2,3] In countries where economic development has been rapid and more widespread, many people have been exposed to urban air pollution due to changes in lifestyles.[4] Studies have shown that exposure to air pollution brings about respiratory[5] and cardiovascular diseases.[6] Exposure to air pollution, in particular, causes oxidative stress, which leads to the production of free radicals and as a result, damages the respiratory system by reducing resistance to bacterial and viral infections.[3] As shown in Figure 15.1, airborne transmission is defined as the spread of infectious agents by aerosols evaporated from droplets and the spread of droplet nuclei that remain suspended in the air for long distances and durations.[7,8]

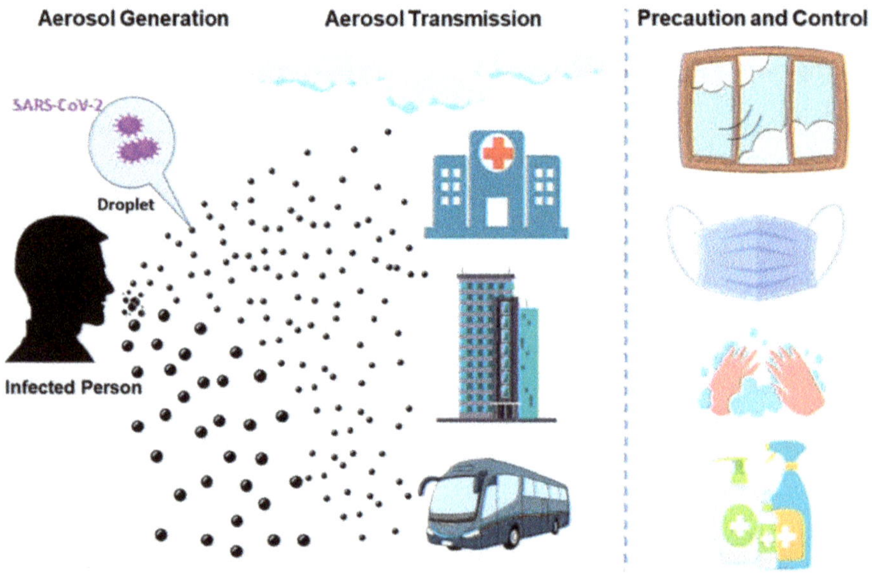

Figure 15.1 Formation of viruses through droplets, their transmission and precautionary measures. Reproduced from ref. 8, https://doi.org/10.1016/j.envint.2020.106039, under the terms of the CC BY 4.0 license, https://creativecommons.org/licenses/by/4.0/.

The significant impact of the pandemic caused by severe acute respiratory syndrome coronavirus 2 (SARS-CoV-2), which emerged in 2019, on health and daily and socioeconomic lives has created significant interest in all issues related to this serious infection.[9] Nowadays, air quality control has become an important issue to be investigated, especially after coronavirus disease (COVID)-19. In addition to the transmission of bioaerosol droplets in closed areas through human contact, other factors have been found to have an effect on the deaths caused in detected COVID-19 cases.[10] These include indoor and outdoor air pollution,[11] weather,[12] meteorological data (temperature, humidity, precipitation, *etc.*),[13] personal health, biological properties of the virus, and certain climatic conditions.[14] Environmental impressions have become an important topic of interest worldwide due to the relationship between human health and environmental pollution. There is a need for innovative, portable and low-cost new detection devices that aim to minimize environmental pollution. Existing strategies have many disadvantages, such as being difficult to use due to being generally heavy and large, consuming high levels of energy and funds, being time-consuming, and requiring trained professional staff. Even if air quality measurement stations detect air pollution correctly in pandemic situations such as COVID-19, they have become difficult to use due to the disadvantages mentioned.[1] In addition to air pollution, which significantly threatens human health, water pollution is also an important environmental problem caused by the increasing population, unplanned urbanization, and technological advancement in the world. With the increasing population, it has become difficult to prevent the pollution of water resources, such as seas, lakes, and rivers, due to chemical or environmental factors. Water is an essential part of human life, as it is used in agriculture, factories, and entertainment sectors, and it is a basic human need. It has become important to protect this basic need from all kinds of contaminants that cause diseases.[15] Although the purification systems used today have achieved good results in removing physical, chemical, and some microbiological contaminants,[16] they are inadequate in removing enteric viruses that accumulate in the intestines of both animals and humans and spread to the environment *via* faeces.[17] Viruses that are persistent and nano-sized settle in underground water resources faster than bacteria found in the soil. Since these viruses are difficult to remove, the risk of infectious diseases caused by enteric viruses increases.[18] According to the drinking water guide, waterborne viruses of great importance to human health include enteroviruses,[19] adenovirus, rotavirus, norovirus and other caliciviruses, astrovirus, hepatitis A (HAV), polioviruses, and coxsackieviruses.[20] Irrigation in the agricultural sector accounts for over 80% of overall water use.[21] Irrigation water used in the agricultural sector is obtained from water sources such as ponds, lakes, and rivers and may contain various viruses that threaten human health. It has been reported that enteric viruses are the most difficult pathogens to detect in water sources.[22,23] In recent years, various methods have been developed for the detection of these viruses in the aquatic environment.

Although biosensors are still under development, they have the potential to become the mainstream method for detecting viruses in aquatic environments.

15.2.1 Water Quality

Water is an important need for the continuation of the human race, energy production, the survival of plants and animals, and consumption of resources.[24] Increasing population and water pollution parallel each other. Society's production methods bring water pollution with them. It is estimated that water sources containing contaminated water cause serious diseases affecting human health and the death of 485 000 people annually worldwide as a result of these diseases.[25] Pathogenic microorganisms in drinking water sources cause epidemic diseases, such as gastroenteritis, cholera, giardiasis, cryptosporidiosis, *etc.*[26] The presence of pathogens in drinking water raises concerns about the spread of bacteria resistant to antibiotics.[27] Evaluation of water quality is essentially based on physiological, biological, and heavy metal concentrations.[28] The quality of water is reduced by pollutants due to the organic, inorganic, biological, thermal, or radiological substances they contain.[29] The survival rate of viruses found in water depends on different factors, such as temperature and pH. Some viruses can survive for long periods in drinking water, such as HAV, where 99% inactivation occurs after approximately 56 days.[30] The quality of water in a source is affected by industrial, agricultural, human exploitation, and natural processes, such as precipitation, abrasion, erosion, mineral deposits, and other geological events.[31] As seen in Figure 15.2, the risk of contamination increases due to irrigation methods and for other reasons, such as being affected by rain and discharges of dirty cooling water.[32]

Surface waters are the most sensitive and vulnerable water resources to pollution when exposed to different types of waste and runoff.[33] Groundwater is better protected against direct runoff, but once contaminated, it remains contaminated for longer.[34] Common parameters used to determine water quality include pH value, total dissolved solids, biological and chemical oxygen demand, ammonia and ammonium value, nitrate, heavy metal values, and bacteriological contamination. The pH of water resources is an important parameter for water quality. Chemical processes in water systems, such as acid–base reactions, solubility reactions, and oxidation–reduction reactions, are all affected by pH value. Water resources around the mining industry are susceptible to metal accumulation due to landfill leaks and other waste discharges from this industry.[35] Dangerous viruses that cause bacteriological contamination are included in the group of enteric viruses, which are non-enveloped viruses that can multiply in the human gastrointestinal tract. These viruses can cause gastrointestinal diseases in animals and humans, the most important complication of which is diarrhoea.[36] The most important enteric viruses that are encountered in water sources belong to the families Picornaviridae (enterovirus and HAV) and

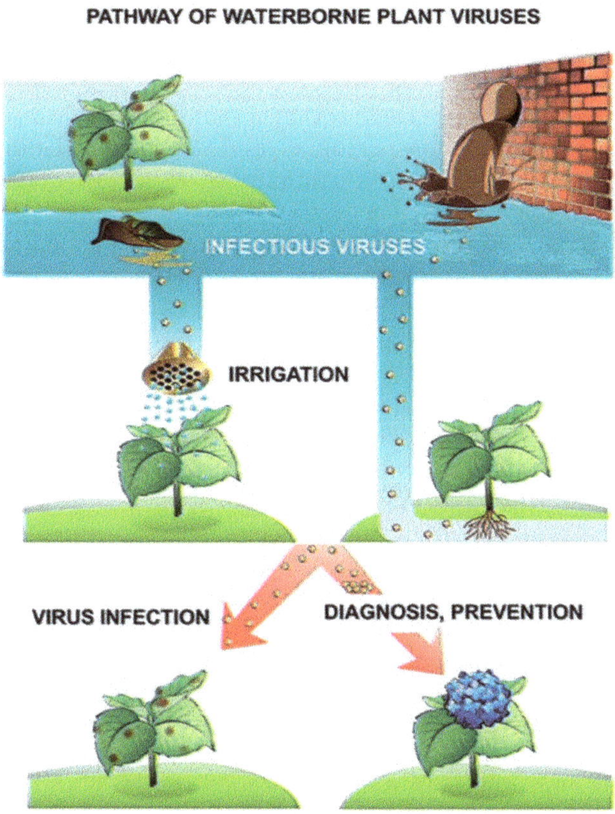

Figure 15.2 Transmission modes of plant viruses in aquatic environments. Reproduced from ref. 32 with permission from Elsevier, Copyright 2012.

Adenoviridae (adenovirus).[37] These viruses, which are detected in almost all water sources (seawater, freshwater, wastewater, drinking water), are usually excreted in high concentrations in the faeces of infected people.[38] Enteric viruses are the leading cause of gastroenteritis, which costs billions of dollars in healthcare costs each year.[39] Waterborne gastroenteritis is estimated to cause 3.4 million deaths annually worldwide.[40] Many types of pathogenic bacteria can be found in water, including *Vibrio cholerae*, *Escherichia coli*, *Salmonella typhi*, *etc.*, and these bacteria can cause waterborne diseases like diarrhoea, cholera, typhoid, *etc.*[41]

15.2.1.1 Detection of Viruses in Water

In recent years, biosensors used in the health sector, and food and water analysis have attracted attention with their powerful and innovative technology to detect different biomarkers in both health and environmental quality control contexts. Biosensors have the potential to be sensitive, fast, and selective detection devices with low sample volumes and high signal strengths.[42]

Optical and electrochemical detection methods are the most widely used biosensor technologies, with low detection limits.[43] Today, traditional methods still used to detect pathogenic microorganisms in the aquatic environment are very effective and sensitive, but they retain their disadvantages, such as high cost, the need for many personnel, and long periods before results are obtained.[44] Table 15.1 presents a summary of the viruses detected in water with biosensors.

In their study, Yıldırım *et al.* used the transient wave optical immunosensor method to detect adenovirus in wastewater samples containing adenovirus.[45] This process, which was completed in less than an hour, was successfully completed with a detection limit of 100 plaque forming units (pfu) per ml.[45] Manzano *et al.* developed an electrochemical biosensor that utilized DNA hybridization with a gold electrode to successfully detect HAV in food and water samples containing pathogenic microorganisms. The cyclic voltammetry (CV) technique was used to monitor the oxidative hybridization of the indicator tripropylamine (TPA), as shown in Figure 15.3. A detection limit of 0.65 pM for HAV in freshwater was achieved.[46]

In a study by Hai *et al.*, an organic electrochemical transistor (OECT) with a trisaccharide-grafted conductive polymer channel was developed for the detection of human influenza A virus in aqueous media.[47] The platform developed offers a potential application for diagnosing different types of influenza viruses with H1 protein on their surfaces. Detection was achieved with a lower limit of detection of 3.54 $\mu g\ ml^{-1}$.[47] Another study used methylene blue as an electrochemical detection redox intercalator probe to generate a diffusion-controlled current that encodes the presence and concentration of reverse transcription loop-mediated isothermal amplification (RT-LAMP) products, specifically amplicons or double-stranded DNA.[48] The sensor has been tested on real wastewater samples for proper operation. SARS-CoV-2 was detected at below the electrochemical detection limit of approximately $2.5 \times 10^{-6}\ ng\ \mu l^{-1}$.[48]

Table 15.1 Total records for biosensor applications in water.

Antigen	Sample	Method	Limit of detection (LOD)	Reference
Adenovirus	Wastewater	Optical fibre sensor	100 plaque forming units (pfu) per ml	45
Hepatitis A virus	Freshwater	Electrochemical DNA biosensor	0.65 pM	46
Influenza A virus	Freshwater	Electrochemical biosensor	3.54 $\mu g\ ml^{-1}$	47
SARS-CoV-2	Wastewater	Loop-mediated isothermal amplification-based electrochemical sensor	$2.5 \times 10^{-6}\ ng\ \mu l^{-1}$	48

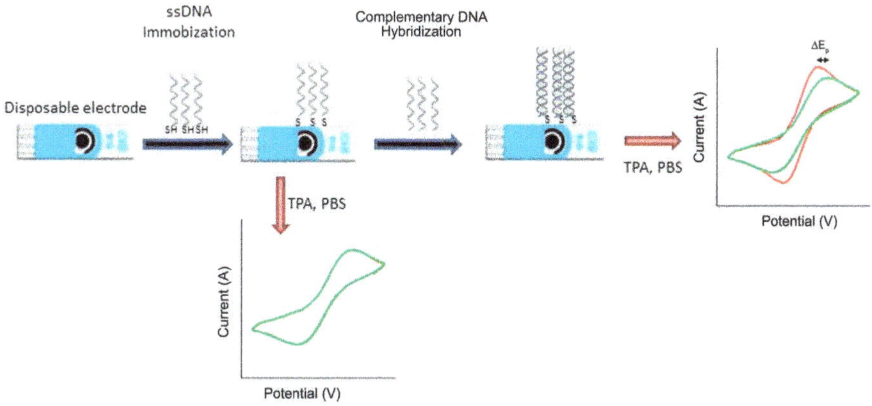

Figure 15.3 A schematic representation summarizing the steps in hepatitis virus detection. Abbreviations: single stranded DNA (ssDNA), tripropylamine (TPA), phosphate buffered saline (PBS). Reproduced from ref. 46 with permission from Elsevier, Copyright 2018.

15.2.2 Air Quality

Air quality is reduced by environmental sources, such as pets and cleaning products, pollutants released into the air by burning coal, wood and tobacco products inside the building, as well as outdoor pollutants that can be produced mostly through vehicle traffic, industrial activities, and natural and mechanical ventilation systems.[49] Abiotic events, such as dust storms, as shown in Figure 15.4, increase and trigger the spread of microbes while also affecting their cycles in the air.[50]

Non-communicable diseases have become a worrying situation, accounting for 70% of deaths caused by air pollution.[51] In 2019, the number of deaths caused by air pollution with an uncertainty rate of 95% was 6.67 million, accounting for 11.62% of deaths due to environmental factors.[52] Long-term exposure to air pollution containing particulate matter and heavy metals increases the risk of chronic cardiovascular[53] and respiratory diseases, type 2 diabetes and cancers, and premature death.[54] The World Health Organization detected a novel coronavirus, SARS-CoV-2, in Wuhan, China, in late 2019, this virus caused the COVID-19 outbreak, which became a pneumonia epidemic. Pollutants and respiratory viruses in the air environment interact with each other. It is known that particulate matter can carry microorganisms such as viruses, and SARS-CoV-2 RNA can be detected in particulate matter.[55] Detection of pollution has been identified as an important criterion for controlling and analysing pollutants. Fast, accurate, low-cost and sensitive sensors that can determine pollutant detection limits can increase the possibility of protecting the environment and human health.[56] As shown in Figure 15.5, pathogenic microorganisms in the air can be detected by nanosensors that interact with biomolecules on their surfaces.[57]

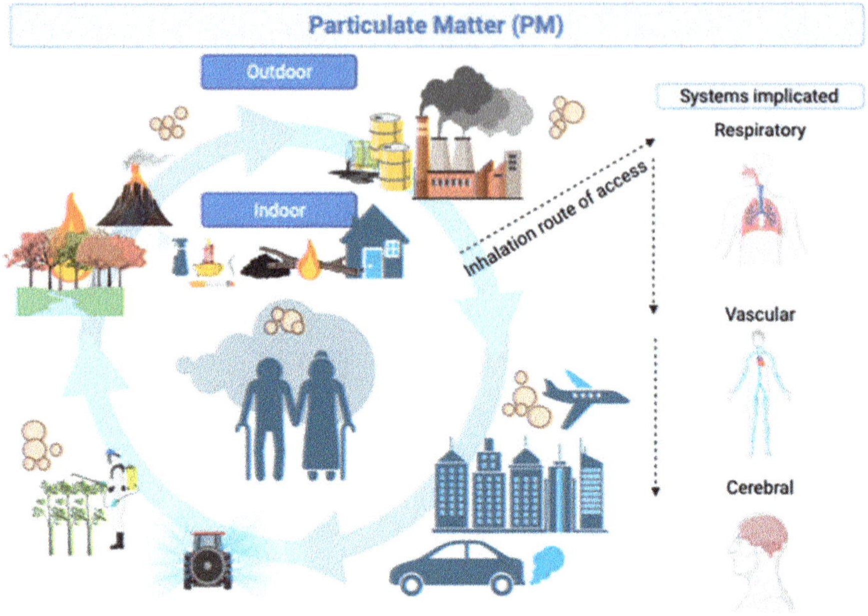

Figure 15.4 Factors causing air quality to decline and diseases caused by it. Reproduced from ref. 50, https://doi.org/10.3390/biom14010035, under the terms of the CC BY 4.0 license, https://creativecommons.org/licenses/by/4.0/.

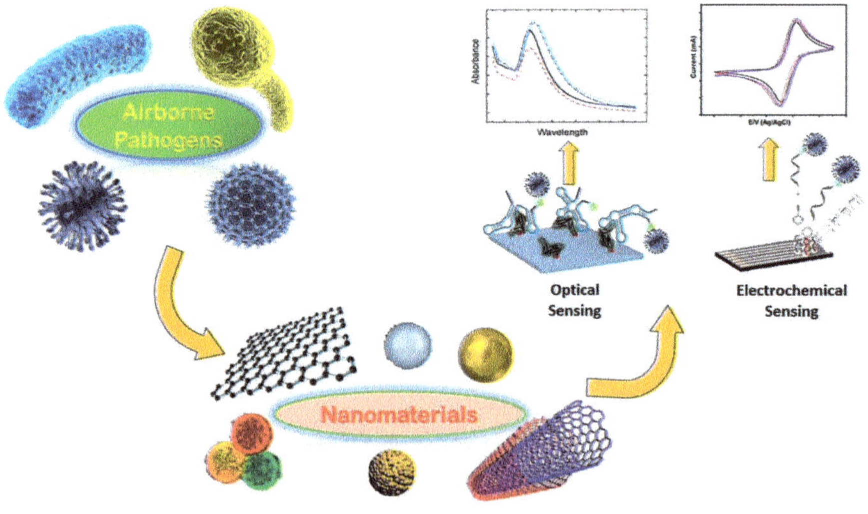

Figure 15.5 Schematic representation of nano-biosensors used in air quality measurement. Reproduced from ref. 57, https://doi.org/10.1016/j.envint.2020.106183, under the terms of the CC BY 4.0 license, https://creativecommons.org/licenses/by/4.0/.

15.2.2.1 Detection of Viruses in Air

Among the current and traditional methods for detecting viruses in the air that threaten human health, reverse transcriptase polymerase chain reaction (RT-PCR)[58] and enzyme-linked immunosorbent assay (ELISA) based methods are most frequently used.[59] However, the ELISA technique causes serious problems due to expensive antibodies, longer detection time, and the need for extra personnel. In order to prevent and detect the spread of these viruses, fast and selective biosensors have been investigated. These detection tools can be classified as optical methods, electrochemistry, and acoustic wave technologies.[60] As shown in Table 15.2, many researchers have investigated the detection of viruses in the air using biosensors and a summary of these studies is presented.

Lee *et al.* studied a paper-based immunosensor that supports label-free measurement of multiple avian influenza virus (H5N1, H7N9, and H9N2) antigens using flexible screen-printed carbon nanotube–polydimethylsiloxane electrodes. Detection limits of 55.7 pg ml^{-1} (0.95 pM) were achieved for H5N1 haemagglutinin (HA), 99.6 pg ml^{-1} (1.69 pM) for H7N9 HA, and 54.0 pg ml^{-1} (0.72 pM) for H9N2 HA antigens in phosphate-buffered saline. The sensors exhibited good selectivity and reproducibility. These paper-based sensors are

Table 15.2 Studies demonstrating biosensors for bioaerosol detection.

Antigen	Method	Working electrode	LOD	Reference
H5N1 haemagglutinin (HA)	A paper-based immunosensor	Flexible screen-printed carbon nanotube-polydimethylsiloxane electrodes	55.7 pg ml^{-1}	61
H7N9 HA			99.6 pg ml^{-1}	
H9N2 HA			54.0 pg ml^{-1}	
Micrococcus serum 2 (MS2)	A wet paper-based electrochemical immunosensor	A modified electrostatic particle concentrator (EPC)	16 plaque forming units (PFU) ml^{-1}	62
SARS-CoV-2	A paper-based electrochemical immunosensor incorporating paper working electrodes (PWEs)	Carbon nanotube coated porous paper working electrodes	65.7 PFU ml^{-1}	63
Respiratory syncytial virus (RSV)	Electrochemical impedance spectroscopy (EIS)	Screen printed gold carbon electrode	1.10×10^3 PFU ml^{-1}	64
H1N1	Electrochemical sensors based on modified screen-printed electrodes	Gold electrode	0.002 ng μl^{-1}	65

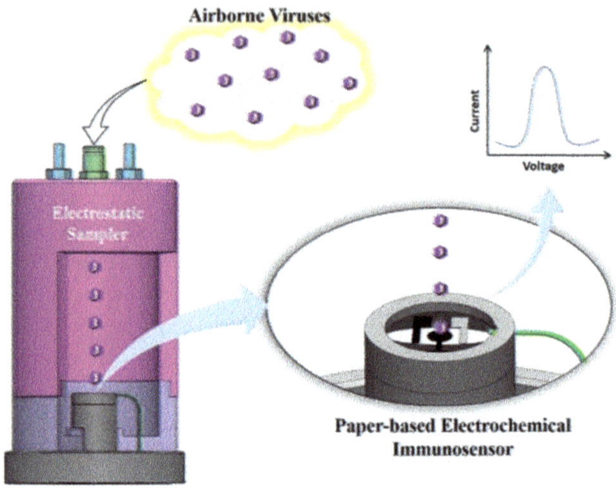

Figure 15.6 Schematic representation of a wet paper based electrochemical immunosensor. Reproduced from ref. 62 with permission from Elsevier, Copyright 2023.

economical, flexible, durable, easy to fabricate, and capable of detecting a variety of avian influenza viruses.[61] In another study, it was demonstrated that airborne Micrococcus serum 2 (MS2) viruses could be measured with accurate time and sensitivity using a wet paper-based electrochemical immunosensor and a modified electrostatic particle concentrator (EPC) (Figure 15.6). It was stated that this biosensor detected MS2 viruses even at very low concentrations, such as 16 plaque forming units (PFU) ml^{-1}.[62]

Lee, *et al.* studied a carboxylated single-walled carbon nanotubes (SWCNT)-modified porous paper-based wireless electrochemical biosensor for the extremely rapid and sensitive detection of all coronaviruses in liquids and air.[63] In another study, a screen-printed gold electrode was used to detect respiratory syncytial virus (RSV) and a successful result was obtained in terms of sensitivity and LOD. In the study, the measurement of the gold electrode extended to the analysis of electrochemical impedance spectroscopy spectra. A good analytical result was recorded for RSV detection with a detection limit of 1.10×10^3 PFU ml^{-1}.[64] Ravina and colleagues developed a screen-printed electrode using a gold electrode modified with an amine group due to its advantages, such as high biocompatibility and high conductivity. In this study, 0.002 ng μl^{-1} H1N1 was detected using the differential pulse voltammetry (DPV) analysis method.[65]

In conclusion, the transmission of viruses through environmental factors, such as air and water, presents a significant public health risk. This section has explored how airborne viruses, particularly those responsible for respiratory diseases, interact with environmental factors, such as temperature, humidity, airflow, and water quality. Airflow and ventilation are key factors in viral spread; poorly ventilated indoor spaces allow viruses to linger in the air for longer periods, facilitating wider transmission. In particular, the rapid and effective airborne transmission of viruses, as seen in the cases of COVID-19, influenza, and other respiratory infections, highlights the importance of these factors. Waterborne viruses also contribute to substantial health challenges worldwide, particularly in relation to water contamination. Viruses in irrigation, drinking water, and wastewater systems pose direct risks to human health. Pollution in water sources can extend the survival of viruses, allowing them to spread more broadly. Moreover, the physical and chemical properties of water play a crucial role in the stability and infectivity of viruses. In this research, the quality of both air and water is essential in the transmission of viruses through the environment. Factors such as air pollution, poor ventilation, high humidity, and elevated temperatures can significantly enhance the spread of viruses. Similarly, water pollution increases the risk of public health threats by promoting the proliferation of harmful microorganisms. Implementing effective environmental policies and hygiene practices is critical to controlling virus transmission and safeguarding public health. Looking ahead, developing strategies to reduce transmission through both air and water by gaining a deeper understanding of how viruses interact with environmental factors will be crucial in preventing future global health crises. Similarly, water pollution increases the risk of public health threats by supporting the proliferation of harmful microorganisms. Therefore, it is of great importance to control viral contamination and protect public health by giving importance to effective environmental factors and hygiene. In order to prevent these risks, biosensors have been developed to detect viruses that pollute the environment. A biosensor should be able to provide low levels of detection at a low cost and with a fast analysis time. It should also be a device that is easy to operate, portable, and can be used to make on-site measurements. When today's technologies are examined, devices that can fulfil these goals have been designed and are still being developed.

15.3 Surveillance of Public Spaces

Environmental monitoring is essential for public health since it serves as an early warning system for viral outbreaks, such as COVID-19, by detecting viral RNA in environmental samples prior to the emergence of clinical cases.[66] This proactive strategy facilitates prompt responses, especially in regions with managed transmission or when public health controls are being eased.[66] Moreover, environmental surveillance is proven to be economically efficient, particularly wastewater monitoring, due to the widespread availability and relative affordability of quantitative reverse transcription PCR (qRT-PCR) testing for viral RNA.[67] Environmental surveillance overcomes the limitations of clinical testing by avoiding obstacles, such as informed consent, sample collection difficulties, and accessibility issues, allowing for targeted interventions at various levels.[67] This approach has demonstrated its use not just for COVID-19 but also for surveillance of poliovirus, detecting new virus importations, and verifying the presence of vaccine-related viruses after vaccination campaigns.

Environmental monitoring employs a number of strategies and tactics, including air sampling, surface swabs, and wastewater sampling.[67,68] Cell culture for viability evaluation, intact virus characterisation, and nucleic acid based approaches are the main detection techniques. Nucleic acid based methods are the gold standard for both qualitative and quantitative analysis.[68] Nonetheless, obstacles, such as low viral load, matrix interference, and problems with structural integrity, need to be addressed.[68] To get over these restrictions, better pretreatment, analysis, and sampling protocols have been proposed. These include improved viral recovery and higher detection sensitivity.[68] Environmental surveillance has proven useful for SARS-CoV-2 detection, with studies detecting viral RNA with portable air samplers and in heating, ventilation and air conditioning (HVAC) systems.[67] Finally, it has been pivotal in the diagnosis of zoonotic diseases, providing insights into the transmission of infection as well as strengthening public health responses.[69] These advances enable environmental monitoring to mitigate global readiness against outbreaks of viruses, providing valuable insights for making informed public health decisions.[66,67,70,71]

Public space surveillance is essential for preventing virus epidemics as it can help in identifying suspected health threats earlier. Surveillance systems (*e.g.*, environmental sensors, wastewater analysis, and digital tracking technologies) assist in identifying the presence of viruses in crowded areas (*e.g.*, transport hubs, marketplaces, and hospitals).[72] Such systems provide near real-time data that can inform public health officials, allowing them to implement targeted actions, such as quarantine and disinfection regimes, to prevent widescale transmission.[52] Moreover, the integration of artificial intelligence (AI) and machine learning in monitoring enhances predictive modelling, enabling the prediction of disease trends and optimizing resource allocation for outbreak prevention.[73]

Also, public space surveillance assists with policy-making and assessment of risks in virus epidemics. Surveillance data can inform government strategies to mitigate this risk of infection, such as social distancing efforts and travel policies.[74,75] Surveillance technology (*e.g.*, thermal imaging and contact tracking programs) could also be useful to identify asymptomatic carriers and limit silent transmission.[75] Still, ongoing ethical challenges tied to privacy and data security are surfacing, making it crucial to develop regulatory structures that align public health interests with personal rights.[76] Notwithstanding these challenges, the proficient surveillance of public spaces constitutes an essential strategy for alleviating the repercussions of viral epidemics and safeguarding public health.

Sensor technologies have emerged as pivotal instruments in the realm of virus detection, facilitating expedited, more efficient, and increasingly sensitive methodologies for identifying viral infections across diverse contexts. These technologies leverage an array of approaches, encompassing physical sensors, such as electrochemical and optical sensors, and molecular biosensors designed to identify viral genetic material. For instance, the popularity of electrochemical sensors arises from their economic advantages, portability, and greater sensitivity, rendering them highly appropriate for point-of-care diagnostic functionalities. Typically, these sensors ascertain the presence of viruses by monitoring variations in electrical characteristics upon the binding of virus particles to specific receptor sites.[77] Similarly, optical sensors, including surface plasmon resonance (SPR) and fluorescence-based detectors, facilitate the real-time observation of viral interactions, thereby providing exceptional specificity and the capability to identify low viral concentrations with minimal sample preparation.[78]

Moreover, advancements in wearable sensor technology and biosensor arrays have rendered it possible to detect viral biomarkers in biological fluids, such as saliva or sweat, in real-time, thus enabling continuous monitoring of viral infections in both clinical and public health contexts.[79] This section examines how combining AI with sensor technologies in order to enhance virus detection allows for more accurate data processing, prompt detection, and predictive epidemic modelling. These technological developments have greatly improved the ability to identify viruses such as COVID-19, influenza, and other infections, offering vital instruments for disease management and monitoring. A variety of detection techniques, such as biosensors, environmental monitoring, non-invasive technologies, and AI-enhanced systems, will be examined emphasizing their benefits, drawbacks, and practical uses. Future advancements, including downsizing, multidisciplinary partnerships, regulatory obstacles, and the development of multimodal detection techniques, in order to build a more effective and easily accessible worldwide disease monitoring system, will also be covered.

15.3.1 Recent Advances in Environmental Detection of SARS-CoV-2: Methods and Applications

15.3.1.1 Real-time Monitoring Systems in Public Spaces

Yang's paper described an air quality checking system for finding SARS-CoV-2 pathogens in closed spaces. The system uses a special kind of light effect with tiny chips treated with specific antibodies. It has an air collecting part, which uses a suction method, a spotting part with small light sensors, and a phone app for showing data and controlling the device through Bluetooth. The chip was tried out with different amounts of dead virus germs showing a strong link between virus amount and changes in light intensity ($R^2 = 0.99$),[80] as shown in Figure 15.7.[80]

Viral aerosols are efficiently detected by the integrated air quality monitoring system (iAQMS) throughout a broad dynamic range, and findings remain consistent. The technology is a promising tool for portable, real-time airborne infection surveillance, despite obstacles such as possible signal interference from water droplets or unsteady airflow. Through modification of the

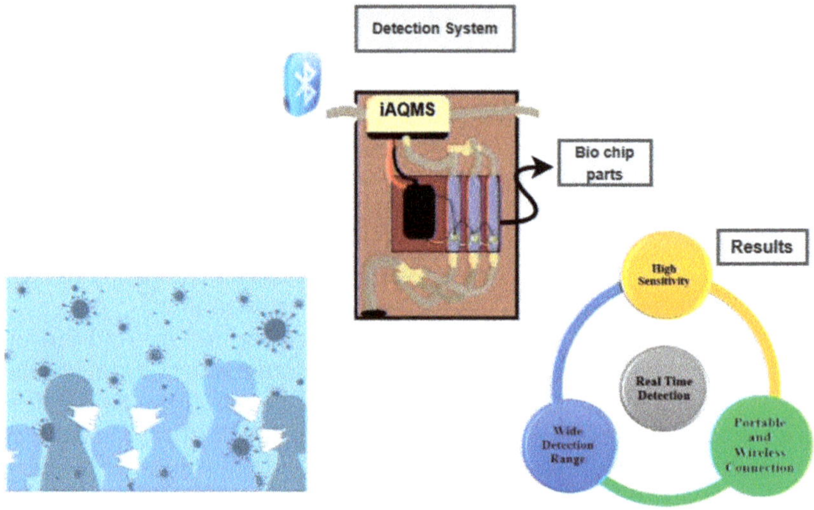

Figure 15.7 A portable, wireless detection device for real-time viral surveillance [integrated air quality monitoring system (iAQMS)].

antibody-functionalized gold nanoparticles, the method provides a scalable and adaptable platform for the detection of various respiratory viruses.[80]

Alorda-Ladaria *et al.* elucidated that the utilization of public Wi-Fi networks for the analysis of probe requests facilitates a cost-efficient and non-intrusive approach for monitoring urban mobility and crowd density, thereby enhancing urban management and social distancing protocols, particularly in the context of the COVID-19 pandemic.[81]

Hany and Khaled investigated the ways in which internet of things (IoT) technology augments the management of public spaces and tourism by fostering real-time data transmission for effective urban solutions, such as intelligent lighting systems and waste management, ultimately promoting its extensive adoption in the development of intelligent and sustainable urban environments.[82]

15.3.1.2 *Virus Detection on Public Transport*

Hoffman *et al.* investigated a passive monitoring strategy for detecting SARS-CoV-2 RNA in public transport using air filtration devices. Given the challenges of widespread clinical testing for airborne transmission, the study explored whether air filters in public buses could serve as cost-effective sensors for viral RNA, offering a scalable approach to community surveillance. The authors proposed that viral particles exhaled by infected individuals accumulate in bus HVAC systems, enabling systematic detection.[83]

Environmental sampling conducted on 37 public buses in Seattle during the years 2020 to 2021 revealed the presence of SARS-CoV-2 RNA in 14% of air filters and 42% of handrail swabs, indicating that bus air filters can

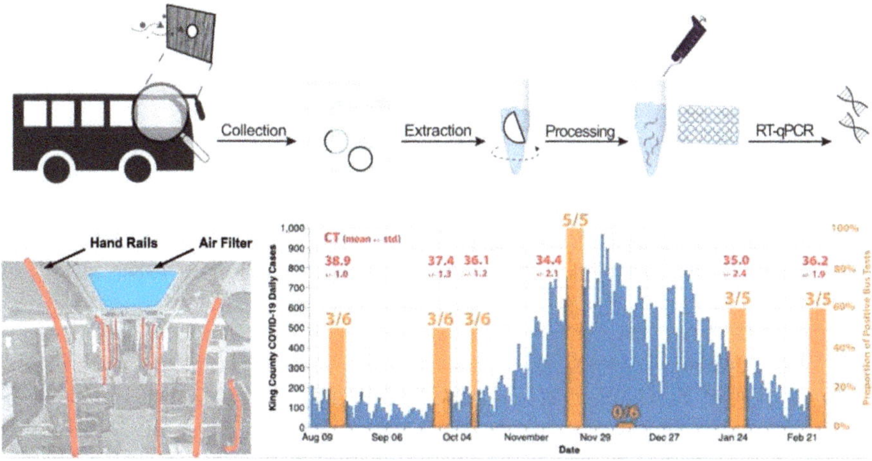

Figure 15.8 Detection of materials recovered from metro busses. Reproduced from ref. 83 with permission from Elsevier, Copyright 2022.

passively accumulate viral RNA, albeit with lower efficacy compared with surface sampling (Figure 15.8). Public transport systems present a significant opportunity for passive viral surveillance, as air filters and surface swabs can capture viral RNA, thereby yielding insights into community infection patterns. Nonetheless, the optimization of filter design, RNA extraction methodologies, and testing intervals is imperative for enhancing the reliability and scalability of this technique within urban transportation networks.[83]

In order to assess the viability of wastewater monitoring in public transportation environments, Medrano-Félix *et al.* employed wastewater-based epidemiology (WBE) to detect SARS-CoV-2 RNA in tourist bus drainage systems. During a peak COVID-19 epidemic between February and April 2022, researchers sampled wastewater from domestic and international buses at a bus stop in Culiacan, Sinaloa, Mexico. They isolated and analysed viral RNA using the polyethylene glycol (PEG) concentration and filtration method, a well-known technology for enhancing RNA recovery from wastewater. This approach made it possible to thoroughly examine the presence of SARS-CoV-2 in wastewater holding tanks, yielding useful insights into viral circulation among passengers and in transit surroundings.[84]

The findings demonstrated that WBE is a dependable and effective tool for monitoring SARS-CoV-2 in public transportation networks, emphasizing its potential for early epidemic identification. Researchers discovered detectable quantities of viral RNA in wastewater from tourist buses, indicating that this technology might act as an early warning system for COVID-19 transmission hazards among passengers. This method provides a scalable, non-invasive method for tracking viral transmission at transit hubs, eliminating individual clinical testing. According to the study, incorporating wastewater surveillance into public health programs might enhance risk assessments, increase outbreak preparation, and promote continuous epidemiological monitoring for infectious disease management.[84]

Pilipenco *et al.* used a rapid acoustic biosensor with an antifouling nano-coating to assess the risk of SARS-CoV-2 surface transmission in public transportation. Over three days in April 2021, 482 surface swabs were taken from regularly touched surfaces in Prague's public transportation system, including trams, buses, metro trains, and platforms. To identify viral RNA, the samples were examined using a biosensor as well as qRT-PCR. This dual-analysis technology enabled the precise detection of viral traces on high-touch surfaces.[85]

The study discovered a very low risk of surface transmission, with only 10 of the 482 swabs (2.1%) testing positive for SARS-CoV-2, and none of these contained live virus. The findings indicated that the chance of infection by surface contact is low, even when viral RNA is present. The study helped to enhance cleaning methods for the Prague transportation system by concentrating sanitation efforts more efficiently. Furthermore, the study emphasized the biosensor's quick viral detection capabilities, which promoted its application in public health and epidemic surveillance. The data support the notion that airborne transmission is the principal route of SARS-CoV-2 dissemination in public places.[85]

15.3.1.3 Environmental Surveillance in Healthcare Settings

Dietz *et al.* compared surface and bioaerosol collection techniques in hospital rooms with COVID-19 patients in order to examine an integrated environmental viral surveillance approach. The study, which was carried out at Oregon Health & Science University (OHSU) between November and December 2020, sought to determine how well surface swabbing, active air sampling, and passive air settling plates detected SARS-CoV-2. High-touch surface swabs, passive air settling plates, and the AerosolSense air sampler were used to gather environmental samples from 32 hospital rooms (Figure 15.9). Targeting the SARS-CoV-2 nucleocapsid (N), spike (S), and open reading frame 1 a and b (ORF1ab) genes, RNA was collected and subjected to qRT-PCR analysis. The results demonstrate how airborne virus surveillance might be used as an extra line of defence for public health in medical facilities.[86]

The AerosolSense active air sampler had the greatest detection rate, with 53.8% of samples positive for SARS-CoV-2 RNA, *vs.* 12.1% for passive air settling plates and 14.8% for surface swabs. Overall, 69% (22/32) of hospital rooms had at least one positive environmental sample. Notably, 32% of positive rooms were found only by active air sampling, demonstrating the value of bioaerosol surveillance.[86] The study also discovered lower cycle threshold (Ct) values in air samples *vs.* surface swabs, indicating a larger virus load in airborne samples. These findings indicate that integrated environmental surveillance, which includes active air sampling, is a desirable addition to hospital infection control efforts. However, the study admits limitations, such as

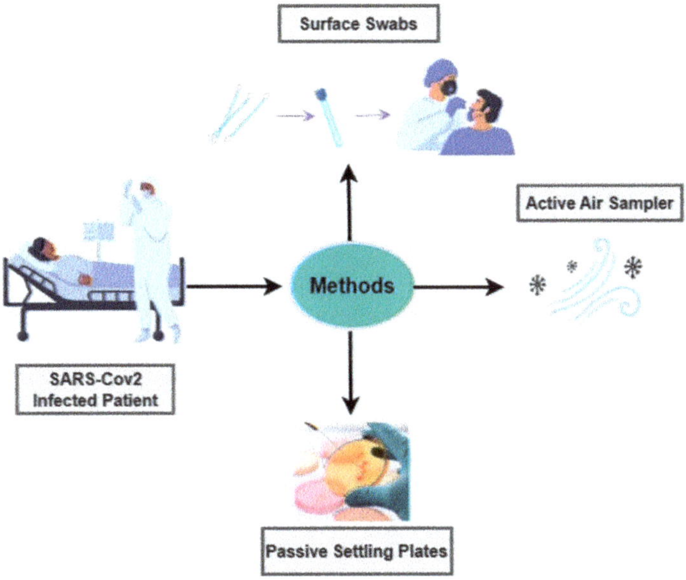

Figure 15.9 Techniques to identify surface and airborne SARS-CoV-2 contamination.

sampling time inconsistency and the inability to discriminate between live and non-viable virus particles. Despite these obstacles, using bioaerosol surveillance in hospital settings is recommended to improve real-time infection monitoring and outbreak prevention.[86]

The rise of next-generation sequencing (NGS) is marking it as a vital asset for environmental oversight in healthcare settings, demonstrating greater precision in detecting bacterial contamination than older microbiological approaches. A study conducted in an Italian paediatric hospital evaluated NGS in comparison to conventional culture-based techniques by examining environmental samples from high-risk surfaces across various hospital wards. The results indicated that NGS was capable of identifying bacterial levels surpassing anticipated thresholds, thereby providing a comprehensive profile of microbial communities. However, it was unable to differentiate between viable and non-viable bacteria or detect fungal contamination, which traditional methodologies successfully captured, along with data on drug resistance. The integration of NGS with conventional microbiological practices can enhance hospital surveillance by refining assessments of microbial contamination and fortifying infection control strategies.[87]

Ray and collaborators demonstrated the efficacy of environmental surveillance as a proactive alert system for COVID-19 outbreaks by detecting SARS-CoV-2 RNA on high-touch surfaces within a tertiary care facility. Through systematic sampling employing sterile swabs and replicate organism detection and counting (RODAC) plates, followed by RT-PCR analysis, the investigators identified contamination in patients' rooms and communal spaces, frequently prior to the emergence of clinical cases. This highlights the critical role of environmental monitoring in infection control, indicating that the integration of regular testing into hospital protocols can enhance early contamination detection, reduce transmission risks, and bolster safety for both patients and healthcare personnel.[88]

15.3.2 Exhaled Breath Analysis with Electronic Nose Technology

Exhaled breath analysis has gained attention as a non-invasive, convenient, and safe method for health diagnosis, offering a pain-free alternative to traditional invasive techniques. The diagnostic value of breath analysis lies in the complex composition of breath, which includes volatile organic compounds (VOCs) and inorganic chemicals that can serve as biomarkers for various physiological and pathological conditions. While traditional techniques, like mass spectrometry and gas chromatography, are effective, their high costs, operational complexity, and lack of portability limit their clinical use. As a result, electronic nose (eNose) technology (Figure 15.10)[89] has emerged as a promising alternative, offering high sensitivity, rapid response times, ease of use, and portability for non-invasive disease detection through exhaled breath analysis.[90]

A prospective study was conducted at public testing facilities, where individuals seeking SARS-CoV-2 testing were grouped into training, validation, replication, and asymptomatic categories. After undergoing throat/nasopharyngeal swabs for transcription-mediated amplification (TMA) or RT-PCR,

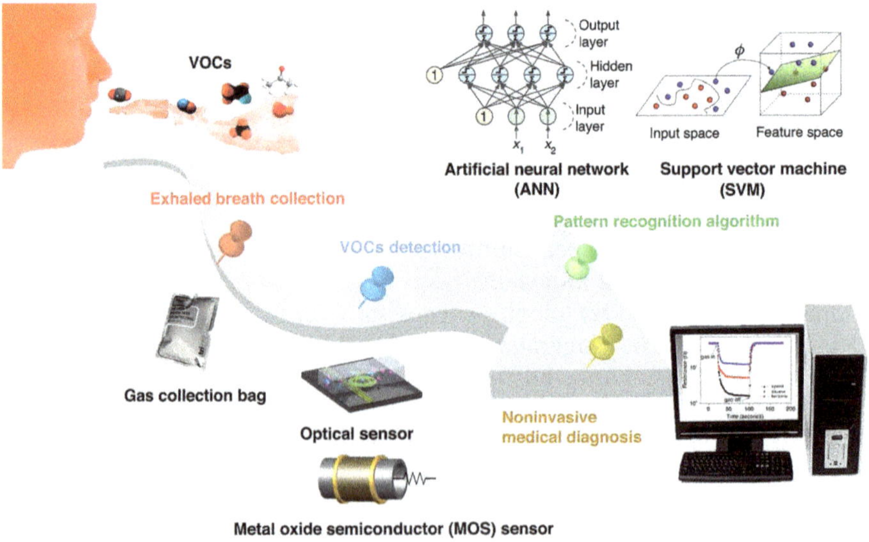

Figure 15.10 Schematic representation of non-invasive breath detection using the eNose system. Reproduced from ref. 89, https://doi.org/10.1038/s41378-023-00594-0, under the terms of the CC BY4.0 license, https://creativecommons.org/licenses/by/4.0/.

exhaled breath samples were collected using a cloud-connected eNose device developed by Breathomix BV. Without naming specific sensor types, this gadget uses a variety of cross-reactive nanosensors that are intended to use probabilistic pattern recognition to identify VOCs in exhaled breath. Statistical and signal processing techniques, such as receiver operating characteristic (ROC) analysis, linear discriminant analysis (LDA), and *t*-tests, were used to examine the data. The training data was used to build the diagnosis model, which was subsequently assessed using the validation, replication, and asymptomatic sets.[90] A total of 4510 participants in the study were used to confirm that eNose technology successfully differentiated between people who were positive and those who were negative for SARS-CoV-2. The eNose sensors generated a composite biomarker in the training set with a ROC-area under the curve (AUC) of 0.947 [confidence interval (CI): 0.928–0.967], which was validated in the subsequent sets: validation (0.957; CI: 0.942–0.971), replication (0.937; CI: 0.926–0.947), and asymptomatic (0.909; CI: 0.879–0.938) sets.[90] The efficacy of eNose as a quick SARS-CoV-2 screening tool was proven by its excellent sensitivity and specificity across all datasets, using a cut-off value of 0.30 in the training set.[90]

Nurputra *et al.* devised a non-invasive, rapid eNose apparatus for the detection of COVID-19 *via* exhaled breath specimens. This methodology is predicated on the identification of distinctive breath print patterns associated with the virus, which are detected by a sensor array. An eNose analyses VOCs detected in breath, functioning as biomarkers for SARS-CoV-2 infection. The investigation revealed that the eNose could accurately differentiate COVID-19 patients with considerable sensitivity and specificity, surpassing conventional diagnostic techniques in terms of speed and convenience. This technique presents a promising alternative for rapid, extensive COVID-19 screening without necessitating invasive procedures.[91]

Ghazaly *et al.* assessed the efficacy of an eNose in the identification of COVID-19 infection through breath analysis. The eNose employed a sensor array specifically designed to detect VOCs that are indicative of respiratory infections. The research entailed the collection of breath samples from both COVID-19-positive and COVID-19-negative patients, followed by VOC analysis utilizing the eNose. The findings indicated that the eNose could accurately differentiate between COVID-19-positive and COVID-19-negative individuals with significant precision. The results imply that eNose technology could serve as a valuable instrument for rapid, non-invasive COVID-19 detection, presenting potential benefits over traditional diagnostic methods, such as PCR testing, owing to its rapidity and user-friendliness.[92]

15.3.3 AI and Nanoparticle-based Virus Detection

Serious hazards to world health are posed by new and reemerging viral infections, underscoring the necessity for quick and precise detection techniques. Particularly in places with limited resources, traditional lab-based methods can be sluggish, expensive, and less accessible. Mobile health (mHealth) diagnostics, leveraging the widespread availability and enhanced capabilities of smartphones, offers promising solutions for point-of-care viral detection. Draz *et al.* introduced a novel nanoparticle-enabled smartphone (NES) system that used platinum nanoparticle (PtNP)-based signal amplification and convolutional neural network (CNN)-based image processing to detect viruses in a short time and with high sensitivity, without requiring external optical attachments. In their work, Draz *et al.* developed a CNN–NES system for viral detection. The virus is captured on a microchip and labelled with PtNPs, which enhance gas bubble formation from hydrogen peroxide, creating a visual signal.[93] A CNN on the smartphone analyses bubble patterns to detect the virus. PtNPs were conjugated with monoclonal antibodies (mAbs) specific to the virus's envelope protein. The microchip surface was functionalized with mAbs to improve virus capture. Following the addition of a hydrogen peroxide-containing catalyst solution, gas bubbles formed, and the CNN algorithm on the smartphone assessed the bubble patterns to identify the infection. Images of microchips with bubbles created from virus-spiked and PtNP-spiked samples, which represented varying viral quantities, were used to train the system. Hepatitis B (HBV), hepatitis C (HCV), and Zika virus (ZIKV)[93] were among the viruses that were examined. HBV-spiked serum, ZIKV-spiked blood, HCV-infected patient plasma and serum, and ZIKV-infected patient serum samples were among the samples used to test the CNN–NES system. On the basis of a viral concentration threshold of 250 particles per ml, the results indicated that the system could accurately categorize samples as either positive or negative. With a confidence range of 94.39 to 99.97%, the system's sensitivity for identifying virally infected samples at the clinically relevant concentration of 250 copies per ml was 98.97%. These results indicate that the CNN–NES system offers a rapid, precise, and reasonably priced alternative for point-of-care viral detection, particularly in settings with constrained resources.[93]

An AI-powered nanopore system for detecting coronaviruses, such as SARS-CoV-2, without RNA extraction is presented in a paper by Taniguchi *et al.* The technology successfully identified viruses with an *F*-value of 0.66 by using machine learning, more especially a random forest technique, to analyse current waveforms. Compared with conventional RT-PCR, this technique improves sensitivity and accelerates detection, providing a potential step toward quicker and more effective viral diagnoses.[94]

Shiaelis *et al.* combined CNN with single-particle imaging to present a deep learning-based method for viral identification. The CNN was taught to recognize and categorize viral particles using transmission electron microscopy (TEM) pictures, eliminating the need for laborious post-processing or manual feature extraction. By converting TEM pictures into probabilistic maps, the technology made it possible to accurately detect and classify viruses.[95]

The CNN-based approach drastically reduced false positives and was an improvement on existing virus detection methods, according to the results. It had a low error rate and correctly recognized well-known viruses, such as feline calicivirus. Because of its versatility, the model may be tailored to identify different viruses by training on annotated datasets. This method's adaptability makes it suitable for use in medical diagnostics, environmental monitoring, and virology research. This study demonstrated the promise of more accurate, efficient, and automated viral detection by fusing cutting-edge imaging methods like TEM with AI-driven analysis. Viral diagnostics might be significantly improved by such technology, enabling more rapid diagnosis, improved epidemic control, and more successful public health initiatives. The technique's success also sets the stage for future developments in the integration of artificial intelligence and imaging technologies to tackle complex diagnostic problems.[95]

15.3.4 Applications of Fast Virus Detection Techniques for Public Environments

The SARS-CoV-2 pandemic necessitates rapid, accessible testing methods for detecting infected individuals in public spaces. A novel approach addresses this need by offering ultra-fast detection suitable for point-of-entry screening in settings like workplaces and schools. This method prioritizes speed, simplicity, and cost-effectiveness, with the aim of enabling decentralized and regular testing to curb viral transmission. The strategy sacrifices some sensitivity to achieve practical usability and affordability. The diagnostic test employs displaceable probe loop amplification, "digital PCR reverse transcription loop-mediated isothermal amplification" (DP-RT-LAMP), designed to be used with dry nasal swabs, eliminating the need for complex sample preparation or specialized equipment. Primer and probe design focuses on conserved regions of the SARS-CoV-2 genome, minimizing cross-reactivity with other coronaviruses (Figure 15.11). The test uses a simplified reaction mixture, including WarmStart LAMP master mix, primers, strand-displaceable probes, and a thermostable uracil-DNA glycosylase (UDG) enzyme, to enable rapid RNA amplification and signal generation. The test's performance was evaluated using synthetic RNA, heat-inactivated viral isolates, and clinical nasal swabs spiked with the virus, all obtained with ethical approval (protocol number 2020001).[96] The DP–RT–LAMP assay demonstrated the ability to detect SARS-CoV-2 RNA directly from nasal swab samples with minimal processing, providing results in just 30 minutes at a cost of under $3 US per test. The assay has a detection limit of approximately 200 viral RNA copies per response. This point-of-care diagnostic tool simplifies the testing procedure, making it ideal for rapid SARS-CoV-2 identification in public spaces, such as workplaces, without the need for specialized staff or equipment.[96]

15.3.5 Rapid Detection Using SERS Aptasensors

The ongoing COVID-19 pandemic has raised concerns about environmental transmission of SARS-CoV-2, detected in dwellings and wastewater. Traditional detection methods, such as PCR and immunodiagnostic kits, face

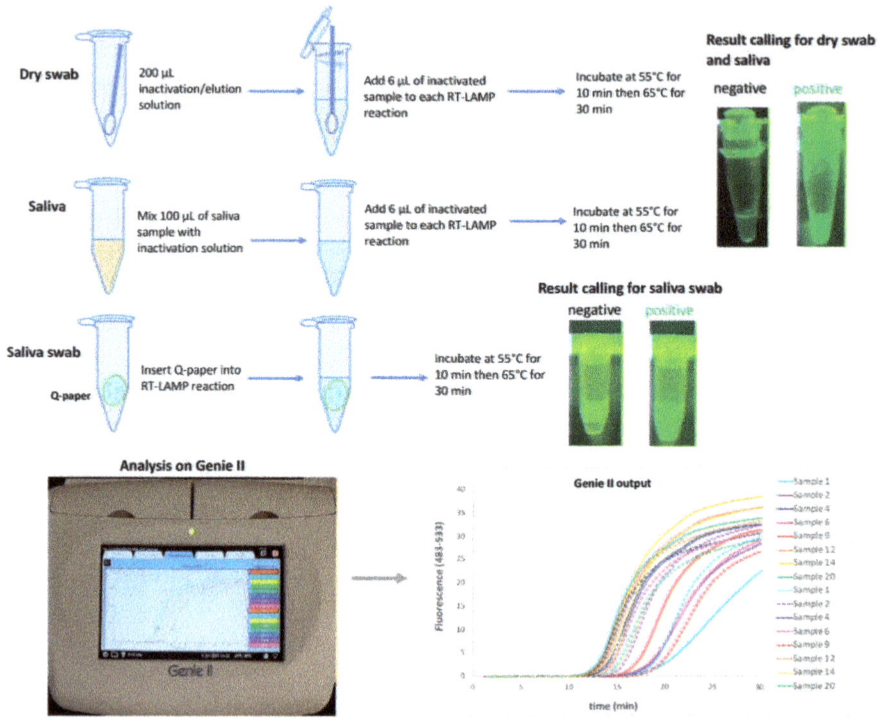

Figure 15.11 Workflow sampling and output results. Reproduced from ref. 96, https://doi.org/10.1371/journal.pone.0240524, under the terms of the CC BY 4.0 license, https://creativecommons.org/licenses/by/4.0/.

challenges related to complexity, time consumption, and sensitivity, hindering effective environmental monitoring. This necessitates the development of rapid, sensitive, and user-friendly biosensors capable of detecting trace amounts of the virus.[97]

A surface enhanced Raman scattering (SERS) aptasensor with a unique "locking amplifier" (Figure 15.12) method has been developed for the quick and ultrasensitive detection of SARS-CoV-2 virus particles. The aptasensor uses dual-function aptamers for viral identification and binding to trigger DNA, creating a "locked trigger" condition that reduces background noise. The test uses a one-pot reaction at 37 °C, simplifying the process and conserving resources. Magnetic beads are included for ease of use and potential interaction with biosensing robots. The SERS aptasensor can detect SARS-CoV-2 with a LOD of 260 transduction units (TU) μl^{-1} in 40 minutes and a linear detection range of 625–10 000 TU μl^{-1}. Its simplicity, speed, sensitivity, and selectivity make it a promising tool for monitoring environmental virus sources.[97]

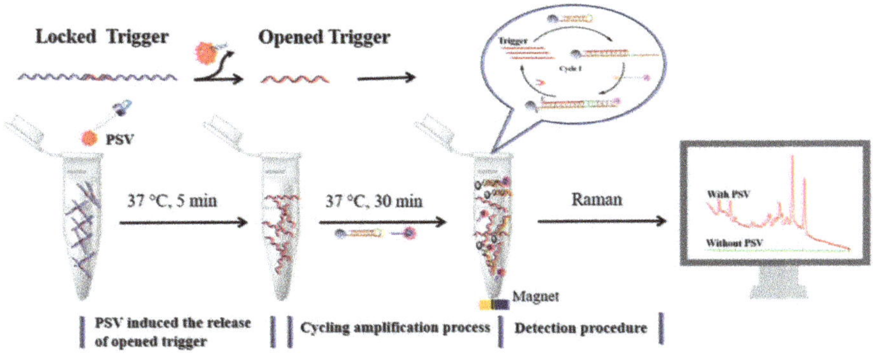

Figure 15.12 SERS aptasensor approach for fast detection of SARS-CoV-2 virus particles. Abbreviation: pseudotyped SARS-CoV-2 virus (PSV). Reproduced from ref. 97 with permission from Elsevier, Copyright 2022.

15.3.6 Cutting-edge Biosensing Technologies for Real-time Virus Monitoring

With thousands of fatalities each year, viruses represent a major danger to global public health and the economy. It is difficult to identify them due to their quick mutation and spread, necessitating quick and precise diagnostic instruments. Multiplexed tests are required for accurate identification since clinical diagnosis is frequently challenging. Timely treatment and illness control are made more difficult by the absence of easily available and sophisticated diagnostic facilities. Biosensors are sensitive analytical instruments that employ transducers to identify biomolecules, such as viruses and proteins, for quick and precise diagnosis.[98]

Respiratory viruses persist as significant contributors to mortality on a global scale, primarily due to their facile transmission *via* aerosolized particles, thereby affecting individuals across the entire age spectrum—from neonates to the geriatric population.[99] Regrettably, a substantial number of fatalities are attributable to the inadequacies in precise and prompt diagnostic measures, particularly in the context of emergent viral variants that elude current detection protocols.[100] In response to these exigencies, scholars have been engaged in the advancement of more sensitive and dependable biosensors. For example, Zhou *et al.* engineered a novel electrochemical sensor integrating immunomagnetic separation, enzymatic amplification, and DPV for the detection of H9N2 avian influenza.[101] Notably, this sensor demonstrated the capability to identify the virus at concentrations as low as 1 ng ml^{-1}. Subsequently, Singh *et al.* unveiled a label-free sensor designed for the detection of the H1N1 influenza virus.[102] Their approach involved the utilization of SWCNTs functionalized with biotinylated antibodies, culminating in a highly sensitive platform for viral detection devoid of labelling requirements. Building upon this concept, Huang *et al.* formulated a graphene-based sensor adorned with silver and gold nanoparticles, which facilitated the sensitive identification of H7 influenza strains.[103] Their astute incorporation of nanomaterials markedly enhanced the sensor's capacity to capture and detect viral entities.

In yet another innovative strategy, Tian and colleagues amalgamated carbon nanotubes (CNTs) and molybdenum sulphide (MoSx) into a three-dimensional aerogel configuration, thereby augmenting the surface area available for viral detection.[104] This sensor exhibited exceptional sensitivity in detecting H7 influenza, achieving detection limits as low as 0.43 ng ml^{-1}. Similarly, Lee *et al.* developed a label-free biosensor for H5N1 utilizing porous gold nanoparticles in conjunction with a specialized DNA three-way junction, simplifying the production process while attaining an ultra-sensitive detection threshold of 1 pM.[105] Amid the COVID-19 pandemic, wherein the demand for rapid and accurate testing became paramount, Raziq *et al.* introduced a molecularly imprinted polymer (MIP)-based electrochemical sensor proficient in detecting the SARS-CoV-2 nucleoprotein at exceedingly low concentrations—as low as 27 fM—even within clinical specimens.[106] This underscores the potential of biosensors to play a pivotal role in mitigating pandemics through the facilitation of rapid and reliable viral detection. According to the Centers for Disease Control and Prevention (CDC), a similar narrative emerges with flaviviruses, including ZIKV, dengue virus (DENV), yellow fever virus (YFV), and Japanese encephalitis virus (JEV), which are predominantly transmitted by mosquitoes and represent substantial health risks in numerous regions globally.[107] The diagnosis of these viruses is frequently complicated due to their overlapping symptoms and structural similarities, rendering cross-reactivity a prevalent challenge. To address this issue, Faria and Zucolotto devised a label-free sensor employing DNA strands affixed to gold electrodes that detect ZIKV through alterations in electrical signals contingent upon the presence of the target.[108] In an alternative approach, Moço *et al.* constructed a graphite-based genosensor augmented with reduced graphene oxide and polythyramine (PTyr), which enhanced both sensitivity and electrical conductivity, albeit necessitating a more intricate fabrication process.[109] For the detection of dengue virus serotype 2, Palomar *et al.* developed an electrochemical immunosensor that combined multi-walled carbon nanotubes (MWCNTs) with gold nanoparticles and specific antibodies.[110] This innovative design significantly enhanced both the sensitivity and selectivity for the detection of DENV; however, the prohibitive cost of antibodies continues to pose a significant obstacle to the scaling of production efforts.

Middle East respiratory syndrome (MERS) constitutes a grave viral respiratory affliction attributable to MERS-CoV, a novel coronavirus initially recognized in Saudi Arabia in 2012 .[111] MERS generally manifests with clinical symptoms such as fever, cough, and dyspnoea, with a substantial number of patients contracting pneumonia and, on occasion, gastrointestinal complications, such as diarrhoea.[112] Notably, asymptomatic instances have been identified, particularly during contact tracing initiatives.[113] Despite the absence of symptoms in certain patients, MERS-CoV has been verified as the aetiological agent responsible for mortality in approximately 35% of fatal cases,[114] thereby underscoring its considerable health threats. In light of the pressing necessity for rapid and dependable diagnostic modalities, researchers have concentrated their efforts on the development of biosensors for the detection of MERS-CoV.[115] A particularly promising methodology involves the application of colorimetric assays utilizing gold nanoparticles (AuNPs), which are user-friendly and yield visual results, rendering them appropriate for point-of-care diagnostics. For example, Kim *et al.* devised a label-free colorimetric assay that employed self-assembled double-stranded DNA (dsDNA)-shielded AuNPs for MERS-CoV detection. By monitoring alterations in localized surface plasmon resonance (LSPR) and colorimetric changes in the AuNP solution, this technique facilitated the identification of the virus at exceedingly low concentrations (1.0 pmol l^{-1}).[116] The DNA probes were meticulously designed to selectively hybridize with conserved regions of the MERS-CoV genome, thereby stabilizing the AuNPs and inhibiting aggregation, which permitted unequivocal identification of the virus. In a separate investigation, Teengam *et al.* introduced a paper-based colorimetric sensor employing aminocyclopentanecarboxylic acid peptide nucleic acid (acpcPNA)-induced silver nanoparticle (AgNP) aggregation for the detection of MERS-CoV.[117] This system attained a detection threshold of 1.53 nM, with the acpcPNA probe exhibiting remarkable specificity, effectively discriminating between complementary and mismatched DNA sequences, thus ensuring accurate and reliable virus detection.

In conclusion, with the growing demand for effective viral detection methods, researchers have created a number of technologies to improve speedy identification and monitoring. These developments are intended to offer early identification, minimize transmission risks, and enhance public health interventions. However, each strategy has advantages and disadvantages that must be considered for practical application.

Advancements in viral detection technology provide substantial advantages in terms of speedy detection and real-time monitoring. SERS aptasensors offer very sensitive and specific detection with quick results, making them ideal for early-stage identification. However, their reliance on specialist equipment and vulnerability to environmental influences restricts their practical applications.

The benefits and drawbacks of many biosensor-based methods for identifying viruses in public areas have been examined in the preceding sections. Air quality monitoring and real-time, non-invasive techniques, like eNose, provide rapid detection, however, they might not be very specific. Although they demand more resources, invasive techniques like RT-PCR and wastewater analysis offer more accuracy. Continuous detection is possible with passive monitoring and air filtration, but further research is still required. The tracking of airborne SARS-CoV-2 particles is aided by environmental monitoring methods, such as bioaerosol sampling and air quality assessment; nonetheless, issues like as false positives and varying circumstances in public transportation remain. IoT-based solutions improve the administration of public spaces, but they are not virus-detection tools. Early detection and outbreak management may be improved by a hybrid strategy that combines biosensors with AI-driven analytics.

Lastly, portable virus detection systems have been revolutionized by the convergence of nanotechnology and AI, which has increased their mobility and accuracy. AI-driven deep learning models can enhance prediction skills, while nanoparticles increase biosensor sensitivity. However, many technologies require rigorous evaluation and regulatory approval before they can be widely adopted. Overall, while each detection strategy has distinct advantages, their limits must be carefully considered in order to optimize their usefulness in real-world applications.

15.3.7 Future Prospects

Looking ahead, virus detection technologies are expected to develop significantly. The downsizing of biosensors and the integration of AI-driven analysis will allow for real-time, at-home diagnostic solutions, decreasing the need for centralized laboratories. Continued advances in nanomaterial engineering are projected to increase the sensitivity and specificity of detection systems, enabling earlier and more precise identification of viral infections. In addition, multidisciplinary partnerships among virologists, engineers, and data scientists will be important in improving detection approaches. The creation of cost-effective and scalable solutions will be critical to ensure worldwide accessibility, especially in low-resource situations. Further study into multi-modal detection methods, such as breath analysis, surface sampling, and AI-based evaluations, might lead to more complete monitoring strategies for infectious disease epidemics. Regulatory developments will also be necessary to facilitate the use of innovative detection methods. Streamlined regulatory processes and worldwide standardization of biosensor performance indicators will speed up adoption and integration into healthcare systems. Furthermore, real-world data collection from deployed systems will enable ongoing optimization and enhancement, assuring long-term efficacy and dependability. Overall, the future of virus detection is fuelled by innovation and cross-disciplinary collaboration. By tackling present limits and using upcoming technology, the field is progressing toward a more robust, efficient, and accessible global disease monitoring system.

15.4 Monitoring Virus Transmission in Wildlife

Nearly every year zoonotic diseases increase in frequency and continue to negatively affect public health because they move between animal reservoirs and human populations. Animals in the wild have crucial functions in disease transmission because they serve as refuges for numerous pathogens. Zoonotic viruses, which include SARS, MERS and COVID-19, have emerged, exposing the biosecurity threats linked with human contact with wildlife and environmental destruction together with illegal wildlife commerce.[118] Higher spillover events in which viruses spread between wildlife and people occur due to increased human contact with nature as a result of deforestation and expanding urban development.[119] The illegal wildlife market, together with human consumption of wild animals, facilitates the transmission of zoonotic pathogens. Three major groups involved in virus transmission are bats, rodents, and primates, because these species can maintain viral infections without exhibiting symptoms.[120] The prevention of future pandemics requires better wildlife monitoring, stronger wildlife preservation work, and enhanced regulations for wildlife commerce. Preventing zoonotic spillover requires resolution of its origin factors for successful outbreak prevention and concurrent protection of both animals and humans.

Identifying viruses in wildlife during their initial stages remains essential to stopping the spread of outbreaks and preventing global pandemics. Wildlife acts as the source for several emerging infectious diseases, which require immediate surveillance with intervention steps to stop health risks. Scientists possess the ability to assess threats and develop preventive measures for viruses to be discovered before human spillovers occur.[118] The surveillance programs actively target bats and rodents because these species serve as known carriers of zoonotic viruses. The programs utilize viral sampling alongside genetic analysis to monitor how viruses change and spread.[121] Early discovery by health authorities enables them to design vaccines, establish biosecurity measures and develop strategies for managing human–wildlife interaction to decrease disease spillover threats.[119] Human and animal health and environmental health need the one health approach to effectively detect and respond to new viral outbreaks. Advanced global surveillance systems coupled with wildlife disease studies form the basis for preventing another pandemic, such as COVID-19.[122] The protection of public health relies on detecting newly arising viruses before human encroachment onto wildlife habitats becomes widespread.

Modern electrochemical sensor methods create essential tools for wildlife surveillance by providing non-intrusive techniques to measure physiological characteristics of animals. Scientists have created implantable electrochemical microsensors for detecting biomarkers, including glucose, lactate, and dopamine, which generate immediate health information about animals and aid in detecting early diseases. The sensors operate for *in vivo* inspection with the purpose of avoiding causing discomfort to animals as they generate precise physiological results.[123] The field has advanced into creating electrochemical sensors that operate wirelessly but also remain flexible and bio-responsive while specifically detecting nitric oxide in real-time. These sensitive sensors hold selectivity capabilities that permit round-the-clock wildlife health observation through non-invasive procedures on the animals.[124] Researchers use electrochemical biosensors as new technology to find viruses in wildlife populations. These sensors possess high sensitivity combined with quick response times and excellent selectivity, which enables them to identify viral pathogens in animal communities, thus enabling quick detection of impending zoonotic disease transmission.[125] The implementation of advanced electrochemical sensor technology in wildlife surveillance systems provides better methods for monitoring animal health while speeding up disease detection. Wildlife conservation, together with zoonotic disease prevention, requires this proactive method as an essential step.

15.4.1 The Link Between Wildlife and Emerging Viruses

Different zoonotic viruses transmitted by animals to humans have led to various significant outbreaks throughout human history. The list of significant outbreaks due to zoonotic viruses includes severe acute respiratory syndrome coronavirus (SARS-CoV) and Ebola virus, as well as different versions of avian influenza. These pathogens start in wildlife, and bats along with birds, function as their main habitat. SARS-CoV is believed to have first emerged in bats before it was transmitted to humans through animal bridges that included civet cats. Human outbreaks of Ebola virus start when humans come into contact with wild animals that host the virus, including bats and non-human primates. The primary place where avian influenza viruses exist is among wild birds, which serve as a natural reservoir, while occasional spread to domestic poultry and human populations triggers disease outbreaks. Scientists report that zoonotic emerging human infectious diseases represent 60.3% of all cases and wildlife specifically is involved in 71.8% of such diseases. The ever-expanding interactions between humans and wildlife result from habitat encroachment and human-caused factors that increase the possibility of zoonotic virus transmission events. The knowledge of how viruses originate from wildlife plays an essential role in creating preventive measures for future outbreaks.[126]

The main sources of zoonotic viruses that jump from animals to humans include bats, birds, and rodents, serving as the viruses' natural reservoirs. One example of virus carriers can be found in bats, which transmit the Ebola virus together with coronaviruses such as SARS-CoV and MERS-CoV. The specific makeup of their immune systems and social patterns helps these organisms retain and spread these pathogens.[127]

Wild waterfowl and their family members, among birds, function as the natural hosts of species-specific influenza viruses. Laboratory tests have revealed that crossing of viral species barriers occasionally allows these viruses to spread among domestic poultry populations and sometimes result in human infections. The worldwide dissemination of avian influenza strains happens through the migratory activities performed by these birds.[128]

The pathogens known as hantaviruses remain in rodents before causing serious respiratory infections in human bodies. Human beings become infected with microorganisms after they breathe in excreta-based airborne particles from rodents. The proximity of specific rodent species to human urban areas intensifies their potential to transmit disease.[129]

Knowledge of wildlife repositories along with their virus connections matters for forecasting as well as stopping upcoming zoonotic disease outbreaks. The observation of virus transmission in migratory animal populations spread across distant regions has proven to be a complex task. Above all else, the remote habitats of these animals are difficult to monitor, which creates extensive gaps in surveillance efforts. The migration patterns of birds lead to transmission of pathogens, such as avian influenza, through worldwide spread, which hinders outbreak tracking and control efforts during epidemics. Migratory species create obstacles for detecting newly emerging diseases because their erratic movement patterns make disease detection difficult. The restrictions of modern technology, including inadequacy of tracking tools and real-time monitoring devices, cause these issues to intensify. Sampling and analysis in distant regions face barriers because of insufficient resources as well as logistical problems. Improving global zoonotic disease early warning depends on creating new surveillance approaches and global cooperation to strengthen disease detection before zoonotic events occur.[130]

15.4.2 Current Techniques for Virus Detection in Wildlife

Wildlife virus monitoring requires effective methods for early disease detection because it helps prevent zoonotic disease outbreaks. Virus detection in wildlife populations requires the implementation of multiple detection techniques.

RT-PCR functions to multiply viral RNA, which provides detailed virus detection and exact virus identification with a high level of sensitivity and precision. RT-PCR plays a vital role in studying Bluetongue virus and African horse sickness virus epidemiology among wildlife populations.[131]

ELISA detects pathogen-specific antibodies in wildlife serum samples, which indicate exposure to particular viruses. This approach makes surveillance screening of many samples possible through its efficient design.[131]

Metagenomic next-generation sequencing (mNGS) supports the identification of existing and entirely new viruses because it analyses all genetic information present in the tested sample. Technical equipment using mNGS in mobile setups has been utilized for field-based RNA and DNA virus detection from zoo animals, which improves rapid surveillance operations.[132]

Clustered regularly interspaced short palindromic repeats (CRISPR)–CRISPR associated (Cas)-based biosensors provide fast viral nucleic acid detection through their particular CRISPR–Cas system specificity. Scientists have created CRISPR–Cas biosensors for detecting the African swine fever virus, which now provides testing methods that work reliably under field conditions.[133]

The surveillance techniques present specific benefits through which field workers can make selections prompted by surveillance requirements and operational limitations.

The surveillance of virus transmission between distant and migratory animal populations faces multiple obstacles during monitoring activities. The massive unapproachable habitats that host these animals create obstacles to regular data collection, thus resulting in surveillance gaps. Migratory birds transmit avian influenza pathogens, which creates extensive difficulties in outbreak tracking and control measures.[134] The irregular migration behaviour of species makes it difficult to promptly recognize new diseases when they emerge. Technological restrictions that limit the effectiveness of trackable devices and real-time surveillance systems cause existing challenges to grow worse. The analysis becomes restricted because of resource scarcity together with limited logistical capabilities in remote locations. Zoonotic disease emergence needs better early warning systems, which require innovative surveillance approaches and international cooperation for their development.[135]

15.4.3 Types of Sensors for Monitoring Virus Transmission in Wildlife

There are several types of sensors for monitoring pathogens in wildlife; these sensors are categorised on the basis of the method used to detect viruses. Nanoparticle (NP)-based biosensors have been used for detection of the canine distemper virus (CDV) disease affecting carnivores. Early detection and control strategies can be achieved using NP-based biosensors and species extinction can be avoided. NPs with optical properties, such as gold NPs and quantum dots have been used.[136] The thin film transistor (TFT) is a promising biosensor system with sensitivity, label-free detection, and quick response time. Its limitations include light, noise, vibration, and limited usage. A fully packaged, portable TFT electrochemical biosensor was developed to address these issues. The biosensor was used in a biosafety level 3 laboratory setting to analyse highly pathogenic avian influenza virus samples. The device showed an increasing voltage signal corresponding to increasing virus concentration, demonstrating its potential in clinical applications.[137] There is also an immunosensor designed for non-invasive and quantitative analysis of saliva. This sensor features a fluid control mechanism with both vertical and lateral flow, and can measure analyte concentrations within 35 minutes. The sensor is easy to use and can be reused, demonstrating its potential for non-invasive, point-of-care measurement.[138] Wearable sensors, environmental sensors, and real-time biosensors are discussed later in this chapter.

Figure 15.13 illustrates biosensor-based virus detection in wildlife, for detecting biomarkers present in biological specimens. The collection of biological fluids containing biomarkers, such as antibodies, proteins, lipids, nucleic acids, and other substances, from human or animal subjects is the first step in the analytical process. When biomarkers interact with sophisticated nanomaterials like graphene in combination with metal NPs and CNTs, the sensitivity of detection increases. A transducer, which can be

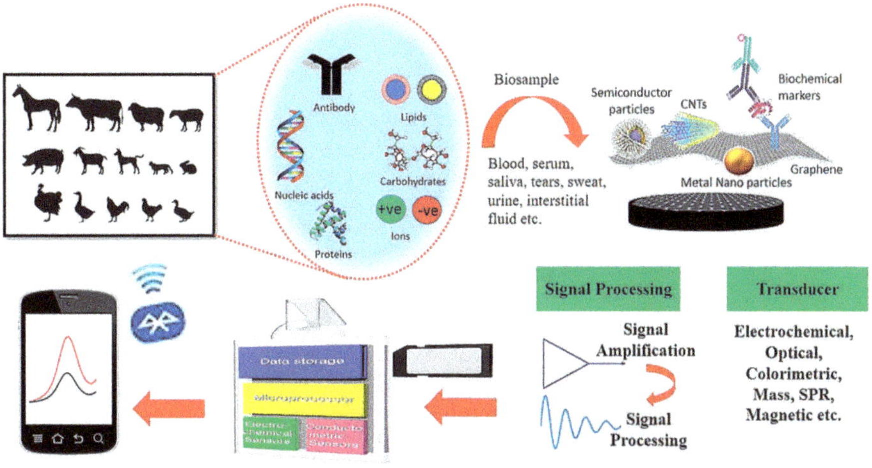

Figure 15.13 Recent advancements in biosensor technology for animal and livestock health management. Reproduced from ref. 139 with permission from Elsevier, Copyright 2017.

electrochemical, optical, colorimetric, or magnetic, transforms the biological signal into an electrical signal. When processed electrical impulses from this signal transmission reach mobile devices wirelessly, real-time monitoring takes place. Biosensors, which are biological detection technologies, are widely used in veterinary medicine and healthcare diagnosis.

15.4.3.1 Wearable Sensors

The use of wearable sensors that combine global position system (GPS) collars and biosensors has become more widespread for tracking wildlife virus spread among migratory birds and bats as well as mammals in the natural environment. The technology allows researchers to obtain instant data about animal displacements and biological measurements while permitting disease research in original habitats. Scientists use GPS collars with accelerometers to track animal behaviours in their environments (Figure 14.14), which reveals how infectious agents disseminate between different species in the wild.[140]

GPS devices lightweight enough for migratory birds enable researchers to collect flight path information along with data about possible avian influenza virus exposure, thus helping with understanding disease transmission through migration patterns. The analysis of bat movement patterns and roosting behaviours depends on GPS collar systems with built-in sensing capabilities to evaluate virus transmission risks to human populations. GPS collars with biosensors attached are used to monitor wild boars by measuring their movements and detecting symptoms of African swine fever early, which helps prevent the disease from spreading.[135]

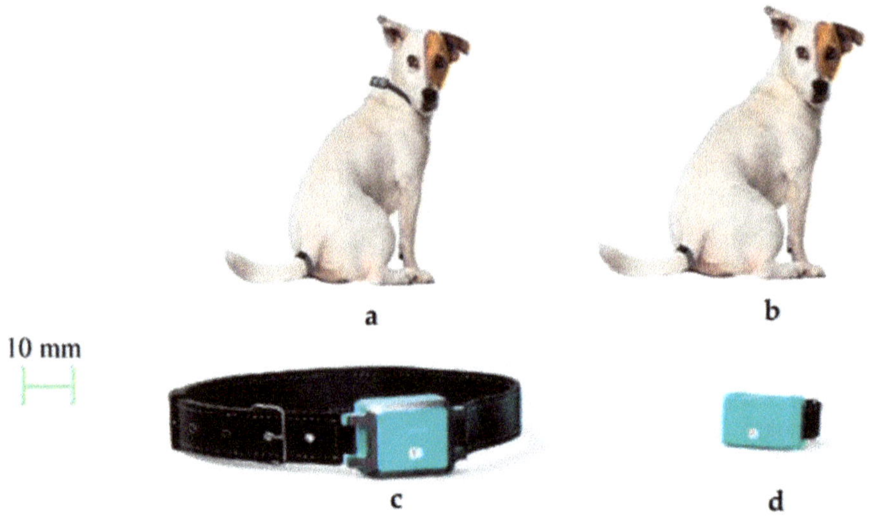

Figure 15.14 Wearable sensor on a dog's neck and tail. (a) Neck-wearable and tail-wearable devices placed on the dog for activity detection. (b) The tail-wearable device is placed for emotion detection. (c) Neck-wearable device. (d) Tail-wearable device. Reproduced from ref. 141, https://doi.org/10.3390/app9224938, under the terms of the CC BY 3.0 license, https://creativecommons.org/licenses/by/3.0/.

15.4.3.2 Habitat Based Sensors

Wildlife habitat inspections depend on environmental sensors to search for viral traces in soil samples and water bodies as well as air circulation samples. The surveillance of the avian influenza virus relies on environmental sampling by collecting various specimens, including faeces, together with water and air samples and surface swabs and mud samples. Remote sensing technologies work together with these methods to supply information about wildlife habitats that cannot be reached through traditional means. Habitat conditions, animal movement patterns, and disease dissemination agents can be monitored through aerial assessments and satellite imagery data, since these processes need no direct human handling. The methods have supported tracking of wild bird migration patterns alongside agricultural developments related to avian influenza transmission dynamics.[130] Environmental sensing platforms used together with remote sensing data improve the capacity to monitor viral transmission between wildlife populations, thus providing complete ecological information about disease origins.

15.4.3.3 Biosensors for Real-time Pathogen Detection

Biosensors made with nanomaterials, aptamers and antibodies serve as real-time detectors of pathogens in wildlife populations as well as their environmental surroundings. Biosensors built with nanomaterials provide both sensitivity and precision that enable the detection of small amounts of viral markers in biological solutions.[142,143]

The use of synthetic oligonucleotides in aptamer-based biosensors results in aptasensors that detect viruses through specific protein binding interactions, leading to prompt correct virus identification. Different aptasensors have proven their ability to detect SARS-CoV-2 alongside other viral pathogens and therefore demonstrated their potential to track emerging wildlife pathogens.[144]

Real-time pathogen detection takes place through immunosensors when antibodies serve as their biorecognition elements (Figure 15.15). Health and safety depend on these sensors to decrease pandemic-related contaminations.[145]

Integrating biosensors into portable devices makes field operation possible because they enable immediate viral pathogen detection in wild animal populations and their natural habitats. Early disease interventions become achievable through real-time monitoring space, which further stops the transmission of diseases between wild animals and humans.

A pathogen detection method based on luminescence through the use of antibody-bound magnetic beads is shown in Figure 15.1.5. Several steps are

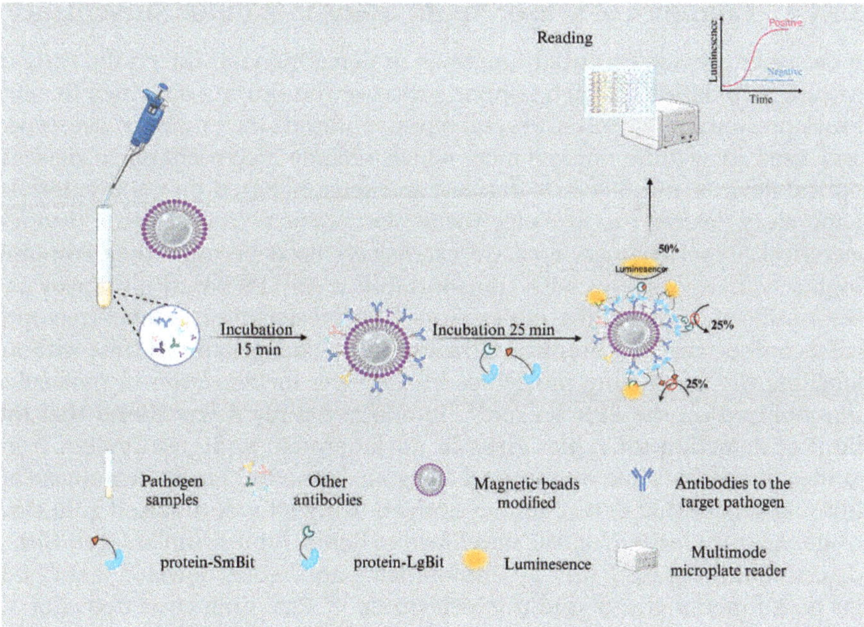

Figure 15.15 Schema illustrating lysis–NanoLuc luciferase complementation assay for the detection of an antibody of interest. Abbreviations: small luciferase fragment (SmBit), large luciferase fragment (LgBit). Reproduced from ref. 146, https://doi.org/10.3390/vetsci12010023, under the terms of the CC BY 3.0 license, https://creativecommons.org/licenses/by/3.0/.

followed in the process. The analysis starts when researchers put pathogen specimens with potential target viruses and bacteria into a reaction tube. The specimen includes different antibodies that will bind to particular pathogens among them. This is followed by incubation where the specific antibodies attached to magnetic beads allow the beads to create complexes with detected pathogens and a second incubation where the pathogen-antibody interaction occurs between the small luciferase fragment (SmBit) and large luciferase fragment (LgBit) protein fragments. The protein fragments contribute to the mechanism that produces luminescence. The biochemical mechanism produces luminescence from the detection solution only when target pathogens exist. The pathogen detection depends on both the intensity of light emission and pathogen concentration levels. The pathogen detection occurs through a multimode microplate reader that quantifies the amount of emitted light. A luminescence–time diagram displaying signal output enables the separation of negative from positive detection results. The diagram presents information about pathogen detection efficiency where magnetic beads serve target isolation purposes and luminescence functions as the output detection system. The diagnostic method enables fast, precise pathogen monitoring with high processing capacity.

15.4.4 Examples of Sensor Applications in Wildlife Surveillance

Biosensors provide essential functions in detecting viral infections early in wildlife populations, which supports disease monitoring activities for outbreak prevention purposes. Several types of biosensors have been developed and used in wildlife surveillance, which include electrochemical devices, optical devices, piezoelectric devices, and sensors based on nanomaterials. Laboratory devices that use electrochemistry detect viral elements through electrical signal changes and are extensively used because they combine high sensitivity features with transportability. An SPR-based biosensor has been utilised to detect the rabies virus. SPR biosensing is a label-free optical technique that can monitor refractive index changes in real time without labelling the interacting molecules in a sample for detection of molecules immobilised on the SPR sensor.[147] In another study it was found that the limit of detection for rabies virus in nucleoprotein and glycoprotein from guinea pig brain tissue was around 70 pg ml^{-1}. Another group introduced an immunosensor that detects rabies antigen using a functionalised gold electrode. An immunosensor was developed to detect rabies viral antigen with a detection limit of 0.5 g ml^{-1} by immobilising anti-rabies polyclonal antibodies on a functionalised gold microelectrode.[148] Real-time virus detection at field sites becomes achievable with fluorescence and SPR sensors because they offer high specificity in addition to their ability to detect targets in real-time.[149] The mass changes from viral binding are detected by piezoelectric biosensors in label-free measurements, while nanomaterial-enhanced biosensors, that include graphene-based and gold nanoparticle sensors, give improved detection stability and sensitivity.[150] Biosensing technologies play a vital role in wildlife zoonotic virus detection by identifying avian influenza virus (AIV), Ebola virus (EBOV), and coronaviruses, thus preventing their spread to human populations.

Wildlife surveillance increasingly relies on advanced sensor technologies to detect, monitor, and check animals' health in real-time. Environmental sensor technologies and biosensors are used to detect disease in wildlife populations, prevent disease outbreaks, and maintain biodiversity, in addition to protecting human health. Avian influenza detection in wild bird populations and coronavirus surveillance in bat populations are two instances in which sensor technology has been used to a great extent.[151] Both instances indicate ways in which sensors detect disease, offer better warning systems, and assist conservation efforts.

15.4.4.1 Detection of Avian Influenza in Wild Bird Population

Avian influenza is a highly contagious viral disease that affects migratory and domesticated birds. Population surveillance of migratory birds is crucial for forecasting and pre-empting disease outbreaks.[152] Environmental sensors have revolutionized the ability to monitor the dissemination of avian influenza viruses using two primary methods: environmental sensors in primary migratory bird flyways and wearable biosensors in vulnerable species.[153]

Environmental sensors in primary migratory bird flyways detect viral RNA fragments, microbial structure of the community, and other environmental indicators of the avian influenza virus's presence. Airborne particle detectors capture airborne viruses shed by sick birds, providing a heads-up on likely disease outbreak potential. Remote sensing and integration of satellite data enable scientists to ascertain habitat health, mobility of waterfowl, and climatic variables that enable disease outbreak potential.[154] The application of environmental sensors in bird migratory patterns has greatly improved the forecasting and management of avian influenza outbreaks. By identifying high-risk areas, wildlife health specialists apply precision biosecurity to restrict disease transfer between wildlife and farm birds. Wearable biosensors have facilitated disease surveillance in vulnerable species of birds, such as Asian bar-headed geese, a primary vector of avian influenza dissemination.[155] GPS biosensors send live-location signals, combined with body heat, heart rate, and mobility patterns surveillance, while miniature biosensors monitor infection biomarkers in real-time. Automated remote monitoring allows live signals to be sent to distant observation centers, allowing monitoring of disease progression, infection patterns, and virus dissemination between groups of birds.[156]

15.4.4.2 Monitoring Bat Population for Coronaviruses

Bats are natural reservoirs of a large number of viruses, including coronaviruses that have caused earlier SARS, MERS, and COVID-19 outbreaks. Population surveillance of bats is important to detect the spillover potential of viruses in order to avert possible zoonosis transmission. Sensors have been utilized in such instances where bat colonies have been monitored using real-time biosensors and sensor systems in caves and natural areas.[157] Biosensors are used to monitor bat colonies and diagnose viral infections in real-time. Miniaturized biosensors that can be attached to bats monitor body temperatures and breathing patterns, potential indicators of active infection. Saliva and faecal sensors in a non-invasive process take bat samples and identify coronavirus, providing warning signals of imminent outbreaks. Biosensors using radio frequency identification (RFID) to monitor individual bat movement patterns and bat interactions in colonies can be used to predict infection between animals and high-risk modes of transfer.[158] Wireless sensor networks in bat colonies send data in real-time to laboratories, reducing stress on animals and human interference. Biosensor applications have been used to detect new coronaviruses and to better identify their patterns of transfer, guiding public health decisions and disease control measures to prevent zoonosis. Environmental sensors in bat caves and associated ecosystems can be used to detect viral particles and estimate disease risk indicators.[159]

Viral RNA fragment-detecting sensors in caves employ advanced air filtering and molecular detection to detect RNA in air. Temperature, humidity, and airflow sensors take readings of cave temperatures, humidity, and airflow, to estimate the optimal survivability of viruses. Surface samplers take readings of cave walls, accumulations of guano, and droplets of water, identifying environmental DNA to detect the genetic material of viruses and other disease organisms. Natural exposure to natural light in bat habitats is measured using UV light sensors to better estimate the potential decomposition of viruses. Environmental surveillance methods, in particular, have been helpful in determining areas of high circulation of viruses and human–wildlife interaction risk assessment.

15.4.5 Integration of Sensor Data with Big Data and AI

The integration of sensor data and AI has revolutionized wildlife disease surveillance to allow the forecasting of diseases that are yet to become prevalent. With AI analysis combined with machine learning (ML), wildlife health experts, ecologists, and virologists can better understand disease dynamics and take preventive action to control disease outbreaks. Sensor technologies are employed on a large scale in wildlife areas to monitor environmental indicators, track animals' movement patterns, and detect biomarkers of disease.[160] The sensors generate large streams of live data that, when combined with AI tools, unveil underlying patterns of disease dissemination. AI analysis enhances predictive modelling to provide knowledge of patterns of disease dissemination in wildlife hosts and even human hosts, as shown in Figure 15.16.

Figure 15.16 takes the form of a circular infographic that demonstrates AI's contributions to zoonotic disease management. The main topic is "Artificial

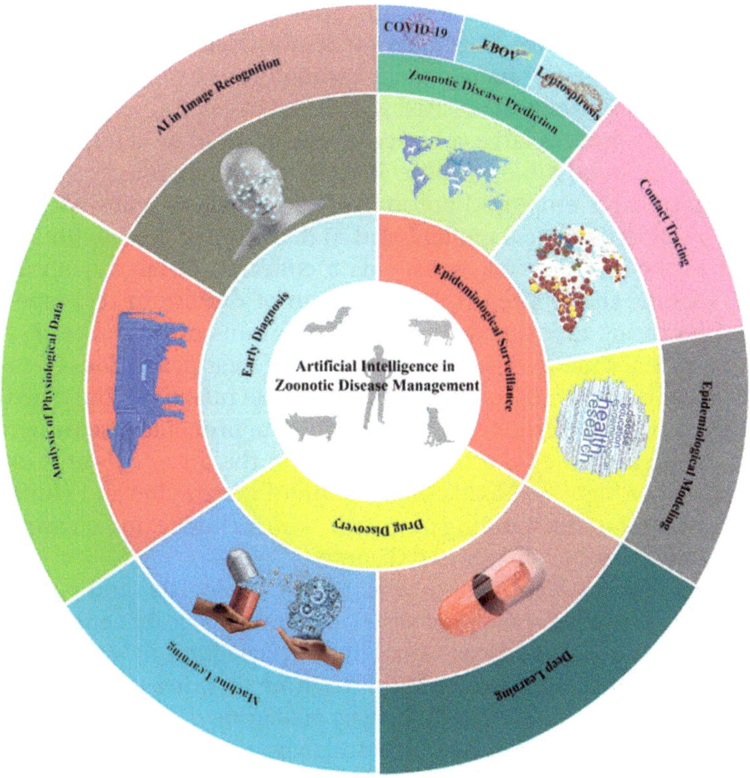

Figure 15.16 Artificial Intelligence in wildlife disease monitoring. Reproduced from ref. 160, https://doi.org/10.1016/j.soh.2023.100045, under the terms of the CC BY 4.0 license, https://creativecommons.org/licenses/by/4.0/.

Intelligence in Zoonotic Disease Management" and it has received further support from different AI applications in this domain.

Figure 15.16 contains different sections that demonstrate that AI technology enables zoonotic disease forecasting for COVID-19 alongside EBOV, leptospirosis and other infectious diseases. The healthcare industry uses AI-based methods called contact tracing for tracking the spread of zoonotic diseases. AI systems create models that forecast disease courses together with their expected impact on populations. ML delivers algorithms that evaluate disease patterns and outbreak developments. The discovery of drugs for zoonotic diseases becomes possible through AI-assisted studies during the drug development process. AI is involved in detecting zoonotic diseases early through diagnostic systems. AI enables the analysis of health data through interpretation systems for disease diagnostics purposes. AI in image recognition is used in image processing and pattern recognition for disease diagnosis. In epidemiological surveillance AI-based monitoring of disease outbreaks can be employed.

Figure 15.16 displays zoonotic categories as different coloured sections within circular structures. This diagram demonstrates the effectiveness of AI technologies in disease management through forecasting and disease control procedures for zoonotic diseases.

AI programs process large volumes of sensor input to determine patterns that would be impossible or difficult to determine manually. Some of the key AI-based tools employed are predictive modelling using ML, forecasting using epidemiological models, and AI detection of disease hotspots. ML models assist in processing sensor data collected in large quantities and deliver instant analysis results.[161] Detection of disease in real-time is crucial to preventing disease effectively. ML programs process wildlife biosensor data in real-time to detect unnatural body heat peaks, loss of mobility, or untypical heart rhythm patterns that indicate infection. The detection of patterns in animals permits the identification of probable disease spreading events, while airborne detection of viruses in the air or in water indicates areas of risk of high viral loads. Integration of geographic and climate data permits a better understanding of ways in which climate and environmental forces allow viruses to survive and disperse.[160]

Data sharing between wildlife researchers, ecologists, and virologists is a crucial aspect of disease-prevention programs. The benefits of sharing data include integrated disease surveillance, detection in a timely manner, and rapid response, improved AI and machine learning models, and cross-disciplinary collaboration. The challenges of sharing data range from security of data to confidentiality of data to standardization of data formats to overcoming interdisciplinary differences.[160,162] There have been various attempts towards developing frameworks of data sharing, such as the Global Virome Project (GVP), Wildlife Health Information Sharing Partnership (WHISPers), and EcoHealth Alliance. By sharing data to facilitate cooperation, researchers can build more accurate disease prediction and prevention models.

15.4.6 Using Sensors in Wildlife Monitoring

Wildlife monitoring is hindered by a set of challenges, such as environmental challenges, technical challenges, transmission or storage challenges, or logistical or ethical challenges. Environmental challenges such as bad weather, power or energy shortages, maintenance or repairs in remote areas, or false positives or negatives hinder it. The technical challenges range from precision or calibration challenges to false positives or negatives Calibration keeps devices from mismeasuring environmental or physiological variables, whereas extraneous variables such as ambient noise or interference from other devices result in misreadings.[163] Biosensors that detect disease biomarkers must discriminate between background microbial noise and the target pathogens. Transmitting or storing challenges include cloud storage or processing capacity to process in real-time, and bandwidth in remote areas. The ethical challenges arise when working with vulnerable animals, where it is a question of keeping human impact to a bare minimum.

The use of GPS collars, biosensors, or radio tags to mark animals in the wild is stressful or discomforting, and animals chew or dislodge devices and may harm themselves in the process. The welfare of animals demands that animals not be handled repeatedly to attach or insert sensors, resulting in stress, social groups should not be separated, or animals exposed to human-diseases. Invasive devices, such as implantable biosensors, demand veterinary knowledge and pose infection or rejection risks.[164] The last challenge is that of security and privacy when GPS tracking of vulnerable animals may become accessible to poachers or illicit wildlife traders. The sharing of data between researchers demands confidentiality agreements to restrict unapproved access to sensitive data. Conservation efforts need to be transparent enough to enable reporting of results without compromising vulnerable animals to human predators.

15.4.7 Future Prospects and Innovations in Sensor-based Wildlife Monitoring

Sensor technology is improving, offering new possibilities in wildlife surveillance. Miniaturization, satellite, drone technologies, and collaborative studies across the globe are making surveillance systems more efficient and responsible. Smaller sensor devices that weigh less are making wildlife surveillance less invasive. Nanotechnology innovations allow rice-grain sized biosensors to be implanted or added to feathers, hair, or scales and enable GPS tags to be used on smaller animals, such as birds and bats, without behavioural effects. Smart wearable wildlife sensor devices, such as smart ear tags for wild mammals to monitor mobility and health non-invasively, are in development.[165]

Integration of next-generation biosensors in satellite and drone technologies facilitates large-scale wildlife surveillance. Satellite-assisted disease surveillance leverages AI-based satellites to monitor environmental conditions to detect warning signs of disease outbreaks. Drones carrying infrared cameras and biosensors can monitor groups of animals, detect stress signals, and obtain environmental samples without human effort.[166] AI-based processing of multiple sources of data plays a crucial role in early warning systems in outbreak prediction based on dynamic environmental conditions. Cooperative studies play a key role in disseminating the global surveillance system of viruses in wildlife, such as open-source data platforms, conservation cooperation across the globe, and citizen science and engagement of the community.

In conclusion, the increasing concern regarding zoonotic disease reveals wildlife's role in transmitting viruses to human beings. With the expansion of human settlements, habitat loss, human–wildlife interaction, and climate change offer a fertile ground for disease spread. The outbreak of such contagious diseases as COVID-19, Ebola, and avian influenza reveals a need for more surveillance and control systems. Wildlife surveillance, enabled by sensor technologies, big data, and AI, is a fundamental instrument in monitoring patterns of disease spread. Nevertheless, technical, logistical, and ethical challenges need to be met *via* cooperation and innovations at a global scale. Mitigation of zoonotic disease risk requires a One Health approach that unites the services of virologists, ecologists, practitioners in public health, and conservation practitioners. Global networks to monitor disease need to be strengthened, investments in early warning systems need to be made, and responsible human–wildlife coexistence needs to be ensured to prevent pandemics in the future.

15.5 Conclusions and Future Perspectives

In conclusion, the transmission of viruses through environmental factors like air and water poses a substantial risk to public health. This chapter has explored the interactions of airborne viruses with environmental factors, including temperature, humidity, airflow, and water quality, emphasizing the importance of environmental quality and hygiene practices in controlling viral transmission. The development of biosensors has been crucial in detecting viruses in these environments, with ongoing efforts to improve detection capabilities in terms of sensitivity, cost-effectiveness, and speed.

The chapter has also addressed the surveillance of public spaces and wildlife, discussing advanced detection methods, including biosensors, environmental monitoring, non-invasive technologies, and AI-enhanced systems. These technologies are vital for early detection, outbreak management, and preventing transmission of zoonoses. The integration of AI and big data has further enhanced surveillance capabilities, enabling predictive modelling and better understanding of disease dynamics.

Looking ahead, the field is poised for significant advancements. Future trends include the development of real-time, at-home diagnostic solutions, enhanced sensitivity and specificity of detection systems through nanomaterial engineering, and multidisciplinary collaborations to improve detection approaches. The development of cost-effective and scalable solutions will be crucial for global accessibility, along with streamlined regulatory processes to facilitate the adoption of innovative detection methods. These advancements promise a more robust, efficient, and accessible global disease monitoring system, essential for preventing future pandemics.

References

1. E. F. Mohamed and G. Awad, *Compr. Anal. Chem.*, 2022, **99**, 163–188.
2. A. P. Jones, *Atmos. Environ.*, 1999, **33**, 4535–4564.
3. J. Ciencewicki and I. Jaspers, *Inhalation Toxicol.*, 2007, **19**, 1135–1146.
4. F. W. Lipfert, *Risk Anal.*, 1997, **17**, 137–146.
5. E. A. F. Simoes, *Lancet*, 1999, **354**, 847–852.

6. N. S. Hendren, M. H. Drazner, B. Bozkurt and L. T. Cooper, *Circulation*, 2020, **141**, 1903–1914.
7. M. Ciotti, M. Ciccozzi, A. Terrinoni, W. C. Jiang, C. Bin Wang and S. Bernardini, *Crit. Rev. Clin. Lab Sci.*, 2020, **57**, 365–388.
8. S. Tang, Y. Mao, R. M. Jones, Q. Tan, J. S. Ji, N. Li, J. Shen, Y. Lv, L. Pan, P. Ding, X. Wang, Y. Wang, C. R. MacIntyre and X. Shi, *Environ. Int.*, 2020, **144**, 106039.
9. J. L. Domingo and J. Rovira, *Environ. Res.*, 2020, **187**, 109650.
10. WHO Director-General's opening remarks at the media briefing on COVID-19 - 11 March 2020. https://www.who.int/director-general/speeches/detail/who-director-general-s-opening-remarks-at-the-media-briefing-on-covid-19---11-march-2020. Accessed 27 Feb 2025.
11. A. M. Elsaid and M. S. Ahmed, *Environ. Res.*, 2021, **199**, 111314.
12. S. Kumar, *Sci. Total Environ.*, 2020, **745**, 141021.
13. S. Lolli, Y. C. Chen, S. H. Wang and G. Vivone, *Sci. Rep.*, 2020, **10**, 1–15.
14. M. F. F. Sobral, G. B. Duarte, A. I. G. da Penha Sobral, M. L. M. Marinho and A. de Souza Melo, *Sci. Total Environ.*, 2020, **729**, 138997.
15. M. Célia da Silva Lanna, A. Viancelli, W. Michelon, S. V. Castro Carvalho, D. de Almeida dos Reis, L. A. Fernandez de Salles, I. H. Sant'Anna, L. T. Resende, C. de Souza Ferreira, I. Aparecido das Chagas, M. Hernández, H. Treichel, D. Rodríguez-Lázaro and G. Fongaro, *Environ. Pollut.*, 2019, **252**, 8–13.
16. S. Tandukar, S. P. Sherchan and E. Haramoto, *Sci. Rep.*, 2020, **10**, 1–8.
17. A. M. Gall, B. J. Mariñas, Y. Lu and J. L. Shisler, *PLoS Pathog.*, 2015, **11**, e1004867.
18. C. P. Gerba, *Adv. Appl. Microbiol.*, 1984, **30**, 133–168.
19. *Guidelines for drinking-water quality, 4th edition, incorporating the 1st addendum*, G. W. H. Organization. G. S. WHO, World Health Organization.
20. B. Adelodun, F. O. Ajibade, J. O. Ighalo, G. Odey, R. G. Ibrahim, K. Y. Kareem, H. O. Bakare, A. G. O. Tiamiyu, T. F. Ajibade, T. S. Abdulkadir, K. A. Adeniran and K. S. Choi, *Environ. Res.*, 2021, **192**, 110309.
21. M. R. Riley, C. P. Gerba and M. Elimelech, *J. Biol. Eng.*, 2011, **5**, 1–10.
22. Surveillance for Waterborne Disease and Outbreaks Associated with Recreational Water—United States, 2003–2004. https://www.cdc.gov/mmwr/preview/mmwrhtml/ss5512a1.htm. Accessed 27 Feb 2025.
23. M. Pilevar, K. T. Kim and W. H. Lee, *J. Hazard. Mater.*, 2021, **410**, 124656.
24. M. T. H. van Vliet, E. R. Jones, M. Flörke, W. H. P. Franssen, N. Hanasaki, Y. Wada and J. R. Yearsley, *Environ. Res. Lett.*, 2021, **16**, 024020.
25. M. Lackner, I. Ouattara, N. Gharae and R. Abolhassani, *J. Water Sci. Eng.*, 2020, **1**, 1–19.
26. S. S. Buehler, B. Madison, S. R. Snyder, J. H. Derzon, N. E. Cornish, M. A. Saubolle, A. S. Weissfeld, M. P. Weinstein, E. B. Liebow and D. M. Wolk, *Clin. Microbiol. Rev.*, 2015, **29**, 59–103.
27. C. Xi, Y. Zhang, C. F. Marrs, W. Ye, C. Simon, B. Foxman and J. Nriagu, *Appl. Environ. Microbiol.*, 2009, **75**, 5714–5718.

28. S. S. Mahapatra, M. Sahu, R. K. Patel and B. N. Panda, *Water Qual., Exposure Health*, 2012, **4**(2), 2012, 93–104.
29. A. Mustapha, A. Z. Aris, F. M. Yusoff, M. P. Zakaria, M. F. Ramli, A. M. Abdullah, N. U. Kura and T. S. Narany, *Water Qual., Exposure Health*, (3), 2014, **6**, 127–142.
30. H. Wang, I. Kjellberg, P. Sikora, H. Rydberg, M. Lindh, O. Bergstedt and H. Norder, *Water Res.*, 2020, **168**, 115141.
31. M. C. Onojake, S. O. Ukerun, G. Iwuoha and S. O. Ukerun, *Water Qual., Exposure Health*, 2011, **3**(2), 91–99.
32. N. Mehle and M. Ravnikar, *Water. Res.*, 2012, **46**, 4902–4917.
33. K. P. Singh, A. Malik and S. Sinha, *Anal. Chim. Acta*, 2005, **538**, 355–374.
34. V. I. Pye and R. Patrick, *Science*, 1983, **221**, 713–718.
35. R. Vinodhini and M. Narayanan, *Int. J. Environ. Sci. Technol.*, 2008, **5**, 179–182.
36. G. La Rosa, L. Bonadonna, L. Lucentini, S. Kenmoe and E. Suffredini, *Water Res.*, 2020, **179**, 115899.
37. Guidelines for drinking-water quality, Volume 2 – Health Criteria and Other Supporting Information, 2nd edn. https://www.who.int/publications/i/item/9241544805. Accessed 27 Feb. 2025.
38. M. Rusiñol and R. Girones, *Water and Sanitation for the 21st Century: Health and Microbiological Aspects of Excreta and Wastewater Management (Global Water Pathogen Project)*, 2017, pp. 1–120, DOI: 10.14321/WATERPATHOGENS.19.
39. K. Farkas, E. M. Adriaenssens, D. I. Walker, J. E. McDonald, S. K. Malham and D. L. Jones, *Food Environ. Virol.*, 2019, **11**, 113–119.
40. *P. diarrhoea through better water, sanitation and hygiene: exposures and impacts in low-and middle-income countries WHO*, World Health Organization.
41. M. A. Al-Abdan, M. N. Bin-Jumah, D. Ali and S. Alarifi, *J. King Saud Univ., Sci.*, 2021, **33**, 101355.
42. Z. Kotsiri, A. Vantarakis, F. Rizzotto, D. Kavanaugh, N. Ramarao and J. Vidic, *Appl. Sci.*, 2019, **9**, 5392.
43. Z. Kotsiri, J. Vidic and A. Vantarakis, *J. Environ. Sci.*, 2022, **111**, 367–379.
44. J. Vidic, M. Manzano, C. M. Chang and N. Jaffrezic-Renault, *Vet. Res.*, 2017, **48**(1), 1–22.
45. N. Yildirim, D. Li, F. Long and A. Z. Gu, *Proc. Sens.*, 2013, 1–5, DOI: 10.1109/ICSENS.2013.6688604.
46. M. Manzano, S. Viezzi, S. Mazerat, R. S. Marks and J. Vidic, *Biosens. Bioelectron.*, 2018, **100**, 89–95.
47. W. Hai, T. Goda, H. Takeuchi, S. Yamaoka, Y. Horiguchi, A. Matsumoto and Y. Miyahara, *Sens. Actuators, B*, 2018, **260**, 635–641.
48. R. G. Ramírez-Chavarría, E. Castillo-Villanueva, B. E. Alvarez-Serna, J. Carrillo-Reyes, R. M. Ramírez-Zamora, G. Buitrón and L. Alvarez-Icaza, *J. Environ. Chem. Eng.*, 2022, **10**, 107488.
49. A. Cincinelli and T. Martellini, *Int. J. Environ. Res. Public Health*, 2017, **14**, 1286.

50. C. López-Granero, L. Polyanskaya, D. Ruiz-Sobremazas, A. Barrasa, M. Aschner and M. Alique, *Biomolecules*, 2024, **14**, 35.
51. M. H. Forouzanfar, A. Afshin, L. T. Alexander, S. Biryukov, M. Brauer, K. Cercy, F. J. Charlson, A. J. Cohen, L. Dandona, K. Estep, A. J. Ferrari, J. J. Frostad, N. Fullman, W. W. Godwin, M. Griswold, S. I. Hay, H. H. Kyu, H. J. Larson, S. S. Lim, P. Y. Liu, A. D. Lopez, R. Lozano, L. Marczak, A. H. Mokdad, M. Moradi-Lakeh, M. Naghavi, M. B. Reitsma, G. A. Roth, P. J. Sur, T. Vos, J. A. Wagner, H. Wang, Y. Zhao, M. Zhou, R. M. Barber, B. Bell, J. D. Blore, D. C. Casey, M. M. Coates, K. Cooperrider, L. Cornaby, D. Dicker, H. E. Erskine, T. Fleming, K. Foreman, E. Gakidou, J. A. Haagsma, C. O. Johnson, L. Kemmer, T. Ku, J. Leung, F. Masiye, A. Millear, M. Mirarefin, A. Misganaw, E. Mullany, J. E. Mumford, M. Ng, H. Olsen, P. Rao, N. Reinig, Y. Roman, L. Sandar, D. F. Santomauro, E. L. Slepak, R. J. D. Sorensen, B. A. Thomas, S. E. Vollset, H. A. Whiteford, B. Zipkin, C. J. L. Murray, C. N. Mock, B. O. Anderson, N. D. Futran, H. R. Anderson, Z. A. Bhutta, M. I. Nisar, N. Akseer, H. Krueger, C. C. Gotay, N. Kissoon, J. A. Kopec, F. Pourmalek, R. Burnett, A. A. Abajobir, L. D. Knibbs, J. L. Veerman, R. Lalloo, J. G. Scott, N. K. M. Alam, H. N. Gouda, Y. Guo, J. J. McGrath, P. Jeemon, R. Dandona, S. Goenka, G. A. Kumar, P. W. Gething, D. Bisanzio, A. Deribew, S. C. Darby, R. Ali, D. A. Bennett, V. Jha, Y. Kinfu, M. McKee, G. V. S. Murthy, N. Pearce, H. Stöckl, L. Duan, Y. Jin, Y. Li, S. Liu, L. Wang, P. Ye, X. Liang, P. Azzopardi, G. C. Patton, A. Meretoja, K. Alam, R. Borschmann, S. M. Colquhoun, R. G. Weintraub, C. E. I. Szoeke, Z. Ademi, H. R. Taylor, T. Wijeratne, C. Batis, S. Barquera, I. R. Campos-Nonato, A. G. Contreras, L. Cuevas-Nasu, V. De, H. Gomez-Dantes, I. B. Heredia-Pi, C. Medina, F. Mejia-Rodriguez, J. C. Montañez Hernandez, C. A. Razo-García, J. A. Rivera, S. Rodríguez-Ramírez, T. G. Sánchez-Pimienta, E. E. Servan-Mori, T. Shamah, G. A. Mensah, H. J. Hoff, B. Neal, T. R. Driscoll, A. H. Kemp, J. Leigh, A. B. Mekonnen, S. Bhatt, T. Fürst, F. B. Piel, A. Rodriguez, S. J. Hutchings, A. Majeed, M. Soljak, J. A. Salomon, A. L. Thorne-Lyman, O. N. Ajala, T. Bärnighausen, L. E. Cahill, E. L. Ding, M. S. Farvid, S. Khatibzadeh, G. R. Wagner, M. G. Shrime, J. R. A. Fitchett, G. M. Aasvang, M. Savic, K. H. Abate, T. T. Gebrehiwot, A. T. Gebremedhin, C. Abbafati, K. M. Abbas, F. Abd-Allah, A. M. Abdulle, S. F. Abera, Y. A. Melaku, G. Y. Abyu, B. D. Betsu, G. B. Hailu, D. Y. Tekle, A. Z. Yalew, B. Abraham, L. J. Abu-Raddad, A. O. Adebiyi, I. A. Adedeji, A. K. Adou, J. C. Adsuar, E. E. Agardh, J. Rehm, A. Badawi, S. Popova, A. Agarwal, A. Ahmad, T. F. Akinyemiju, D. C. Schwebel, J. A. Singh, Z. Al-Aly, S. F. Aldhahri, K. A. Altirkawi, A. S. Terkawi, R. W. Aldridge, T. Tillmann, Z. A. Alemu, T. K. Tegegne, A. Alkerwi, F. Alla, F. Guillemin, P. Allebeck, R. H. S. Rabiee, S. M. Fereshtehnejad, M. Kivipelto, J. J. Carrero, E. Weiderpass, R. Havmoeller, S. Sindi, U. Alsharif, E. Alvarez, N. Alvis-Guzman, A. T. Amare, L. G. Ciobanu, B. W. Taye, A. Amberbir, A. K. Amegah, H. Amini, C. K. Karema, W. Ammar, H. L. Harb, S. M. Amrock, H. H. Andersen, C. A. T. Antonio, E. J. A. Faraon, P. Anwari, J. Ärnlöv,

A. Larsson, A. Artaman, H. Asayesh, R. J. Asghar, R. Assadi, S. Atique, E. F. G. A. Avokpaho, A. Awasthi, B. P. Ayala, U. Bacha, M. C. Bahit, K. Balakrishnan, A. Barac, S. L. Barker-Collo, B. del Pozo-Cruz, S. Mohammed, L. Barregard, M. Petzold, L. H. Barrero, S. Basu, L. C. Del, S. Bazargan-Hejazi, J. Beardsley, N. Bedi, E. Beghi, K. N. Sheth, M. L. Bell, J. J. Huang, A. K. Bello, I. S. Santos, I. M. Bensenor, P. A. Lotufo, A. Berhane, C. D. Wolfe, E. Bernabé, H. S. Roba, A. S. Beyene, T. A. Hassen, Y. M. Mesfin, N. Bhala, A. Bhansali, S. Biadgilign, B. Bikbov, E. Bjertness, A. S. Htet, S. Boufous, L. Degenhardt, S. Resnikoff, B. Calabria, R. R. A. Bourne, M. Brainin, A. Brazinova, M. Majdan, J. Shen, N. J. K. Breitborde, H. Brenner, B. Schöttker, D. M. Broday, T. S. Brugha, B. Brunekreef, H. Kromhout, Z. A. Butt, A. van Donkelaar, R. V. Martin, R. Cárdenas, D. O. Carpenter, C. A. Castañeda-Orjuela, J. Castillo, R. E. Castro, F. Catalá-López, J. Chang, P. P. Chiang, M. Chibalabala, O. Chimed-Ochir, Y. Jiang, K. Takahashi, V. H. Chisumpa, C. C. Mapoma, A. A. Chitheer, J. J. Choi, H. Christensen, D. J. Christopher, L. T. Cooper, J. A. Crump, R. G. Poulton, A. Damasceno, P. I. Dargan, J. das Neves, A. C. Davis, J. N. Newton, N. Steel, K. Davletov, E. F. de Castro, D. De, R. P. Dellavalle, D. C. Des, S. D. Dharmaratne, P. K. Dhillon, D. K. Lal, S. Zodpey, C. Diaz-Torné, E. R. Dorsey, K. E. Doyle, M. Dubey, M. H. U. Rahman, U. Ram, A. Singh, A. K. Yadav, B. B. Duncan, C. Kieling, M. I. Schmidt, I. Elyazar, A. Y. Endries, S. P. Ermakov, B. Eshrati, F. Farzadfar, A. Kasaeian, M. Parsaeian, A. Esteghamati, N. Hafezi-Nejad, S. Sheikhbahaei, S. Fahimi, R. Malekzadeh, G. Roshandel, S. G. Sepanlou, M. S. Hassanvand, P. Heydarpour, V. Rahimi-Movaghar, M. Yaseri, T. A. Farid, A. R. Khan, C. S. E. S. Farinha, A. Faro, V. L. Feigin, J. G. Fernandes, F. Fischer, N. Foigt, I. Shiue, F. G. R. Fowkes, R. C. Franklin, A. L. Garcia-Basteiro, J. M. Geleijnse, T. Jibat, B. D. Gessner, W. Tefera, A. Z. Giref, D. Haile, W. A. A. Manamo, M. Giroud, M. D. Gishu, J. Martinez-Raga, M. C. Gomez-Cabrera, P. Gona, A. Goodridge, S. V. Gopalani, A. Goto, M. Inoue, H. C. Gugnani, R. Gupta, R. A. Gutiérrez, R. Orozco, Y. A. Halasa, E. A. Undurraga, R. R. Hamadeh, S. Hamidi, A. J. Handal, G. J. Hankey, Y. Hao, S. Harikrishnan, J. M. Haro, N. F. Hernández-Llanes, H. W. Hoek, A. K. Tura, M. Horino, N. Horita, H. D. Hosgood, D. G. Hoy, M. Hsairi, G. Hu, A. Husseini, I. Huybrechts, K. M. Iburg, B. T. Idrisov, G. F. Kwan, B. V. Ileanu, A. Pana, N. Kawakami, K. Shibuya, T. A. Jacobs, K. H. Jacobsen, N. Jahanmehr, M. B. Jakovljevic, H. A. F. Jansen, S. K. Jassal, M. B. Stein, M. Javanbakht, S. P. Jayaraman, A. U. Jayatilleke, S. H. Jee, J. B. Jonas, Z. Kabir, Y. Kalkonde, R. Kamal, J. She, H. Kan, A. Karch, C. Karimkhani, A. Kaul, D. S. Kazi, P. N. Keiyoro, C. D. Parry, A. P. Kengne, R. Matzopoulos, C. S. Wiysonge, D. J. Stein, B. M. Mayosi, A. Keren, Y. S. Khader, E. A. Khan, G. Khan, Y. H. Khang, S. Won, S. Khera, M. Tavakkoli, T. A. M. Khoja, J. Khubchandani, C. Kim, D. Kim, R. W. Kimokoti, Y. Kokubo, P. A. Koul, A. Koyanagi, M. Kravchenko, Y. Y. Varakin, B. Kuate, R. S. Kuchenbecker, B. Kucuk, E. J. Kuipers, T. Lallukka, R. Shiri, T. J. Meretoja, Q. Lan, A. A. Latif, A. E. B. Lawrynowicz, J. L. Leasher,

M. Levi, X. Li, J. Liang, B. K. Lloyd, G. Logroscino, R. Lunevicius, D. Pope, M. Mahdavi, D. C. Malta, W. Marcenes, K. Matsushita, J. B. Nachega, B. X. Tran, P. A. Meaney, A. Mehari, B. A. Tedla, Z. A. Memish, W. Mendoza, G. B. M. Mensink, F. A. Mhimbira, T. R. Miller, E. J. Mills, A. Mohammadi, G. L. D. Mola, L. Monasta, L. Morawska, R. E. Norman, R. Mori, D. Mozaff, P. Shi, A. Werdecker, U. O. Mueller, A. J. Paternina, R. Westerman, S. Seedat, A. Naheed, V. Nangia, N. Nassiri, Q. L. Nguyen, P. M. Nkamedjie, O. F. Norheim, B. Norrving, L. Nyakarahuka, C. M. Obermeyer, F. A. Ogbo, I. Oh, O. Oladimeji, B. Sartorius, B. O. Olusanya, P. R. Olivares, J. O. Olusanya, J. N. Opio, E. Oren, A. Ortiz, E. Ota, P. A. Mahesh, E. Park, T. Patel, S. T. Patil, S. B. Patten, J. Wang, D. M. Pereira, M. Cortinovis, G. Giussani, N. Perico, G. Remuzzi, K. Pesudovs, M. R. Phillips, J. D. Pillay, D. Plass, M. Tobollik, S. Polinder, C. D. Pond, C. A. Pope, N. M. Prasad, M. Qorbani, A. Radfar, A. Rafay, S. M. Rana, M. Rahman, S. U. Rahman, S. Rajsic, R. K. Rai, M. Raju, K. Ranganathan, A. H. Refaat, C. D. Rehm, A. L. Ribeiro, D. Rojas-Rueda, A. Roy, M. Satpathy, N. Tandon, D. Rothenbacher, M. M. Saleh, J. R. Sanabria, L. Sanchez-Riera, M. D. Sanchez-Niño, R. Sarmiento-Suarez, M. Sawhney, J. Schmidhuber, I. J. C. Schneider, A. E. Schutte, D. A. S. Silva, S. Shahraz, M. Shin, A. Shaheen, M. A. Shaikh, R. Sharma, M. Shigematsu, S. Yoon, K. Shishani, I. D. Sigfusdottir, P. K. Singh, D. G. A. Silveira, J. I. Silverberg, Y. Yano, S. Soneji, S. Stranges, N. Steckling, C. T. Sreeramareddy, V. Stathopoulou, K. Stroumpoulis, B. F. Sunguya, S. Swaminathan, B. L. Sykes, R. Tabarés-Seisdedos, R. T. Talongwa, D. Tanne, E. M. Tuzcu, J. Thakur, G. Shaddick, M. L. Thomas, A. G. Thrift, G. D. Thurston, A. J. Thomson, R. Topor-Madry, F. Topouzis, J. A. Towbin, O. A. Uthman, R. Tobe-Gai, N. Tsilimparis, Z. Tsala, S. Tyrovolas, K. N. Ukwaja, J. van Os, T. Vasankari, N. Venketasubramanian, F. S. Violante, S. G. Waller, C. J. Uneke, Y. Wang, S. Weichenthal, A. D. Woolf, D. Xavier, G. Xu, B. Yakob, P. Yip, C. N. Kesavachandran, M. Montico, L. Ronfani, C. Yu, Z. Zaidi, N. Yonemoto, M. Z. Younis, M. Wubshet, L. J. Zuhlke, M. E. Zaki and J. Zhu, *Lancet*, 2016, **388**, 1659–1724.
52. E. C. Clark, S. Neumann, S. Hopkins, A. Kostopoulos, L. Hagerman and M. Dobbins, *JMIR Pub. Health Surveill*, 2024, **10**(1), e49185.
53. G. Hoek, R. M. Krishnan, R. Beelen, A. Peters, B. Ostro, B. Brunekreef and J. D. Kaufman, *Environ. Health*, 2013, **12**, 1–16.
54. R. Beelen, O. Raaschou-Nielsen, M. Stafoggia, Z. J. Andersen, G. Weinmayr, B. Hoffmann, K. Wolf, E. Samoli, P. Fischer, M. Nieuwenhuijsen, P. Vineis, W. W. Xun, K. Katsouyanni, K. Dimakopoulou, A. Oudin, B. Forsberg, L. Modig, A. S. Havulinna, T. Lanki, A. Turunen, B. Oftedal, W. Nystad, P. Nafstad, U. De Faire, N. L. Pedersen, C. G. Östenson, L. Fratiglioni, J. Penell, M. Korek, G. Pershagen, K. T. Eriksen, K. Overvad, T. Ellermann, M. Eeftens, P. H. Peeters, K. Meliefste, M. Wang, B. Bueno-De-Mesquita, D. Sugiri, U. Krämer, J. Heinrich, K. De Hoogh, T. Key, A. Peters, R. Hampel, H. Concin, G. Nagel, A. Ineichen, E. Schaffner, N. Probst-Hensch, N. Künzli, C. Schindler, T. Schikowski, M. Adam,

H. Phuleria, A. Vilier, F. Clavel-Chapelon, C. Declercq, S. Grioni, V. Krogh, M. Y. Tsai, F. Ricceri, C. Sacerdote, C. Galassi, E. Migliore, A. Ranzi, G. Cesaroni, C. Badaloni, F. Forastiere, I. Tamayo, P. Amiano, M. Dorronsoro, M. Katsoulis, A. Trichopoulou, B. Brunekreef and G. Hoek, *Lancet*, 2014, **383**, 785–795.
55. T. Bourdrel, I. Annesi-Maesano, B. Alahmad, C. N. Maesano and M. A. Bind, *Eur. Respir. Rev.*, 2021, **30**, 1–18.
56. S. Hassani, S. Momtaz, F. Vakhshiteh, A. S. Maghsoudi, M. R. Ganjali, P. Norouzi and M. Abdollahi, *Arch. Toxicol.*, (1), 2016, **91**, 109–130.
57. S. K. Bhardwaj, N. Bhardwaj, V. Kumar, D. Bhatt, A. Azzouz, J. Bhaumik, K. H. Kim and A. Deep, *Environ. Int.*, 2021, **146**, 106183.
58. K. Dziabowska, E. Czaczyk and D. Nidzworski, *Biosensors*, 2018, **8**, 94.
59. S. Hassanpour, B. Baradaran, M. Hejazi, M. Hasanzadeh, A. Mokhtarzadeh and M. de la Guardia, *TrAC, Trends Anal. Chem.*, 2018, **98**, 201–215.
60. R. L. Caygill, G. E. Blair and P. A. Millner, *Anal. Chim. Acta*, 2010, **681**, 8–15.
61. D. Lee, J. Bhardwaj and J. Jang, *Sci. Rep.*, 2022, **12**(1), 2022, 1–11.
62. J. Bhardwaj, N. D. Ngo, J. Lee and J. Jang, *J. Hazard. Mater.*, 2023, **442**, 130006.
63. D. Lee, J. Jang and J. Jang, *J. Hazard. Mater.*, 2023, **458**, 131972.
64. W. Białobrzeska, D. Firganek, M. Czerkies, T. Lipniacki, M. Skwarecka, K. Dziąbowska, Z. Cebula, N. Malinowska, D. Bigus, E. Bięga, K. Pyrć, K. Pala, S. Żołędowska and D. Nidzworski, *Biosensors*, 2020, **10**, 175.
65. A. D. Ravina, P. S. Gill, J. Narang, M. Prasad and H. Mohan, *Proc. Biochem.*, 2020, **98**, 262–268.
66. World Health Organization, *WHO Guidelines for Environmental Surveillance of Poliovirus Circulation Vaccines and Biologicals*, 2003.
67. R. Zuniga-Montanez, D. A. Coil, J. A. Eisen, R. Pechacek, R. G. Guerrero, M. Kim, K. Shapiro and H. N. Bischel, *PLoS One*, 2022, **17**, e0267212.
68. L. Yao, W. Zhu, J. Shi, T. Xu, G. Qu, W. Zhou, X. F. Yu, X. Zhang and G. Jiang, *Chem. Soc. Rev.*, 2021, **50**, 3656–3676.
69. S. Carver, A. M. Kilpatrick, A. Kuenzi, R. Douglass, R. S. Ostfeld and P. Weinstein, *J. Environ. Monit.*, 2010, **12**, 2048–2055.
70. K. Mihajlovski, M. P. Buttner, P. Cruz, B. Labus, B. S. Pierre Schneider and E. Detrick, *PLoS One*, 2022, **17**, e0278061.
71. Global Polio Eradication Initiative, *Guidelines On Environmental Surveillance For Detection Of Polioviruses*, 2015.
72. Z. Liu, K. Skowron, K. Grudlewska-Buda and N. Wiktorczyk-Kapischke, *Sci. Total Environ.*, 2021, **795**, 148949.
73. K. Leung, J. T. Wu and G. M. Leung, *Nat. Commun.*, 2021, **12**(1), 2021, 1–8.
74. A. Ferretti and E. Vayena, *Epidemics*, 2022, **41**, 100652.
75. M. Verani, A. Pagani, I. Federigi, G. Lauretani, N. T. Atomsa, V. Rossi, L. Viviani and A. Carducci, *Viruses*, 2024, **16**, 482.
76. G. T. Igwama, J. A. Olaboye, C. C. Maha, M. D. Ajegbile and S. Abdul, *Int. J. Appl. Res. Soc. Sci.*, 2024, **6**, 1449–1460.

77. M. Banakar, M. Hamidi, Z. Khurshid, M. S. Zafar, J. Sapkota, R. Azizian and D. Rokaya, *Biosensors*, 2022, **12**, 927.
78. H. Maddali, C. E. Miles, J. Kohn and D. M. O'Carroll, *ChemBioChem*, 2021, **22**, 1176–1189.
79. C. J. Goergen, M. J. Tweardy, S. R. Steinhubl, S. W. Wegerich, K. Singh, R. J. Mieloszyk and J. Dunn, *Annu. Rev. Biomed. Eng.*, 2022, **24**, 1–27.
80. H. Yang, 2022, arXiv:2204.14064.
81. B. Alorda-Ladaria, M. Ruiz-Pérez and V. Ramos, *Electronics*, 2022, **11**, 2897.
82. E. Hany and N. M. Khaled, *Eng. Res. J.*, 2024, **53**, 271–278.
83. J. S. Hoffman, M. Hirano, N. Panpradist, J. Breda, P. Ruth, Y. Xu, J. Lester, B. H. Nguyen, L. Ceze and S. N. Patel, *Sci. Total Environ.*, 2022, **821**, 152790.
84. J. A. Medrano-Félix, J. D. Lira-Morales, I. González-L´ópez, J. P. G. Gómez, N. C. Campo, C. I. Martínez-Rodríguez, B. G.-G. Rodríguez-Sala and C. Chaidez-Quiroz, *Rev. Bio Cienc.*, 2024, **11**, e1631.
85. A. Pilipenco, M. Forinová, H. Mašková, V. Hönig, M. Palus, N. S. Lynn, I. Víšová, M. Vrabcová, M. Houska, J. Anthi, M. Spasovová, J. Mustacová, J. Štěrba, J. Dostálek, C. P. Tung, A. S. Yang, R. Jack, A. Dejneka, J. Hajdu and H. Vaisocherová-Lísalová, *J. Travel. Med.*, 2023, **30**, 1–10.
86. L. Dietz, D. A. Constant, M. Fretz, P. F. Horve, A. Olsen-Martinez, J. Stenson, A. Wilkes, R. G. Martindale, W. B. Messer and K. G. Van Den Wymelenberg, 2021, medRxiv:03.26.21254416.
87. M. Comar, M. D'accolti, C. Cason, I. Soffritti, G. Campisciano, L. Lanzoni, M. Bisi, A. Volta, S. Mazzacane and E. Caselli, *Microorganisms*, 2019, **7**, 708.
88. P. Ray, B. Lim, K. Zorcic, J. Johnstone, A. Hinz, A. M. A. Hicks, A. Wong, D. R. MacFadden, C. Nott, L. Castellani, R. Kassen and M. Fralick, *Infect. Prevent. Pract.*, 2024, **6**, 100353.
89. Y. Li, X. Wei, Y. Zhou, J. Wang and R. You, *Microsyst. Nanoeng.*, 2023, **9**(1), 2023, 1–22.
90. R. de Vries, R. M. Vigeveno, S. Mulder, N. Farzan, D. R. Vintges, J. J. Goeman, S. Bruisten, B. van den Corput, J. J. M. Geelhoed, L. G. Visser, M. van der Lubben, P. J. Sterk, J. C. C. M. in't Veen and G. H. Groeneveld, 2021, medRxiv:02.14.21251712.
91. D. K. Nurputra, A. Kusumaatmaja, M. S. Hakim, S. N. Hidayat, T. Julian, B. Sumanto, Y. Mahendradhata, A. M. I. Saktiawati, H. S. Wasisto and K. Triyana, *npj Digital Med.*, 2022, **5**(1), 2022, 1–17.
92. C. Ghazaly, K. Biletska, E. A. Thevenot, P. Devillier, E. Naline, S. Grass-in-Delyle and E. Scorsone, *J. Breath Res.*, 2023, **17**, 026006.
93. M. S. Draz, A. Vasan, A. Muthupandian, M. K. Kanakasabapathy, P. Thirumalaraju, A. Sreeram, S. Krishnakumar, V. Yogesh, W. Lin, X. G. Yu, R. T. Chung and H. Shafiee, *Sci. Adv.*, 2020, **6**, 5354–5370.
94. M. Taniguchi, S. Minami, C. Ono, R. Hamajima, A. Morimura, S. Hamaguchi, Y. Akeda, Y. Kanai, T. Kobayashi, W. Kamitani, Y. Terada, K. Suzuki, N. Hatori, Y. Yamagishi, N. Washizu, H. Takei, O. Sakamoto, N. Naono,

K. Tatematsu, T. Washio, Y. Matsuura and K. Tomono, *Nat. Commun.*, 2021, **12**(1), 2021, 1–8.
95. N. Shiaelis, A. Tometzki, L. Peto, A. McMahon, C. Hepp, E. Bickerton, C. Favard, D. Muriaux, M. Andersson, S. Oakley, A. Vaughan, P. C. Matthews, N. Stoesser, D. W. Crook, A. N. Kapanidis and N. C. Robb, *ACS Nano*, 2023, **17**, 697–710.
96. O. Yaren, J. McCarter, N. Phadke, K. M. Bradley, B. Overton, Z. Yang, S. Ranade, K. Patil, R. Bangale and S. A. Benner, *Plos One*, 2021, **16**(2), e0240524.
97. C. Tian, L. Zhao, G. Qi, J. Zhu and S. Zhang, *Sens. Actuators, B*, 2022, **371**, 132445.
98. Y. Saylan, Ö. Erdem, S. Ünal and A. Denizli, *Biosensors*, 2019, **9**, 65.
99. M. Nasrollahzadeh, M. Sajjadi, G. J. Soufi, S. Iravani and R. S. Varma, *Nanomaterials*, 2020, **10**, 1072.
100. R. Misra, S. Acharya and N. Sushmitha, *Rev. Med. Virol.*, 2022, **32**, e2267.
101. C. H. Zhou, Y. M. Long, B. P. Qi, D. W. Pang and Z. L. Zhang, *Electrochem. Commun.*, 2013, **31**, 129–132.
102. R. Singh, A. Sharma, S. Hong and J. Jang, *Analyst*, 2014, **139**, 5415–5421.
103. J. Huang, Z. Xie, Z. Xie, S. Luo, L. Xie, L. Huang, Q. Fan, Y. Zhang, S. Wang and T. Zeng, *Anal. Chim. Acta*, 2016, **913**, 121–127.
104. J. Tian, D. Wang, Y. Zheng and T. Jing, *Int. J. Electrochem. Sci.*, 2017, **12**, 2658–2668.
105. T. Lee, G. H. Kim, S. M. Kim, K. Hong, Y. Kim, C. Park, H. Sohn and J. Min, *Colloids Surf., B*, 2019, **182**, 110341.
106. A. Raziq, A. Kidakova, R. Boroznjak, J. Reut, A. Öpik and V. Syritski, *Biosens. Bioelectron.*, 2021, **178**, 113029.
107. About Viral Hemorrhagic Fevers|Viral Hemorrhagic Fevers (VHFs)|CDC. https://www.cdc.gov/viral-hemorrhagic-fevers/about/index.html?CDC_AAref_Val=https://www.cdc.gov/vhf/virus-families/flaviviridae.html. Accessed 12 Mar 2025.
108. H. A. M. Faria and V. Zucolotto, *Biosens. Bioelectron.*, 2019, **131**, 149–155.
109. A. C. R. Moço, P. H. Guedes, J. M. R. Flauzino, H. S. da Silva, J. G. Vieira, A. C. H. Castro, É. V. R. Gomes, F. M. Tolentino, M. M. C. N. Soares, J. M. Madurro and A. G. Brito-Madurro, *Electroanalysis*, 2019, **31**, 1580–1587.
110. Q. Palomar, X. X. Xu, C. Gondran, M. Holzinger, S. Cosnier and Z. Zhang, *Microchim. Acta*, 2020, **187**, 1–10.
111. A. Zumla, D. S. Hui and S. Perlman, *Lancet*, 2015, **386**, 995–1007.
112. A. A. Elkholy, R. Grant, A. Assiri, M. Elhakim, M. R. Malik and M. D. Van Kerkhove, *J. Infect. Pub. Health*, 2020, **13**, 418–422.
113. J. A. Al-Tawfiq and P. Gautret, *Travel Med. Infect. Dis.*, 2019, **27**, 27–32.
114. N. Ramadan and H. Shaib, *Germs*, 2019, **9**, 35.
115. L. A. Layqah and S. Eissa, *Microchim. Acta*, 2019, **186**, 1–10.
116. H. Kim, M. Park, J. Hwang, J. H. Kim, D. R. Chung, K. sung Lee and M. Kang, *ACS Sens.*, 2019, **4**, 1306–1312.

117. P. Teengam, W. Siangproh, A. Tuantranont, T. Vilaivan, O. Chailapakul and C. S. Henry, *Anal. Chem.*, 2017, **89**, 5428–5435.
118. P. Daszak, K. J. Olival and H. Li, *Biosaf Health*, 2020, **2**, 6–8.
119. K. E. Jones, N. G. Patel, M. A. Levy, A. Storeygard, D. Balk, J. L. Gittleman and P. Daszak, *Nature*, 2008, **451**(7181), 2008, 990–993.
120. M. E. J. Woolhouse and S. Gowtage-Sequeria, *Emerging Infect. Dis.*, 2005, **11**, 1842–1847.
121. S. J. Anthony, J. H. Epstein, K. A. Murray, I. Navarrete-Macias, C. M. Zambrana-Torrelio, A. Solovyov, R. Ojeda-Flores, N. C. Arrigo, A. Islam, S. A. Khan, P. Hosseini, T. L. Bogich, K. J. Olival, M. D. Sanchez-Leon, W. B. Karesh, T. Goldstein, S. P. Luby, S. S. Morse, J. A. K. Mazet, P. Daszak and W. I. Lipkin, *mBio*, 2013, **4**, e00598-13.
122. D. Carroll, P. Daszak, N. D. Wolfe, G. F. Gao, C. M. Morel, S. Morzaria, A. Pablos-Méndez, O. Tomori and J. A. K. Mazet, *Science*, 2018, **359**, 872–874.
123. J. Zhou, S. Zhou, P. Fan, X. Li, Y. Ying, J. Ping and Y. Pan, *Nano-Micro Lett.*, 2023, **16**(1), 1–29.
124. R. Li, H. Qi, Y. Ma, Y. Deng, S. Liu, Y. Jie, J. Jing, J. He, X. Zhang, L. Wheatley, C. Huang, X. Sheng, M. Zhang and L. Yin, *Nat. Commun.*, 2020, **11**(1), 2020, 1–11.
125. X. He, S. Wang, C. Ma, G. R. Xu, J. Ma, H. Xie, W. Zhu, H. Liu, L. Wang and Y. Wang, *Animals*, 2023, **13**, 3141.
126. H. Chai, Q. Liu, N. N. Gaudreault and W. Zhang, *Front. Microbiol.*, 2023, **14**, 1165365.
127. L. F. Wang and B. T. Eaton, *Curr. Top Microbiol. Immunol.*, 2007, **315**, 325–344.
128. B. Olsen, V. J. Munster, A. Wallensten, J. Waldenström, A. D. M. E. Osterhaus and R. A. M. Fouchier, *Science*, 2006, **312**, 384–388.
129. C. B. Jonsson, L. T. M. Figueiredo and O. Vapalahti, *Clin. Microbiol. Rev.*, 2010, **23**, 412–441.
130. D. J. Prosser, P. Cui, J. Y. Takekawa, M. Tang, Y. Hou, B. M. Collins, B. Yan, N. J. Hill, T. Li, Y. Li, F. Lei, S. Guo, Z. Xing, Y. He, Y. Zhou, D. C. Douglas, W. M. Perry and S. H. Newman, *PLoS One*, 2011, **6**, e17622.
131. A. L. Michel, H. van Heerden, D. Prasse, V. Rutten, S. Al Dahouk and B. M. Crossley, *OIE Revue Sci. Tech.*, 2021, **40**, 105–118.
132. A. S. Fomsgaard, S. A. Tahas, K. Spiess, C. Polacek, J. Fonager and G. J. Belsham, *Viruses*, 2023, **15**, 1399.
133. A. A. Pérez, G. Vazquez-Meves and M. E. Hunter, *CRISPR J.*, 2024, **7**(6), 327–342.
134. A. L. Michel, R. G. Bengis, D. F. Keet, M. Hofmeyr, L. M. De Klerk, P. C. Cross, A. E. Jolles, D. Cooper, I. J. Whyte, P. Buss and J. Godfroid, *Vet. Microbiol.*, 2006, **112**, 91–100.
135. K. Morelle, J. A. Barasona, J. Bosch, G. Heine, A. Daim, J. Arnold, T. Bauch, A. Kosowska, E. Cadenas-Fernández, M. M. Aviles, D. Zuñiga, M. Wikelski, J. M. Vizcaino-Sanchez and K. Safi, *Proc. R. Soc. B*, 2023, **290**(2005), 20231396.

136. C. S. Cheng, B. R. Ou and F. Di Lung, *J. Pers. Med.*, 2022, **12**, 594.
137. J. Choi, M. Jeun, S. S. Yuk, S. Park, J. Choi, D. Lee, H. Shin, H. Kim, I. J. Cho, S. K. Kim, S. Lee, C. S. Song and K. H. Lee, *ACS Nano*, 2019, **13**, 812–820.
138. M. Yamaguchi, Y. Matsuda, S. Sasaki, M. Sasaki, Y. Kadoma, Y. Imai, D. Niwa and V. Shetty, *Biosens. Bioelectron.*, 2013, **41**, 186–191.
139. S. Neethirajan, S. K. Tuteja, S. T. Huang and D. Kelton, *Biosens. Bioelectron.*, 2017, **98**, 398–407.
140. R. N. Handcock, D. L. Swain, G. J. Bishop-Hurley, K. P. Patison, T. Wark, P. Valencia, P. Corke and C. J. O'Neill, *Sensors*, 2009, **9**, 3586–3603.
141. S. Aich, S. Chakraborty, J. S. Sim, D. J. Jang and H. C. Kim, *Appl. Sci.*, 2019, **9**, 4938.
142. H. Yoo, H. Jo and S. S. Oh, *Mater. Adv.*, 2020, **1**, 2663–2687.
143. W. Zhou, P. J. Jimmy Huang, J. Ding and J. Liu, *Analyst*, 2014, **139**, 2627–2640.
144. R. Yuan, J. Cai, H. Ma, Y. Luo, L. Wang and S. Su, *Chemosensors*, 2023, **11**, 488.
145. E. Janik-Karpinska, M. Ceremuga, M. Niemcewicz, M. Podogrocki, M. Stela, N. Cichon and M. Bijak, *Sensors*, 2022, **22**, 9757.
146. Y. Zhao, L. Zhang, A. Wang and D. Zhou, *Vet. Sci.*, 2025, **12**, 23.
147. J. Xu, J. yu Wan, S. tao Yang, S. feng Zhang, N. Xu, N. Li, J. ping Li, H. ying Wang, X. Bai and W. sen Liu, *Acta Vet. Brno*, 2012, **81**, 107–111.
148. M. Hnaien, M. F. Diouani, S. Helali, I. Hafaid, W. M. Hassen, N. J. Renault, A. Ghram and A. Abdelghani, *Biochem. Eng. J.*, 2008, **39**, 443–449.
149. S.-H. Lee, T.-H. Park, S.-A. Kim, S.-G. Kim, K.-M. Byun, M. L. Shuler and S.-J. Kim, *J. Opt. Soc. Korea*, 2009, **13**(3), 392–397.
150. M. Sharma, P. Mahajan, A. S. Alsubaie, V. Khanna, S. Chahal, A. Thakur, A. Yadav, A. Arya, A. Singh and G. Singh, *Mater. Today Sustainability*, 2025, **29**, 101068.
151. A. Macdonald, L. A. Hawkes and D. K. Corrigan, *Philos. Trans. R. Soc., B*, 2021, **376**(1831), 20200228.
152. B. J. Hoye, V. J. Munster, H. Nishiura, M. Klaassen and R. A. M. Fouchier, *Emerging Infect. Dis.*, 2010, **16**, 1827–1834.
153. Y. Si, Q. Xin, H. H. T. Prins, W. F. de Boer and P. Gong, *Sci. Bull.*, 2015, **60**, 1984–1993.
154. B. C. Easterday, D. O. Trainer, B. Tůmová and H. G. Pereira, *Nature*, 1968, **219**(5153), 523–524.
155. M. Geldenhuys, M. Mortlock, J. H. Epstein, J. T. Pawęska, J. Weyer and W. Markotter, *Viruses*, 2021, **13**, 936.
156. M. Hemnani, P. G. da Silva, G. Thompson, P. Poeta, H. Rebelo and J. R. Mesquita, *Ecohealth*, 2024, **21**, 125–140.
157. T. G. Ksiazek, D. Erdman, C. S. Goldsmith, S. R. Zaki, T. Peret, S. Emery, S. Tong, C. Urbani, J. A. Comer, W. Lim, P. E. Rollin, S. F. Dowell, A.-E. Ling, C. D. Humphrey, W.-J. Shieh, J. Guarner, C. D. Paddock, P. Rota, B. Fields, J. DeRisi, J.-Y. Yang, N. Cox, J. M. Hughes, J. W. LeDuc, W. J. Bellini and L. J. Anderson, *N. Engl. J. Med.*, 2003, **348**, 1953–1966.

158. C. H. Calisher, J. E. Childs, H. E. Field, K. V. Holmes and T. Schountz, *Clin. Microbiol. Rev.*, 2006, **19**, 531–545.
159. G. Mashura, B. Maburutse, V. Chidoti, T. R. Zinyakasa, E. Porovha, R. B. Nhara, E. Mwandiringana and E. Gori, *Trop. Anim. Health Prod.*, 2025, **57**(2), 1–12.
160. W. Guo, C. Lv, M. Guo, Q. Zhao, X. Yin and L. Zhang, *Sci. One Health*, 2023, **2**, 100045.
161. S. Neethirajan, *Sens. Biosens. Res.*, 2020, **29**, 100367.
162. Z. Allam and Z. A. Dhunny, *Cities*, 2019, **89**, 80–91.
163. J. Linchant, J. Lisein, J. Semeki, P. Lejeune and C. Vermeulen, *Mamm Rev.*, 2015, **45**, 239–252.
164. P. J. Stephenson, *Curr. Opin Environ. Sustainable*, 2020, **45**, 36–41.
165. P. K. Mathur, B. Habib and P. Mathur, *Human Bond Communication*, 2017, pp. 115–130.
166. I. Nandutu, M. Atemkeng and P. Okouma, *Sensors*, 2022, **22**, 2478.

Section V: Challenges and Solutions

CHAPTER 16

Sensor Technologies for Virus Detection: Challenges and Perspectives

KUMSAL KOCADAL[a] AND DILEK BATTAL[a,b*]

[a]Department of Toxicology, Faculty of Pharmacy, Near East University, Nicosia, North Cyprus; [b]Department of Toxicology, Faculty of Pharmacy, Mersin University, Mersin, Turkey
*E-mail: dilekbattal@mersin.edu.tr

16.1 Introduction

Viruses are capable of infecting all forms of living organisms. Starting in the 20th century, people have been fighting viral outbreaks, including dengue fever, Zika, Ebola, hepatitis, human immunodeficiency virus (HIV), and the latest outbreak of corona virus disease (COVID)-19. These viral pandemics have resulted in extensive morbidity and mortality worldwide.[1,2]

Efficient diagnostic methods are crucial for the rapid and precise detection of viral infections as a preventive measure. The prompt and precise detection of viral infections is vital for appropriate treatment selection, cost minimisation, and disease management. It also facilitates the progression of illness prevention and treatment strategies, encompassing the development of antiviral vaccines and novel pharmaceuticals.[3]

Virus detection has significance in terms of protecting health. Many approaches are being used to identify and/or detect viruses. These approaches

Detection Science Series No. 32
Advances in Virus Detection with Sensors
Edited by Adil Denizli
© The Royal Society of Chemistry 2026
Published by the Royal Society of Chemistry, www.rsc.org

can involve biochemical and microbiological tests, genetic engineering, and immunological approaches.[4] Currently, sensors are extensively used for the detection of viral particles due to their user-friendly applications and superior sensitivity compared with traditional methods.

Sensors are frequently used to detect viral particles due to their relative simplicity compared with traditional methods and their superior sensitivity. Various sensor technologies are being used for virus detection, and they can be listed as optical, electrochemical, piezoelectric, fluorescence, thermal, magnetic, and micromechanical biosensors.[5] Sensor technologies are frequently classified according to their structures. Biosensors constitute the paramount category and include several subtypes, such as immunosensors, DNA- or RNA-based sensors, and enzyme-linked sensors. Additionally, there are optical sensors with varying functionalities, including surface plasmon resonance (SPR), fluorescence, and colorimetric detection. Electrochemical sensors exemplify a category of sensors, encompassing voltametric, amperometric, and impedimetric types. In addition to these technologies, several novel sensor types are being developed. Clustered regularly interspaced short palindromic repeats (CRISPR)-based sensors, nanomaterial-based sensors, and paper-based microfluidics exemplify new sensor technologies.[6]

Owing to their exceptional sensitivity and selectivity, polymerase chain reaction (PCR)-based tests are presently considered the gold standard for diagnosing viral diseases. Nonetheless, PCR relies on costly apparatus and intricate sample preparation prior to nucleic acid amplification; in contrast, biosensors offer direct detection, rapid response, and significant versatility for use to analyse many viruses.[7]

Significant concerns have arisen due to increased viral outbreaks, as the viruses might be transmitted rapidly and initiate a pandemic. Thus, viral detection sensors must exhibit high sensitivity and specificity towards a particular virus, or multiple viruses when detecting more than one simultaneously, for optimal functionality and precision. Rapid response is essential as well. Analytes including viral nucleic acids (DNA and RNA), viral proteins, intact viral particles, and antibodies produced by the patient's immunological reaction against the virus have been employed for virus detection with the intention of clinical application. These analytes are identified by many conventional techniques, including PCR, viral culture, enzyme-linked immunosorbent assay (ELISA), western blotting, and serological antibody detection methods.[8–10] Traditional methods are frequently inadequate for fast on-site analysis, as they necessitate virus isolation and biocontainment, the capacity to cultivate cultured cells for cytopathology-related assays, and/or complex and costly laboratory equipment that is challenging to transport and utilise at the point of care. Conventional laboratory assays are time-consuming, labour-intensive, and may exhibit relative insensitivity; furthermore, they necessitate the transportation of samples to centralised diagnostic laboratories for analysis.[10–12] Consequently, portability and user-friendliness are two essential criteria for effective virus detection. In comparison to conventional methods, innovative virus detection sensors are expected to provide a more cost-effective solution. Moreover, this cost-effectiveness should be applied to large-scale development as well. Ultimately, sensors must be appropriate for mass deployment, and all sensors produced must demonstrate uniform effectiveness, sensitivity, and selectivity. An optimal sensor for detecting infectious diseases should possess simplicity, rapidity, precision, selectivity, sensitivity, and cost-effectiveness.[13]

This chapter aims to provide an up-to-date perspective on the problems and potential solutions associated with the application of sensor technology in viral detection. This chapter is intended to explain the challenges associated with sensitivity, detection limits, selectivity, cross-reactivity, sample preparation, matrix interference, stability, shelf life, scalability, mass production, cost, accessibility, and data analysis and integration while proposing potential solutions for those challenges.

16.2 Sensor Technologies Used for Virus Detection and Analysis

Sensors and biosensors are sophisticated analytical instruments that integrate biological and non-biological recognition components or biomimetics, utilised for the rapid detection and real-time monitoring of molecules pertinent to medicine, security, food, and environmental sectors. Sensors and biosensors identify the interaction between the target molecule and the receptor or bioreceptor in the detection of a target analyte. A physical or chemical binding reaction results in a change that can be detected by a transducer and quantified by a detector.

Analytical systems consist of a recognition element and a transducer that identify a target molecule for sample analysis and subsequently evaluate the composition, structure, and function of target molecules by adjusting their responses to the signals.[14,15] The target molecule binds to the receptor, and the detecting component identifies the presence of the target molecule through a binding process. The transducer converts the changes in the signal detected by the sensor into an output.[15,16] An optimal biosensor possesses several attributes such as selectivity, sensitivity, rapid reaction time, disposability, user-friendliness, extended shelf life, cost-effectiveness, and multifunctional sensing capabilities.[15,17] Figure 16.1 illustrates multi-virus biosensing platforms.

16.2.1 Optical Sensors

Optical transducers represent the most basic form of biosensor, detecting the analyte by assessing alterations in the fluorescence, absorption, or reflectance of the sensing material. This technique utilises the colour alteration of the sensing materials resulting from variations in their size or concentration

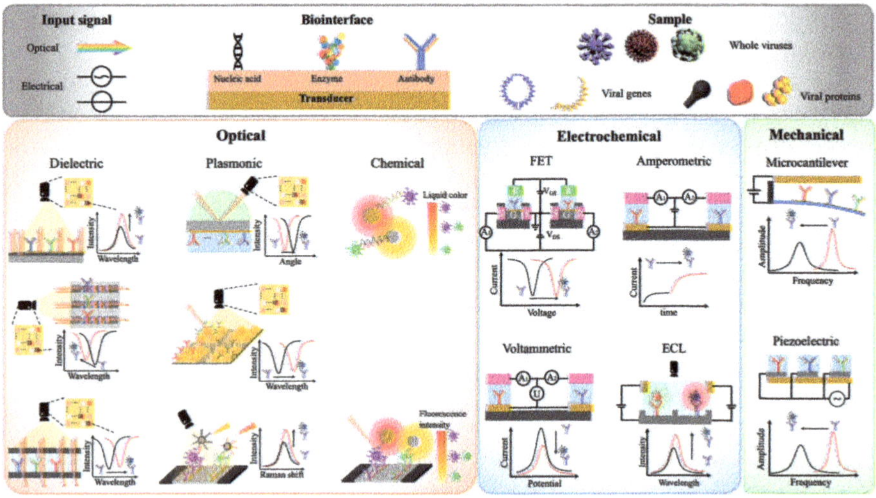

Figure 16.1 Multi-virus biosensing platforms detect viruses, genes, proteins, and nucleic acids, enzymes, or antibodies using electrical or optical input signals, and can be categorized as optical, mechanical, or electrochemical. Abbreviations: field effect transistor (FET), electrochemiluminescence (ECL). Reproduced from ref. 7 with permission from Elsevier, Copyright 2023.

owing to interaction with the analyte.[15,18] Optical biosensors assess changes in reflection, absorption, refractive index, Raman spectra, fluorescence, infrared spectrum, chemiluminescence, or energy transfer upon the formation of a complex between the recognition element and the target molecule.[19–28] Researchers have enhanced biosensor sensitivity by concentrating on fabrication quality, increasing the affinity among various surface chemical alterations, and employing polymers for signal amplification investigations.[29–33]

Research on optical biosensors commenced several years ago, with a multitude of scientists-including biologists, biochemists, chemists, biophysicists, physicists, bioengineers, and medical doctors employing these biosensors in novel applications across various disciplines, including medical diagnosis,[34] environmental studies,[35] pharmaceutics,[36] and many more. Medical diagnosis has been studied as a significant area of application due to the persistent global threat that diseases, particularly infectious diseases, pose to human health.[37,38] Significant diseases can proliferate due to inadequate access to treatment modalities, disparate therapeutic strategies, and limited access to well-equipped healthcare facilities.[39] In response to the demands for improved monitoring capabilities and swift analysis, many recognition elements and platforms have been utilised for recognition. The detection efficacy of optical biosensors can be enhanced through their utilisation in various combinations. These attributes render them suitable for medical diagnosis, facilitating multi-molecule analysis and expedited monitoring.[40–43]

16.2.2 Electrochemical Sensors

Electrochemical biosensors quantify the electrical potential difference across nanopores resulting from the interaction between an analyte and the membrane or sensor surface. There is a proportionate relationship between the electrical potential difference and the logarithm of the electrochemically active concentration of the substance. The current through the system or the potential difference between the electrodes resulting from the redox reactions involving the analyte is utilised for its measurement in the sample. These sensors exhibit enhanced stability in output, elevated sensitivity, rapid responsiveness, and resistance to interference. Electrochemical measurements are predominantly used for sensing applications.[44–47] Electrochemical biosensors can be further categorised into distinct categories according to the electrical characteristics measured. These encompass conductimetric, amperometric, potentiometric, and impedimetric sensors.[48,49]

16.2.3 Piezoelectric Sensors

Piezoelectric biosensors are micromechanical transducers that operate on the basis of mass sensitivity. Their mass is augmented by molecular interactions, and the frequency is assessed using alternating current and voltage reductions. Piezoelectric, mass-sensitive, or acoustic instruments provide rapid and label-free identification of molecules on the basis of their characteristic molecular masses. These transducers comprise: (a) surface acoustic wave (SAW) transducers;[50,51] (b) bulk acoustic wave (BAW) devices, such as quartz crystal microbalances (QCMs);[52–54] and (c) shear transverse wave (STW) devices,[55] commonly utilised in conjunction with diverse bioreceptor components for biomedical diagnostics and molecular recognition.[56] Piezoelectric biosensors are widely utilised in food safety monitoring,[57] drug delivery mechanisms,[58,59] environmental pollution monitoring,[60] *etc.* Piezoelectric biosensors, specifically mass-based chemical sensors like QCMs, possess significant attributes, including high sensitivity and selectivity, user-friendliness, cost-efficiency, stability, portability, and simplicity. Diverse methodologies can be employed to engineer piezoelectric biosensor surfaces for various application domains.[15,58]

16.3 Challenges and Solutions for Sensor Technologies for Virus Detection and Analysis

The challenges of virus detection include the following. The detection of low quantities of viruses in complex biological samples, achieving low limits of detection with a high regression coefficient, and developing sensing platforms suitable for inter- and intra-laboratory use. Table 16.1 summarises the challenges and their solutions with examples from the literature.

Table 16.1 Summary of challenges and their solutions with examples from the literature.

Challenge	Solution	Example	Reference
Sensor sensitivity	Using functional monomers. Using AI during design	MIP-Based sensors: functional monomers enhance chemical recognition, while nano-patterned electrodes serve as transducers, generating additional cavities with high efficacy.	13
Identifying viruses in low quantities (low limit of detection)	Using innovative compounds and exploring new electrode materials.	A novel nanocomposite film with gold nanoparticles, carbon nanotubes, and propolis has been used for HIV-1 p24 antigen detection, offering high surface area and minimal detection limits.	63
		The early detection of AIDS was achieved using fluorescent quantum dots and testing strips, with a detection limit of 0.76 pmol l^{-1} and a range of 1 pmol l^{-1} to 10 nmol l^{-1}.	65
Selectivity	Using nanomaterials and/or MIP-based sensors	Electrochemical polymer composite biosensor to detect SARS-CoV-2 particles in the environment, achieved higher signalling than negative controls and exhibited linearity of three log scales in buffer and water samples.	117
Cross-reactivity	Using a specific biomarker	Researchers have developed a multiplex electrochemical biosensor for simultaneous detection of Zika and dengue infections on a single chip, using NS1 proteins from Zika and dengue as target molecules due to limited cross-reactivity.	73 and 74
Sample preparation	Novel sensor designs	A biosensor with a specialized cuvette design accommodates a saliva sample and electrode, allowing for direct placement and minimizing contamination interference. The downward-facing electrode design allows debris and particles to accumulate at the bottom, ensuring the electrode is exposed to the target viral RNA sample.	86

Table 16.1 (*Continued*)

Challenge	Solution	Example	Reference
Integration of multiple sensor types	Mathematical biosensors	Advanced designs use four square split-ring resonators and a central graphene ring, enhancing detection capabilities and demonstrating potential in next-generation diagnostic systems.	87
Data interpretation	The integration of advanced data analytics and machine learning algorithms	Machine learning algorithms are used in image analysis for LFDs to reduce human error and bias in rapid virus detection, despite variability in visual perception.	100
Ethical issues	Informed consent. Clear delineation of ownership rights	In sensor-based virus detection, clear information about the nature of the data, intended use, and risks is required. Transparent communication is essential for voluntary consent, especially in digital disease detection technologies. Rigid consent protocols protect individual rights and facilitate responsible implementation of biosensing.	112

16.3.1 Sensitivity and Specificity

Sensor technologies have been designed to identify various viruses. Prior to gaining entry to the market for therapeutic applications, comprehensive studies must be conducted to address the challenges. Even though it is accepted as the most important feature of a biosensor,[49] sensitivity is equally important as a challenge to overcome.[13] The sensitivity of a biosensor is defined as the relationship between the variation in analyte concentration and the intensity of the signal generated by the transducer. A biosensor should ideally emit a signal in response to slight fluctuations in the concentration of the target analyte. Biosensors are required to detect analytes within the ng ml^{-1} or fg ml^{-1} concentration levels, depending on the application. This is generally important for medical applications and environmental monitoring.[49]

Molecularly imprinted polymer (MIP)-based sensors are extensively used to detect viruses and to overcome sensitivity issues, functional monomers can be used to enhance chemical recognition, while nano-patterned electrodes may serve as transducers; these electrodes can activate MIPs to generate additional cavities with high efficacy and exceptional spatial matching effectiveness. Moreover, machine learning can be utilised within artificial intelligence (AI) to develop MIPs with enhanced recognition abilities.[13]

Advanced virus detection methods are essential in order to implement effective measures for preventing the transmission of viral infections. This necessitates the advancement of quick virus detection technologies to identify viruses in low quantities, including in the biological fluids of patients in the early stages of illness or in environmental samples. In the advancement of viral detection technology, efforts have mostly focused on two areas: lowering the limit of detection (LOD) and enhancing the speed of detection.[61] Thus, the detection limit of a sensor must be equal to or lower than that of traditional methods. Achieving a low limit of detection is a must and a challenge in the development of virus-detecting sensors. Recent work on amperometric immunosensors has focussed on the development of electrode surfaces utilising novel chemicals and the study of new electrode materials to overcome this challenge. This also facilitates the identification of ultra-low quantities of viral particles.[62] Kheiri *et al.* fabricated a novel nanocomposite film consisting of gold nanoparticles (AuNPs), carbon nanotubes (CNTs), and propolis for the detection of the HIV-1 p24 antigen. This combination, along with a p24 antibody fixed on their electrode surface, offers a high surface area, resulting in an efficient amperometric immunosensor with a minimal limit of detection (0.0064 ng ml^{-1}) and a rapid runtime of 18 seconds.[63] Silva *et al.*[64] have recently created an economical electrochemical immunosensor (EI sensor) for the detection of non-structural proteins associated with the dengue virus (DENV). The authors immobilised anti-non-structural protein 1 (NS1) antibodies onto an electrode surface *via* poly(allylamine) on carboxylated carbon nanotubes and assessed the responses by amperometry. This innovative sensor has an exceptional limit of detection (0.035 μg ml^{-1}) with remarkable repeatability, sensitivity, and specificity.[64]

Deng *et al.*[65] created fluorescent quantum dots (QDs) utilising testing strips alongside strand displacement amplification (SDA) for the early detection of AIDS. This assay exhibited a low detection limit of 0.76 pmol l^{-1}, with a detection range spanning from 1 pmol l^{-1} to 10 nmol l^{-1}. The strength of the signal in the test zone was a significant determinant in the operation of this device. QDs possess the capability to aggregate more biomolecules, hence enhancing the precision and sensitivity of strips.[65,66]

The viral load is contingent upon the number of days following the onset of sickness. Consequently, the identification of intact viral particles can furnish clinicians with insights on the stage of the infection or the response to therapy. The concentration of viral particles is often assessed by plaque-based tests, wherein patient samples are inoculated into cultured cell lines, and cell death is indicated by the quantifiable plaques in the confluent cell layer.[10] These tests detect viruses using a slow method, correlated to viral load, calibration curves, and cut-off values. SPR biosensors use unsupervised data analysis techniques to identify virus adhesions from spatio-temporal data, distinguishing positive from negative cases.[67] Silicon-based capacitance sensors facilitate swift viral detection by contrasting sense and dummy capacitors, where variations in capacitance signify the existence of living material.[68] Plasmonic biosensors employ phenomena such as propagating SPR, localised SPR, and surface-enhanced spectroscopic approaches for viral diagnosis.[69] These technologies seek to enhance the speed and efficiency of virus detection while minimising bioprocessing processes, potentially utilising mass-producible integrated circuit platforms.[68] The advancement of these sensors facilitates early identification and disease management, tackling global biosecurity issues highlighted by the COVID-19 pandemic.[69] Ymeti *et al.* produced a quick, ultrasensitive sensor capable of detecting numerous viruses, including the herpes virus, at minimal concentrations by a combination of integrated optical interferometric sensing and lab-on-a-chip nanotechnology (antibody–antigen recognition).[70]

Selectivity describes the biosensor's capacity to exclusively bind and react to the target analyte amidst other molecules or chemicals. A false positive result occurs when a signal or reaction is produced by interactions with an analyte that differs from the target analyte. This is prevalent in biosensors with inadequate selectivity, hence hindering their therapeutic applications. Selectivity is a crucial characteristic, particularly in medical applications, because the test sample or matrix, typically blood or urine, has several molecules that closely resemble the target analyte and compete for attachment to the biological receptor.[49] The primary problems in sensor fabrication are the specificity and selectivity of the sensor. To improve the selectivity of the sensor, various nanomaterials have been employed to construct innovative sensors, including gold-based thin films, gold nanoparticles, graphene oxide, reduced graphene oxide, cobalt-functionalized TiO_2 nanotubes, palladium nano-thin films, polyaniline, activated graphene oxide, and Au nanostars.[71] Additionally, MIPs provide exquisite opportunities to improve sensor selectivity. During the design of a MIP, machine learning and AI can be used to improve the recognition capabilities.[13] Quantum mechanics can be utilised to calculate the affinity energy between functional monomers and template molecules, resulting in functional monomers with exceptional affinity characteristics. Molecular dynamics simulation can be utilised to model the production mechanisms of imprinted cavities and binding sites. Furthermore, numerous interactions (*via* the incorporation of multiple binding sites) should be systematically engineered to enhance the selectivity and recognition capabilities of MIP-based systems for the targeted molecules. Extensive investigations and analytical research are essential to enhance the shape recognition capabilities and selectivity of MIP-based sensors.[13] Sukjee *et al.* fabricated an electrochemical polymer composite biosensor in order to detect SARS-coronavirus (CoV)-2 whole-virus particles in the environment. The fabricated MIP composites were formed using inactivated SARS-COV-2. As a result, sensor signalling was higher for SARS-Cov-2 than for negative controls, H5N1 influenza A virus, and non-imprinted polymers This sensor identified SARS-CoV-2 at concentrations as low as 0.1 fM in buffer and samples derived from reservoir water, exhibiting a linearity of three log scales.[72]

Assay cross-reactivity can be broadly described as the creation of signals by analogous analytes. For example, Zika virus (ZIKV) and DENV are well known for their cross-reactivity. A group of researchers have found that NS1 may represent an efficient biomarker for discriminating DENV infection from ZIKV infection, providing limited cross-reactivity due to group-specific epitopes.[73] In the light of that information, another group designed and developed a multiplex electrochemical biosensor for the simultaneous detection of Zika and dengue infections on a single chip, utilising NS1 proteins from Zika (NS1ZV) and Dengue (NS1DV) as target molecules.[74] Thus, the best way to minimise cross-reactivity is therefore identifying an efficient biomarker and using that specific biomarker for the determination of viruses. Addressing these challenges results in the elimination of false negative and false positive results.

Sample preparation and matrix interference are critical processes that influence sensor selectivity and specificity. The sample matrices and assays must be meticulously chosen for the specific virus being targeted. Besides the identification of targets collected by particles, motion-based sensing has garnered significant interest owing to its capacity for real-time detection and elevated spatial resolution.[61,75,76] Nevertheless, it was challenging to regulate and analyse the movements of numerous particles in high resolution within the bulk solution, leading to reduced sensitivity of these systems compared with PCR assays, specifically 10^3 virus particles per ml. In order to address this issue a group of researchers fabricated an external force-assisted near-field illumination (EFA-NI) biosensor.[77,78] This sensor employs magnetic force to drive the beads, hence minimising false positives by the concurrent detection of optical signals from markers and particle movements. The EFA-NI biosensor eliminates the need for a washing procedure, allowing for the measurement of samples with several contaminants by merely combining the sample with a detection solution (comprising antibody-modified magnetic particles and optical markers) and injecting it into the device.[61,77,78]

Portable optical sensors for rapid detection of SARS-COV-2 have been researched and described in the literature. Minimising the number of sample preparation steps is one of the challenges in improving a virus detection sensor. Huang *et al.* developed a rapid optical biosensor chip for the detection of SARS-COV-2 without extensive sample preparation in a single step to overcome this problem. Using a nanoplasmonic resonance sensor specific for the virus spike protein, they detected concentrations as low as 370 virus particles (vp) per ml in 15 minutes (Figure 16.2a).[79] Song *et al.* also developed an optical sensor using a smartphone and point-of-care up conversion luminescence diagnosis (PULD) to identify SARS-CoV-2. The use of up conversion probes modified with complementary oligonucleotides and gold probes tailored to bind the target nucleocapsid (N)-gene quenched the fluorescence intensity. This platform achieved 11.46 fM detection sensitivity without target amplification (Figure 16.2b).[80] Draz *et al.* developed a smartphone-based system known as nanoparticle-enabled smartphone (NES), which uses convolutional neural network (CNN) technology to provide a simple and

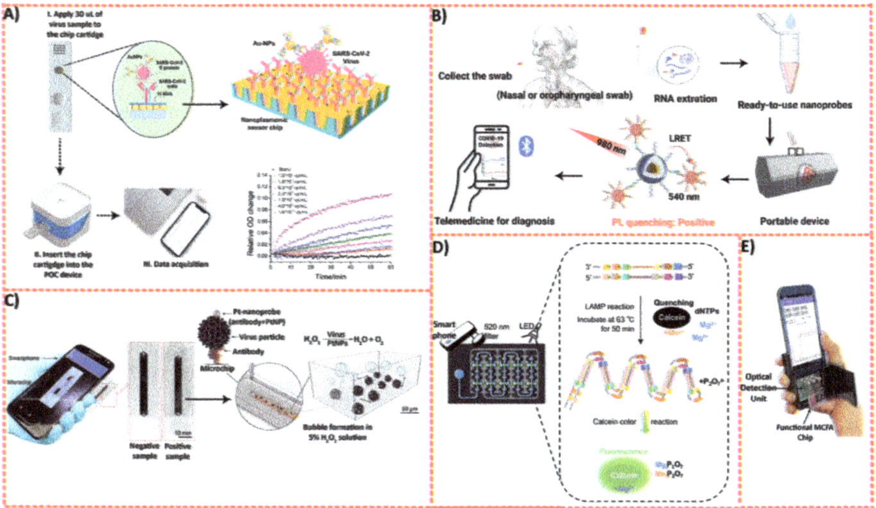

Figure 16.2 (a) Nanoplasmonic sensor chip cartridge for the unique detection process and dynamic binding curves of SARS-CoV-2 antibodies interacting with varying quantities of pseudovirus.[79] (b) A mobile phone-controlled portable platform for COVID-19 diagnostics featuring a ready-to-use, target amplification-free upconversion luminescence sandwich assay. Abbreviation: photoluminescence (PL).[80] (c) A CNN-nanoparticle-enabled smartphone for virus detection, consisting of three primary steps: virus capture and labelling with Pt-nanoprobes (PtNPs), washing with phosphate buffer, and a catalytic reaction with hydrogen peroxide.[81] (d) Principle of the LAMP-based autonomous chip for multiplex pathogen detection.[82] (e) Microfluidic MCFA chip linked to a smartphone for point-of-care testing.[83] Reproduced from ref. 84 with permission from Elsevier, Copyright 2024.

sensitive method for the early detection of hepatitis B virus (HBV), as shown in Figure 16.2c.[81] Xie *et al.* developed a self-operating microfluidic chip using loop-mediated isothermal amplification (LAMP) technology and a cell phone camera to perform multiple HBV detections, but this method required complex sample preparation (Figure 16.2D).[82] Smartphones can use their built-in cameras as detectors in lab-on-a-chip (LOC) devices. However, their inherent sensitivity may not be sufficient for demanding immunodiagnostics. Therefore, in order to improve immunodiagnostic capabilities, it is necessary to integrate more sensitive optical detectors with smart phone analysers. Ghosh *et al.* developed a mobile-connected LOC sensor using a microchannel capillary flow assay (MCFA) platform as shown in Figure 16.2e. The device uses freeze-dried chemiluminescent materials to detect the malaria biomarker histidine rich protein II (HRP-II). The LOC sensor was demonstrated to be reliable for malaria diagnosis.[83]

Methods for preparing clinical samples have been refined for diagnostic testing across several clinical scenarios, such as the extraction, concentration, and purification of respiratory virus particles. These techniques provide a high degree of purity and concentration of target samples; however, they necessitate substantial equipment and highly skilled personnel, which are challenging to attain in resource-constrained settings where prompt testing and diagnostics are essential for effective management of respiratory viruses.[85] Upon acquisition of the biological sample by the designated method, the samples undergo preprocessing to facilitate various concentration and purification steps. Samples, including blood, faeces, and other bigger or more viscous specimens, undergo pretreatment to facilitate processing. This encompasses the utilisation of standard anticoagulants during blood draws (often present in collection tubes) or dilution in buffers, as observed with stools, to facilitate pipetting. This encompasses, but is not restricted to, a lysing phase *via* chemical or physical methods employing various buffers, concentration and purification stages utilising filter membranes or centrifugation, and lysing and/or isolation through thermal variations. These are essential steps in the procedure as they release and isolate viral samples from cells and inactivate nucleases that would otherwise destroy the sample.[85] The biosensor is engineered for minimal sample preparation, rendering it appropriate for point-of-care testing. The impedance measurements exhibit significant sensitivity to alterations in the electrical characteristics of the sample, facilitating the identification of viral RNA even at minimal concentrations.

Wityk *et al.* developed a biosensor featuring a specialised cuvette design that accommodates a saliva sample and the electrode. On account of the complexities of saliva as a medium, particularly contaminants that may disrupt biosensor results, a cuvette with a downward-facing electrode was created. The cuvette design permits the direct placement of the saliva sample into the container, with the electrode surface orientated downward at the bottom. This orientation allows bigger debris, epithelial cells, and other unwanted particles to accumulate in the bottom of the cuvette, furthest from the active sensing surface. This reduces contamination interference and guarantees that the electrode is exposed to the liquid phase of the sample containing the target viral RNA.[86]

16.3.2 Integration of Multiple Sensor Types

The integration of multiple sensor technologies into virus detection systems significantly improves diagnostic accuracy while expanding the range of pathogens that can be detected. By combining the complementary strengths of different sensor technologies, this multi-sensor approach mitigates the limitations of each detection method, resulting in a more robust, sensitive and reliable diagnostic framework. The synergistic combination of different sensor types facilitates improved signal accuracy, reduces false positives and negatives, and enables a comprehensive assessment of viral presence, thereby optimising detection efficiency in both clinical and field applications.

Metamaterial biosensors have emerged as a powerful tool for the simultaneous detection of multiple viral agents, offering a compact and highly sensitive diagnostic platform. Researchers have developed advanced designs that incorporate four square split-ring resonators coupled to a central graphene ring, achieving near-perfect absorption at 1.354 THz. This sophisticated configuration enables precise and selective virus identification, exploiting the unique electromagnetic properties of metamaterials to enhance detection capabilities. The ability of these biosensors to simultaneously detect different viral pathogens highlights their potential in next-generation diagnostic systems, facilitating rapid and accurate multi-virus screening in clinical and field settings.[87]

The integration of microfluidic technology with biosensors has created a revolution in virus detection, enabling highly sensitive and specific identification of viral antigens and nucleic acids. These advanced platforms allow precise manipulation of minute volumes of fluid, minimising sample consumption while maximising analytical efficiency. Recent innovations have led to the development of microfluidic devices capable of detecting viral DNA, RNA, and protein markers, providing a versatile and scalable tool for comprehensive viral diagnostics. By streamlining assay workflows and improving detection accuracy, microfluidic-based biosensors hold significant potential for rapid, point-of-care, and high-throughput diagnostic applications.[88]

Optical resonator-based sensors, particularly subwavelength micro-ring resonators, have been instrumental in the development of highly sensitive point-of-care biosensors for rapid virus differentiation, such as distinguishing SARS-CoV-2 from influenza A H1N1. These advanced sensors use virus-specific antibodies that bind to target antigens and induce measurable shifts in resonance peaks, enabling precise and reliable detection. In addition, their seamless integration with microfluidic chips facilitates the simultaneous analysis of multiple analytes, significantly increasing diagnostic throughput. With detection limits as low as 100 fg ml^{-1} in a remarkably short timeframe of 15 minutes, these optical biosensors provide an efficient and scalable solution for real-time virus detection in clinical and field settings.[89]

Integrated air quality monitoring systems have been innovatively designed to enable real-time detection of airborne viruses, including SARS-CoV-2, representing a major advance in public health surveillance and infection control. These sophisticated systems incorporate air sampling mechanisms coupled with high-precision optical detection units, where biochips functionalised with gold nanoparticles and virus-specific antibodies facilitate the selective capture and identification of viral aerosols. By exploiting the unique optical and biochemical interactions of nanomaterial-based biochips, these monitoring platforms provide highly sensitive and rapid virus detection, making them invaluable tools for mitigating airborne disease transmission in both healthcare and community settings.[90]

The widespread availability and technological sophistication of smartphones have spurred the development of innovative diagnostic frameworks that leverage their embedded sensors for virus detection. AI-driven methods have been proposed to harness the computational power, imaging capabilities and sensor functionalities of modern smartphones to diagnose infections such as COVID-19. By integrating advanced machine learning algorithms with smartphone-based detection platforms, these approaches offer a low-cost, portable, and accessible diagnostic solution that has the potential to transform point-of-care testing and expand access to healthcare, particularly in resource-limited settings.[91]

In the context of globalisation, individuals face a heightened risk of simultaneous infections by several viruses, underscoring the necessity for the development of multiplexed devices. Besides adequate sensitivity and swift response, multi-virus sensing approaches are anticipated to provide further benefits, such as high throughput, single-sample analysis for. The development of multi-virus detection techniques, including parallel evaluation and data visualization, is crucial for controlling virus spread and protecting public health. These biosensors consist of a viral receptor, transducer, signal detector, processor, and display.[7,92] Bioreceptors enhance selectivity, while bio-interface design enhances binding efficiency. Biosensors can generate response signals and integrate multiple functions for maximum detection performance.[7,93] In contrast to biosensing for individual viruses and multiplexed illness indicators, multi-virus biosensing necessitates heightened scrutiny in a more stringent setting. Multi-virus biosensing technology has advanced significantly in recent years. The meticulous design of transducers, cutting-edge nanofabrication, and advancements in bio-interface technology have enhanced the sensitivity, accuracy, and selectivity of multi-virus biosensors. The sample volume required for detection has been reduced from the microlitre to the picolitre scale, while improvements in AI enable the prediction of virus species. Presently, multiplexed detection predominantly depends on the replication of detection units and various receptors, which increases throughput but lacks a focused design for a particular virus. For instance, respiratory viruses exhibit varying characteristics, including viral particle diameters that range from 60 to 250 nm,[7,94] and one transducer design may not be suitable for all types of viral antigens. Bioreceptors enhance selectivity, while bio-interface design enhances binding efficiency. Biosensors can generate response signals and integrate multiple functions for maximum detection performance.[7] Han *et al.* presented a multi-detectable and nano-flow immunosensor using ZnO nanorods (NRs) grown on the inner surface of polydimethylsiloxane (PDMS) sensors to detect H1N1, H5N1, and H7N9 influenza viruses simultaneously. The nanostructured ZnO NRs interact with proteins with lower isoelectric points (IEP), such as H1N1, H5N1, and H7N9 antibodies. The sensors showed high selectivity towards these viruses, with a low limit of detection of 1 pg ml^{-1}. The steady-state oxidation current output increased linearly with virus concentrations.[95]

16.3.3 Data Interpretation and Validation in Sensor-based Virus Detection

Sensor-based virus detection relies on accurate data interpretation and validation to ensure reliable, timely, and actionable results. The appropriate validation ensures that sensor output is meaningful and error-free. In addition, data interpretation and validation are critical in sensor-based virus detection to ensure accuracy, reliability, and clinical usability. Integrating modern signal processing, AI-driven validation, and cross-referencing with established procedures can allow sensor-based detection to become a crucial instrument for swift and effective virus identification. The following paragraphs present an overview of how the data is interpreted and validated in such systems.

16.3.3.1 Data Interpretation

Sensor systems (optical, electrochemical, piezoelectric, *etc.*) convert raw signals into measurable outputs, such as fluorescence intensity, electrical resistance changes, or impedance shifts. These outputs need to be processed and interpreted in order to obtain meaningful data.[96] Accurate and precise data interpretation is critical for extracting meaningful information from sensor-based virus detection systems and involves a variety of approaches and technologies. Interpreting data from sensor-based virus detection systems presents several challenges, including managing complex datasets, ensuring data accuracy, and integrating diverse data sources. Addressing these issues is crucial for reliable diagnostics and effective disease monitoring.

Sensor-based virus detection systems frequently generate vast, multidimensional datasets that pose significant challenges for traditional analytical methodologies. The complexity of these data arises from the intricate signal patterns produced by advanced biosensing platforms, such as plasmonic and metamaterial-based sensors, which rely on subtle variations in optical properties for virus detection. Extracting meaningful insights from such sophisticated signal outputs necessitates the application of advanced computational techniques, including machine learning algorithms and multivariate statistical analyses, to enhance interpretability and diagnostic accuracy. Without these refined analytical approaches, the potential of these cutting-edge biosensors remains constrained by the limitations of conventional data processing methods.[97] The integration of advanced data analytics and machine learning algorithms significantly augments the analysis of complex datasets by uncovering patterns and correlations that conventional methods may overlook. These AI-driven frameworks, particularly those leveraging smartphone sensors, offer a promising avenue for improving diagnostic accuracy, as demonstrated by their application in COVID-19 detection. By harnessing machine learning models to process sensor data, these systems enable more nuanced interpretations and real-time assessments, thus enhancing the precision and efficiency of virus detection. The ability of AI to adapt and learn from new data further strengthens its potential in transforming sensor-based diagnostic tools into more robust and intelligent solutions.[91]

Ensuring the accuracy and reliability of sensor data is critical, as even minor inaccuracies can result in false positives or false negatives, which could have significant implications for public health interventions. Variability in environmental factors, such as temperature, humidity, and electromagnetic interference, as well as issues related to sensor calibration, can all influence the quality of the data collected. Therefore, maintaining stringent calibration protocols and accounting for environmental conditions are essential to minimize errors and ensure that the sensor data remains consistent and trustworthy for diagnostic purposes. Without addressing these factors, the integrity of sensor-based detection systems could be compromised, undermining their effectiveness in real-world applications.[98]

The integration of heterogeneous data sources from diverse sensor platforms presents a considerable challenge due to variations in data formats, measurement standards, and technological specifications. This inherent heterogeneity complicates seamless data fusion, often leading to inconsistencies that hinder comprehensive analysis and interpretation. Addressing these challenges requires the development of sophisticated data harmonization techniques and interoperable frameworks that facilitate the synthesis of multi-source sensor data, ensuring coherence and reliability in virus detection and monitoring applications.[99]

The implementation of advanced data fusion techniques is crucial for the seamless integration of heterogeneous data sources, enabling a more comprehensive and accurate interpretation of sensor outputs. By consolidating information from multiple sensing modalities, these methodologies enhance the reliability and depth of virus detection systems. One such approach, tuneable resistive pulse sensing (TRPS), facilitates high-throughput single-particle measurements, allowing for precise assessments of sample stability and viral particle concentration. By leveraging these sophisticated fusion strategies, sensor-based detection platforms can achieve greater diagnostic precision, ultimately improving their efficacy in real-world biomedical applications.

Various advanced approaches and technologies are currently being developed to effectively process and analyse sensor data and are described in the literature. Image analysis techniques for lateral flow devices (LFDs) are widely used for rapid virus detection, such as for SARS-CoV-2. Interpreting LFD results can be challenging due to variability in visual perception. To address this, image analysis techniques employing machine learning algorithms have been developed. These methods analyse visual features of the LFD's control region to automatically classify results as positive, negative, or inconclusive, reducing human error and bias.[100] Raman spectroscopy provides molecular fingerprints of viruses, which can be complex to interpret. Machine learning models, particularly CNNs, have been applied to analyse Raman spectra for accurate virus identification. These models achieve high accuracy in classifying virus types and subtypes and offer interpretability by highlighting spectral regions critical for identification.[101] Plasmonic metamaterials enhance the sensitivity of biosensors through unique optical properties. These sensors detect viruses by monitoring changes in refractive index upon virus binding. Advancements in this field include integrating machine learning and AI to improve detection capabilities, offering rapid and cost-effective virus identification.[102] The interferometric reflectance imaging sensor (IRIS) utilizes optical interference principles to detect biomolecular interactions. It measures changes in interferometric signatures as biomass accumulates on a silicon–silicon oxide substrate. This label-free technique has been adapted for various applications, including virus and pathogen detection, offering a quantitative approach to monitor biomolecular interactions in real-time.[103] TRPS measures individual virus particles as they pass through a size-tuneable nanopore, allowing simultaneous determination of particle size and concentration. This high-throughput method provides rapid assessments of viral samples without the need for pre-staining, making it efficient for evaluating sample stability and aggregate contributions as well as total viral particle concentration (vp per ml).[104] Nanowire arrays coated with specific antibodies can detect multiple viruses simultaneously. Interpreting the complex data from these arrays, especially with multiple viruses docking, requires advanced modelling. Hidden semi-Markov models have been employed to analyse patterns of conductance changes in nanowire arrays, facilitating real-time and accurate virus detection.[105] These approaches and technologies exemplify the integration of advanced computational methods with sensor-based platforms to enhance the accuracy and efficiency of virus detection. Ongoing research continues to refine these methods, contributing to improved public health responses. Unsupervised methods, such as non-negative matrix factorization and *k*-Shape clustering, have shown promise in detecting viruses without labelled data.[67,106] Supervised machine learning techniques, including extreme gradient boosting (XGBoost), have also been successfully applied to classify virus types using sensor data.[106] Surface-enhanced Raman scattering (SERS) sensors combined with machine learning algorithms offer the potential for improved virus detection and identification.[106,107] Other sensor technologies, such as SPR and capacitance-based sensors, have demonstrated effectiveness in virus detection.[67,68] A number of challenges (*e.g.* environmental noise, sample contamination, and random variation) can affect the efficiency of sensor technologies used for virus analysis. However, various approaches and technologies can be used to assist in processing raw sensor signals, reducing noise and ensuring reliable virus identification, *etc.* For example, filtering techniques (*e.g.* machine learning algorithms, Fourier transform analysis) aid in distinguishing true virus signals from background noise. The main methods used for this purpose are shown in Figure 16.3.

16.3.3.2 Validation

Ensuring the accuracy and reliability of sensor-based virus detection systems requires rigorous validation methodologies. These methodologies encompass a range of approaches and technologies designed to comprehensively assess the performance of detection assays.

Validation of sensor-based virus detection systems is a multifaceted process that integrates analytical assessments, comparative analyses, robustness

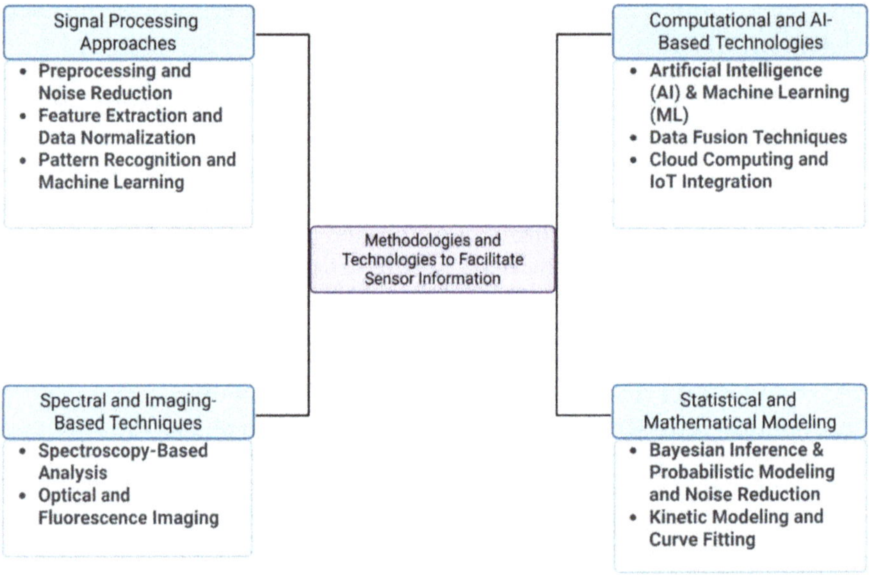

Figure 16.3 Different methodologies and technologies facilitate the processing of raw sensor information, mitigate noise, and ensure dependable virus identification. Abbreviation: internet of things (IoT).

testing, specificity evaluations, and clinical trials. Adherence to these rigorous validation protocols ensures that these diagnostic tools provide accurate, reliable, and reproducible results, thereby enhancing their utility in public health and clinical diagnostics.[108]

Analytical validation of sensor-based virus detection involves a comprehensive evaluation of sensitivity and specificity, as well as the accurate determination of detection and quantification limits. Sensitivity, defined as the assay's ability to accurately detect the presence of virus (true positive rate), and specificity, which ensures correct classification in the absence of virus (true negative rate), are critical performance parameters. Their evaluation requires controlled testing with known concentrations of viral particles and non-target organisms to establish detection thresholds and assess potential cross-reactivity. LOD is the minimum viral concentration that can be reliably identified, while the limit of quantification (LOQ) is the lowest viral load that can be measured with acceptable precision. These parameters are determined by systematic serial dilution analysis, ensuring robust detection capabilities and enhancing the diagnostic confidence of sensor-based technologies.[108]

Comparative validation of sensor-based assays requires rigorous benchmarking against gold-standard diagnostic methods, such as PCR and ELISA, to establish their clinical reliability and diagnostic efficacy. This process involves a systematic evaluation of the concordance between the new and conventional techniques to ensure that the new assay demonstrates equivalent or superior accuracy while offering potential advantages such as improved speed, sensitivity, or ease of use. For example, the results of studies evaluating the performance of recombinase polymerase amplification (RPA) have demonstrated its ability to provide rapid and highly sensitive virus detection, positioning it as a viable alternative to traditional PCR-based diagnostics.[109] Such comparative assessments not only demonstrate the diagnostic robustness of next-generation biosensors, but also facilitate their seamless integration into routine virological surveillance and clinical diagnostic workflows.

Robustness and reproducibility testing are critical components of sensor-based assay validation, ensuring consistent performance across diverse operational settings. Inter-laboratory studies play a pivotal role in assessing reproducibility by subjecting identical samples to the same assay protocols across multiple laboratories, thereby identifying potential sources of variability and standardizing methodologies to enhance diagnostic reliability. In addition, environmental stress testing evaluates the assay's resilience under fluctuating external conditions, such as temperature variations and humidity changes, to ascertain its stability and accuracy in real-world applications. These rigorous validation measures collectively ensure that sensor-based detection systems maintain high diagnostic precision and accuracy, regardless of environmental or procedural inconsistencies.[108]

Cross-reactivity and interference studies are essential in validating the specificity of sensor-based assays, ensuring that the detection mechanism precisely identifies the target virus without erroneous interactions with non-target pathogens or interfering substances. This validation process involves systematically exposing the sensor to a diverse panel of non-target organisms and potential interferents to rigorously assess its selectivity and minimize false-positive results. For instance, nanomaterial-based biosensors developed for SARS-CoV-2 detection have undergone extensive specificity evaluations, demonstrating minimal cross-reactivity with other respiratory viruses. Such assessments are crucial in establishing the assay's diagnostic fidelity, reinforcing its clinical applicability, and preventing misinterpretations that could compromise public health responses.[110]

Clinical validation is a crucial phase in assessing the real-world applicability of sensor-based assays, ensuring their efficacy in practical diagnostic settings. Field testing with clinical specimens involves evaluating the assay's diagnostic performance by analysing samples from both infected and non-infected individuals, thereby determining its sensitivity, specificity, and overall reliability in diverse patient populations. Beyond initial validation, post-market surveillance plays a pivotal role in continuously monitoring the assay's performance after deployment, enabling the identification of potential issues related to accuracy, reliability, and user accessibility. This ongoing assessment facilitates iterative refinements, ensuring sustained diagnostic precision and optimizing the assay's integration into routine clinical and public health applications.[108]

16.4 Ethical Considerations in Sensor-based Virus Detection

The integration of sensor-based technologies for virus detection offers significant advances in public health surveillance and disease management. However, its use raises a number of ethical considerations (*e.g.* fairness, privacy, and responsible distribution) that need to be carefully addressed to balance technological benefits with individual rights and societal values.

The implementation of sensor-based virus detection systems inherently involves the collection and transmission (*e.g. via* wearable biosensors or mobile health apps) of sensitive health data, necessitating stringent data privacy and security measures to safeguard against unauthorized access and breaches. As these systems often rely on digital communication channels, including wireless networks and cloud-based storage, they are particularly vulnerable to cyber threats such as data breaches, hacking, and unauthorized surveillance. The potential misuse of such data—ranging from identity theft to discriminatory practices based on an individual's health status—underscores the critical need for robust encryption protocols, access control mechanisms, and regulatory frameworks that ensure data integrity and confidentiality. Establishing comprehensive security measures is paramount to fostering public trust and ensuring the ethical deployment of these advanced diagnostic technologies.[111]

Informed consent is a fundamental pillar of ethical medical practice, ensuring that individuals retain autonomy over their personal health data. Before participating in sensor-based virus detection, individuals must be provided with clear and comprehensive information about the nature of the data being collected, its intended use, access parameters, and any potential risks associated with its processing. Transparent communication is essential to ensure that consent is given voluntarily and with full understanding of the implications. This ethical obligation is particularly critical in the context of digital disease detection technologies, where the integration of new data sources and sophisticated analytical tools may not be readily understood by the general public. By promoting clarity and trust, rigorous informed consent protocols help protect individual rights while facilitating the responsible implementation of advanced biosensing technologies.[112]

The issue of data ownership and control is a critical ethical consideration in sensor-based virus detection, as it directly impacts individual autonomy and trust in these technologies. Clear delineation of ownership rights is essential to ensure that individuals retain authority over their personal health information, including the ability to access, amend inaccuracies, and regulate data sharing and usage parameters. Without well-defined policies governing data stewardship, there is a risk of unauthorized exploitation or misuse, undermining public confidence in these diagnostic systems. Therefore, the establishment of robust regulatory frameworks that explicitly address data ownership, access rights, and control mechanisms is imperative to safeguard individual privacy while fostering ethical and transparent deployment of sensor-based health monitoring technologies.[113]

The utilization of data from sensor-based virus detection systems carries the inherent risk of unintended discrimination if not managed with stringent ethical safeguards. Misuse of health data–such as by employers or insurance providers–could result in prejudiced decision-making, leading to stigmatization, social marginalization, or inequitable access to opportunities and services. To mitigate these risks, it is imperative to establish and enforce comprehensive data governance policies that explicitly prohibit discriminatory practices and ensure that health information is utilized solely for its intended medical and public health purposes. Upholding these ethical standards not only protects individuals from bias but also fosters public trust in the responsible deployment of sensor-based diagnostic technologies.[107]

The implementation of sensor-based virus detection systems must be grounded in rigorous validation to ensure their accuracy and reliability, as diagnostic inaccuracies can lead to significant consequences. False positives may result in unnecessary distress, medical interventions, or social and economic disruptions, while false negatives could create a false sense of security, facilitating the unchecked spread of infection. To uphold diagnostic integrity, these systems require continuous performance monitoring, stringent quality control measures, and robust validation protocols that assess sensitivity, specificity, and reproducibility. Ensuring high diagnostic precision is paramount for the credibility and effectiveness of sensor-based detection technologies in both clinical and public health settings.[114]

The integration of AI into sensor-based virus detection systems enhances analytical precision and predictive capabilities, significantly improving disease surveillance and diagnostic efficiency. However, the deployment of AI-driven methodologies also presents complex ethical challenges, necessitating stringent oversight to ensure transparency in decision-making processes, mitigate algorithmic bias, and uphold accountability for AI-generated conclusions. Unchecked biases within AI models could lead to disparities in diagnostic outcomes, disproportionately affecting certain populations and undermining healthcare equity. To address these concerns, the implementation of responsible AI frameworks-incorporating fairness, interpretability, and regulatory compliance is essential to promote ethical AI usage while maximizing its potential in sensor-based disease detection.[115]

In the context of public health crises, the deployment of sensor-based virus detection technologies necessitates a delicate balance between safeguarding collective well-being and preserving individual privacy rights. While these technologies play a crucial role in enhancing disease surveillance, outbreak response, and epidemiological monitoring, their implementation must be carefully regulated to prevent undue encroachments on personal freedoms. Striking this equilibrium requires the establishment of comprehensive ethical guidelines and robust legal frameworks that delineate the scope of data collection, usage, and retention. By ensuring that public health interventions remain proportionate and non-intrusive, policymakers can uphold both societal health imperatives and the fundamental rights of individuals, fostering public trust and ethical compliance in the use of sensor-driven diagnostic systems.[116]

16.5 Conclusions and Future Aspects

The field of virus detection *via* sensor technology is poised for significant advancements, driven by the persistent need for rapid, accurate, and accessible diagnostic tools. Despite existing challenges, the solutions being formulated provide the foundation for a transformative future in healthcare and beyond. This chapter has delineated the current problems and their solutions during the production and application of biosensors for viral detection.

In the near future, nanomaterial-based sensors, due to their remarkable surface-to-volume ratios, should attain enhanced sensitivity, enabling the detection of minuscule quantities of virus particles. Improvements in biorecognition components, like modified antibodies and aptamers, will increase specificity, hence reducing false positives. Additionally, the integration of AI will facilitate the processing of intricate data sets and enhance the capacity to differentiate between analogous infections. Furthermore, multiplexing capabilities will facilitate the concurrent detection of multiple viruses, essential for differential diagnosis and recognising co-infections. The advancement of reliable, user-friendly point-of-care (POC) sensors will transform diagnostics in resource-constrained environments, permitting swift on-site testing. Sensor networks implemented in public areas, transportation centres, and wastewater systems will facilitate real-time surveillance of viral transmission, allowing for pre-emptive public health measures. Machine learning and AI will be vital tools in assessing sensor data and identifying trends that will facilitate the prediction of potential epidemics.

In conclusion, innovative solutions are being developed to address the challenges of virus detection using sensors, including limitations in sensitivity, concerns regarding specificity, and issues related to scalability. The integration of nanomaterials, sophisticated biorecognition components, and AI is propelling the advancement of exceptionally sensitive, precise, and user-centric sensors.

References

1. A. Baskar, K. Madhivanan, R. Atchudan, S. Arya and A. K. Sundramoorthy, Nanoparticle electrochemical biosensors for virus detection, *Clin. Chim. Acta*, 2025, **566**, 120054.
2. S. I. Kaya, L. Karadurmus, G. Ozcelikay, N. K. Bakirhan and S. A. Ozkan, Electrochemical virus detections with nanobiosensors, in *Nanosensors for Smart Cities*, 2020, pp. 303–326
3. T. P. Harsh, Medical viruses: diagnostic techniques, *Virol. J.*, 2023, **20**(1), 143.

4. O. I. Guliy, B. D. Zaitsev, O. S. Larionova and I. A. Borodina, Virus detection methods and biosensor technologies, *Biophysics*, 2019, **64**(6), 890–897.
5. S. Akgönüllü, Y. Aslan, S. Celik, U. Celik, A. Denizli, E. Derin, *et al.*, in *Biosensors for Virus Detection*, ed. A. Denizli, and Y. Saylan, IOP Publishing, 2021.
6. A. M. Rheima, Z. T. Al-Sharify, A. A. Mohaimeed, M. A. A. H. Kazem, J. M. Dhabab and D. M. Athair, *et al.*, Nano biosensors: classification, electrochemistry, nanostructures, and optical properties, *Results Eng.*, 2024, **24**, 103428.
7. Y. Zheng, X. Song, Z. Fredj, S. Bian and M. Sawan, Challenges and perspectives of multi-virus biosensing techniques: a review, *Anal. Chim. Acta*, 2023, **1244**, 340860.
8. K. R. Bandi, A. K. Singh and A. Upadhyay, Construction and performance characteristics of polymeric membrane electrode and coated graphite electrode for the selective determination of Fe^{3+} ion, *Mater. Sci. Eng.: C.*, 2014, **36**(1), 187–193.
9. M. A. Lifson, M. O. Ozen, F. Inci, S. Q. Wang, H. Inan and M. Baday, *et al.*, Advances in biosensing strategies for HIV-1 detection, diagnosis, and therapeutic monitoring, *Adv. Drug Delivery Rev.*, 2016, **103**, 90–104.
10. T. Ozer, B. J. Geiss and C. S. Henry, Review—chemical and biological sensors for viral detection, *J. Electrochem. Soc.*, 2020, **167**(3), 037523.
11. A. Ahmed, J. V. Rushworth, N. A. Hirst and P. A. Millner, Biosensors for whole-cell bacterial detection, *Clin. Microbiol. Rev.*, 2014, **27**(3), 631–646.
12. V. Velusamy, K. Arshak, O. Korostynska, K. Oliwa and C. Adley, An overview of foodborne pathogen detection: In the perspective of biosensors, *Biotechnol. Adv.*, 2010, **28**(2), 232–254.
13. G. Jamalipour Soufi, S. Iravani and R. S. Varma, Molecularly imprinted polymers for the detection of viruses: challenges and opportunities, *Analyst*, 2021, **146**(10), 3087–3100.
14. R. Antiochia, Developments in biosensors for CoV detection and future trends, *Biosens. Bioelectron.*, 2021, **173**, 112777.
15. S. Akgönüllü, Y. Aslan, S. Celik, U. Celik, A. Denizli, E. Derin, *et al.*, in *Biosensors for Virus Detection*, ed. A. Denizli, and Y. Saylan, IOP Publishing, 2021.
16. S. A. Abid, M. A. Ahmed, I. M. S. Al-Kadmy, A. A. Sattar, A. M. Beshbishy and G. E. S. Batiha, *et al.*, Biosensors as a future diagnostic approach for COVID-19, *Life Sci.*, 2021, **273**, 119117.
17. Y. Saylan and A. Denizli, Virus detection using nanosensors, in *Nanosensors for Smart Cities*, 2020, pp. 501–511.
18. F. Narita, Z. Wang, H. Kurita, Z. Li, Y. Shi and Y. Jia, *et al.*, A review of piezoelectric and magnetostrictive biosensor materials for detection of COVID-19 and other viruses, *Adv. Mater.*, 2021, **33**(1), 2005448.
19. D. Y. Li, Y. P. Qin, H. Y. Li, X. W. He, W. Y. Li and Y. K. Zhang, A "turn-on" fluorescent receptor for detecting tyrosine phosphopeptide using the

surface imprinting procedure and the epitope approach, *Biosens. Bioelectron.*, 2015, **66**, 224–230.
20. H. Duan, L. Li, X. Wang, Y. Wang, J. Li and C. Luo, A sensitive and selective chemiluminescence sensor for the determination of dopamine based on silanized magnetic graphene oxide-molecularly imprinted polymer, *Spectrochim. Acta, Part A*, 2015, **139**, 374–379.
21. S. E. Diltemiz and O. Uslu, A reflectometric interferometric nanosensor for sarcosine, *Biotechnol. Prog.*, 2015, **31**(1), 55–61.
22. M. Mehrzad-Samarin, F. Faridbod, A. S. Dezfuli and M. R. Ganjali, A novel metronidazole fluorescent nanosensor based on graphene quantum dots embedded silica molecularly imprinted polymer, *Biosens. Bioelectron.*, 2017, **92**, 618–623.
23. P. Weber, B. R. Riegger, K. Niedergall, G. E. M. Tovar, M. Bach and G. Gauglitz, Nano-MIP based sensor for penicillin G: Sensitive layer and analytical validation, *Sens. Actuators, B*, 2018, **267**, 26–33.
24. C. He, U. H. Ledezma, P. Gurnani, T. Albelha, K. J. Thurecht and R. Correia, *et al.*, Surface polymer imprinted optical fibre sensor for dose detection of dabrafenib, *Analyst*, 2020, **145**(13), 4504–4511.
25. J. Wang, J. Li, C. Zeng, Q. Qu, M. Wang and W. Qi, *et al.*, Sandwich-like sensor for the highly specific and reproducible detection of rhodamine 6G on a surface-enhanced raman scattering platform, *ACS Appl. Mater. Interfaces*, 2020, **12**(4), 4699–4706.
26. B. Regan, F. Boyle, R. O'Kennedy and D. Collins, Evaluation of molecularly imprinted polymers for point-of-care testing for cardiovascular disease, *Sensors*, 2019, **19**(16), 3485.
27. S. M. Kim, T. Lee, Y. G. Gil, G. H. Kim, C. Park and H. Jang, *et al.*, Fabrication of bioprobe self-assembled on au–te nanoworm structure for sers biosensor, *Materials*, 2020, **13**(14), 3234.
28. F. Inci, Y. Saylan, A. M. Kojouri, M. G. Ogut, A. Denizli and U. Demirci, A disposable microfluidic-integrated hand-held plasmonic platform for protein detection, *Appl. Mater. Today*, 2020, **18**, 100478.
29. F. T. T. Cavalcante, I. R. Italo, J. E. José, T. G. Rocha, I. G. de Sousa and A. L. G. Cavalcante, *et al.*, Designing of nanomaterials-based enzymatic biosensors: synthesis, properties, and applications, *Electrochem*, 2021, **2**(1), 149–184.
30. D. Li, Q. He, Y. He, M. Xin, Y. Zhang and Z. Shen, Molecular imprinting sensor based on quantum weak measurement, *Biosens. Bioelectron.*, 2017, **94**, 328–334.
31. S. Liu, R. Cheng, Y. Chen, H. Shi and G. Zhao, A simple one-step pretreatment, highly sensitive and selective sensing of 17B-estradiol in environmental water samples using surface-enhanced Raman spectroscopy, *Sens. Actuators, B*, 2018, **254**, 1157–1164.
32. S. Sharma and B. D. Gupta, An efficient and selective sensing of creatinine based on fiber optic SPR technique exploiting the advantages of molecular imprinting technique, in *Optical Sensing and Detection V*, SPIE, 2018, vol. 10680, p. 106801T.

33. X. Zhang, S. Yang, R. Jiang, L. Sun, S. Pang and A. Luo, Fluorescent molecularly imprinted membranes as biosensor for the detection of target protein, *Sens. Actuators, B*, 2018, **254**, 1078–1086.
34. Y. Saylan, F. Yilmaz, E. Özgür, A. Derazshamshir, H. Yavuz and A. Denizli, Molecular imprinting of macromolecules for sensor applications, *Sensors*, 2017, **17**(4), 898.
35. Ö Erdem, Y. Saylan, N. Cihangir and A. Denizli, Molecularly imprinted nanoparticles based plasmonic sensors for real-time Enterococcus faecalis detection, *Biosens. Bioelectron.*, 2019, **126**, 608–614.
36. J. Luo, Q. Ma, W. Wei, Y. Zhu, R. Liu and X. Liu, Synthesis of water-dispersible molecularly imprinted electroactive nanoparticles for the sensitive and selective paracetamol detection, *ACS Appl. Mater. Interfaces*, 2016, **8**(32), 21028–21038.
37. Ö Erdem, I. Eş, G. A. Akceoglu, Y. Saylan and F. Inci, Recent advances in microneedle-based sensors for sampling, diagnosis and monitoring of chronic diseases, *Biosensors*, 2021, **11**(9), 296.
38. E. N. Primo, M. J. Kogan, H. E. Verdejo, S. Bollo, M. D. Rubianes and G. A. Rivas, Label-free graphene oxide-based surface plasmon resonance immunosensor for the quantification of galectin-3, a novel cardiac biomarker, *ACS Appl. Mater. Interfaces*, 2018, **10**(28), 23501–23508.
39. A. E. Bodoki, B. C. Iacob and E. Bodoki, Perspectives of molecularly imprinted polymer-based drug delivery systems in cancer therapy, *Polymers*, 2019, **11**(12), 2085.
40. P. Li, G. H. Lee, S. Y. Kim, S. Y. Kwon, H. R. Kim and S. Park, From diagnosis to treatment: recent advances in patient-friendly biosensors and implantable devices, *ACS Nano*, 2021, **15**(2), 1960–2004.
41. N. Van Sau, N. Q. Minh, T. Bach Phan, N. Quyen Tran and T. Tai Nguyen, Optical biosensor using near infrared laser for enhancement of detection accuracy, *J. Electron. Mater.*, 2020, **49**(12), 7420–7426.
42. M. K. Esentürk, S. Akgönüllü, F. Yılmaz and A. Denizli, Molecularly imprinted based surface plasmon resonance nanosensors for microalbumin detection, *J. Biomater. Sci. Polym. Ed.*, 2019, **30**(8), 646–661.
43. Y. Saylan and A. Denizli, Molecularly imprinted polymer-based microfluidic systems for point-of-care applications, *Micromachines*, 2019, **10**(11), 766.
44. J. Wang, G. Rivas, X. Cai, E. Palecek, P. Nielsen and H. Shiraishi, *et al.*, DNA electrochemical biosensors for environmental monitoring: a review, *Anal. Chim. Acta*, 1997, **347**(1–2), 1–8.
45. O. Lazcka, F. J. D. Campo and F. X. Muñoz, Pathogen detection: a perspective of traditional methods and biosensors, *Biosens. Bioelectron.*, 2007, **22**(7), 1205–1217.
46. N. A. Mungroo and S. Neethirajan, Biosensors for the detection of antibiotics in poultry industry—a review, *Biosensors*, 2014, **4**(4), 472–493.
47. A. Koyun, E. Ahlatcıoğlu, Y. K. İpek, A. Koyun, E. Ahlatcıoğlu and Y. K. İpek, Biosensors and their principles, a roadmap of biomedical engineers and milestones, 2012. Available at https://www.intechopen.com/chapters/37345.

48. A. C. Huet, P. Delahaut, T. Fodey, S. A. Haughey, C. Elliott and S. Weigel, Advances in biosensor-based analysis for antimicrobial residues in foods, *TrAC, Trends Anal. Chem.*, 2010, **29**(11), 1281–1294.
49. P. Tetyana, P. M. Shumbula and Z. Njengele-Tetyana, Biosensors: design, development and applications, *Nanopores*, 2021, DOI: 10.5772/intechopen.97576.
50. Y. Pan, L. Zhang, B. Cao, X. Xue, W. Liu and C. Zhang, *et al.*, Effects of temperature and humidity on the performance of a PECH polymer coated SAW sensor, *RSC Adv.*, 2020, **10**(31), 18099–18106.
51. M. J. Jeng, M. Sharma, Y. C. Li, Y. C. Lu, C. Y. Yu and C. L. Tsai, *et al.*, Surface acoustic wave sensor for C-reactive protein detection, *Sensors*, 2020, **20**(22), 6640.
52. F. Temel, One novel calix[4]arene based QCM sensor for sensitive, selective and high performance-sensing of formaldehyde at room temperature, *Talanta*, 2020, **211**, 120725.
53. E. Haghighi and S. Zeinali, Formaldehyde detection using quartz crystal microbalance (QCM) nanosensor coated by nanoporous MIL-101(Cr) film, *Microporous Mesoporous Mater.*, 2020, **300**, 110065.
54. X. Jin, Y. P. Zhang, D. M. Li, D. Ma, S. R. Zheng and C. H. Wu, *et al.*, The interaction of an amorphous metal–organic cage-based solid (aMOC) with miRNA/DNA and its application on a quartz crystal microbalance (QCM) sensor, *Chem. Commun.*, 2020, **56**(4), 591–594.
55. A. Mujahid, A. Afzal, G. Glanzing, A. Leidl, P. A. Lieberzeit and F. L. Dickert, Imprinted sol-gel materials for monitoring degradation products in automotive oils by shear transverse wave, *Anal. Chim. Acta*, 2010, **675**(1), 53–57.
56. A. Afzal, A. Mujahid, R. Schirhagl, S. Z. Bajwa, U. Latif and S. Feroz, Gravimetric viral diagnostics: QCM based biosensors for early detection of viruses, *Chemosensors*, 2017, **5**(1), 7.
57. M. Pan, R. Li, L. Xu, J. Yang, X. Cui and S. Wang, Reproducible molecularly imprinted piezoelectric sensor for accurate and sensitive detection of ractopamine in swine and feed products, *Sensors*, 2018, **18**(6), 1870.
58. D. Battal, S. Akgönüllü, M. S. Yalcin, H. Yavuz and A. Denizli, Molecularly imprinted polymer based quartz crystal microbalance sensor system for sensitive and label-free detection of synthetic cannabinoids in urine, *Biosens. Bioelectron.*, 2018, **111**, 10–17.
59. S. Akgönüllü, D. Battal, M. S. Yalcin, H. Yavuz and A. Denizli, Rapid and sensitive detection of synthetic cannabinoids JWH-018, JWH-073 and their metabolites using molecularly imprinted polymer-coated QCM nanosensor in artificial saliva, *Microchem. J.*, 2020, **153**, 104454.
60. L. T. Lee, M. A. Mohamed, I. Yahya, J. Kulothungan, M. Muruganathan and H. Mizuta, Comparison of piezoelectric energy harvesting performance using silicon and graphene cantilever beam, *Microsyst. Technol.*, 2018, **24**(9), 3783–3789.

61. M. Yasuura, Z. L. Tan, Y. Horiguchi, H. Ashiba and T. Fukuda, Improvement of sensitivity and speed of virus sensing technologies using nm- and μm-scale components, *Sensors*, 2023, **23**(15), 6830.
62. N. Manring, M. M. N. Ahmed, N. Tenhoff, J. L. Smeltz and P. Pathirathna, Recent advances in electrochemical tools for virus detection, *Anal. Chem.*, 2022, **94**(20), 7149–7157.
63. F. Kheiri, R. E. Sabzi, E. Jannatdoust, E. Shojaeefar and H. Sedghi, A novel amperometric immunosensor based on acetone-extracted propolis for the detection of the HIV-1 p24 antigen, *Biosens. Bioelectron.*, 2011, **26**(11), 4457–4463.
64. M. M. S. Silva, A. C. M. S. Dias, B. V. M. Silva, S. L. R. Gomes-Filho, L. T. Kubota and M. O. F. Goulart, *et al.*, Electrochemical detection of dengue virus NS1 protein with a poly(allylamine)/carbon nanotube layered immunoelectrode, *J. Chem. Technol. Biotechnol.*, 2015, **90**(1), 194–200.
65. X. Deng, C. Wang, Y. Gao, J. Li, W. Wen and X. Zhang, *et al.*, Applying strand displacement amplification to quantum dots-based fluorescent lateral flow assay strips for HIV-DNA detection, *Biosens. Bioelectron.*, 2018, **105**, 211–217.
66. S. Mudenkattil, A. Girigoswami, T. Jayaprakash and K. Girigoswami, Development of nanobiosensors for human immunodeficiency virus detection a mini review, *Open Explorat.*, 2019, **3**(5), 479–493.
67. D. Siedhoff, M. Strauch, V. Shpacovitch and D. Merhof, Unsupervised data analysis for virus detection with a surface plasmon resonance sensor, International Conference on Image Processing Theory Tools and Applications, 2017.
68. A. Balasubramanian, B. Bhuval, R. Memaugh and F. R. Haselton, Si-based sensor for virus detection, *IEEE Sens. J.*, 2003, **5**(3), 340–344.
69. A. M. Shrivastav, U. Cvelbar and I. Abdulhalim, A comprehensive review on plasmonic-based biosensors used in viral diagnostics, *Commun. Biol.*, 2021, **4**(1), 1–12.
70. A. Ymeti, P. H. J. Nederkoorn, A. Dudia, V. Subramaniam, J. S. Kanger, Rapid, ultrasensitive detection of microorganisms based on interferometry and lab-on-a-chip nanotechnology, in *Optics and Photonics in Global Homeland Security V and Biometric Technology for Human Identification VI*, 2009, vol. 7306, pp. 105–111
71. N. Kumar, N. P. Shetti, S. Jagannath and T. M. Aminabhavi, Electrochemical sensors for the detection of SARS-CoV-2 virus, *Chem. Eng. J.*, 2021, **430**, 132966.
72. W. Sukjee, A. Thitithanyanont, S. Manopwisedjaroen, S. Seetaha, C. Thepparit and C. Sangma, Virus MIP-composites for SARS-CoV-2 detection in the aquatic environment, *Mater. Lett.*, 2022, **315**.
73. M. da R. Q. Lima, T. Chouin-Carneiro, E. Azeredo, L. S. Barbosa, T. M. A. Souza and J. B. C. da Silva, *et al.*, The inability of a dengue NS1 ELISA to detect Zika infections, *Sci. Rep.*, 2019, **9**(1), 1–7.

74. I. Sampaio, F. D. Quatroni, J. N. Yamauti Costa and V. Zucolotto, Electrochemical detection of Zika and Dengue infections using a single chip, *Biosens. Bioelectron.*, 2022, **216**, 114630.
75. E. W. A. Visser, L. J. Van Ijzendoorn and M. W. J. Prins, Particle motion analysis reveals nanoscale bond characteristics and enhances dynamic range for biosensing, *ACS Nano*, 2016, **10**(3), 3093–3101.
76. M. S. Draz, N. K. Lakshminaraasimulu, S. Krishnakumar, D. Battalapalli, A. Vasan and M. K. Kanakasabapathy, *et al.*, Motion-based immunological detection of Zika virus using Pt-nanomotors and a cellphone, *ACS Nano*, 2018, **12**(6), 5709.
77. M. Yasuura and M. Fujimaki, Detection of extremely low concentrations of biological substances using near-field illumination, *Sci. Rep.*, 2016, **6**(1), 39241.
78. M. Yasuura, H. Shirato, K. Higo-Moriguchi and M. Fujimaki, Detection of norovirus-like particles with an external force-assisted near-field illumination biosensor, *Jpn. J. Appl. Phys.*, 2019, **58**(7), 071005.
79. L. Huang, L. Ding, J. Zhou, S. Chen, F. Chen and C. Zhao, *et al.*, One-step rapid quantification of SARS-CoV-2 virus particles via low-cost nanoplasmonic sensors in generic microplate reader and point-of-care device, *Biosens. Bioelectron.*, 2021, **171**, 112685.
80. M. Song, M. C. Wong, L. Li, F. Guo, Y. Liu and Y. Ma, *et al.*, Rapid point-of-care detection of SARS-CoV-2 RNA with smartphone-based upconversion luminescence diagnostics, *Biosens. Bioelectron.*, 2023, **15**, 222.
81. M. S. Draz, A. Vasan, A. Muthupandian, M. K. Kanakasabapathy, P. Thirumalaraju and A. Sreeram, *et al.*, Virus detection using nanoparticles and deep neural network-enabled smartphone system, *Sci. Adv.*, 2020, **6**, 5354–5370.
82. C. Xie, S. Chen, L. Zhang, X. He, Y. Ma and H. Wu, *et al.*, Multiplex detection of blood-borne pathogens on a self-driven microfluidic chip using loop-mediated isothermal amplification, *Anal. Bioanal. Chem.*, DOI: 10.1007/s00216-021-03224-8.
83. S. S. Ghosh, T. Hebbeker, A. Meyer and T. Pook, Motivation for general model independent search for new physics, 2020, DOI: 10.1007/978-3-030-53783-8_2.
84. S. Rasheed, T. Kanwal, N. Ahmad, B. Fatima, M. Najam-ul-Haq and D. Hussain, *TrAC, Trends Anal. Chem.*, Advances and challenges in portable optical biosensors for onsite detection and point-of-care diagnostics, 2024, **173**.
85. R. Zenhausern, C. H. Chen and J. Y. Yoon, Microfluidic sample preparation for respiratory virus detection: A review, *Biomicrofluidics*, 2021, **15**(1), 011503.
86. P. Wityk, A. Terebieniec, R. Nowak, J. Łubiński, M. Mroczyńska-Szeląg and T. Wityk, *et al.*, Reusable biosensor for easy RNA detection from unfiltered saliva, *Sensors*, 2025, **25**(2), 360.

87. P. Upender, S. Prasanna Bharathi, Sukriti, K. Kumba and A. Kumar, A compact metamaterial biosensor for multi-virus detection with tunability and high incidence angle absorption, *IEEE Access*, 2017, **11**, 131915–1311925.
88. K. T. L. Trinh, H. D. K. Do and N. Y. Lee, Recent advances in molecular and immunological diagnostic platform for virus detection: a review, *Biosensors*, 2023, **13**(4), 490.
89. S. Ning, H. C. Chang, K. C. Fan, P. Y. Hsiao, C. Feng and D. Shoemaker, *et al.*, A point-of-care biosensor for rapid detection and differentiation of COVID-19 virus (SARS-CoV-2) and influenza virus using subwavelength grating micro-ring resonator, *Appl. Phys. Rev.*, 2023, **10**(2), 021410.
90. H. Yang, An integrated air quality monitoring system for COVID-19 virus detection. Available at https://blynk.io/.
91. M. S. Halgurd, K. Region-FRIraq FirstLast, Z. G. Kayhan, S. S. Ali, C. Kevin and R. Khaled, *A Novel AI-Enabled Framework to Diagnose Coronavirus COVID-19 Using Smartphone Embedded Sensors: Design Study*.2020.
92. D. Kadadou, L. Tizani, V. S. Wadi, F. Banat, H. Alsafar and A. F. Yousef, *et al.*, Recent advances in the biosensors application for the detection of bacteria and viruses in wastewater, *J. Environ. Chem. Eng.*, 2022, **10**(1), 107070.
93. M. X. Li, W. Zhao, G. S. Qian, Q. M. Feng, J. J. Xu and H. Y. Chen, Distance mediated electrochemiluminescence enhancement of CdS thin films induced by the plasmon coupling of gold nanoparticle dimers, *Chem. Commun.*, 2016, **52**(99), 14230–14233.
94. L. E. Breshears, B. T. Nguyen, S. M. Robles, L. Wu and J. Y. Yoon, Biosensor detection of airborne respiratory viruses such as SARS-CoV-2, *SLAS Technol.*, 2022, **27**(1), 4–17.
95. J. H. Han, D. Lee, C. H. C. Chew, T. Kim and J. J. Pak, A multi-virus detectable microfluidic electrochemical immunosensor for simultaneous detection of H1N1, H5N1, and H7N9 virus using ZnO nanorods for sensitivity enhancement, *Sens. Actuators, B*, 2016, **228**, 36–42.
96. M. van der Voort, D. Jensen, C. Kamphuis, I. N. Athanasiadis, A. De Vries and H. Hogeveen, 2021, **104**, Invited review: Toward a common language in data-driven mastitis detection research, *J. Dairy Sci.*, 10449–10461.
97. J. Wang, Z. Xu and D. G. Kotsifaki, Plasmonic and metamaterial biosensors: a game-changer for virus detection, *Sens. Diagn.*, 2023, **2**, 600.
98. E. Seymour, F. Ekiz Kanik, S. Diken Gür, M. Bakhshpour-Yucel, A. Araz and Ü. N. Lortlar, *et al.*, Solid-phase optical sensing techniques for sensitive virus detection, *Sensors*, 2023, **23**(11), 5018.
99. C. Lauber and S. Seitz, Opportunities and Challenges of Data-Driven Virus Discovery, *Biomolecules*, 2022, **12**, 1073.
100. N. Vashistha, SARS-CoV-2 Result Interpretation based on Image Analysis of Lateral Flow Devices, 2022. Available at http://arxiv.org/abs/2205.13311.

101. J. Ye, Y. T. Yeh, Y. Xue, Z. Wang, N. Zhang and H. Liu, *et al.*, Accurate virus identification with interpretable Raman signatures by machine learning, *Proc. Natl. Acad. Sci. U. S. A.*, 2022, **119**(23), e2118836119.
102. M. M. Hassan, F. S. Sium, F. Islam and S. M. Choudhury, *Sens. Bio-Sens. Res.*, 2021, **33**, A review on plasmonic and metamaterial based biosensing platforms for virus detection.
103. G. G. Daaboul, A. Yurt, X. Zhang, G. M. Hwang, B. B. Goldberg and M. S. Ünlü, High-throughput detection and sizing of individual low-index nanoparticles and viruses for pathogen identification, *Nano Lett.*, 2010, **10**(11), 4727–4731.
104. G. S. Roberts, S. Yu, Q. Zeng, L. C. L. Chan, W. Anderson and A. H. Colby, *et al.*, Tunable pores for measuring concentrations of synthetic and biological nanoparticle dispersions, *Biosens. Bioelectron.*, 2012, **31**(1), 17–25.
105. S. Ghosh, P. Lincoln, C. Petersen and A. Valdes, Virus detection in multiplexed nanowire arrays using hidden semi-markov models, 2014.
106. Z. F. Pitafi, W. Song, Z. Tse, Y. Yang, Y. Zhao, J. Murray, *et al.*, Intelligent surface-enhanced raman scattering sensor system for virus identification, in *Proceedings - IEEE Global Communications Conference*, GLOBECOM, 2022, pp. 554–559
107. M. Punia, A. Choudhary, S. Agarwal and V. Shukla, Ethical Considerations and legal frameworks for biometric surveillance systems: the intersection of AI, Soft biometrics, and human surveillance, *Lecture Notes Networks Syst.*, 2024, **918**, 659–674.
108. EFPIA, Considerations for Validation and Implementation of Next Generation Sequencing for Adventitious Virus Detection for Biological Medicinal Products, 2024. Available at www.efpia.eu.
109. S. Wu, W. Yu, X. Fu, X. Yu, Z. Ye and M. Zhang, *et al.*, Advances in virus detection techniques based on recombinant polymerase amplification, *Molecules*, 2024, **29**, 4972.
110. A. Bisht, A. Mishra, H. Bisht and R. M. Tripathi, Nanomaterial based biosensors for detection of viruses including SARS-CoV-2: a review, *J. Anal. Test*, 2021, **5**(4), 327–340.
111. J. Loscalzo, *Biosensors and Privacy: Navigating Ethical Challenges in Health Monitoring*, 2024.
112. K. Denecke, An ethical assessment model for digital disease detection technologies, *Life Sci. Soc. Policy*, 2017, **13**(1), 1–11.
113. K. Montgomery, J. Chester and K. Kopp, Ensuring fairness, preventing discrimination, and promoting equity in an emerging internet-of-things environment, *J. Inf. Policy*, 2018, **8**, 34–77.
114. R. Bouderhem, Ethical and regulatory challenges for AI biosensors, *Healthcare*, 2024, **37**.
115. A. Borda, A. Molnar, C. Neesham and P. Kostkova, Ethical issues in ai-enabled disease surveillance: perspectives from global health, *Appl. Sci.*, 2022, **12**(8), 3890.

116. M. M. Mello and C. J. Wang, Ethics and governance for digital disease surveillance, *Science*, 2020, **368**(6494), 951–954.
117. W. Sukjee, A. Thitithanyanont, S. Manopwisedjaroen, S. Seetaha, C. Thepparit and C. Sangma, Virus MIP-composites for SARS-CoV-2 detection in the aquatic environment, *Mater. Lett.*, 2022, **315**, 131973.

Section VI: Case Studies

CHAPTER 17

Case Studies in Virus Detection Using Sensor Technologies

MONIREH BAKHSHPOUR-YÜCEL[a*], SINEM DIKEN-GÜR[b,c], NESE LORTLAR ÜNLÜ[d,e] AND BILGEN OSMAN[a]

[a]Department of Chemistry, Faculty of Arts and Science, Bursa Uludag University, Bursa, Turkey; [b]Department of Biology, Hacettepe University, Ankara, Turkey; [c]Department of Electrical Engineering, Boston University, Boston, MA, USA; [d]Department of Biomedical Engineering, Boston University, Boston, MA, USA; [e]Faculty of Medicine, Histology and Embryology, Atlas University, Istanbul, Turkey
*E-mail: myucel@uludag.edu.tr

17.1 Introduction

Influenza viruses are known for their rapid mutation rates, a phenomenon called antigenic drift, which allows them to adapt and evolve. These changes can alter their antigenic properties as the viruses accumulate small genetic mutations, particularly in their surface proteins. This can make previously acquired immunity less effective, potentially leading to new disease outbreaks.[1,2] This constant mutation is one of the key reasons why influenza remains a persistent global health threat, as it can quickly adapt to immune defences and spread through susceptible populations. Currently used techniques for the diagnosis and laboratory confirmation of viruses, such as conventional PCR, loop-mediated isothermal amplification (LAMP),

Detection Science Series No. 32
Advances in Virus Detection with Sensors
Edited by Adil Denizli
© The Royal Society of Chemistry 2026
Published by the Royal Society of Chemistry, www.rsc.org

enzyme-linked immunosorbent assay (ELISA), and whole-virus isolation and cell-culture-based methods are insufficient to detect the pathogen in the early stage because of the requirements for trained personnel, non-portable and specialized equipment, maintenance costs, and results only becoming available after a period ranging from days to weeks. As a result, there is a need to develop rapid, ultra-sensitive, robust, user-friendly, and point-of-care diagnostic methods capable of detecting viruses to achieve infection control. In the last three decades, as a result of technological advances, new biosensors have been developed by incorporating photonics, electrochemical detection, microfluidics, magnetic and acoustic technologies, and mass spectroscopy that can detect viruses with a sensitivity beyond the boundaries of traditional techniques.[3–5] Interest is growing in developing rapid and reliable influenza sensors requiring minimal sample handling and laboratory expertise. One approach is to create single-step direct sensing methods that avoid separation, incubation, or signal-reporting agents. Techniques like surface plasmon resonance (SPR), quartz crystal microbalance (QCM), surface-enhanced Raman scattering (SERS), and fluorescence resonance energy transfer (FRET) have shown promise in viral detection, offering detection limits in the sub-nanogram range. Another promising method is colorimetric sensors using functional polymers, which provide simple, litmus-test-like detection.[6,7]

17.2 Successful Applications of Sensor Technologies in Real-world Scenarios

Animal disease outbreaks are an increasing challenge for the food industry in affected countries due to economic consequences, such as food trade disruption and increased food prices in the domestic market. Among the causative pathogens of animal diseases, viruses emerge as a result of the lack of hygiene, and biosecurity protocols are the most difficult to manage due to the lack of effective vaccines, limited treatments, and their high transmission potential. As a result, implementing strict quarantine procedures or culling infected herds, which are the main ways to control the epidemic, lead to significant economic losses. Viruses are microscopic parasites that require a host to replicate and can cause various diseases. Traditional viral detection methods like PCR and antibody assays are time-consuming and costly, highlighting the need for faster, more efficient diagnostic tools.

On the other hand, following the corona virus disease (COVID)-19 pandemic, the first pandemic of this century, the significance of infectious disease control and diagnosis has become increasingly prominent, especially for airborne pathogens within global health systems. This increased awareness highlights the urgent need for improved strategies and frameworks to address and manage infectious diseases effectively in the future. COVID-19 case reports and data demonstrate that early detection of the causative agent prevents its spread. Even if the effect of COVID-19 seems to have weakened, infectious diseases are still responsible for almost 15% of all deaths worldwide. Access to rapid, cost-effective, and easily implemented diagnostic tools is vital for preventing and controlling outbreaks. The time-consuming and laborious features of traditional diagnostic techniques have led to delays in diagnosis and initiation of treatment during the COVID-19 pandemic. Thus, biosensors, revolutionary diagnostic tools that can be enhanced by integrating technological innovations, are attracting attention for detecting pathogens and improving infection management. Thanks to their most attractive feature—the capability to be miniaturized and integrated into portable devices–they have become prominent in the development of point-of-care and real-time diagnostic tools for clinical and field use.[8] In this chapter, the role and significance of sensors in monitoring several viral outbreaks seen in animals and humans are highlighted, followed by a discussion of studies where sensors have been successfully used to detect viral agents and provide rapid diagnostic results.

17.2.1 Sensor Applications for Avian Influenza Virus (AIV) Detection

Sensor technologies have significantly transformed the monitoring of bird populations and tracking of potential avian influenza outbreaks. Early detection is vital for preventing the spread of bird flu, given its rapid transmission among avian species and the risk it poses to human health. Biosensors and environmental sensors are deployed in poultry farms, wetlands, and migratory bird habitats to identify early indicators of infection, such as alterations in temperature, behaviour, and vital signs of birds.[9] Sensors are a fast and effective technology for early detection of avian influenza in birds and diagnosis in humans, which have played a crucial role in controlling avian influenza (AIV) outbreaks. AIV, commonly known as bird flu, is an infectious disease caused by the influenza A virus, primarily affecting birds. However, it can also occasionally infect mammals, including humans. Wild aquatic birds are the primary hosts of the virus, which remains endemic (enzootic) in various bird populations. Humans contract avian flu through prolonged, close contact with infected birds. Human cases are rare and generally involve symptoms ranging from mild to severe, including fever, diarrhoea, and respiratory issues. The virus is spread primarily through the excretions of infected birds, such as saliva, mucus, and faeces, and can infect both poultry flocks and wild bird populations. The highly pathogenic strain H5N1 has caused significant mortality in domestic poultry, leading to the culling of hundreds of millions of birds globally to control outbreaks.[10]

Influenza A is an RNA virus characterized by a segmented genome that encodes 11 viral proteins. Two key surface glycoproteins, hemagglutinin (H) and neuraminidase (N) are critical for the virus's ability to bind to host cells and release new viral particles. These proteins are the primary targets of immune responses, and their variations result in different virus subtypes.

The influenza A subtypes are named according to the hemagglutinin and neuraminidase proteins they possess, such as H5N1, which refers to a virus with hemagglutinin type 5 and neuraminidase type 1.[11–13] Various subtypes, with different combinations of H and N proteins, have been identified in wild birds, the natural reservoir of the virus, and these variations can influence the virus's pathogenicity and ability to spread.

The ability of influenza A viruses to undergo genetic mutations and antigenic shifts, particularly in the hemagglutinin and neuraminidase proteins, increases the risk of new strains emerging that could pose a more significant threat to both animal and human health. This situation has raised concerns about the potential for a new pandemic strain to emerge, especially in cases where viruses like H5N1 mutate to become more easily transmissible among humans.

In the early stages of AIV detection, traditional methods, such as PCR, viral culture, and serological assays, were the primary tools for identifying and diagnosing the virus. In the early 2000s, researchers began exploring the potential of sensors to provide faster, more efficient, on-site diagnostic tools for AIV detection. One of the first sensor-based approaches was the development of immunosensors that used an interferometric biosensor immunoassay for direct and label-free detection of AIV through whole virus capture on a planar optical waveguide. The assay relies on refractive index changes when virus particles bind to antigen-specific (hemagglutinin) antibodies on the waveguide surface. Using monoclonal and polyclonal antibodies, the sensor detected two virus subtypes (H7 and H8) in buffer solution. The real-time antigen–antibody interaction response was concentration-dependent, with detection limits as low as 0.0005 hemagglutination units per ml. The system used a Mach–Zehnder interferometer, where light propagates through a waveguide core with a high refractive index, generating an evanescent field sensitive to refractive index changes. The binding of the target antigen alters the refractive index, measured through interference patterns. This approach offered high sensitivity, with refractive index changes measurable to 10^{-6} levels.[14]

As faster detection technologies were needed, advancements in microfluidic-based biosensors were made. An integrated microfluidic system was developed for efficiently screening aptamers specific to the influenza A/H1N1 virus. The system automated the entire systematic evolution of ligands by exponential enrichment (SELEX) process, including aptamer incubation, extraction, and nucleic acid amplification. The selected aptamers were tested in a magnetic-bead assay, successfully detecting influenza A/H1N1 in biological samples like throat swabs, demonstrating high specificity and sensitivity. This study was the first to demonstrate microfluidic SELEX for screening influenza-specific aptamers, offering a promising tool for rapid diagnostic assays and potential expansion to other pathogens.[15]

Over time, these optical systems became more advanced, incorporating microfluidic devices that simultaneously detected multiple pathogens from small biological samples. Integrating nano-materials, such as gold nanoparticles or carbon nanotubes, further enhanced the sensitivity and specificity of these sensors.

Today, sensor systems for AIV detection have evolved to incorporate various advanced technologies. More recently, the integration of smartphone technology with sensor-based systems has emerged as a game-changer in the field of avian influenza detection. The ability to use smartphones as a platform for analysing sensor data not only makes the technology more accessible but also improves the speed and efficiency of AIV surveillance.[16,17]

Kazemi *et al.* proposed an innovative terahertz (THz) metamaterial biosensor designed for the highly sensitive detection of AIVs, including subtypes like H1N1, H5N2, and H9N2. The sensor employed spoof surface plasmon polaritons (SSPP), utilizing a grating structure to support spoof plasmon modes with high-quality factors. This enabled detection at a frequency of 1.93 THz with an impressive quality factor (Q-factor) of 690. This design achieved strong field confinement and resonance, making it highly sensitive to minute variations in the properties of the analyte, such as viral proteins.[18]

The sensor was tested with several AIV subtypes, including H1N1, H5N2, and H9N2. The results demonstrated that the sensor could detect virus proteins, even at low concentrations. The sensor's average sensitivity (Savg) was up to 8% in the refractive index range of 1.1 to 1.5, indicating its ability to accurately detect small amounts of viral material. This level of sensitivity is crucial for early detection and monitoring of viral infections.

The results of this study demonstrated the potential of terahertz metamaterials to overcome the diffraction limits of traditional sensing methods, providing rapid diagnosis of viral infections and offering valuable applications in healthcare and diagnostics. Further development of such sensors could play a pivotal role in early-stage virus detection, thus improving the speed and accuracy of disease diagnosis and control.

A highly sensitive and multiplexed detection platform for AIV biomarkers was presented by Tian *et al.*, which used a microfluidic paper-based analytical device (μPAD) combined with chemiluminescence (CL) technology. Using a time-resolved CL assay, this platform was designed to rapidly and accurately detect three AIV strains: H1N1, H7N9, and H5N1. The platform incorporated DNAzyme/Luminol (Lum)/polyethyleneimine (PEI)/$CaCO_3$-based functionalized microparticles, where the mesoporous $CaCO_3$ microspheres played a dual role as both carriers for luminol (the chemiluminescent reagent) and as amplification platforms *via* catalytic hairpin assembly (CHA) reactions and DNAzyme catalysis.[19]

The μPAD-based chemiluminescence sensor was built to provide multiplex detection, allowing for the simultaneous detection of three AIV biomarkers in a single test. By leveraging the chemiluminescence reaction with DNAzyme-based amplification, the platform generated three distinct time-resolved chemiluminescence signals, which were sequentially detected, effectively preventing interference between the signals. The system was

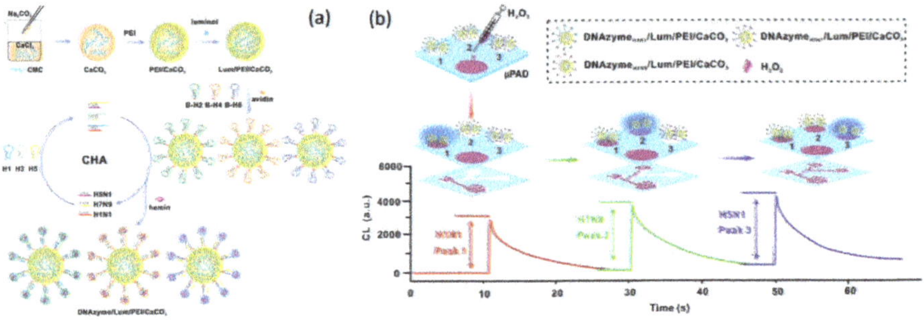

Figure 17.1 Graphical representation of (a) the synthesis process for DNAzyme/Lum/PEI/$CaCO_3$, (b) the multiplex chemiluminescence assay for simultaneous detection of three DNA targets. Reproduced from ref. 19 with permission from Elsevier, Copyright 2024.

designed to detect minute amounts of virus biomarkers, with a 0.32 pM limit of detection (LOD) value for H1N1, 0.34 pM for H7N9, and 0.29 pM for H5N1, making it highly sensitive. Figure 17.1 shows the DNAzyme/Lum/PEI/$CaCO_3$ synthesis process and the multiplex chemiluminescence assay for simultaneous detection of three DNA targets. Figure 17.2 shows the characterization of microparticles.

The μPAD-based CL platform offered impressive time-resolved multiplex detection within 70 seconds, with a high temporal resolution of 20 seconds. The system's sensitivity and selectivity were validated through rigorous testing, with detection limits in the picomolar range and a broad detection range from 1.0 pM to 10 nM. The CL signals were highly specific, demonstrating excellent immunity to interference from nonspecific DNA, and the assay displayed accurate mismatch discrimination.

The platform offered low cost, high sensitivity, excellent selectivity, and rapid detection capabilities. Using a smartphone to record and visualize the chemiluminescent signals enabled straightforward interpretation of results, even at low concentrations (as low as 10 pM). The ability to perform rapid multiplex detection significantly improved diagnostic accuracy and efficiency compared with traditional single-target detection methods.

Another study presented a rapid method for fabricating a microfluidic chip integrated with electrodes for quantitatively detecting H5N1-DNA sequences, a highly virulent and variable influenza subtype with epidemic and pandemic potential.[19] The detection process utilized a catalytic redox-recycling signal through a Fe_3O_4@TMU-8 nanocomposite, which enhanced the electrochemical signal by facilitating the turnover of $[Ru(NH_3)_6]^{3+}$, a molecule that binds to the phosphate backbone of H5N1 DNA. The aptasensor's redox signal, influenced by the DNA hybridization, served as the electrochemical response, with a detection range from 1 fM to 1 nM and a 0.16 fM LOD value (Figure 17.3). The microfluidic device was fabricated using

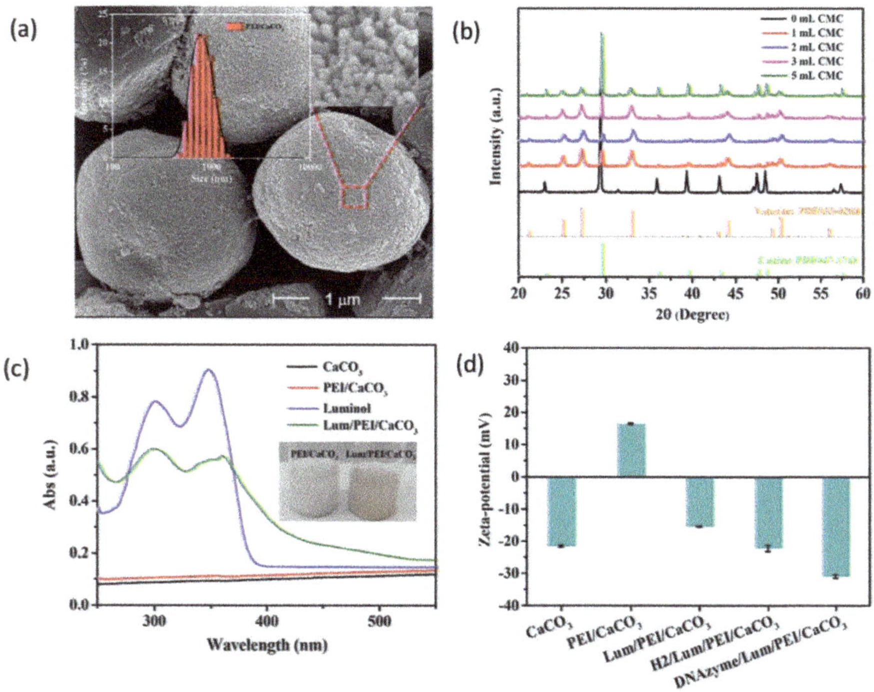

Figure 17.2 (a) Scanning electron microscope (SEM) image of Lum/PEI/$CaCO_3$; (b) X-ray diffraction (XRD) patterns of $CaCO_3$ synthesized with varying amounts of carboxymethyl cellulose (CMC) added; (c) UV–visible absorption (Abs) spectra of $CaCO_3$, PEI/$CaCO_3$, luminol, and Lum/PEI/$CaCO_3$ (D) Zeta potential measurements of $CaCO_3$, PEI/$CaCO_3$, Lum/PEI/$CaCO_3$, H2/Lum/PEI/$CaCO_3$, and DNAzyme/Lum/PEI/$CaCO_3$. Reproduced from ref. 19 with permission from Elsevier, Copyright 2024.

poly(methyl methacrylate) (PMMA) and a laser system to create microfluidic patterns, ensuring ease of manufacturing. The sensor showed excellent long-term stability, reproducibility, and applicability in real-world samples, making it a promising tool for sensitive and quantitative detection of H5N1-DNA in diverse settings.

In another study, Gong *et al.* designed a monoclonal molecular imprinted polymer sensor for the specific recognition of H5N1 AIV, showing the challenge of distinguishing between similar subtypes of influenza A viruses (IAV),[21] such as H1N1, H7N9, and H9N2. The process began by immobilizing H5N1 onto magnetic microspheres to create H5N1-magnetic nanoparticles (MagNPs), followed by the preparation of high-affinity H5N1-molecular-imprinted particles (MIPs) using a solid-phase imprinting technique. The sensor's detection capability was based on competitive substitution, where varying concentrations of H5N1 were added to the sensor, and the resulting changes in fluorescence intensity were measured in the supernatant. The

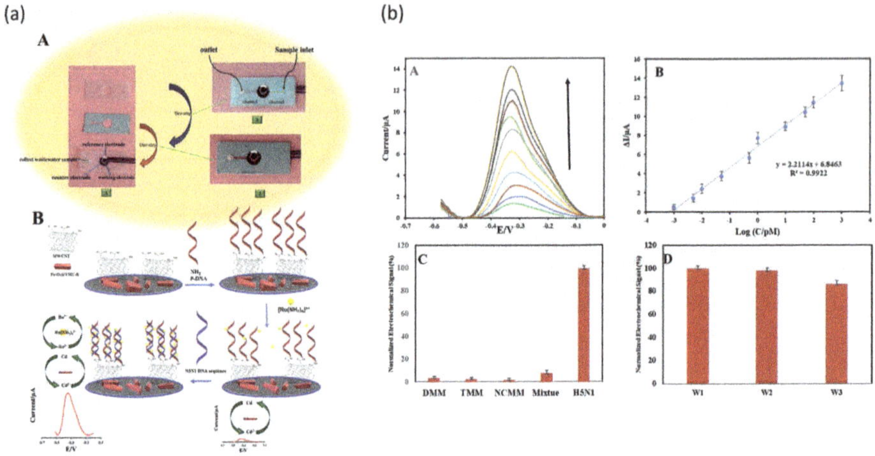

Figure 17.3 (a) (I). Diagram of the μPAD sensor fabrication process. (II). Illustration depicting the sensing procedure. (b) (I) Differential pulse voltammograms (DPVs) of the DNA-sensor-μPAD across a concentration range from 1 fM to 1.0 nM. (II). Plot of change in current (ΔI) in μA *versus* the target DNA concentration (C) in pM. (III) Normalized DPV responses of DNA-sensor-μPAD to double-base mismatched (DMM), triple-base mismatched (TMM), and non-complementary DNA (NCMM), and H5N1 DNA sequences. (IV) Normalized DPV responses of DNA-sensor-μPAD over three weeks. Reproduced from ref. 20 with permission from Elsevier, Copyright 2025.

sensor demonstrated excellent selectivity, accurately identifying H5N1 even in the presence of closely related IAV subtypes. It was highly sensitive, with a 0.58 fM LOD value, and offered a selectivity factor similar to those of other small MIP-based sensors. This monoclonal sensor provided a promising approach for large-scale screening and precise treatment of H5N1 infections, offering a valuable tool for viral detection and public health applications. Figure 17.4 shows the process of developing a monoclonal molecularly imprinted sensor for the highly selective recognition of H5N1, that can distinguish it from similar IAV subtypes. A silica layer was first coated onto magnetic nanoparticles using tetraethyl orthosilicate (TEOS), followed by surface modification with amino groups *via* 3-aminopropyltriethoxysilane (APTES) and glutaraldehyde to introduce aldehyde groups. These aldehyde groups attached the template virus particles to the magnetic nanoparticles. Using solid-phase imprinting, the particles with fixed template viruses underwent polymerization with functional monomers [*N*-isopropyl acrylamide (NIPAm), acrylic acid (AA), 3-aminopropyl methacrylate (APMA), and *N*-tertbutyl acrylate (TBA)] and crosslinking agents (*N*,*N*-methylene bisacrylamide (BIS), catalysed by ammonium persulphate (APS) and *N*,*N*,*N*′,*N*′-tetramethyl ethylenediamine (TEMED)] to form H5N1-specific molecular-imprinted

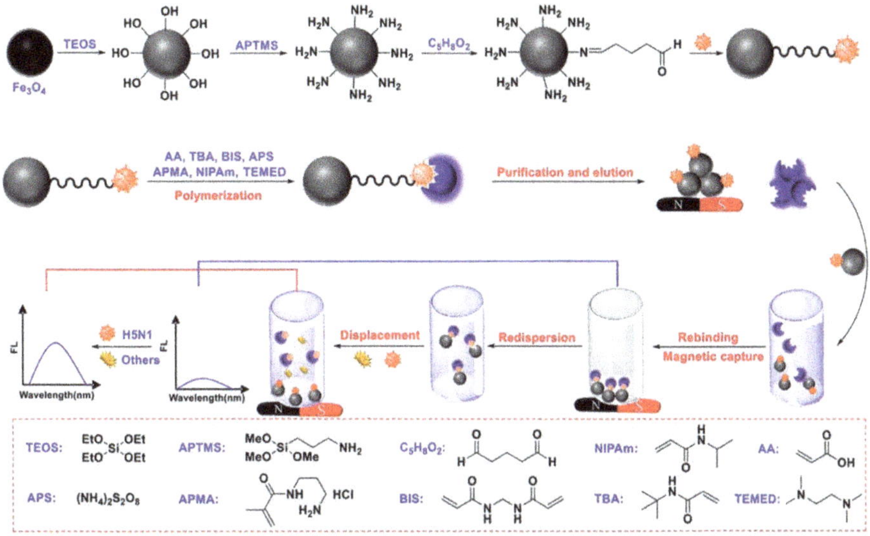

Figure 17.4 Schematic illustration of the development process for a monoclonal molecularly imprinted sensor. Reproduced from ref. 21 with permission from Elsevier, Copyright 2024.

nano gels (H5N1-MIPs). After washing and removal of unreacted materials, the imprinted nanogels were eluted at 60 °C. The sensor demonstrated high sensitivity with a detection limit of 0.58 fM and low cost, making it suitable for large-scale screening and targeted treatment.

In another study, the authors presented an intensity-based SPR sensor designed to rapidly and accurately detect the H5N1 AIV, which continues to cause highly fatal human infections. The sensor utilized near-infrared (NIR) wavelength excitation in combination with a precisely engineered NIR-optimized organic photodetector (NIR-OPD). The active layer of the OPD consisted of a poly[[2,6′-4,8-di(5-ethylhexylthienyl)benzo[1,2-*b*;3,3-*b*]dithiophene][3-fluoro-2[(2-ethylhexyl)carbonyl]thieno[3,4-*b*]thiophenediyl]] (PTB7-Th) and 2,2′-((2*Z*,2′*Z*)-(((4,4-bis(2-ethylhexyl)-4*H*-cyclopenta[2,1-*b*:3,4-*b*′]dithiophene-2,6-diyl)bis(4-(2-ethylhexyloxy)thiophene-5,2-diyl))bis(methanylylidene))bis(5,6-difluoro-3-oxo-2,3-dihydro-1*H*-indene-2,1-diylidene))dimalononitrile (COTIC-4 F) blend, which was optimized for excitation at 980 nm.

This SPR sensor was employed to detect H5N1 AIV using a custom polyclonal antibody targeting the viral hemagglutinin (HA) envelope protein. The sensor detected H5N1 in culture medium samples within 12 minutes, with a detection limit of approximately 4.3×10^4 copies per ml in phosphate buffered saline (PBS)-diluted samples. Compared with a commercial H5-antigen lateral flow test kit, the biosensor demonstrated 37 times higher sensitivity. The developed SPR biosensor proved to be an efficient and cost-effective tool for on-site detection of H5N1 AIV, offering significant advancements in diagnostic technologies for influenza outbreaks.[22]

Wang *et al.* developed a digital microfluidics (DMF) platform integrated with SERS for the rapid, automated, and sensitive detection of disease biomarkers, including H5N1. DMF provided significant advantages for immunoassays, particularly those with multiple steps, due to its low reagent consumption and high automation. SERS was used to enhance the sensitivity of the detection method.[23]

The proposed method utilized SERS tags labelled with the Raman reporter 4-mercaptobenzoic acid (4-MBA), which were synthesized with a core–shell nanostructure, offering strong signals, good uniformity, and high stability. A sandwich immunoassay design was used, where magnetic beads coated with antibodies captured antigens from the sample to form an immunocomplex. The complex was then labelled with SERS-tagged detection antibodies, allowing sensitive antigen detection through the enhanced SERS signal.

By automating the immunoassay procedure, the DMF–SERS system reduced the risk of exposure to hazardous samples and simplified the process. The system demonstrated excellent sensitivity with a 74 pg ml^{-1} LOD for H5N1 in buffer and human serum samples. The detection process was completed in less than an hour, using approximately 30 μl of reagents, substantially lower than traditional ELISA methods. The integration of DMF with SERS holds significant potential for advancing diagnostic technologies, especially in high-throughput or point-of-care applications.

17.2.2 Sensor Applications for the Detection of Swine Fever Viruses

For the pig industry, diseases caused by African swine fever viruses (ASFV) and classical swine fever viruses (CSFV) result in significant financial losses due to high morbidity and mortality rates. Although they are different types of viruses, ASFV, a large, icosahedral, and enveloped DNA virus from the Asfarviridae family, and CSFV, a small RNA virus, icosahedral, enveloped, and belonging to the Flaviridae family, both commonly cause similar diseases characterized by haemorrhagic fever in pigs and piglets and become endemic in affected areas due to their fast transmission ratios. Furthermore, the lack of commercial antivirals or vaccines against these viruses exacerbates disease control. On the basis of their significant impact on the swine industry due to their transmission dynamics, high virulence, and economic consequences, they are classified as the most significant viral diseases for pigs in Europe by the World Organization for Animal Health [formerly the Office International des Epizooties (OIE)]. To overcome the challenges posed by the rapid spread of the disease, establishing effective disease control strategies that rely on early and accurate diagnosis of the virus is important globally.

In 1998, a QCM system was integrated into a flow injection analysis (FIA) system based on immune sensing technology to rapidly detect ASVF. In this study, the quartz crystal was covered with physically adsorbed viral protein 73 (73VP), a major envelope protein of ASVF, to obtain an ultra-specific sensor that relies on monitoring the resonant frequency change of the quartz crystal for the detection of serum anti-ASVF in pigs. A reusable sensor was fabricated to reduce the cost of each test. The high viscosity of the pig serum caused a significant drop in resonant frequency. This loss was approximately ten times greater than the frequency shift triggered by actual analyte binding, making it difficult to distinguish the true signal. The discrimination of the negative and positive serum samples could be achieved even with different titres of serum; thus, more sensitive results were obtained with the QCM-integrated immunosensor than with the other immunosensor techniques used in this study.[24]

Another study used similar sensor technology, although there were differences due to preparing the receptor layer by coating the gold electrode of the quartz crystal. Two different methods were employed, using monoclonal antibody 18BG3 (a peptide-specific antibody) and 73VP as receptors to detect the presence of each other in the serum sample. While antibodies were immobilized to gold electrode covalently in order to avoid the washout of receptors, the immobilization of 73VP depended on physisorption *via* thiol–gold and hydrophobic interactions. Regarding the mass-sensitive behaviour of this sensor system, the bioaccumulation of the analytes on the covered quartz crystal led to a shift in the resonant frequency. With this sensor technology, the detection limit of virus protein was determined for spiked buffer and serum as 0.31 and 1 μg ml^{-1}, respectively. Furthermore, the limit of detection values for antibodies were 0.1 and 0.2 g ml^{-1} in the spiked buffer and serum, respectively. A comparison of different techniques to obtain more specific receptor layers was conducted to achieve a highly sensitive QCM-integrated FIA system.[25]

Cho and Park used a commercial SPR system to detect anti-CSFV in infected pig serum, utilizing g55 recombinant protein as a receptor.[26] To compare the techniques in terms of LOD, 117 pig serum samples were evaluated with SPR and ELISA. The SPR proved its capacity by outperforming the ELISA by a factor of 10.

Innovations in the diagnostic field are being developed, including DNA-based receptors, molecular imprinting-based artificial receptors, and various sensor technologies, to create more robust, cost-effective, easily applied, and rapid point-of-care systems that increase sensitivity to accurately detect both viruses. In this way, diagnostic tools that can be used in laboratories may now be replaced with portable devices that enable on-site diagnosis in field cases and outbreaks. In a study by Biagetti *et al.*, a label-free biosensor based on SPR technology that used a single-strand, 25-*mer*-long DNA–locked nucleic acid (LNA) probe as a receptor for the vp72 gene region of the ASFV genome was developed to screen pig blood samples for ASF disease (Figure 17.5). Validation of the biosensor was performed by comparing the method in terms of specificity, rapidity, and sensitivity with real-time PCR, defined as a gold standard method for virus detection by the OIE. In this method, probe capture relied on a strong streptavidin–biotin coupling interaction, a biotinylated DNA–LNA probe designed to be complementary to the ASFV gene

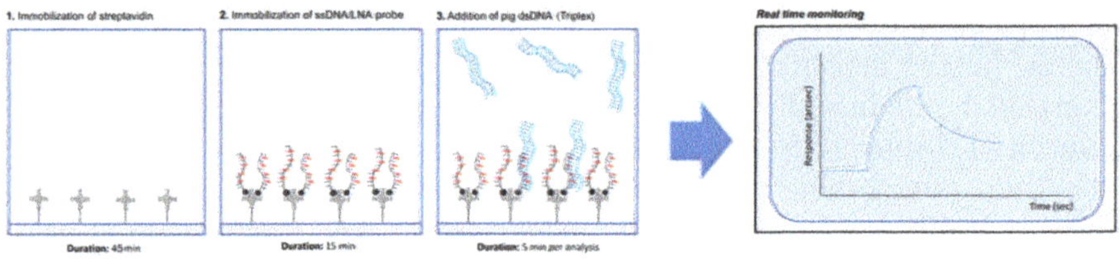

Figure 17.5 Schematic representation of the developed biosensor. Reproduced from ref. 27 with permission from Elsevier, Copyright 2018.

region was immobilized onto the surface-located streptavidin. In this proposed biosensor technology, it was possible to eliminate the pretreatment required for denaturation and amplification steps by capturing the unamplified double-stranded DNA of ASFV through the formation of a triplex structure.[27] In this study, the LOD and limit of quantification (LOQ) values were determined to be 178 and 245 copies per μl of viral DNA, respectively. Moreover, the proposed biosensor and the OIE-recommended real-time PCR method showed perfect correlation in discriminating between infected and non-infected animals by ASFV.

Recently, two promising studies based on colorimetric sensing were proposed owing to the *trans*-cleavage activity of the clustered regularly interspaced short palindromic repeats (CRISPR)–CRISPR-associated (Cas) protein system, which is activated by the recognition of single or double-stranded target DNA. In one of them, CRISPR-associated protein 12a (Cas12a), a target recognition system, was combined with an approach defined by the absorbance shift induced by the cascade reaction in the presence of the urease enzyme. The sensing was achieved by leveraging the advantages of the double enzyme, eliminating the need for amplification processes or advanced equipment. In this method, Cas12a recognized target viral DNA was conjugated with the urease enzyme, which was bound to magnetic beads for easy separation at the end of the reaction. The working mechanism of the detection platform was based on the cleavage of magnetic beads bound to urease-conjugated single-stranded DNA by CRISPR–Cas12a, which is activated in the presence of ASFV double-stranded DNA (dsDNA). Then, free urease was separated from the solution magnetically, and colorimetric change due to the presence of viral DNA could be measured to confirm the result[28] (Figure 17.6).

In another colorimetric sensor study, a combination of a paper-based fluidic device, the CRISPR–Cas14a system, and the G-Quadruplex deoxyribozyme (DNAzyme) was utilized on a wax-based hydrophobic platform to detect ASFV genes in porcine plasma, enabling rapid and cost-effective virus diagnosis.[29] This paper platform exhibited better sensitivity from the above-mentioned colorimetric method, with LOD as low as five copies of viral DNA per μl. However, this method requires additional steps, such as amplification of the ASFV gene region and the digestion of the amplicons by T7 exonuclease, to form single-stranded DNA (ssDNA) targets that trigger the *trans*-cleavage activity of the CRISPR–Cas14a system. As a result, this colorimetric biosensor strategy requires specialized equipment for the PCR step. Upon the activation of the CRISPR–Cas14a system, the G-Quadruplex DNAzyme lost its ability to catalyse the colorimetric 2,2′-azino-bis(3-ethylbenzothiazoline-6-sulfonic acid) diammonium salt (ABTS)–H_2O_2 reaction in the presence of hemin. This impairment in catalytic activity resulted from the cleavage of the DNAzyme by the Cas14a system (Figure 17.7). This innovative approach enhances the sensitivity of the detection process.

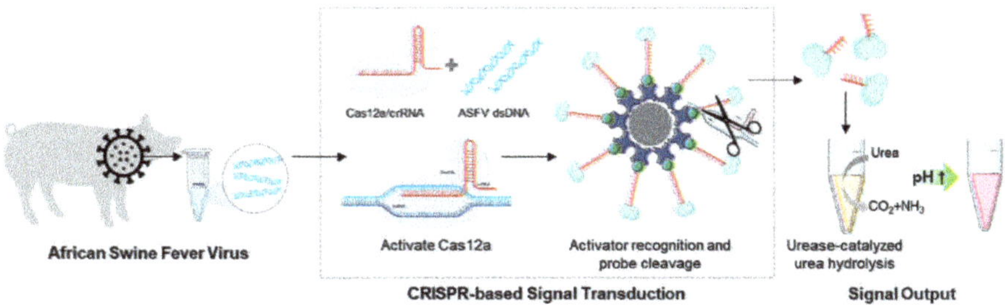

Figure 17.6 Schematic representation of the colorimetric sensor system that can be used to detect dsDNA of ASFV directly by utilizing the CRISPR–Cas12a recognition system. At the end of the reaction, the appearance of the magenta colour confirms the presence of ASFV in the infected animal. Reproduced from ref. 28 with permission from American Chemical Society, Copyright 2022.

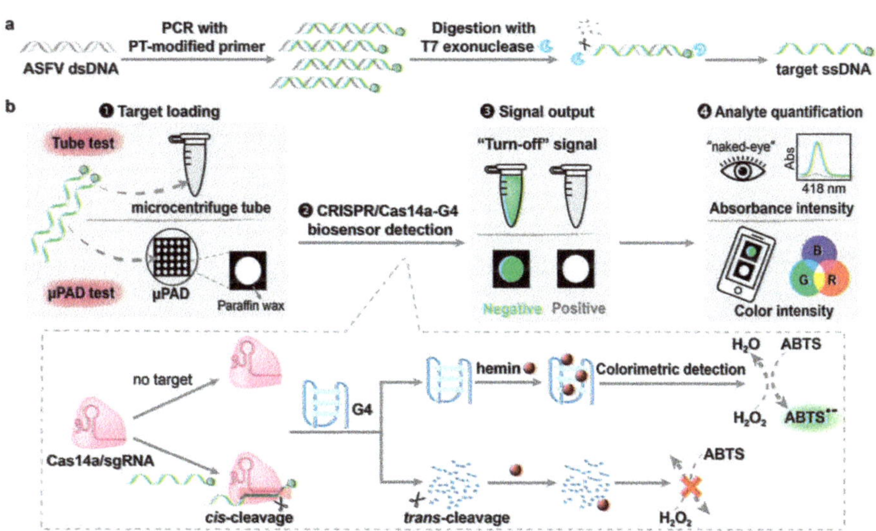

Figure 17.7 Schematic illustration of a paper-based colorimetric method that integrates the key advantages of the CRISPR–Cas system and DNAzyme. (a) The preparation of ASVF ssDNA. (b) The detection of ASVF by using the CRISPR/Cas 14a-G4 biosensor system. Reproduced from ref. 29 with permission from American Chemical Society, Copyright 2024.

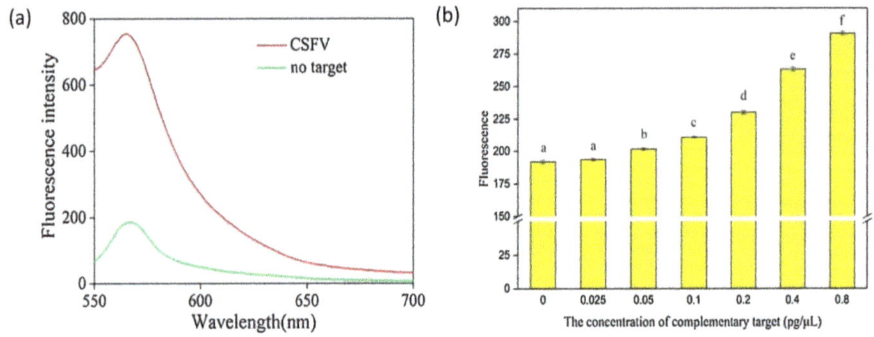

Figure 17.8 (a) Fluorescence spectra of nanoflares, demonstrating the absence of a target (green line) compared with the presence of the CSFV target (red line). (b) Fluorescence intensity of the nanoflares (1 nM) across a range of complementary target concentrations (0 to 0.8 pg μl^{-1}). The annotations a through f indicate statistically significant differences ($P < 0.05$). Reproduced from ref. 30 with permission from Elsevier, Copyright 2018.

Recently, several countries within the European Union have demonstrated success in managing the CSF epidemic. However, the ongoing threat of CSFV infection remains a considerable concern for the pig-breeding industries in developing nations, particularly China. In light of the challenges posed by existing diagnostic techniques for CSVF, including PCR and immunodiagnostic tools that rely on antibodies produced through costly processes, researchers have focused on developing advanced detection methods. These innovative approaches aim to enable direct and sensitive identification of virions, thereby enhancing diagnostic capabilities. Ning and colleagues developed a nanocomposite material featuring gold nanoparticles conjugated with DNA sequences complementary to the viral RNA of CSFV. This method enables the detection of CSFV at concentrations as low as 50 pg μl^{-1} through generating a fluorescent signal (Figure 17.8).[30]

In a different study, the molecular imprinting method was used to synthesize template-directed cavities capable of recognizing CSFV, leading to the development of the MIP-based QCM sensor. Surface imprinting of the whole virus on a copolymer, including acrylamide (AAM), methacrylic acid (MAA), methyl methacrylate (MMA), and *N*-vinylpyrrolidone (VP), was performed on a QCM electrode to create binding cavities for the target. The main benefit of this method is the use of bulk chemicals to produce artificial receptors, which allows for the cost-effective manufacturing of the biosensor system. Integrating MIP and QCM technologies significantly enhanced sensitivity, yielding a LOD of 1.7 $\mu g\ ml^{-1}$ and LOQ of 5.1 $\mu g\ ml^{-1}$.[31]

17.2.3 Sensor Applications for the Detection of Influenza Viruses

Among infectious agents, influenza, belonging to the Orthomyxoviridae family, is a formidable respiratory virus that triggers annual epidemics and has been at the heart of four significant pandemics since 1918. Its impact on public health is profound, making it a crucial concern for global health agencies and communities alike. Isolates of this virus are categorized into three groups: A, B, and C, on the basis of differences in the nucleoprotein and matrix proteins. Type A influenza is not only the most dangerous but also the most contagious strain of the virus. The H1N1 subtype, generally known as swine flu, played a pivotal role in the recent pandemic, highlighting the potential severity of this threat. A range of innovative biosensing platforms has been meticulously designed to meet the urgent need for early, rapid, precise, and on-site detection of the influenza virus.[32,33]

A study by Yang *et al.* focused on developing an electrochemical biosensor to detect type A influenza (H1N1) virus using duplex-specific nuclease (DSN) and rolling circle amplification (RCA) technologies. A ssDNA sequence as a probe was designed to include a complementary sequence for the target H1N1-specific RNA sequence, and a primer sequence for initiating RCA. Following the cyclical DSN cleavage of the probe triggered with a specific sequence of target RNA, the released primer sequence initiated RCA to create detectable ssDNA-labelled multiple biotins. Then, ssDNA-labelled multiple biotins were captured onto the electrode surface *via* surface-located streptavidin. Additionally, a free streptavidin–horseradish peroxidase (SA–HRP) complex was used to label immobilized ssDNA with multiple biotins. As the concentration of target viral RNA increased, greater recruitment of the signal probe SA–HRP occurred, leading to the generation of a stronger electrical signal, which was formed *via* the oxidation of 3,3′,5,5′-tetramethylbenzidine (TMB) to quinondiimine catalysed by HRP. As a result, the H1N1 influenza virus was detected by measuring changes in the signal on the electrode surface with a LOD of 0.44 pM.[34]

The influenza virus, characterized by its rapid mutation rates and significant adaptability across various hosts, has led to both historically significant pandemics and seasonal infections. Ongoing monitoring of antigenic evolution in humans is crucial for preventing the threat posed by influenza viruses originating from animals at early stages. Understanding the mechanism of HA receptor-binding specificity against host receptors is crucial for comprehending how animal-sourced influenza viruses adapt to humans. Since the specificity of HA receptors is the primary factor determining the infection, transmission, and adaptation processes of viruses, researchers have focused on the interspecific differences in bonds between the sialic acid groups of complex glycans that constitute the host receptor.

While human-adapted influenza binds to the α2,6 linked sialic acid moieties present on the epithelial cells of the human upper respiratory tract *via* HA receptors, animal-specific influenza viruses target the α2,3 linked sialic acid moieties of the epithelial cells in the intestine or respiratory system. Thus, biosensors that can rapidly identify the specificity of the HA receptors, particularly in distinguishing between the influenza HA proteins that bind to α2,6-and α2,3-linked sialic acids, are of the essence for evaluating the adaptation potential of newly emerged animal influenza viruses.

Given this information, Zhang *et al.* focused on developing a rapid sensor technology to discriminate viral isolates on the basis of variations in HA proteins, enabling early prevention of animal-sourced outbreaks (Figure 17.9). Synthetic glycans, or receptor-binding analogues, were used to obtain microarrays prepared on the arrayed imaging reflectometry (AIR) sensing platform. This microarray-based sensing platform revealed the potential for differentiating between influenza viruses that can infect humans and birds and for understanding their potential interrelations.[35]

17.3 Lessons Learned from Past Implementations

Viruses significantly affect public health and economic development, with thousands of deaths worldwide and growing concerns about their potential use as biological weapons. Current diagnostic methods, though effective, have limitations, like long processing times and complexity, often requiring confirmatory tests and centralized labs, which delay treatment. Sensors,

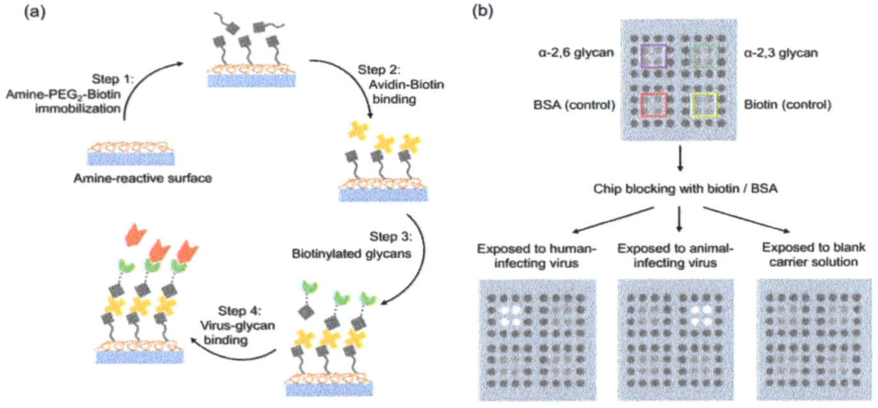

Figure 17.9 (a) Schematic illustration for the preparation of a microarray chip and virus-receptor analogues binding detection. Firstly, the amine-reactive Si/SiO_2 chip surface is coated with amine–polyethylene glycol $(PEG)_2$–biotin (Step 1), and then avidin is immobilized *via* the biotin–avidin interaction (Step 2). Then, biotinylated synthetic glycans are spotted (Step 3), and incubated with the virus (Step 4). (b) Designing a microarray with glycan receptors for the discrimination of human (α2,6-sialic acid binding) and avian (α2,3-sialic acid binding) influenza viruses. Reproduced from ref. 35 with permission from American Chemical Society, Copyright 2021.

which can provide rapid, high-performance detection by monitoring real-time molecular interactions, offer a promising alternative. Advances in smartphone integration and microfluidic devices are improving accessibility and efficiency in diagnosing viral infections. Over the years, sensor technologies for virus detection have witnessed significant advancements, provided valuable insights, and contributed to more effective surveillance systems. Despite the successes, earlier implementations have highlighted several key challenges that must be addressed to improve the design and deployment of virus detection systems, especially for highly contagious pathogens like influenza, swine fever, and avian influenza. These lessons emphasize the need for enhanced accuracy, speed, scalability, and adaptability in future virus detection strategies.

One of the most critical lessons learned is the importance of early detection in controlling the spread of viruses. A significant challenge encountered with earlier sensor technologies was the lack of specificity. Virus subtypes, particularly those within the IAV family, often share similar structural characteristics, making it difficult to differentiate between them using traditional sensor technologies. This limitation has led to innovations in molecular imprinting techniques and template-assisted strategies, such as the use of magnetic nanoparticles and MIPs. These approaches enhance the specificity and selectivity of sensors, particularly in recognizing target viruses like H5N1, thereby improving detection capabilities and reducing the risk of false positives or negatives.[19,36] Another challenge seen with past technologies is stability and robustness. Many biosensors utilize fragile biological components, which are susceptible to environmental variations (temperature, pH), affecting their performance and shelf-life. Using artificial recognition elements instead of fragile biological components leads to the development of more robust sensors that can withstand real-world conditions. Developing robust sensors that can withstand real-world conditions is vital. This may involve using more stable recognition elements, incorporating protective coatings, preparing artificial cavities with MIP, or designing self-referencing systems to compensate for environmental changes.

Another critical lesson from past implementations is the need for highly sensitive detection techniques to identify viral markers at very low concentrations, particularly in complex sample matrices. Early sensor systems struggled to detect viruses accurately when their abundance was low, frequently resulting in false negatives and delayed detection. However, recent advancements in sensor technologies, such as redox-recycling amplification and fluorescent signalling, have significantly enhanced the sensitivity of detection systems.[37,38]

These innovations are particularly valuable in the context of early diagnosis and containment of AIV outbreaks. By detecting viral markers at extremely low concentrations, nanoparticle-based and redox-recycling amplification sensors allow identification of infections in their earliest stages, when interventions are most effective. This capability is vital in preventing the spread of viruses, especially when dealing with rapidly evolving and highly contagious strains.

Current sensor technologies often involve expensive materials, sophisticated fabrication techniques, and complex readout instrumentation, which limits their accessibility, especially in resource-limited settings. However, cost-effectiveness is the key factor for widespread adaptation of the sensors. Utilizing low-cost materials and scalable manufacturing processes is critical to make these sensors affordable and accessible for point-of-care diagnostics and mass screening. Research on paper-based microfluidics, screen-printed electrodes, and other low-cost fabrication techniques addresses this challenge.[39,40]

The evolution of virus detection systems has been shaped by the lessons learned from past implementations. Early detection, specificity, sensitivity, scalability, and adaptability are all critical factors that must be prioritized in future sensor designs. By addressing these challenges, researchers and public health officials can develop more effective and efficient tools for detecting and controlling viral outbreaks, ultimately improving global health security.

17.4 Impact on Public Health and Preventive Measures

17.4.1 Public Health Impacts of Influenza, Avian Flu, Swine Flu, and COVID-19

The public health impacts of influenza, avian flu, swine flu, and COVID-19 have been profound, underscoring the vulnerabilities of global health systems to respiratory viral pandemics. Influenza, which affects millions annually, is a significant cause of morbidity and mortality, particularly among vulnerable populations. Seasonal flu, although manageable through vaccination, still leads to substantial numbers of hospitalizations and deaths each year, while pandemic strains, such as the 1918 Spanish flu and the 2009 H1N1, can cause widespread devastation due to genetic mutations that lead to novel, highly transmissible viruses. Similarly, avian influenza (H5N1, H7N9) and swine flu (H1N1) highlight the risks of zoonotic transmission, where viruses originating in animals can spill over into human populations, sometimes resulting in severe disease and high mortality rates.

COVID-19, caused by the novel severe acute respiratory syndrome coronavirus 2 (SARS-CoV-2), further demonstrated the rapid spread and overwhelming impact a respiratory virus can have on global health systems, with devastating consequences for both public health and the economy. The pandemic led to millions of deaths, stressed healthcare infrastructures, and disrupted economies worldwide. These events have emphasized the critical need for effective surveillance, early detection, rapid response measures, and vaccine development to mitigate the impact of future outbreaks. The lessons learned from these pandemics will guide global health preparedness in the face of emerging viral threats.

17.4.1.1 Influenza as a Major Global Health Concern

Influenza is a highly contagious viral infection that remains a significant global health challenge. Each year, it affects millions of people, leading to hospitalizations, severe complications, and fatalities, particularly among vulnerable populations, such as the elderly, young children, and individuals with preexisting conditions. While seasonal influenza is generally manageable, periodic pandemics have demonstrated the virus's potential for widespread devastation. In addition to its direct health burden, influenza imposes significant economic costs, straining healthcare systems and disrupting economies worldwide.[41,42] Influenza occurs in seasonal epidemics and/or pandemic forms. While influenza viruses cause both, their behaviour, severity, and impact differ.

Seasonal influenza occurs annually, primarily in colder months, and is caused by influenza A and B viruses that undergo antigenic drift, small genetic changes that enable the virus to evade immune defences while remaining partially recognizable. Due to prior immunity, vaccination programs, and antiviral treatments, seasonal flu outbreaks are generally predictable and manageable. Nevertheless, they still result in an estimated three to five million severe cases and between 290 000 and 650 000 deaths worldwide each year.[43,44]

Unlike seasonal flu, pandemic influenza arises when a novel influenza A strain emerges through antigenic shift, resulting in significant genetic changes. With little to no prior immunity, the global population is highly vulnerable to rapid and widespread transmission. Major influenza pandemics in history include the 1918 Spanish flu (H1N1), which infected one-third of the world's population and caused an estimated 50 million deaths;[45] the 1957 Asian flu (H2N2), which originated in East Asia and resulted in over 1 million deaths worldwide;[46] the 1968 Hong Kong flu (H3N2), which led to approximately 1–4 million deaths globally;[47] and the 2009 Swine flu (H1N1), which spread to over 214 countries, causing an estimated 200 000+ deaths, though with milder symptoms than expected.[48]

The COVID-19 pandemic underscored how quickly a respiratory virus can overwhelm healthcare systems. Although caused by a coronavirus rather than influenza, it highlighted key challenges in the pandemic response, including vaccine development, supply chain disruptions, and healthcare system capacity.[49]

Influenza remains a leading cause of acute respiratory infections worldwide, significantly contributing to morbidity and mortality rates. While most healthy individuals recover from influenza within a few weeks, high-risk populations face severe complications, such as pneumonia, organ failure, and exacerbation of chronic conditions. The World Health Organization (WHO) estimates that seasonal flu leads to millions of hospitalizations yearly, with mortality rates varying with strain virulence, vaccination coverage, and healthcare accessibility.[50]

During peak flu seasons, hospitals, emergency departments, and intensive care units (ICUs) experience surges in patient numbers, leading to bed shortages and increased healthcare costs. Influenza pandemics exacerbate these pressures, stretching resources to their limits. The COVID-19 crisis provided a modern example of how a widespread respiratory virus can disrupt well-prepared healthcare systems, emphasizing the importance of pandemic preparedness.[51]

Beyond its health impact, influenza places a considerable economic burden on healthcare systems and global productivity. Influenza-related hospitalizations, intensive care unit (ICU) admissions, and antiviral treatments generate billions of dollars in healthcare expenditures annually. Countries with limited healthcare infrastructure often struggle to manage outbreaks effectively.[52] Influenza outbreaks also cause significant absenteeism in workplaces and schools, reducing productivity. Severe pandemics intensify economic instability through shutdowns and trade restrictions, as demonstrated by COVID-19.[53]

Innovations such as biosensor-based influenza detection are improving early outbreak identification, allowing for faster responses and better containment strategies. Electrochemical biosensors for influenza detection offer promising potential for real-time monitoring, which could significantly aid future pandemic prevention efforts.[54,55]

17.4.1.2 Avian Influenza (H5N1, H7N9) and Zoonotic Transmission

Avian influenza, particularly the H5N1 and H7N9 strains, poses a significant global health threat due to its potential to spread from birds to humans and trigger pandemics. Unlike seasonal flu, which circulates primarily among humans, avian flu originates in wild birds and poultry. In rare cases, it jumps to humans, often resulting in severe disease and high fatality rates. While sustained human-to-human transmission has not been observed, genetic mutations could increase the risk, making these viruses a major focus of global surveillance and pandemic preparedness efforts.[56]

Wild birds, mainly migratory waterfowl, such as ducks and geese, serve as natural reservoirs for avian influenza. These birds often carry the virus asymptomatically, shedding it through saliva, nasal secretions, and faeces. High-density poultry farms amplify the virus, increasing the likelihood of spillover into human populations. Key factors contributing to transmission include long-distance migration, which enables infected birds to introduce the virus to new regions, sometimes spanning thousands of miles.[57] Crowded markets where birds from various sources are kept together create ideal conditions for viral spread and genetic reassortment. The H7N9 strain, for example, has been strongly linked to live bird markets in China. Avian influenza viruses can persist in water, especially in colder climates, facilitating continued transmission in shared water supplies.[58]

Both H5N1 and H7N9 have triggered severe outbreaks among poultry, leading to mass culling efforts to contain the virus. The highest risks are in regions where human activity, farm animals, and wild birds intersect frequently, such as Southeast Asia, China, and Africa. Although avian influenza primarily affects birds, certain strains, especially H5N1 and H7N9, have demonstrated the ability to infect humans, often with fatal consequences. Since its detection in humans in 1997, H5N1 has infected over 800 individuals globally, with a case fatality rate (CFR) ranging from 50% to 60%. First identified in China in 2013, H7N9 has resulted in more than 1500 confirmed infections, with an estimated mortality rate of 40%.[59,60]

Most human infections occur through direct or indirect contact with infected birds. While sustained human-to-human transmission has not been documented, limited cases of family clusters indicate that person-to-person spread can occur under specific conditions. What makes these viruses particularly concerning is their potential for genetic mutation. A global pandemic could occur if H5N1 or H7N9 were to acquire mutations that enable efficient human-to-human transmission, either through spontaneous genetic changes or reassortment with a human flu virus. H7N9 was first reported in China in 2013 and was linked to live poultry markets. Over four epidemic waves, more than 1500 confirmed human infections occurred. In 2016–2017, the emergence of highly pathogenic strains made outbreaks more severe and more challenging to control. Early detection and rapid response are crucial; live poultry markets pose a high risk, and continuous monitoring of virus mutations is essential.[61–65]

17.4.1.3 Swine Flu (H1N1) and Agricultural Biosecurity Challenges

Swine flu, particularly the H1N1 strain, is a subtype of influenza A that affects both pigs and humans, raising significant concerns for public health and agricultural biosecurity. While seasonal flu primarily circulates among humans, swine flu originates in pigs and occasionally crosses species barriers, leading to outbreaks in human populations. The 2009 H1N1 pandemic was a stark reminder of how a virus emerging from animal reservoirs can quickly escalate into a global crisis. Understanding its transmission dynamics, improving surveillance of pig farms, and strengthening biosecurity measures are critical for preventing future outbreaks.[66,67]

Key pathways for pig-to-human transmission involve farmers, veterinarians, and slaughterhouse workers, who face the highest risk of zoonotic transmission due to prolonged exposure to pigs. The H1N1 virus can persist on surfaces in pig farms, spreading through equipment, feed, and workers' clothing. Enclosed spaces with poor ventilation facilitate airborne transmission, increasing infection risks for farm workers. One of the most significant concerns with swine flu is genetic reassortment, where swine and human influenza viruses exchange genetic material, potentially producing a new, highly transmissible strain.[68–70]

The 2009 H1N1 pandemic emerged after such a reassortment event, where a virus containing swine, human, and avian influenza genes spread efficiently among humans. This underscores the importance of continuous influenza surveillance in pigs, as mutations can lead to more severe human outbreaks.[71] Monitoring influenza in pig populations is essential to prevent zoonotic spillover events. Pig farms serve as critical points for surveillance, where early detection of emerging influenza variants can help mitigate potential pandemics. Regular influenza A virus testing in pigs helps in identifying emerging strains before they spread to humans. Health checks for workers with direct pig contact provide early warnings of potential zoonotic infections. Unlike human flu vaccines, swine influenza vaccines are not globally standardized, emphasizing the need for farm-specific or regional vaccine strategies. Measures to limit human movement between farms to prevent cross-contamination, implement rigorous disinfection protocols for equipment and transport vehicles, and encourage farm workers to wear protective gear and minimize direct pig contact during outbreaks are vital.[72]

The Centers for Disease Control and Prevention (CDC) and the WHO emphasize the importance of global surveillance networks in tracking the evolution of swine flu and preventing another H1N1-like pandemic.[73] Historically, swine-origin influenza viruses have triggered several significant outbreaks, with the 2009 H1N1 pandemic being the most notable in recent history. H1N1 was first detected in Mexico and the USA in early 2009, and the virus quickly spread worldwide, prompting the WHO to declare a pandemic in June 2009. Unlike previous influenza pandemics, H1N1 (2009) disproportionately affected younger individuals, including healthy adults and pregnant women, while older populations had partial immunity due to prior exposure to similar flu strains. The pandemic resulted in an estimated 284 000 deaths globally, with millions requiring hospitalization due to severe respiratory complications.[74] The outbreak cost billions of dollars in lost productivity and healthcare costs, while international travel and trade restrictions further disrupted economies. The pandemic highlighted the importance of rapid vaccine development, with an H1N1 vaccine distributed worldwide within six months.[75–77]

17.4.1.4 COVID-19 and Global Pandemics

The COVID-19 pandemic has been one of the most transformative public health crises in modern history, affecting nearly every aspect of human life. Emerging in late 2019, the SARS-CoV-2 virus spread rapidly across the globe, overwhelming healthcare systems, disrupting economies, and reshaping societal norms. The pandemic exposed vulnerabilities in public health infrastructure, highlighted the complexities of infectious disease containment, and emphasized the critical role of rapid vaccine development. Although significant progress has been made in understanding and managing the virus, its continued evolution, the emergence of new variants, and the long-term health implications remain areas of active research.[78]

COVID-19 is caused by SARS-CoV-2, a novel coronavirus believed to have originated in bats, with potential transmission to humans through an intermediate host. Initial cases were linked to a seafood market in Wuhan, China, but the exact origins of the virus remain a topic of scientific investigation. Human-to-human transmission occurs primarily through respiratory droplets, aerosols, and direct contact. Asymptomatic and pre-symptomatic carriers played a significant role in silent community spread, challenging early containment. The virus has an incubation period of 2 to 14 days, during which infected individuals can spread the virus before exhibiting symptom. Variants of concern (VOCs), such as alpha, delta, and omicron, have demonstrated increased transmissibility and immune evasion, complicating public health responses.[79,80]

COVID-19 symptoms range from mild to severe, with risk factors including advanced age, pre-existing conditions (*e.g.*, diabetes, cardiovascular disease), and weakened immune systems. While many individuals recover without complications, severe cases often lead to acute respiratory distress syndrome (ARDS), organ failure, and death.[81] New variants have resulted in differing severity levels and reinfection rates, challenging long-term disease management.[82]

The scale of COVID-19 exposed weaknesses in global pandemic preparedness, from medical supply shortages to inconsistent public health messaging and disparities in healthcare access. During peak outbreak waves, ICUs and hospital capacities were overwhelmed.[83] Supply chain disruptions led to shortages of ventilators, personal protective equipment (PPE), and essential medications. Healthcare worker burnout reached critical levels, exacerbating staffing shortages worldwide.[84] Initial delays in testing and contact tracing allowed the virus to spread unchecked in many countries. Vaccine hesitancy and misinformation hampered immunization efforts, particularly in vulnerable populations. Lockdowns and social distancing measures were necessary but resulted in severe economic and psychological consequences.[85]

Despite these challenges, the global response to COVID-19 showcased the power of scientific collaboration. The rapid development and deployment of mRNA vaccines (Pfizer-BioNTech, Moderna) marked an unprecedented milestone in medical history, significantly reducing severe disease and mortality rates.[86]

As the pandemic progressed, two major ongoing concerns emerged: the long-term health effects of COVID-19, often referred to as long COVID-19, and the continuous evolution of the virus, resulting in new variants with varying degrees of transmissibility and severity. Some individuals experience symptoms for months after recovering from the initial infection, including fatigue, cognitive impairment ("brain fog"), joint pain, and respiratory issues. Results of research indicate that immune system dysregulation and persistent inflammation may contribute to long COVID. The condition affects millions worldwide, affecting quality of life, workforce productivity, and healthcare systems.[87,88]

The delta variant exhibited higher transmissibility and increased disease severity compared with earlier strains. Omicron and its subvariants spread more rapidly but were generally associated with milder symptoms, though their immune evasion properties led to recurrent infection waves.[94] The ongoing evolution of SARS-CoV-2 underscores the need for updated vaccines, continuous surveillance, and adaptive public health strategies.[89]

17.4.2 Preventive Strategies for Viral Control

Preventing the spread of viral diseases requires a comprehensive approach that includes vaccination programs, global surveillance, early detection systems, and public health policies. Past outbreaks, including influenza, avian flu, swine flu, and COVID-19, have demonstrated how rapidly viruses can spread and the severe consequences of inadequate preparedness. Effective mitigation strategies emphasize preventing human and animal infections, limiting transmission, and enhancing early warning systems.[90]

Vaccination is one of the most effective tools in controlling viral infections. However, influenza viruses, particularly influenza A, mutate frequently, requiring continuous updates to existing vaccines. Moreover, achieving universal protection remains a significant scientific challenge. Influenza vaccines are reformulated annually to align with the circulating strains predicted through global surveillance data. The effectiveness of flu vaccines varies yearly, typically between 40% and 60%, due to antigenic drift and mismatches between vaccine strains and circulating viruses. The WHO and the CDC are key in monitoring influenza patterns worldwide and guiding vaccine formulation.[91–93]

Current influenza vaccines provide only short-term immunity and necessitate annual updates. Research efforts are focused on developing a universal influenza vaccine that offers long-term immunity against multiple strains, potentially eliminating the need for yearly immunization. The stem region of HA, which mutates less frequently than the head region, is a promising target for universal vaccine development.[94] H5N1 and H7N9 avian influenza vaccines have been developed primarily for pandemic preparedness, though their use in humans is limited due to infrequent outbreaks. Swine flu vaccines are available for pigs, but only high-risk populations receive H1N1 vaccines as part of seasonal flu immunization efforts. COVID-19 vaccines were developed at an unprecedented pace using mRNA (Pfizer, Moderna) and viral vector (AstraZeneca, Johnson & Johnson) technologies, demonstrating how rapid vaccine development can mitigate pandemic impacts.[95]

Preventing pandemics necessitates continuous monitoring of viral evolution and early identification of emerging threats. International health organizations and advanced technologies are pivotal in detecting pathogens before they become widespread. The WHO, CDC, and OIE conduct global influenza surveillance in humans, birds, and livestock.[96] The WHO Global Influenza Surveillance and Response System (GISRS) collects and analyses flu virus samples globally to forecast future strains. The CDC's Influenza Risk Assessment Tool (IRAT) evaluates new influenza strains for their pandemic potential.

Next-generation sequencing (NGS) technology enables real-time sequencing of viral genomes, allowing researchers to identify mutations and track the emergence of variants of concern.[97] This method has been instrumental in monitoring SARS-CoV-2 mutations, such as delta and omicron, and informing vaccine modifications. NGS also aids in tracking cross-species transmission, which is crucial for pandemic prevention. Electrochemical and optical biosensors can detect viral proteins, RNA, or antibodies in patient samples within minutes, providing a valuable tool for early diagnosis.[98] Researchers are developing portable biosensor devices for use in high-risk settings, such as airports, hospitals, and outbreak zones. Results indicate that biosensors could revolutionize infectious disease diagnostics, particularly for rapidly mutating viruses like influenza and coronaviruses.[99]

Beyond vaccines and surveillance, robust public health measures are essential in containing infectious diseases. These measures include quarantine protocols, hospital infection control strategies, and improved biosecurity in agriculture. Hand hygiene, mask-wearing, and social distancing have proven effective in controlling COVID-19 and influenza outbreaks.[100] Hospitals implement airborne isolation precautions for highly transmissible diseases, utilizing negative pressure rooms to prevent viral spread. Antiviral treatments such as oseltamivir (Tamiflu) and remdesivir help mitigate disease severity but are most effective when administered early.

Border closures, travel bans, and mandatory quarantines were extensively employed during COVID-19, with mixed outcomes.[101] The results of studies on past pandemics [*e.g.*, the 1918 flu, SARS, and Middle East respiratory syndrome (MERS)] indicate that early quarantine enforcement can slow outbreaks; however, prolonged restrictions may negatively impact economies and mental health. Future outbreak response strategies must balance public health protection and socio-economic considerations.

Live animal markets have been linked to SARS, avian flu, and COVID-19 outbreaks, emphasizing the need for stricter regulations. Poultry and pig farms have enhanced biosecurity measures, including livestock vaccination programs and reducing human–animal interactions. Some experts advocate gradually phasing out wet markets or implementing stricter operational controls to minimize zoonotic transmission risks.[102–104]

17.4.3 Emerging Diagnostic and Sensor Technologies for Viral Control

Early and accurate detection of viruses is essential for controlling outbreaks and preventing pandemics. While traditional diagnostic methods, such as PCR and antigen tests, remain widely used, they have limitations in terms of speed, cost, and accessibility. Advances in biosensors, lab-on-a-chip systems, and artificial intelligence (AI)-driven diagnostic tools are revolutionizing virus detection, enabling faster, more portable, and highly sensitive diagnostics. Among these, interferometric reflectance imaging sensor (IRIS) technology has emerged as a powerful tool for label-free, high-throughput viral detection. These innovations are shaping the future of virus surveillance and pandemic preparedness.[105]

17.4.3.1 Biosensor-based Approaches for Virus Detection

Biosensors have transformed viral diagnostics by offering rapid, real-time analysis without extensive laboratory infrastructure. These devices detect viral proteins, nucleic acids, or antibodies through various sensing mechanisms.

Optical biosensors that use SPR[106] and fluorescence can detect viral markers with high sensitivity. Electrochemical biosensors, on the other hand, convert virus–protein interactions into electrical signals, enabling low-cost and portable diagnostics.[107] Additionally, microfluidic biosensors integrate sample handling and detection within a chip-based system, allowing for rapid virus identification within minutes.[108,109]

17.4.3.2 IRIS Technology and Its Role in Virus Detection

IRIS technology provides high-throughput viral detection without fluorescent labelling. This method measures changes in light reflection caused by viral binding events, enabling real-time monitoring. IRIS has been successfully used for detecting influenza viruses and SARS-CoV-2, with accuracy similar to PCR tests but significantly faster turnaround times.[110] This technology can differentiate between virus strains, making it an invaluable tool for monitoring mutations and emerging variants.

17.4.3.3 AI-powered Diagnostics and Digital Detection Systems

AI is increasingly being used in viral detection and outbreak surveillance. AI-driven diagnostic tools are used to analyse genomic data, biosensor outputs, and epidemiological trends to identify patterns in viral evolution. These AI-assisted detection systems provide early warning signals for outbreaks, thereby improving response times for public health officials. Additionally, multiomics approaches, which combine genomics, proteomics, and metabolomics, offer deeper insights into viral interactions with the human body.[111]

17.5 Conclusion

The persistent threat of viral pandemics, including influenza, avian flu, swine flu, and COVID-19, underscores the need for a multi-layered public health strategy that integrates surveillance, vaccination, diagnostic advancements, and pandemic preparedness. The lessons learned from past outbreaks have demonstrated that rapid detection, global collaboration, and technological innovation are fundamental in mitigating the impacts of infectious diseases.

The unpredictable nature of influenza and its ability to mutate require continuous investment in vaccine development, enhanced surveillance systems, and proactive public health measures. Avian and swine influenza remain critical concerns due to their zoonotic potential and the possibility of genetic reassortment leading to novel strains. Improved biosecurity measures in livestock farming, early detection mechanisms, and research into cross-protective vaccines will be crucial in reducing the risk of spillover events. The COVID-19 pandemic has reshaped global health priorities, emphasizing the need for resilient healthcare systems, equitable vaccine distribution, and the integration of AI-powered diagnostics. The rapid development of messenger RNA (mRNA) vaccines and antiviral treatments has highlighted the importance of investing in cutting-edge biomedical research. However, the emergence of new variants and long-term complications, such as long COVID, necessitate sustained vigilance and adaptive response strategies. Advancements in biosensors, lab-on-a-chip technologies, and digital health monitoring are revolutionizing the detection and tracking of viral outbreaks. IRIS technology represents a transformative approach in diagnostics, offering real-time, label-free, and highly sensitive virus detection. With the ability to integrate AI-driven analysis and multiplexing capabilities, IRIS has the potential to become a cornerstone of future pandemic preparedness, providing rapid, cost-effective, and scalable diagnostic solutions. Moving forward, the key to effective pandemic response lies in a collaborative global approach. Strengthening international partnerships, ensuring equitable access to medical resources, and fostering interdisciplinary research will be critical in preventing and managing future outbreaks. By leveraging innovations in biosensing, AI diagnostics, and vaccine technologies, it will be possible to build a more robust and adaptive public health infrastructure capable of addressing emerging infectious threats efficiently.

References

1. D. Bhatia, S. Paul, T. Acharjee and S. S. Ramachairy, Biosensors and their widespread impact on human health, *Sens. Int.*, 2024, **5**, 100257.
2. S. A. Walper, G. Lasarte Aragonés, K. E. Sapsford, C. W. Brown III, C. E. Rowland, J. C. Breger and I. L. Medintz, Detecting biothreat agents: from current diagnostics to developing sensor technologies, *ACS Sens.*, 2018, **3**(10), 1894–2024.
3. M. I. Razzak, M. Imran and G. Xu, Big data analytics for preventive medicine, *Neural Comput. Appl.*, 2020, **32**, 4417–4451.
4. S. K. Metkar and K. Girigoswami, Diagnostic biosensors in medicine – A review, *Biocatal. Agric. Biotechnol.*, 2019, **17**, 271–283.
5. M. Bakhshpour and A. Denizli, Highly sensitive detection of Cd(II) ions using ion-imprinted surface plasmon resonance sensors, *Microchem. J.*, 2020, **159**, 105572.

6. H. Bai, R. Wang, B. Hargis, H. Lu and Y. Li, A SPR aptasensor for detection of avian influenza virus H5N1, *Sensors*, 2012, **12**(9), 12506–12518.
7. C. Wang, C. Wang, X. Wang, K. Wang, Y. Zhu, Z. Rong, W. Wang, R. Xiao and S. Wang, Magnetic SERS strip for sensitive and simultaneous detection of respiratory viruses, *ACS Appl. Mater. Interfaces*, 2019, **11**(21), 19495.
8. Y. Rasmi, X. Li, J. Khan, T. Ozer and J. R. Choi, Emerging point-of-care biosensors for rapid diagnosis of COVID-19: current progress, challenges, and future prospects, *Anal. Bioanal. Chem.*, 2021, **413**(16), 4137–4159.
9. W. W. Hsiao, G. Fadhilah, C.-C. Lee, R. Endo, Y.-J. Lin, S. Angela, C.-C. Ku, H.-C. Chang and W.-H. Chiang, Nanomaterial-based biosensors for avian influenza virus: a new way forward, *Talanta*, 2023, **265**, 124892.
10. J. Astill, R. A. Dara, E. D. G. Fraser and S. Sharif, Detecting and predicting emerging disease in poultry with the implementation of new technologies and big data: a focus on avian influenza virus, *Front. Vet. Sci.*, 2018, **5**, 263.
11. T. Lee, J.-H. Ahn, S. Y. Park, G.-H. Kim, J. Kim, T.-H. Kim, I. Nam, C. Park and M.-H. Lee, Recent advances in AIV biosensors composed of nanobio hybrid material, *Micromachines*, 2018, **9**, 651.
12. L. Byrd-Leotis, R. D. Cummings and D. A. Steinhauer, The interplay between the host receptor and influenza virus hemagglutinin and neuraminidase, *Int. J. Mol. Sci.*, 2017, **18**, 1541.
13. P. S. Wikramaratna, M. Sandeman, M. Recker and S. Gupta, The antigenic evolution of influenza: drift or thrift?, *Philos. Trans. R. Soc., B*, 2013, **368**, 20120200.
14. J. Xu, D. Suarez and D. S. Gottfried, Detection of avian influenza virus using an interferometric biosensor, *Anal. Bioanal. Chem.*, 2007, **389**, 1193–1199.
15. Y. Amano and Q. Cheng, Detection of influenza virus: traditional approaches and development of biosensors, *Anal. Bioanal. Chem.*, 2005, **381**, 156–164.
16. H.-C. Lai, C.-H. Wang, T.-M. Liou and G.-B. Lee, Influenza A virus-specific aptamers screened by using an integrated microfluidic system, *Lab Chip*, 2014, **14**(12), 1953–2132.
17. K. Wu, T. Klein, V. D. Krishna, D. Su, A. M. Perez and J.-P. Wang, Portable GMR handheld platform for the detection of influenza A virus, *ACS Sens.*, 2017, **2**(11), 1594–1601.
18. F. Kazemi, Z. S. Tabatabaeian and F. B. Zarrabi, A terahertz metamaterial biosensor based on spoof surface plasmon polaritons transmission line for avian influenza virus detection, *Opt. Laser Technol.*, 2024, **174**, 110694.
19. Y. Tian, Y. Zhang, X. Lu, D. Xiao and C. Zhou, Microfluidic paper-based chemiluminescence sensing platform based on functionalized $CaCO_3$ for time-resolved multiplex detection of avian influenza virus biomarkers, *Anal. Biochem.*, 2024, **693**, 115583.

20. H. Ehzari, M. Amiri, R. Hallaj and M. Sadeghi, Rapid, flexible fabrication of a microfluidic electrochemical chip nucleic acid target for selective, label-free detection of influenza virus DNA using catalytic redox-recycling, *Anal. Biochem.*, 2025, **700**, 115771.
21. H. Gong, G. Cai, C. Chen, F. Chen and C. Cai, Construction of a monoclonal molecular imprinted sensor with high affinity for specific recognition of influenza A virus subtype, *Talanta*, 2024, **278**, 126568.
22. Y.-C. Huang, S.-F. Wang, B.-C. Chen, Z.-S. Yang, M.-C. Li, X.-Y. Wu, M.-J. Youh, H.-Y. Chou, Y.-X. Lin, W. Assavalapsakul, A. Thitithanyanont and L.-C. Su, Towards cost-effective and lightweight surface plasmon resonance biosensing for H5N1 avian influenza virus detection: Integration of novel near-infrared organic photodetectors, *Sens. Actuators, B*, 2024, **400**(Pt B), 13898.
23. Y. Wang, Q. Ruan, Z.-C. Lei, S.-C. Lin, Z. Zhu, L. Zhou and C. Yang, Highly sensitive and automated surface enhanced raman scattering-based immunoassay for H5N1 detection with digital microfluidics, *Anal. Chem.*, 2018, **90**(8), 5224–5231.
24. E. Uttenthaler, C. Kößlinger and S. Drost, Quartz crystal biosensor for detection of the African swine fever disease, *Anal. Chim. Acta*, 1998, **362**(1), 91–100.
25. E. Uttenthaler, C. Kößlinger and S. Drost, Characterization of immobilization methods for African swine fever virus protein and antibodies with a piezoelectric immunosensor, *Biosens. Bioelectron.*, 1998, **13**(12), 1279–1286.
26. H. S. Cho and N. Y. Park, Serodiagnostic comparison between two methods, ELISA and surface plasmon resonance for the detection of antibodies of classical swine fever, *J. Vet. Med. Sci.*, 2006, **68**(12), 1327–1329.
27. M. Biagetti, M. Cuccioloni, L. Bonfili, V. Cecarini, C. Sebastiani, L. Curcio, M. Giammarioli, G. M. De Mia, A. M. Eleuteri and M. Angeletti, Chimeric DNA/LNA-based biosensor for the rapid detection of African swine fever virus, *Talanta*, 2018, **184**, 35–41.
28. J. Ki, H. K. Na, S. W. Yoon, V. P. Le, T. G. Lee and E. K. Lim, CRISPR/Cas-assisted colorimetric biosensor for point-of-use testing for African swine fever virus, *ACS Sens.*, 2022, **7**(12), 3940–3946.
29. X. Zhao, Y. He, S. Shao, Q. Ci, L. Chen, X. Lu, Q. Liu and J. Chen, CRISPR/Cas14 and G-Quadruplex DNAzyme-driven biosensor for paper-based colorimetric detection of African swine fever virus, *ACS Sens.*, 2024, **9**(5), 2413–2420.
30. P. Ning, Z. Wu, X. Li, Y. Zhou, A. Hu, X. Gong, J. He, Y. Xia, K. Guo, R. Zhang and X. Zhang, Development of functionalized gold nanoparticles as nanoflare probes for rapid detection of classical swine fever virus, *Colloids Surf., B*, 2018, **171**, 110–114.
31. S. Klangprapan, B. Choke-Arpornchai, P. A. Lieberzeit and K. Choowongkomon, Sensing the classical swine fever virus with molecularly imprinted polymer on quartz crystal microbalance, *Heliyon*, 2020, **6**(6), e04211.

32. S. A. Nakhjavani, H. Mirzajani, S. Cararra and M. C. Onbaşlı, Advances in biosensor technologies for infectious diseases detection, *TrAC, Trends Anal. Chem.*, 2024, 117979.
33. P. Pitak-Arnnop, S. Schubert, K. Dhanuthai, K. Sappayatosok, U. Bauer, P. Ngamwannagul, U. G. Liebert and A. Hemprich, Swine-origin H1N1 influenza A virus and dental practice: a critical review, *Clin. Oral Investig.*, 2010, **14**, 11–17.
34. Y. Yang, S. Bu, X. Zhang, Q. Duan, H. Meng, Z. Hao, X. He and J. Wan, Influenza virus detection using an electrochemical biosensor based on DSN and RCA, *Microchem. J.*, 2024, **199**, 109998.
35. H. Zhang, A. M. Klose and B. L. Miller, Label-free, multiplex glycan microarray biosensor for influenza virus detection, *Bioconjugate Chem.*, 2021, **32**(3), 533–540.
36. M. N. Matrosovich, T. Y. Matrosovich, T. Gray, N. A. Roberts and H.-D. Klenk, Human and avian influenza viruses target different cell types in cultures of human airway epithelium, *Proc. Natl. Acad. Sci. U. S. A.*, 2004, **101**, 4620–4624.
37. K. Shinya, M. Ebina, S. Yamada, M. Ono, N. Kasai and Y. Kawaoka, Avian flu: influenza virus receptors in the human airway, *Nature*, 2006, **440**, 435.
38. K. Wu, J. Liu, R. Saha, D. Su, V. D. Krishna, M. C.-J. Cheeran and J.-P. Wang, Magnetic particle spectroscopy for detection of influenza A virus subtype H1N1, *ACS Appl. Mater. Interfaces*, 2020, **12**(12), 13686–13697.
39. Y. Liu, X. Z. Meng, X. Luo, H. W. Gu, X. L. Yin, W. L. Han, H. C. Yi and Y. Chen, Molecularly imprinted polymer combined with MOF-assisted redox recycling amplification: A powerful electrochemical sensing strategy for pathogenic bacteria, *Sens. a Actuators, B*, 2024, **410**, 135682.
40. Y. Shen, Y. Wei, C. Zhu, J. Cao and D. M. Han, Ratiometric fluorescent signals-driven smartphone-based portable sensors for onsite visual detection of food contaminants, *Coord. Chem. Rev.*, 2022, **458**, 214442.
41. V. Faustino, S. O. Catarino, R. Lima and G. Minas, Biomedical microfluidic devices by using low-cost fabrication techniques: A review, *J. Biomech.*, 2016, **49**(11), 2280–2292.
42. S. M. Scott, Ali. Zulfiqur "fabrication methods for microfluidic devices: an overview", *Micromachines*, 2021, **12**(3), 319.
43. World Health Organization (WHO), Influenza (Seasonal), Retrieved from WHO database, https://www.who.int/en/news-room/fact-sheets/detail/influenza-(seasonal).
44. Centers for Disease Control and Prevention (CDC), *Estimated Flu Burden*, Retrieved from CDC website, https://www.cdc.gov/flu/about/burden/index.html.
45. J. K. Taubenberger ve and D. M. Morens, 1918, Influenza: the mother of all pandemics, *Emerging Infect. Dis.*, 2006, **12**, 15.
46. E. D. Kilbourne, Influenza pandemics of the 20th century, *Emerging Infect. Dis.*, 2006, **12**, 9–14.
47. C. Viboud and R. F. Grais, *et al.*, Global mortality impact of the 1957–1959 influenza pandemic, *J. Infect. Dis.*, 2005, **192**, 233.

48. F. S. Dawood and S. Jain, *et al.*, Emergence of a novel swine-origin influenza A (H1N1) virus in humans, *N. Engl. J. Med.*, 2009, **360**, 2605.
49. A. S. Fauci, H. C. Lane and ve R. R. Redfield, COVID-19—Navigating the uncharted, *N. Engl. J. Med.*, 2020, **382**, 1268.
50. G. Forni ve and A. Mantovani, COVID-19 vaccines: where we stand and challenges ahead, *Cell Death Differ.*, 2021, **28**, 626.
51. C. J. Murray ve and A. D. Lopez, Global mortality, disability, and the contribution of risk factors: global burden of disease study, *Lancet*, 1997, **349**, 1436.
52. N. A. Molinari and I. R. Ortega-Sanchez, *et al.*, The annual impact of seasonal influenza in the u.s.: measuring disease burden and costs, *Vaccine*, 2007, **25**, 5086.
53. Y. Yang, S. Bu and X. Zhang, *et al.*, Influenza virus detection using an electrochemical biosensor, *Microchem. J.*, 2024, **183**, 107492.
54. A. Pichlmair ve and C. Reis e Sousa, Innate recognition of viruses, *Nat. Rev. Immunol.*, 2007, **7**, 265.
55. R. G. Webster ve and E. A. Govorkova, H5N1 Influenza—continuing evolution and spread, *N. Engl. J. Med.*, 2006, **355**, 2174.
56. Centers for Disease Control and Prevention (CDC), Avian Influenza in Birds and People, https://www.cdc.gov/flu/avianflu/index.htm.
57. R. Gao and B. Cao, *et al.*, Human infection with a novel avian-origin influenza A (H7N9) virus, *N. Engl. J. Med.*, 2013, **368**, 1888.
58. M. D. Van Kerkhove and S. Vong, *et al.*, Risk factors for human infection with highly pathogenic avian influenza A H5N1, *J. Infect. Dis.*, 2011, **204**, 119.
59. Centers for Disease Control and Prevention (CDC), Avian Influenza in Wild Birds and Poultry, https://www.cdc.gov/flu/avianflu/index.htm.
60. D. E. Stallknecht ve and J. D. Brown, Ecology of Avian Influenza in Wild Birds, *Avian Dis.*, 2009, **53**(1), 391.
61. World Health Organization (WHO), Avian Influenza Overview, https://www.who.int/news-room/fact-sheets/detail/avian-influenza.
62. Y. Chen, W. Liang, *et al.*, Clinical and epidemiological characteristics of a fatal case of avian influenza A H7N9 virus infection with acute respiratory distress syndrome, *Lancet*, 2013, 381, 1091.
63. T. Imai and T. Watanabe, *et al.*, A Highly pathogenic H7N9 influenza virus emergent in China, *Cell Host Microbe*, 2017, **22**, 725.
64. J. Zhou and D. Wang, *et al.*, Biological features of H7N9 influenza viruses and their implications for pandemic risk assessment, *Proc. Natl. Acad. Sci. U. S. A.*, 2017, **114**, 4325.
65. J. S. M. Peiris, M. D. de Jong and Y. Guan, Avian influenza virus (H5N1): A threat to human health, *Clin. Microbiol. Rev.*, 2007, **20**, 243.
66. World Health Organization (WHO), Avian Influenza Reports, Retrieved from WHO website, https://www.who.int/influenza/surveillance_monitoring/avian_influenza_reports/en/.
67. H. Chen and G. J. D. Smith, *et al.*, Establishment of multiple sublineages of H5N1 influenza virus in Asia: implications for pandemic control, *Proc. Natl. Acad. Sci. U. S. A.*, 2006, **103**, 2845.

68. A. L. Vincent, K. M. Lager and T. K. Anderson, *et al.*, Origins and evolution of H1N1 swine influenza viruses, *Trends Microbiol.*, 2014, **22**, 543.
69. W. Ma, R. E. Kahn and J. A. Richt, The pig as a mixing vessel for influenza viruses: human and veterinary implications,*J. Mol. Virol.*, 2008, **82**, 27.
70. Centers for Disease Control and Prevention (CDC), Swine Influenza: How It Spreads, 2023, Retrieved from CDC website, https://www.cdc.gov/flu/swineflu/howitspread.htm.
71. R. G. Webster, W. J. Bean and O. T. Gorman, *et al.*, Evolution and ecology of influenza A viruses, *Microbiol. Rev.*, 1992, **56**, 152.
72. G. J. D. Smith and D. Vijaykrishna, *et al.*, Origins and evolution of the 2009 Swine-origin H1N1 influenza A epidemic, *Nature*, 2009, **459**, 1122.
73. World Health Organization (WHO), Zoonotic Influenza Surveillance in Swine Farms, 2023, Retrieved from WHO website, https://www.who.int/influenza/surveillance_monitoring/zoonotic_influenza/en/.
74. G. Neumann and Y. Kawaoka, Predicting pandemic potential of influenza A viruses, *Emerging Infect. Dis.*, 2015, **21**, 1.
75. F. S. Dawood and S. Jain, *et al.*, The Global burden of the 2009 H1N1 pandemic influenza, *N. Engl. J. Med.*, 2012, **366**, 1859.
76. J. W. McCauley and R. S. Daniels, Influenza virus surveillance and pandemic preparedness, *Virus Res.*, 2013, **178**, 32–38.
77. M. T. Osterholm, The compelling need for game-changing influenza Vaccines,*J. Infect. Dis.*, 2012, **206**, 1652.
78. A. S. Fauci and H. C. Lane, COVID-19 and lessons from influenza pandemics, *N. Engl. J. Med.*, 2020, **382**, 1303–1305.
79. M. Chinazzi and J. T. Davis, *et al.*, The Effect of travel restrictions on COVID-19 spread, *Science*, 2020, **368**, 395.
80. K. G. Andersen and A. Rambaut, *et al.*, The proximal origin of SARS-CoV-2, *Nat. Med.*, 2020, **26**, 450.
81. World Health Organization (WHO), Origins of the SARS-CoV-2 Virus, 2023, Retrieved from WHO website, https://www.who.int/health-topics/coronavirus/origins.
82. Centers for Disease Control and Prevention (CDC), How COVID-19 Spreads, 2022, Retrieved from CDC website, https://www.cdc.gov/coronavirus/2019-ncov/prevent-getting-sick/how-covid-spreads.html.
83. N. G. Davies and S. Abbott, *et al.*, Estimated transmissibility and impact of SARS-CoV-2 variants of concern, *Science*, 2021, **372**(6538), eabg3055.
84. C. Huang and Y. Wang, *et al.*, Clinical features of patients infected with 2019 novel coronavirus in Wuhan, China, *The Lancet*, 2020, **395**(10223), 497–506.
85. S. Richardson and J. S. Hirsch, *et al.*, Presenting characteristics, comorbidities, and outcomes among 5700 patients hospitalized with COVID-19 in the New York City Area,*JAMA*, 2020, **10**(16), e020255.
86. M. Ronco-López, C. Peñafiel-Saiz and L. Echegaray-Eizaguirre, Public information about vaccines and treatments on COVID-19. Approach to the documentary sources of institutions and organizations, *Rev. Lat. Comun. Soc.*, 2024, **82**, 1.

87. F. P. Polack and S. J. Thomas, *et al.*, Safety and efficacy of the BNT162b2 mRNA COVID-19 vaccine, *N. Engl. J. Med.*, 2020, **383**(27), 2603–2615.
88. H. E. Davis and G. S. Assaf, *et al.*, Characterizing long COVID in an international cohort: 7 months of symptoms and their impact, *EClinicalMedicine*, 2021, **38**, 101019.
89. M. Gandhi, D. S. Yokoe and D. V. Havlir, Asymptomatic transmission, the Achilles' heel of current strategies to control COVID-19, *N. Engl. J. Med.*, 2020, **382**(22), 2158–2160.
90. M. Chinazzi and J. T. Davis, *et al.*, The effect of travel restrictions on COVID-19 spread, *Science*, 2020, **368**(6489), 395–400.
91. C. Tymm, J. Zhou, A. Tadimety, A. Burklund and J. X. Zhang, Scalable COVID-19 detection enabled by lab-on-chip biosensors, *Cell. Mol. Bioeng.*, 2020, **13**, 313.
92. A. J. Hay and J. W. McCauley, The WHO global influenza surveillance and response system (GISRS)—a future perspective, *Influenza Other Respir. Viruses*, 2018, **12**(5), 551.
93. Centers for Disease Control and Prevention (CDC), Influenza SARS-CoV-2 (Flu SC2) Multiplex Assay, July 2, 2020, Retrieved from CDC website, https://www.cdc.gov/coronavirus/2019-ncov/lab/flu-sc2.html.
94. Q. Song, X. Sun, Z. Dai, Y. Gao, X. Gong, B. Zhou, J. Wu and W. Wen, Point-of-care testing detection methods for COVID-19, *Lab Chip*, 2021, **21**(9), 1634.
95. G. Kabay, J. DeCastro, A. Altay, K. Smith, H. W. Lu, A. M. Capossela, M. Moarefian, K. Aran and C. Dincer, Emerging biosensing technologies for the diagnostics of viral infectious diseases, *Adv. Mater.*, 2022, **34**(30), 2201085.
96. E. Seymour, F. Ekiz Kanik, S. Diken Gür, M. Bakhshpour-Yucel, A. Araz, N. Lortlar Ünlü and M. S. Ünlü, Solid-phase optical sensing techniques for sensitive virus detection, *Sensors*, 2023, **23**(11), 5018.
97. G. Wang, L. Wang, Z. Meng, X. Su, C. Jia, X. Qiao, S. Pan, Y. Chen, Y. Cheng and M. Zhu, Visual detection of COVID-19 from materials aspect, *Adv. Fiber Mater.*, 2022, **4**(6), 1304–1333.
98. P. Rajpurkar, C. O'Connell, A. Schechter, N. Asnani, J. Li, A. Kiani, R. L Ball, M. Mendelson, G. Maartens, D. J. van Hoving and R. Griesel, CheXaid: deep learning assistance for physician diagnosis of tuberculosis using chest X-rays in patients with HIV, *NPJ Digital Med.*, 2020, **3**(1), 115.
99. X. Wang, S. Stelzer-Braid, M. Scotch and W. D. Rawlinson, Detection of respiratory viruses directly from clinical samples using next-generation sequencing: A literature review of recent advances and potential for routine clinical use, *Rev. Med. Virol.*, 2022, **32**(5), e2375.
100. A. Sharma, R. K. Mishra, K. Y. Goud, M. A. Mohamed, S. Kummari, S. Tiwari, Z. Li, R. Narayan, L. A. Stanciu and J. L. Marty, Optical biosensors for diagnostics of infectious viral disease: A recent update, *Diagnostics*, 2021, **11**(11), 2083.

101. N. Bhalla, Y. Pan, Z. Yang and A. F. Payam, Opportunities and challenges for biosensors and nanoscale analytical tools for pandemics: COVID-19, *ACS Nano*, 2020, **14**(7), 7783–7807.
102. N. C. Chiu, H. Chi, Y. L. Tai, C. C. Peng, C. Y. Tseng, C. C. Chen, B. F. Tan and C. Y. Lin, Impact of wearing masks, hand hygiene, and social distancing on influenza, enterovirus, and all-cause pneumonia during the coronavirus pandemic: retrospective national epidemiological surveillance study, *J. Med. Int. Res.*, 2020, **22**(8), e21257.
103. K. A. Grépin, J. Aston and J. Burns, Effectiveness of international border control measures during the COVID-19 pandemic: a narrative synthesis of published systematic reviews, *Philos. Trans. R. Soc., A*, 2023, **381**(2257), 20230134.
104. S. Mubareka, J. Amuasi, A. Banerjee, H. Carabin, J. Copper Jack, C. Jardine, B. Jaroszewicz, G. Keefe, J. Kotwa, S. Kutz and D. McGregor, Strengthening a One Health approach to emerging zoonoses, *Facets*, 2023, **8**(1), 1.
105. G. Karuppaiah, A. Vashist, M. Nair, M. Veerapandian and P. Manickam, Emerging trends in point-of-care biosensing strategies for molecular architectures and antibodies of SARS-CoV-2, *Biosens. Bioelectron. X*, 2023, **13**, 100324.
106. W. Reis de Araujo, H. Lukas, M. D. T. Torres, W. Gao and C. de la Fuente-Nunez, Low-cost biosensor technologies for rapid detection of COVID-19 and Future pandemics, *ACS Nano*, 2024, **18**(3), 1757–1777.
107. S. Akgönüllü, M. Bakhshpour, A. K. Pişkin and A. Denizli, Microfluidic systems for cancer diagnosis and applications, *Micromachines*, 2021, **12**(11), 1349.
108. Ö. Erdem, E. Derin and K. Sagdic, *et al.*, Smart materials-integrated sensor technologies for COVID-19 diagnosis, *Emergent Mater.*, 2021, **4**, 169–185.
109. O. Avci, N. L. Ünlü, A. Y. Özkumur and M. S. Ünlü, Interferometric reflectance imaging sensor (IRIS)—a platform technology for multiplexed diagnostics and digital detection, *Sensors*, 2015, **15**(7), 17649–17665.
110. Z. Wu, C. Wang, B. Liu, C. Liang, J. Lu, J. Li, X. Tang, C. Li and T. Li, Smartphone-based high-throughput fiber-integrated immunosensing system for point-of-care testing of the SARS-CoV-2 nucleocapsid protein, *ACS Sens.*, 2022, 7(7), 1985–1995.
111. H. Panwar, P. K. Gupta, M. K. Siddiqui, R. Morales-Menendez and V. Singh, Application of deep learning for fast detection of COVID-19 in X-Rays using nCOVnet, *Chaos, Solitons Fractals*, 2020, **138**, 109944.

Section VII: Future Directions

CHAPTER 18

Future Directions and Opportunities

HUMA SHAIKH*, MEHRUNNISA TANZIL AND ASBAH KABEER

National Center of Excellence in Analytical Chemistry, University of Sindh, Jamshoro, Pakistan
*E-mail: huma.hashu@gmail.com

18.1 Promising Research Directions in Sensor-based Virus Detection

The swift detection of viruses is very crucial and has become of the utmost importance in current diagnosis. In this regard, sensing technology is rising as the most eligible candidate due to its robustness, simplicity, and speed. Sensor-based virus detection is a promising tool for improving public health and monitoring diseases. The priority of current research in sensing technology is to develop innovative sensors that are able to promptly detect viral infections more accurately and non-invasively. The progress of "wearable sensors" is one of the major advancements in sensor technology because they can monitor physiological changes, such as heart rate, temperature, blood oxygen level, *etc.*; they are able to provide real time monitoring and insights into viral infections. Material chemistry is also playing its role very efficiently by introducing smart, robust, and cost-effective sensing materials that lead to highly sensitive electrochemical and biosensors for the detection of viral

Detection Science Series No. 32
Advances in Virus Detection with Sensors
Edited by Adil Denizli
© The Royal Society of Chemistry 2026
Published by the Royal Society of Chemistry, www.rsc.org

particles at a molecular level. Using these smart nanomaterials; lab-on-a-chip (LOC) devices have been designed that are capable of on-site and point-of-care detection of viruses. The optical sensors, more specifically surface plasmon resonance (SPR) or fluorescence sensors, help in precise detection of viruses by measuring the minute changes in optical properties of sensor surfaces when viral particles bind to them. Another milestone is the use of clustered regularly interspaced short palindromic repeats (CRISPR) technology for the extremely specific and swift detection of viral RNA. This technique is very helpful in detection of emerging viruses. The field of sensing is transforming with the integration of machine learning and artificial intelligence, as the complex sensor data can be analysed using these tools, which can assist in predicting virus outbreaks and assessing disease progression. The most recent interest in sensing technologists is to devise hybrid sensing systems that employ multiple detection techniques, such as electrochemical, optical, and mechanical sensors *etc.,* to improve the limits of detection and sensitivity of sensors. In case of viral detection, non-invasive sensing is playing a crucial role because it is capable of analysing saliva, sweat or breath to enable early and less disruptive virus detection. These sensing technologies are performing their tasks efficiently; however, challenges of compromised sensitivity and specificity still exist. Moreover, their scalability and affordability are other challenges to be dealt with. However, the continuous revolution and innovation in the field of sensing technology are promising for the reliable, swift and prompt detection of viruses at point of care. This will not only help in mitigating outbreaks but will also improve public health.

The population round the globe is increasing perpetually, driving the consumption of resources, and prompting an increased need to produce smart and environmentally friendly solutions.[1] Millions of people suffer from several diseases due to encounters with pathogens.[2] There is a noteworthy rise globally in infectious diseases that is affecting all living beings;[3] particularly, in undeveloped countries, due to increases in various diseases, such as human immunodeficiency virus (HIV), malaria, tuberculosis, and so on.[4] Viruses are parasites that require a host cell in order to generate and replicate. However, mammalian cells have the power to boost complex protection mechanism in order to hinder and detect viral replication. In reaction, viruses are capable of controlling and breaking down the host immune response, allowing growth which is proficient at destroying the host's immune system.[5] In recent years, several viruses have caused epidemics, including the avian influenza virus (AIV) in 2013,[6] Zika virus in 2015,[7] Ebola virus (EBOV) in 2018,[8] the African swine fever virus (ASFV),[9] corona virus disease (COVID)-19[10] in 2019, and the influenza A virus (IAV)[11] in 2023. In order to cope up with such diseases, new vaccines have been developed and some bio-security measures have been taken, that have played a significant role in prevention and protection.

Isothermal amplification technology has been the subject of keen consideration due to its robustness and ability to eliminate instrumental complications.[12] In this class, techniques include, recombinase polymerase amplification (RPA),[13] recombinase-aided amplification assay (RAA),[14] loop-mediated isothermal amplification (LAMP),[15] nucleic acid sequence-based amplification (NASBA),[16] strand displacement amplification (SDA),[17] rolling circle amplification (RCA),[18] and isothermal helicase-dependent amplification (HDA).[19] It has been concluded from results of recent studies that RPA in combination with other technologies leads to reliable diagnosis, producing results that are more accurate and sensitive than those obtained with standard RPA alone. These methods include the detection of fluorescence signals,[20] lateral flow devices (LFDs),[21] CRISPR–CRISPR-associated (Cas) systems,[22] flocculation assays detection,[23] electrochemical detection,[24] chemiluminescent detection,[25] surface-enhanced Raman scattering (SERS),[26] surface-enhanced infrared absorption spectroscopy (SEIRA),[27] and microfluidic technology.[28]

Since the first case of identification of COVID-19, worldwide catastrophic effects led to the pandemic. The cumulative count globally has exceeded 769 million verified COVID-19 cases, with documented fatalities of more than 6.9 million.[29] Specific treatments for many viral infections remain elusive. The results emphasize how urgent it is to investigate novel and effective strategies for early detection and viral transmission containment. Biomarkers for COVID-19, quantitative reverse transcription PCR (qRT-PCR) recognition of viral RNA, isothermal detection of RNA, CRISPR-based detections of viral RNA, and lab-on-a-chip devices for nucleic acid testing are among the advancements in sensing technologies and diagnostic techniques that are essential for overcoming the obstacles. Figure 18.1 illustrates some tests used during the COVID-19 outbreak.

Patient monitoring is now more achievable, effective, and responsive thanks to the incorporation of portable biosensors into telemedicine and home healthcare. The gadgets are crucial to facilitate remote patient monitoring (RPM), which allows medical professionals to keep an eye on patients' health indicators outside of traditional clinical settings. Vital indicators, like blood pressure, oxygen saturation, heart rate, and glucose levels, can be continuously monitored with wearable biosensors.[31] Health practitioners may stay informed about the patient's condition and take prompt action if any irregularities are found thanks to this real-time monitoring.[32] When it comes to chronic conditions like hypertension, diabetes, and cardiovascular diseases, wearable biosensors enable people to keep a close eye on their health. To avoid irregularities in data collection by the patient and healthcare provider for suitable action, regular data collection helps patients better manage their diseases and guarantees fewer hospital visits and emergency interventions.[33] Allowing decision-making to be data-driven, wearable biosensors can collect information that can be used to analyse long-term patterns and trends in a patient's health.[34] These would then be used in clinical judgments to help improve health outcomes by further adjusting treatment regimens on the basis of individual needs and responses. Biosensors increase patient participation in health management by enabling remote health monitoring.[35] Additionally, by giving users immediate, direct feedback and health-related

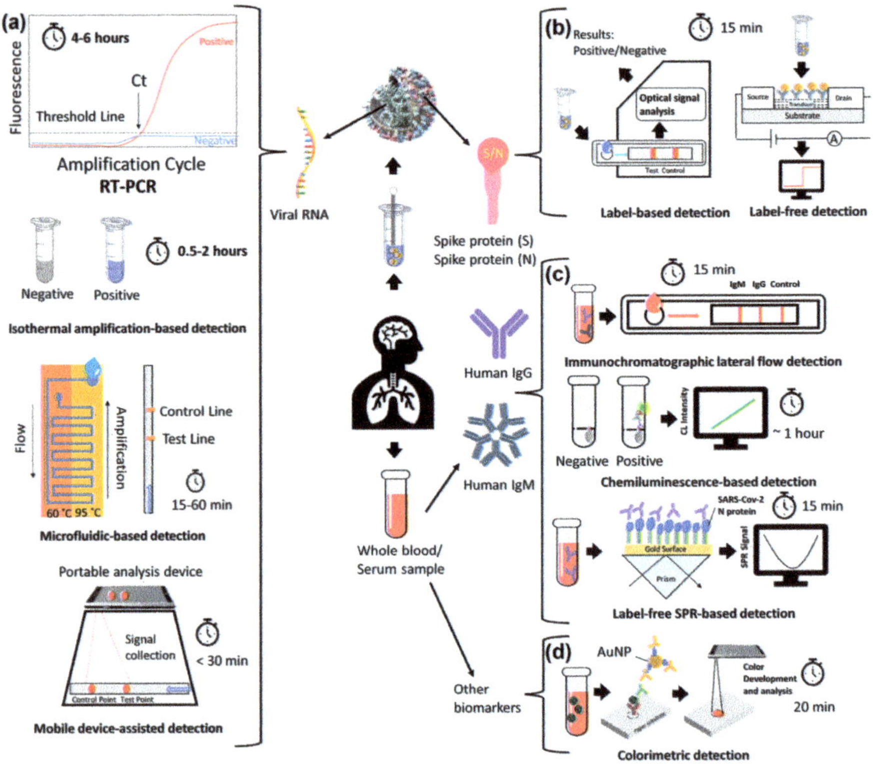

Figure 18.1 General descriptions and approximate analysis times for the diagnosis of COVID-19 based on detection of (a) viral RNA, (b) viral proteins and viral particles, (c) serology tests for IgG/IgM and (d) other markers and biomarkers.[30] Reproduced from ref. 30 with permission from Elsevier, Copyright 2021.

insights, these devices support the development of a treatment-compliant mind set and improved lifestyle. As a result, patients can be held directly accountable for their health. Additionally, by monitoring patients continuously at home, difficulties can be identified considerably sooner, allowing for the required actions to be performed and a reduction in readmissions to the hospital.[36] Patients who have had surgery or even those who have chronic illnesses, when complications are more common, can benefit greatly from this. People who live in rural areas or in remote locations with insufficient medical facilities can now get healthcare thanks to biosensors. These patients can supervise their health and quality of life because they can obtain excellent home-based care.[37] Wearable biosensors can easily connect to telemedicine systems, allowing medical professionals to monitor patients' well-being in real time and conduct virtual consultations.[37–39] Once more, this will increase the efficacy of home visits and yield more reliable evaluations, which will allow provision of individualized treatment plans. Biosensors are at the

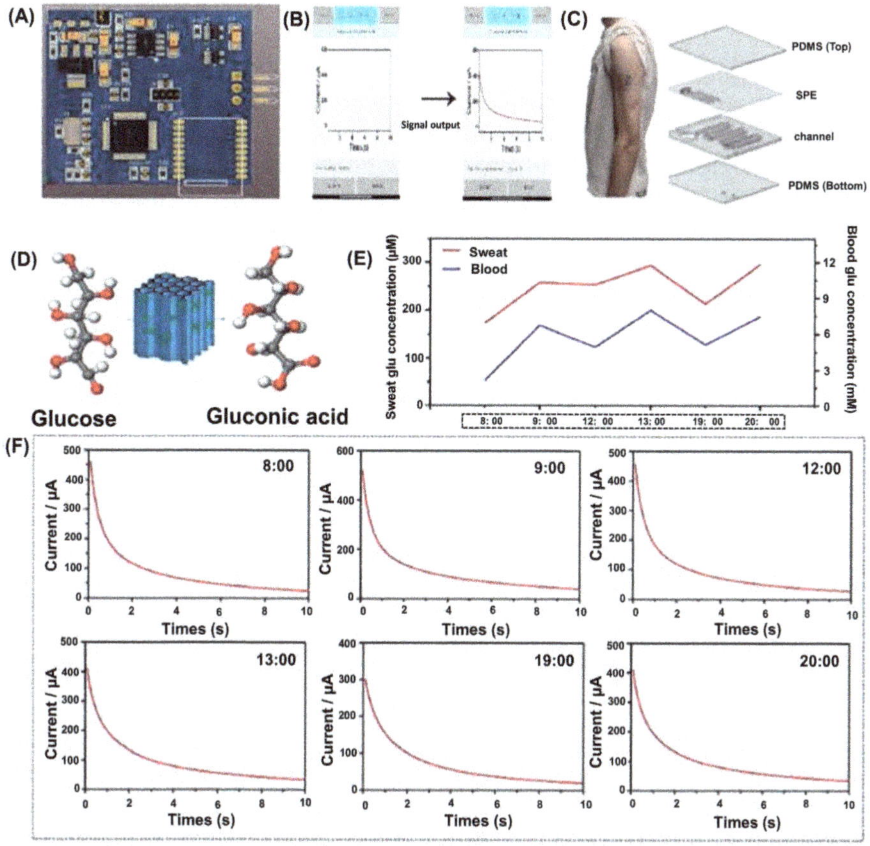

Figure 18.2 (a) Point of care testing device signal transmission circuit diagram. (b) Smartphone app for electrochemical signal reception and processing. (c) Wearable device attached to skin surface. Abbreviations: polydimethylsiloxane (PDMS), screen printed electrodes (SPE). (d) Glucose catalysed into glucuronic acid with the help of Co_3O_4/reduced graphene oxide (rGO)/Pt. (e) Variations in blood glucose (glu) and sweat glucose levels before and after meals. (f) Sweat glucose detection results at various time intervals. Copyright Elsevier (2024).[40] Reproduced from ref. 40, https://doi.org/10.3390/bios14110560, under the terms of the CC BY 4.0 license, https://creativecommons.org/licenses/by/4.0/.

leading position in remote patient analysis, improving home healthcare and the efficacy and accessibility of telemedicine. Through ongoing health monitoring, support for managing chronic diseases, and patient engagement, these tools enhance health outcomes and reduce the strain on healthcare systems. Equipment will continue to be the driving force behind the development of biosensors in RPM, revolutionizing the way healthcare is provided and received (Figure 18.2).

18.2 Collaboration with Academia, Industry and Health Care

The swift progress and adoption of intelligent internet of things (IoT)-based technologies have created a multitude of innovative prospects for technical breakthroughs in all areas of life.[41,42] Digitalization has resulted in "smart" becoming key to today's technological breakthroughs. IoT technologies have enormous potential for innovation and societal advantages, making them one of the primary pillars of the fourth industrial revolution. A new internet-based technology, the IoT promises to link physical items like household appliances and commercial machinery. These gadgets could provide customers with a range of services and useful data by leveraging appropriate communication networks and sensors. Over 125×10^9 IoT devices are expected to be connected in the next ten years, according to projections, as the proliferation of IoT technology gains momentum.[42] The compound annual growth rate is added to over 7.3%, the finances for IoT technologies are substantial and reached over 120×10^9 US dollars in 2021.[43] The current IoT technology market structure is depicted in Figure 18.3, which indicates that the industrial and smart city IoT segments account for the majority of the market. When developing an IoT application, the first step in constructing IoT systems is the choice of IoT components, such as processing, data storage, communication protocols, and sensor devices, must be appropriate for the intended use. For instance, an IoT platform that controls a building's heating, ventilation, and air conditioning (HVAC) system makes use of the right environmental sensors and pertinent communication technology.[44] Figure 18.4 depicts the various modules of the IoT platform and the common IoT connection and network structure.[45] The current global population is estimated to be 8.2 billion, and it is growing at a rate of 1.1% year, according to World Bank statistics. As mentioned previously, cities are densely populated, and according to UN projections, 68% of the world's population will reside in cities by 2050). As a result, cities are projected to experience significant pressures on their infrastructures as a result of growing urbanization; therefore, novel technological solutions would be essential to guaranty the regular operation of cities in these complex and demanding situations. In this case, addressing some of the city's primary infrastructure issues may be greatly aided by the extensive usage of smart technology and the IoT. Additionally, a variety of smart devices, apps, networking, and services have been made possible by IoT technologies, which have proven advantageous for daily life and produce synergistic consequences. Additionally, there are several uses for IoT, ranging from manufacturing, logistics, and construction[46] to power-efficient building management, drone-based services, environmental monitoring co-ordination, and healthcare services.[47–50] According to a number of project findings, the concept of smart cities,[51] the industrial sector,[52] the agriculture sector, power network management,[53] smart energy management,[43] and the transportation sector[54] are the most important and rapidly expanding IoT applications. Furthermore, IoT-based methods offer a completely new

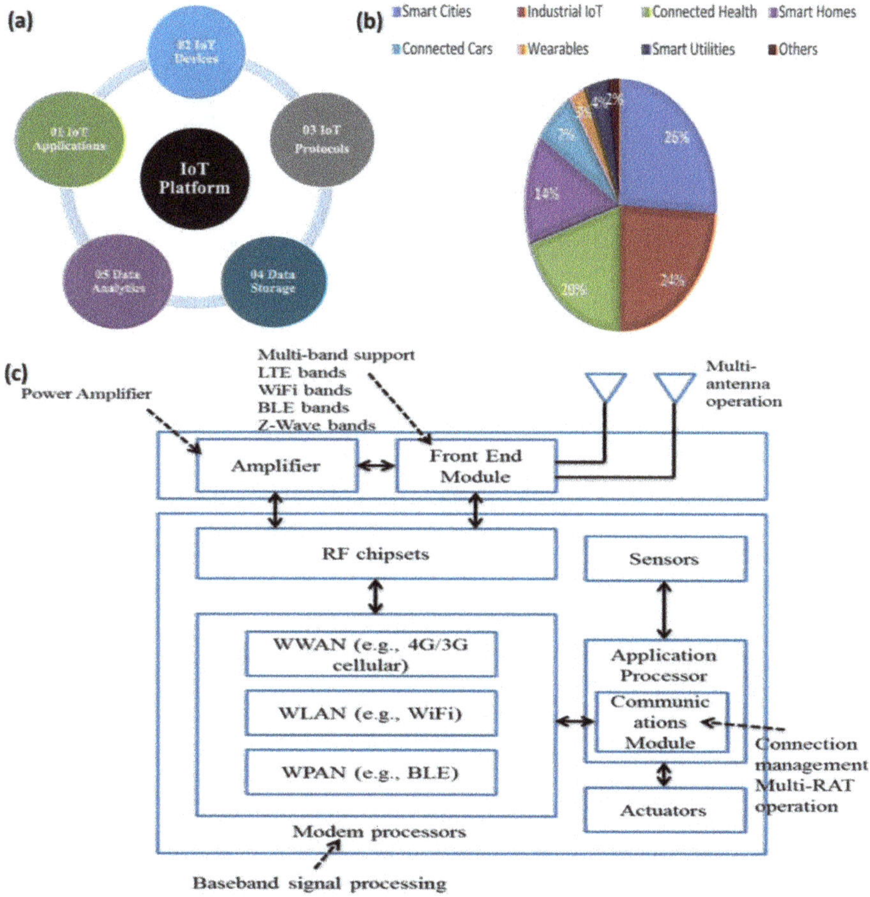

Figure 18.3 The components of an IoT platform are shown schematically in (a). The general market structure of IoT technologies is shown in (b). The general structure of IoT networks and connections is shown in (c).[57] Abbreviations: long term evolution (LTE), Bluetooth low energy (BLE), radio frequency (RF), wireless wide area network (WWAN), wireless local area network (WLAN), wireless personal area network (WPAN), multiple radio access technologies (Multi-RAT). Reproduced from ref. 57, https://doi.org/10.1016/j.biosx.2022.100153, under the terms of the CC BY 4.0 license, https://creativecommons.org/licenses/by/4.0/.

perspective for progress in a variety of fields, including engineering,[55] agriculture,[55] medical,[56] and so on. Certain possible applications of IoT technologies are still unclear or unknown, which means that in order to find new and significant potential benefits for society, in-depth research and analysis should be conducted on this difficult subject. As a result, it should be easier to understand the significance and applicability of IoT techniques in the future.

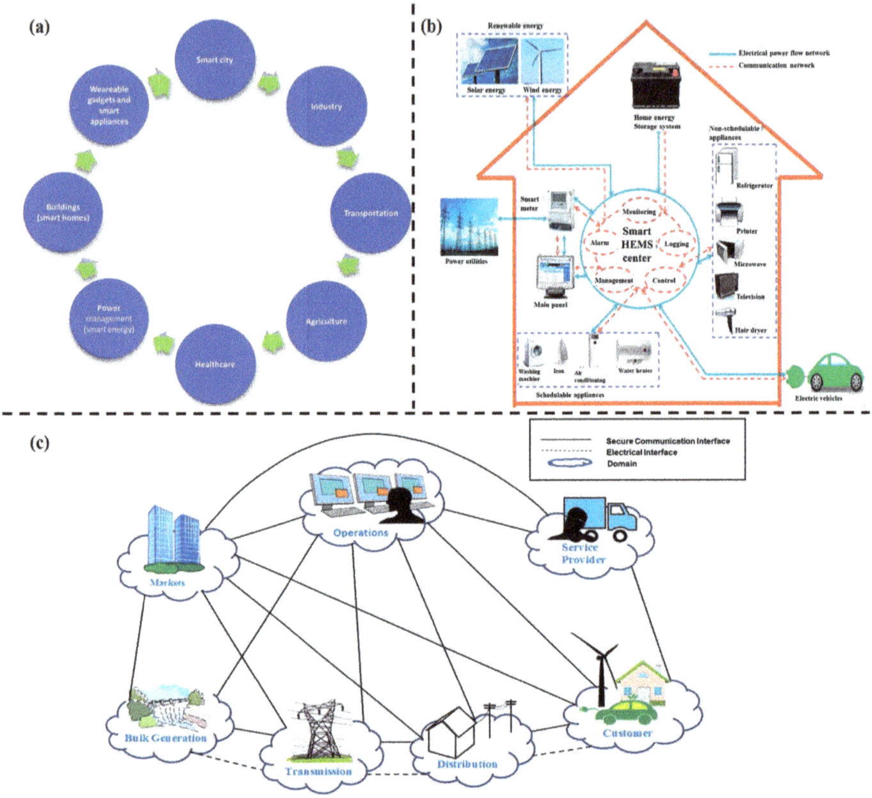

Figure 18.4 (a) Application areas of IoT technologies. (b) Schematic representation of a smart home system using smart home energy management system (HEMS). (c) Smart grids.[57] Reproduced from ref. 57, https://doi.org/10.1016/j.biosx.2022.100153, under the terms of the CC BY 4.0 license, https://creativecommons.org/licenses/by/4.0/.

The IoT has effectively transformed the globe and affected general and working lifestyles.[58] Because of its many uses, ranging from patient-centric to hospital-centric, the most significant IoT domain is healthcare and medicine, which has the potential to completely transform the health maintenance industry after 10 years.[59] Patients are taking more active roles in their care since IoT technology enables them to communicate with their physicians, set appointments, and access their medical records *via* a portal. Additionally, home monitoring devices made it possible for people and medical professionals to check their health in real time from the convenience of their own homes, which was advantageous for old and chronically affected patients.[60] Through a range of devices, such as blood pressure (BP) analysers, glucose meters, mobile X-ray machines, smart beds, ultrasounds, and thermometer units, the IoT offers the potential to improve patient care.[61] Currently, IoT-based portable devices are classified as follows: body apparel (coats, pants, and underwear); head protection (helmets and spectacles); wrist protection (gloves, watches, and bracelets); and somatic sensory modulators, such as sensory control devices for the body and feet. Furthermore, the way the internet functions is shifting from the internet of computers (IoC) to the industrial internet of things (IIoT), since humans and their surroundings make up the majority of interconnected systems, including embedded devices, smart items, and infrastructure. The IoT is at risk due to security and ethical issues because, once all data and devices are linked to set of connections, hackers could contact them and use them for different types of fraud. Artificial intelligence (AI) is widely used in healthcare commerce to improve quality of life in a variety of ways through wearable devices. The ability of AI to transform computers into intelligent agents has made it more and more popular in the current world.,[62] and it can assist with medical image analysis and diagnosis, which could improve patient care and staff efficiency.[63] Additionally, clinicians may now employ an AI-enabled IoT on patients to identify a range of conditions, such as diabetes, heart disease, and gait issues, with the aid of data gathered in the Kaggle repository. The Google Brain initiative's AI-powered diagnosis of diabetic eye illness is an example of AI-based IoT based on deep learning, which employs neural networks to learn and accomplish a particular task through repetition and self-correction. Another example is next-generation wearable devices for the blind, which employ ultrasonic technology to identify and alert users of hazards on their route, allowing them to safely navigate things in their surroundings. AI wearables can also assist those seeking to improve their fitness with their regular training regimens by allowing them to track their progress. If the user reaches 12 000 steps, the wearable device will count and show the steps. In addition to measuring data, wearables with AI capabilities can make recommendations about how to increase one's level of fitness, including what to eat, how much sleep to obtain, and how to train. Fitness tracker headphones that use Bluetooth Smart with AI biosensors to measure heart rate, elevation, motion, proximity, and touch provide an effective means of reaching personal fitness objectives depending on health criteria. In conclusion, wearable AI pushes the envelope by assisting clinicians and patients with remote diagnostics, remote tracking, preventative measures, and patient guidance for progressive decision-making. AI makes it possible for devices in the fitness sector to serve as personal assistants, supporting customers.

18.2.1 Technology and Policies

Although biosensors have made great strides in laboratory settings, a major challenge remains to quickly translate lab research into marketable prototypes and manage the intricate dogmatic requirements obligatory for clinical settings, especially during epidemics. Corporate and government spending on technology transfer is steadily rising. In addition, several countries' governments are starting to make their rules more business-friendly for students. This is exemplified by the fact that many associations are currently categorizing and filling gaps between industrial needs and academic research through seminars and proof-of-concept programs financed by universities. Furthermore, several governments have established limitless specialized urgent funds during infectious illness outbreaks, like the COVID-19 pandemic. However, despite being well-established and developed from an infrastructural perspective, instruments like SPR, atomic force microscopy (AFM), X-ray diffraction (XRD), and electron microscopy (EM) are expensive, which primarily restricts their use to laboratory testing.[64,65] However, morphological traits may be revealed by large-scale enterprises, particularly those engaged in vaccine research and testing.[66] Though developing human resource capability is a major obstacle to the technological transfer of the majority of biosensors and sophisticated instrumentation. The primary issue is that the operator of the sensing equipment frequently needs practical, hands-on training in disciplines that are fundamentally different from one another, like biology and electronics. Many educational institutions are also addressing this limitation by encouraging transdisciplinary projects and courses. The dependability of the sensors and the rapid adoption of biosensor manufacturing methods are two more technological obstacles that still stand in the way of rapid technology transfer during pandemics. Since, most rapidly expanding technologies are adapted for endemic strains; research can take many months to yield numerical data for drawing educated conclusions. In this case, fundamental research and engineering must keep finding naturally stable materials in terms of chemistry and physics and create technologies that will improve the accuracy of measurements. Furthermore, modular sensing systems offer versatility in the application of various analytical biomarkers. Two more technological barriers that continue to impede rapid technology transfer during pandemics are the reliability of the sensors and the quick adoption of biosensor production techniques.

18.2.2 Ethics and Technology Adoption

There are significant ethical concerns around cutting-edge sensing technologies, even while they have the ability to completely transform the healthcare system in times of crisis, like COVID-19. It is quite difficult to resolve disputes pertaining to data privacy, ownership, and security in a short amount of time, which causes community members to utilize the technology less frequently.[67,68] In particular, sensors that are designed to remotely offer information management systems with fast, thorough, and impartial feedback on measurements must be compatible with local data protection laws. Although passing data security quality controls has been easier in recent years, these technologies' acceptability in requisites of dependability is still low. Many medical professionals are doubtful that the technology was created with sufficient understanding of biology to make accurate measurements. Furthermore, in conventional medicine, a practitioner must obtain a patient's agreement before administering a treatment and fairly explain the risks and advantages of the planned course of action. The hazards and advantages of self-use biosensors and mobile health applications are not made clear in this context since clients are typically less inclined to read the conditions and regulations that appear when using an application. When it comes to laboratory-based testing, surface characterization technologies and biosensors remove ethical concerns like those related to testing on humans and animals.[69,70] For instance, extensive testing on high-throughput sensing systems may be possible before a medicine is deemed suitable for testing on humans and animals if it needs to be created quickly during a pandemic. A range of characterizations could be used in large-scale testing to determine the similarity of molecules, their selectivity, as well as cross-reactivity amid putative chemicals inside the body. Hence, many uncertainties related to rapid drug development and testing may be reduced with the growing usage of contemporary characterization methods and biochemical or chemical sensors.

18.3 Potential for Sensor Technologies in Future Pandemic Preparedness

The culture method, CRISPR-circularisation (CRISPR-C), PCR, reverse transcription PCR (RT-PCR), enzyme-linked immunosorbent assay (ELISA), and variations of these biological techniques are examples of traditional virus detection methods. Recently, biosensors have been used to identify different analytes, include bacteria, viruses, proteins, nucleic acids, cancer biomarkers, and toxins. They are widely used in environmental monitoring, food safety, and clinical diagnostics. Electrochemical biosensors attracted a lot of focus because of their great sensitivity, affordability, and quick reaction times. Nanoparticles can be used to further improve the sensitivity and effectiveness of these biosensors. Recent years have seen the development of numerous biosensors employing 2D nanomaterials, like graphene or transition metal and carbon or nitrogen-based MXenes, because of their remarkable electrical conductivity as well as other characteristics, including good reproducibility, ease of synthesis, strength, high specificity, and appropriate band gaps. Because of these characteristics MXene- and graphene-based nanomaterials are special because that can be used to create electrochemical biosensors that can identify a variety of analytes, including bacteria, viruses, cancer antigens, *etc.* (Figure 18.5)

The use of AI has become important for improving the ability to prepare reactions of healthcare systems to pandemics. Because AI can process enormous amounts of health data in real time, hospitals and policymakers may predict spikes in patient demand and change resource allocation accordingly.[72] AI-driven predictive models project hospital bed occupancy and intensive critical care unit capacity requirements by analysing epidemiological data, patient admission trends, and recent infection patterns.[73] AI-based forecasting models helped in anticipating hospital strain during

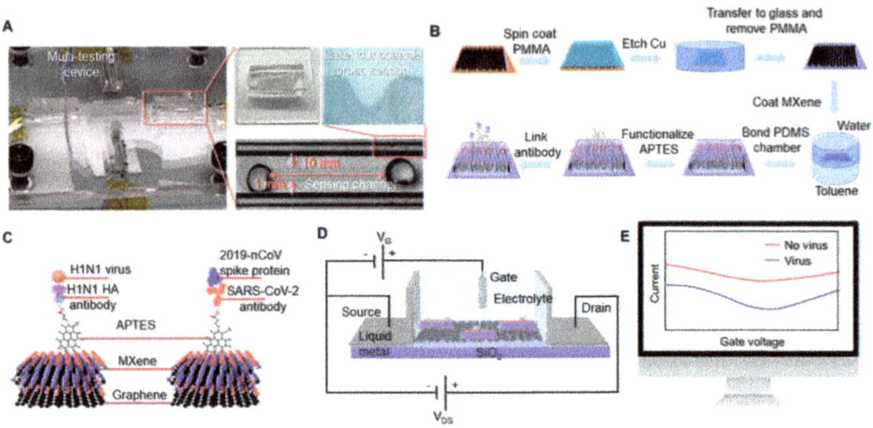

Figure 18.5 (a) Transistor sensor utilizing MXene–graphene field effect. (b) Deposition. Abbreviations: polymethyl methacrylate (PMMA), polydimethylsiloxane (PDMS), (3-aminopropyl)triethoxysilane (APTES) (c) Mechanism of detection. Abbreviations: hemagglutinin (HA), novel coronavirus (nCoV). (d) Circuit. (e) Drain–source current shift.[71] Reproduced from ref. 71 with permission from American Chemical Society, Copyright 2021.

the COVID-19 epidemic and directed decision-makers in directing the supply of vital medications, medical personnel, and ventilators to the areas that required them the most.[74] Additionally, AI systems have been used to automate triage, helping medical professionals in evaluating patients according to comorbidities and the severity of their diseases. Advanced computer vision medical imaging data, including chest X-rays and CT scans, were evaluated by software to assist radiologists in diagnosing viral pneumonia, greatly speeding up the process of diagnosis.[75] Additionally, by examining electronic health records (EHRs), emergency room (ER) logs, and tele-health consultations, AI-driven natural language processing (NLP) algorithms have made it possible to follow illness symptoms and treatment requirements in real time. By offering automated symptom assessment, self-isolation advice, and mental health support for the general public, AI-based chat-bots and virtual assistants significantly lessen the workload for medical professionals.[76] The use of reinforcement learning models to predict different pandemic situations and propose the most effective methods of allocating resources has been a significant development in AI-assisted hospital administration.[77] Dynamic and data-driven decision-making can be supported by AI-driven systems that take into account factors like supply chain disturbances, healthcare facility capacity, and the severity of regional outbreaks.[78] However, even with these advantages there are still issues with data accuracy, interoperability, and ethics with AI-based healthcare planning systems. For AI models to operate effectively, high-quality real-time data is required, yet data integration is usually hampered by uneven reporting of data and privacy regulations. It is also still very important to ensure that AI-driven healthcare actions are impartial and democratic, which calls for constant enhancement of models and supervision.[79,80]

18.4 Artificial Intelligence and Emerging Technologies in Public Health

Rapid developments in emerging technologies are improving early outbreak identification, predictive modelling, and real-time response strategies in AI-driven disease surveillance.[78] Some of the major developments influencing the future of AI in public health include the use of synthetic data, enhanced data integration frameworks, and advanced deep learning algorithms.[79] Federated learning, which enables several healthcare organizations to train AI models together without sharing raw patient data, is one of the most promising developments.[79] By using a variety of universal datasets this method improves privacy protection while increasing the generalizability and accuracy of epidemic prediction models.[81] Federated learning has been especially successful in pandemic surveillance allowing safe cross-border data sharing while adhering to privacy laws like the General Data Protection Regulation (GDPR) and the Health Insurance Portability and Accountability Act (HIPAA).[82] The combination of block chain technology and AI is another innovation that ensures safe, unchangeable sharing and storage of health data.[83] By preventing data breaches and illegal changes, block chain-enhanced AI systems can boost public confidence in AI-powered surveillance systems. Block chain-based vaccine tracking has been successfully utilized by nations to cut down on supply chain inefficiencies, fraud, and misreporting.[84] Neural–symbolic AI, which blends symbolic thinking and machine learning, is also becoming a significant advancement for interpretable AI in illness surveillance. Neural–symbolic AI, as compared with black-box deep learning models, can provide human-understandable explanations for its predictions, enhancing transparency and confidence in public health decision-making.[85] Additionally real time disease diagnosis is being revolutionized by the use of edge AI. Edge AI improves rapid epidemic detection in resource-constrained environments and lessens reliance on cloud computing by processing health data directly on local devices, including smartphones, wearable sensors, and remote diagnostic tools.[86] to help contain such outbreaks more quickly AI-driven mobile diagnostic kits, for example, have been placed in remote areas to analyse pathogen samples in real time. Additionally, public health preparedness is being improved by the combination of AI and genetic surveillance. AI systems examine new infections' genomic alterations to identify potentially problematic variants before they spread.[87] This was seen during the COVID-19 pandemic when SARS-CoV-2 mutations were found *via* AI-assisted genomic analysis, resulting in identification of variations like alpha, delta, and omicron before they produced significant epidemics. These developments are opening the way for a more robust, AI-powered global health system that can identify and treat infectious diseases more quickly and effectively than in the past.[88]

18.5 Early Identification of the Next Pandemic

Early warning signs of new infectious diseases include unusually high rates of illness among young, healthy adults or medical specialists as well as an unusual rise in the use of over-the-counter drugs and traditional medicines. The use of event tracking systems to identify these signs in a proactive approach to epidemic identification have led to its increased popularity in recent years. Early warning techniques using AI attracted a lot of discussion during the COVID-19 pandemic. In order to produce early warning signals for pandemic monitoring, AI systems use technologies, like using machine learning and text mining, to analyse and filter quantities of publicly available data without requiring a lot of human intervention. AI has several benefits for early warning systems. Because the system processes open-source data connected globally to provide epidemic alerts it is quick and works in real time. It can theoretically alleviate the problem of a lack of capability in the sector because it requires little human intervention. It has been shown to be capable of identifying an Ebola outbreak swiftly, even in places with comparatively low adoption of smartphones. It is also appropriate for poor nations.[89] It can be used in conjunction with more risk simulation and analysis software tools to predict the severity of future epidemics and prioritize resources accordingly. The majority of AI systems are freely accessible online, allowing other industries to use them as tools for pandemic preparedness and early warning. Although other systems including Health Map and EPIWATCH, can perform analyses using syndrome-based methods, which makes it possible to identify epidemics early on, even for unidentified illnesses.

Most AI early warning systems, like Epidemic Intelligence from Open Sources (EIOS) and Meta-biota, only investigate particular pathogens. Event-based surveillance's major drawback is that, like indicator-based surveillance, much of the data analysed is non-specific. Unless the data is derived from credible sources, like medical professionals, or is particularly concerning, like a patient's or healthcare worker's unexplained severe acute respiratory illness, it takes a lot of work for public health personnel to confirm every signal found by event-based monitoring and to conduct thorough microbiological surveillance for every occurrence. For the purpose of identifying the new agents for the upcoming pandemic, event-based surveillance might not be as effective on its own. Furthermore, censorship may change AI algorithms, resulting in inaccurate reporting by AI technology, depending on how it is regulated. Rapid outbreak containment should be the goal of AI early warning systems. The integration of multidimensional information, like past travel records, animal and vector distribution across regions, temperature, and humidity, is the future direction for AI research. This will allow for the spatiotemporal analysis of upcoming outbreaks. Officials in charge of public health can gather monitoring samples and implement focused actions to prevent recurrences by combining animal data with spatial data on disease cases. This helps with identifying significant interactions between humans and animals that lead to microbial spillover. However, the news or social media was where SARS-CoV-1 in 2003 and SARS-CoV-2 in 2019 outbreaks were first reported. Specifically, when a physician expressed on social media in 2019 that they were concerned about seeing patients with unusually severe respiratory conditions, or when young, healthy, and health-conscious healthcare workers were admitted to hospital in 2003 due to atypical pneumonia, these are significant examples of soft intelligence that require careful consideration. Event-based surveillance should also focus on major international meetings, as they could be potential sites for outbreaks that could spread globally. Examples include M-pox outbreaks linked to Pride celebrations, Hepatitis A, and arboviruses linked to the Hajj pilgrimage, as well as *Vibrio cholerae* and *Neisseria meningitides*.[90,91] Healthcare specialists who attend international conferences and work at animal–human connection conferences are checked for fever, respiratory symptoms, sick leave, or hospitalization. This is probably where AI early warning systems could be able to detect a helpful signal of the next significant outbreak. Laboratories used viruses that were unknown or non-typable to determine the human cases from the 2009 HINI influenza pandemic in the USA, the Hong Kong 1997 H5N1 avian influenza, Saudi Arabia's 2012 Middle East respiratory syndrome coronavirus (MERS-CoV), Wuhan's SARS-CoV-2 in 2019 and SARS-CoV-1 in Hong Kong in 2003 and the past epidemics caused by new respiratory viruses. Clinical laboratory surveillance has found isolates not recognized using traditional laboratory techniques, local syndromic monitoring and early warning systems were not capable of finding, identifying, or providing notification of these new infections, which may indicate that these methods are not very sensitive. This is caused by acute community acquired pneumonia being an extremely prevalent clinical disease that leads to hospitalization.

18.6 Preventing and Getting Ready for the Next Pandemic

In order to prepare for and avoid the next pandemic and guarantee both productivity and safety, strict control of biosafety level 3 and level 4 laboratories is also necessary, as laboratory mishaps or bioterrorism may be the cause of the next pandemic There were many instances of infections linked to laboratories that caused outbreaks.[92] The 1978 smallpox outbreak appears to have started at the UK's University of Birmingham Medical School. Since a service duct was not properly maintained while the lab was testing the variola virus. With respect to the Russian influenza of 1977, a H1N1 case closely looked like a strain that was circulating worldwide from 1946 to 1957 it was possible that the virus has accidently escaped from a lab or an active vaccination trial.[93] The 2004 SARS outbreaks were probably caused by laboratory-acquired cases, which led to the disease's subsequent spread in hospitals and the community.[94] In 2004 the College of American Pathologists accidently sent live H2N2 influenza A virus samples (A/Japan/305/57) to a number of worldwide laboratories. Accordingly, in order to avoid disastrous outcomes the samples were promptly recalled.[95] In 2019, a Brucella vaccine strain was discharged into the waste gas from a biopharmaceutical plant in Lanzhou killing at least 10 000 nearby individuals. To prepare for the future pandemics or epidemics, development of vaccines is essential for preventing infection stopping the spread of infection and lessening the severity of viral diseases such as COVID-19 and influenza.[96] Additionally, a sufficient oxygen supply, medications for symptom management (antipyretics), along with steroid immunomodulation may all be crucial. When administered during outbreaks, vaccines, particularly those that protect the mucosa, may prevent the virus from spreading further. In order to prevent the spread of possible developing pandemics, it may be essential to have flexible vaccination platforms and enough facilities for producing vaccines that can create large quantities of vaccines quickly or replenish stockpiles. One produced vaccine is the 17DD vaccination against yellow fever. It was used to prevent the transmission of the virus when yellow fever struck Angola and the Democratic Republic of the Congo in 2016.[97]

18.7 Challenges in the Developing Nations

Given the occurrence of human–animal contacts and frequent spillover events in tropical and subtropical climates, conducting microbial surveillance there is beneficial. However, in contrast to clinical testing and medical countermeasures for pandemics, developing nations may face challenges including routine surveillance capabilities and public health interest and investment in surveillance. It is difficult to discuss the problem of incomplete resources in developing nations while simultaneously enhancing microbiological surveillance and pandemic preparedness. Particularly in developing nations, the COVID-19 outbreak has bolstered the establishment of regional sequencing centers for microbial surveillance. However, most nations are reducing their pandemic control expenditures and concentrating on resuming normal economic activity as a result of SARS-CoV-2 strains currently showing a reduction in the intensity of sickness. It is important to convince political leaders of the advantages of improving their pandemic preparation laboratories. Another way to overcome these difficulties is to establish a centralized laboratory for high throughput sequencing at the regional or national level. In order to efficiently employ resources, a centralized laboratory enables targeted funding for the acquisition of sequencing equipment, concentrates expertise in one place, and processes all surveillance specimens centrally. This enables the selection of representative samples. Additionally, the issue of a shortage of personnel can be avoided by utilizing AI early warning systems. Some parties, like biotechnology or pharmaceutical firms, might also assist by keeping prices down when offering antibiotics, vaccines, consumables, and reagents when the diseases in developing countries have disrupted regular socioeconomic activity. Due to their segmented genomes Bunyavirales and Reoviridae may become more powerful pandemic agents as climate change increases.[98] While the ambient temperature that controls arthropod activity severely affects arboviruses. Quiet pandemics caused by agents such as HIV or germs or fungi that are resistant to antibiotics in addition to acute, explosive respiratory or systemic disorders must not be overlooked. On the other hand, the latter may lead to germs that are nearly resistant to carbapenem. An increased susceptibility to several opportunistic agents occurs in patients with acute respiratory illnesses who have fungal (*Candida auris*, Mucorales) or Enterobacteriaceae (pan-drug-resistant *Acinetobacter baumannii*) superinfections.[99]

18.8 Conclusion

In conclusion, the collaboration between academia, industry, and healthcare has significantly advanced virus detection sensors, offering rapid, accurate, and cost-effective solutions for diagnosing viral diseases by integrating emerging technologies, by focusing on user friendly devices and through investments, collaboration, mass manufacturing, and by integration with the (IoT), and ease of use for members of the community. These collaborations highlight the power of interdisciplinary partnerships in driving innovation in virus detection sensors. By combining the strengths of academia, industry, and healthcare, it should be possible to accelerate the development and deployment of advanced diagnostic technologies, ultimately enhancing global health security and preparedness against infectious diseases.

List of Abbreviations

ASFV	African swine fever virus
AFM	Atomic force microscopy
AIV	Avian influenza virus
COVID	Corona virus disease
CRISPR	Clustered regularly interspaced short palindromic repeats
EBOV	Ebola virus
EHRs	Electronic health records
EIOS	Epidemic intelligence from open sources
HDA	Helicase-dependent amplification
IAV	Influenza A virus
IIoT	Industrial internet of things
IoC	Internet of computers
IoT	Internet of things
LOC	Lab-on-a-chip
LAMP	Loop-mediated isothermal amplification

NLP	Natural language processing
NASBA	Nucleic acid sequence-based amplification
RAA	Recombinase-aided amplification assay
RPA	Recombinase polymerase amplification
RPM	Remote patient monitoring
RCA	Rolling circle amplification
SDA	Strand displacement amplification
SEIRA	Surface-enhanced infrared absorption spectroscopy
SERS	Surface-enhanced Raman scattering
SPR	Surface plasmon resonance
XRD	X-ray diffraction

References

1. G. P. Hancke and G. P. Hancke Jr, *Sensors*, 2013, **13**, 393–425.
2. M. M. Alvarez, J. Aizenberg, M. Analoui, A. M. Andrews, G. Bisker, E. S. Boyden, R. D. Kamm, J. M. Karp, D. J. Mooney and R. Oklu, *ACS Nano*, 2017, **11**, 5195–5214.
3. A. A. Malik, C. Nantasenamat and T. Piacham, *Mater. Sci. Eng.: C*, 2017, **77**, 1341–1348.
4. P. J. Hotez, *Microbes Infect.*, 2015, **17**, 539–541.
5. Y. K. Chan and M. U. Gack, *Nat. Rev. Microbiol.*, 2016, **14**, 360–373.
6. B. Danqi, Z. Li, Q. Liu and J. A. Richt, *Expert Rev. Anti-Infect. Ther.*, 2017, **15**, 723–727.
7. D. Tian and W. Chen, *Chin. J. Biotechnol.*, 2017, **33**, 1–15.
8. M. Keita, H. Lucaccioni, M. K. Ilumbulumbu, J. Polonsky, J. Nsio-Mbeta, G. T. Panda, P. C. Adikey, J. K. Ngwama, M. K. Tosalisana and B. Diallo, *Emerging Infect. Dis.*, 2021, **27**, 2988.
9. B. Yang, C. Shen, D. Zhang, T. Zhang, X. Shi, J. Yang, Y. Hao, D. Zhao, H. Cui and X. Yuan, *Virol. J.*, 2021, **18**, 1–16.
10. S. Kannan, P. S. S. Ali, A. Sheeza and K. Hemalatha, *Eur. Rev. Med. Pharmacol. Sci.*, 2020, **24**, 2006–2011.
11. Y. Li, J. Yu, Y. Wang, J. Yi, L. Guo, Q. Wang, G. Zhang, Y. Xu and Y. Zhao, *J. Med. Virol.*, 2024, **96**, e29602.
12. X. Feng, Y. Liu, Y. Zhao, Z. Sun, N. Xu, C. Zhao and W. Xia, *Int. J. Nanomed.*, 2023, 6311–6331.
13. R. Kaur, S. Maan, K. Batra, N. Singh, N. Chahal and A. Kumar, *Mol. Biol. Rep.*, 2024, **51**, 178.
14. X. Li, S. Zhu, X. Zhang, Y. Ren, J. He, J. Zhou, L. Yin, G. Wang, T. Zhong and L. Wang, *Front. Bioeng. Biotechnol.*, 2023, **11**, 1215466.
15. Y. H. Nai, E. H. Doeven and R. M. Guijt, *PLoS One*, 2022, **17**, e0265391.
16. Y. Zhang, A. N. Luck and N. A. Tanner, *BioTechniques*, 2024, **76**, 255–262.
17. Y. Okamura, M. Suemitsu, T. Ishikawa and H. Takahashi, *Int. J. Mol. Sci.*, 2024, **25**, 5089.
18. A. Gulinaizhaer, C. Yang, M. Zou, S. Ma, X. Fan and G. Wu, *Virol. J.*, 2023, **20**, 274.

19. R. K. Daher, G. Stewart, M. Boissinot and M. G. Bergeron, *Clin. Chem.*, 2016, **62**, 947–958.
20. Q.-J. Huang, Y. Chen, H. Liu, S. St-Hilaire, S. Gao, B. MacKinnon, S.-Q. Zhu, Z.-Q. Wen, P. Jia and X.-C. Zheng, *J. Virol. Methods*, 2022, **300**, 114377.
21. P. G. Preena, T. V. A. Kumar, T. K. Johny, A. Dharmaratnam and T. R. Swaminathan, *Aquacult. Int.*, 2022, **30**, 1211–1220.
22. K. Lin, J. Guo, X. Guo, Q. Li, X. Li, Z. Sun, Z. Zhao, J. Weng, J. Wu and R. Zhang, *Anal. Chim. Acta*, 2023, **1248**, 340938.
23. J. Hu, Y. Wang, H. Ding, C. Jiang, Y. Geng, X. Sun, J. Jing, H. Gao, Z. Wang and C. Dong, *Int. J. Food Microbiol.*, 2020, **331**, 108691.
24. B. Y. Ng, W. Xiao, N. P. West, E. J. Wee, Y. Wang and M. Trau, *Anal. Chem.*, 2015, **87**, 10613–10618.
25. C. Kober, R. Niessner and M. Seidel, *Biosens. Bioelectron.*, 2018, **100**, 49–55.
26. S. Schlücker, *Angew. Chem., Int. Ed.*, 2014, **53**, 4756–4795.
27. Z. Yao, Q. Zhang, W. Zhu, M. Galluzzi, W. Zhou, J. Li, A. V. Zayats and X.-F. Yu, *Nanoscale*, 2021, **13**, 10133–10142.
28. X. P. He, B. J. Zou, X. M. Qi, S. Chen, Y. Lu, Q. Huang and G. H. Zhou, *Yi chuan = Hereditas*, 2019, **41**, 611–624.
29. L. E. Dumkow, L. J. Worden and S. N. Rao, *J. Clin. Microbiol.*, 2021, **76**, iii4–iii11.
30. Y. Shen, T. B. Anwar and A. Mulchandani, *Sens. Actuators Rep.*, 2021, **3**, 100025.
31. P. Zhu, H. Peng and A. Y. Rwei, *Med. Novel Technol. Dev.*, 2022, **14**, 100118.
32. Z. Song, S. Zhou, Y. Qin, X. Xia, Y. Sun, G. Han, T. Shu, L. Hu and Q. Zhang, *Biosensors*, 2023, **13**, 630.
33. X. Yuan, C. Li, X. Yin, Y. Yang, B. Ji, Y. Niu and L. Ren, *Biosensors*, 2023, **13**, 313.
34. M. Di Rienzo and R. Mukkamala, *Sensors*, 2021, **21**, 1291.
35. P. Li, G.-H. Lee, S. Y. Kim, S. Y. Kwon, H.-R. Kim and S. Park, *ACS Nano*, 2021, **15**, 1960–2004.
36. A. Haleem, M. Javaid, R. P. Singh, R. Suman and S. Rab, *Sens. Int.*, 2021, **2**, 100100.
37. A. Sonawane, P. Manickam and S. Bhansali, *IEEE Rev. Biomed. Eng.*, 2017, **10**, 174–186.
38. S. Kalasin and W. Surareungchai, *Anal. Chem.*, 2023, **95**, 1773–1784.
39. S. Faham, A. Salimi and R. Ghavami, *Talanta*, 2023, **253**, 123892.
40. D.-K. Vo and K. T. L. Trinh, *Biosensors*, 2024, **14**, 560.
41. C. Zheng, J. Yuan, L. Zhu, Y. Zhang and Q. Shao, *J. Cleaner Prod.*, 2020, **258**, 120689.
42. K. Haseeb, A. Almogren, N. Islam, I. Ud Din and Z. Jan, *Energies*, 2019, **12**, 4174.
43. S. Nižetić, P. Šolić, D. L.-D.-I. Gonzalez-De and L. Patrono, *J. Cleaner Prod.*, 2020, **274**, 122877.
44. A. Ramamurthy and P. Jain, *ADB Sustainable Development Working Paper Series*, Asian Development Bank, Philippines, 2017, pp. 1–36.

45. M. Jia, A. Komeily, Y. Wang and R. S. Srinivasan, *Autom. Constr.*, 2019, **101**, 111–126.
46. J. Höller, V. Tsiatsis, C. Mulligan, S. Karnouskos, S. Avesand and D. Boyle, Chapter 1 – Introduction and book structure, in *From Machine-To-Machine to the Internet of Things*, Academic Press, Oxford, 2014, pp. 3–8.
47. L. Atzori, A. Iera and G. Morabito, *Comput. Networks*, 2010, **54**, 2787–2805.
48. D. Evans, How the Next Evolution of the Internet is Changing Everything, Whitepaper, Cisco Internet Business Solutions Group (IBSG), 2011, 1, 1–12.
49. T. K. Hui, R. S. Sherratt and D. D. Sánchez, *Fut. Generat. Comput. Syst.*, 2017, **76**, 358–369.
50. N. H. Motlagh, M. Bagaa and T. Taleb, *IEEE Internet Things J.*, 2019, **6**, 6523–6536.
51. G. Sivanageswara Rao, K. Raviteja, G. Phanindra and D. Vignesh, *Int. J. Adv. Sci. Technol.*, 2020, **29**, 3691–3704.
52. P. Osterrieder, L. Budde and T. Friedli, *Int. J. Product. Econ.*, 2020, **221**, 107476.
53. M. M. Martín-Lopo, J. Boal and Á. Sánchez-Miralles, *Comput. Networks*, 2020, **171**, 107101.
54. S. Porru, F. E. Misso, F. E. Pani and C. Repetto, *J. Traffic Transp. Eng., Engl. Ed.*, 2020, **7**, 88–97.
55. A. Zaidan and B. Zaidan, *Artif. Intell. Rev.*, 2020, **53**, 141–165.
56. S. Salagare and R. Prasad, *Wireless Personal Commun.*, 2020, **110**, 1345–1371.
57. D. Verma, K. R. Singh, A. K. Yadav, V. Nayak, J. Singh, P. R. Solanki and R. P. Singh, *Biosens. Bioelectron.: X*, 2022, **11**, 100153.
58. Y. Myrka, *IoT For All*, 2020.
59. S. K. Ismail, *Fen Bilimleri Enst.*, 2017.
60. N. Wickramasinghe and F. Bodendorf, *Delivering Superior Health and Wellness Management with IoT and Analytics*, Springer, 2020.
61. C. A. Marino and C. Diaz Paz, *J. Med. Internet Res.*, 2025, **27**, e69465.
62. L. K. Ramasamy, F. Khan, M. Shah, B. V. V. S. Prasad, C. Iwendi and C. Biamba, *Sensors*, 2022, **22**, 1076.
63. A. A. Ibrahim and W. Zhuopeng, *Int. J. Eng. Res. Applicat. ISSN*, 2018, **2248**, 1–3.
64. A. J. Malkin, M. Plomp and A. McPherson, *DNA Viruses*, 2005, 85–108.
65. M. A. Iyer, G. Oza, S. Velumani, A. Maldonado, J. Romero, M. D. L. Munoz, M. Sridharan, R. Asomoza and J. Yi, *Sens. Actuators, B*, 2014, **202**, 1338–1348.
66. L. Pasquina-Lemonche, J. Burns, R. Turner, S. Kumar, R. Tank, N. Mullin, J. Wilson, B. Chakrabarti, P. Bullough and S. Foster, *Nature*, 2020, **582**, 294–297.
67. R. Evans, M. McNamee and O. Guy, *Sci. Eng. Ethics*, 2017, **23**, 1487–1505.
68. A. G. Spagnolo and V. Daloiso, *Bioethics*, 2009, **23**, 394–402.
69. S. K. Doke and S. C. Dhawale, *Saudi Pharm. J.*, 2015, **23**, 223–229.
70. J. M. Wilkinson, in *Organ-on-a-chip*, Elsevier, 2020, pp. 1–11.

71. Y. Li, Z. Peng, N. J. Holl, M. R. Hassan, J. M. Pappas, C. Wei, O. H. Izadi, Y. Wang, X. Dong, C. Wang, Y.-W. Huang, D. Kim and C. Wu, *ACS Omega*, 2021, **6**, 6643–6653.
72. G. Babanejaddehaki, A. An and M. Papagelis, *J. ACM Trans. Comput. Healthcare*, 2025, **6**, 1–40.
73. R. K. Debbadi and O. Boateng, *Int. J. Sci. Res. Arch.*, 2025, **14**, 802–820.
74. V. Jeyakumar, P. Sundaram and N. Ramapathiran, in *System design for epidemics using machine learning and deep learning*, Springer, 2023, pp. 123–152
75. O. Ajayi, Available at SSRN 5044945, 2023.
76. B. Marandi and M. K. Deep, *Mach. Learn.*, 2024, **1**, 3.
77. J. N. Chukwunweike, A. Praise and B. Bashirat, *Int. J. Res. Publ. Rev.*, 2024, **5**, 3608–3617.
78. Q.-V. Pham, D. C. Nguyen, T. Huynh-The, W.-J. Hwang and P. N. Pathirana, *IEEE Access*, 2020, **8**, 130820–130839.
79. S. J. Fong, N. Dey and J. Chaki, *Artificial Intelligence for Coronavirus Outbreak*, Springer, 2021.
80. N. L. Bragazzi, H. Dai, G. Damiani, M. Behzadifar, M. Martini and J. Wu, *Int. J. Environ. Res. Publ. Health*, 2020, **17**, 3176.
81. R. K. Debbadi and O. Boateng, *Int. J. Sci. Res. Arch.*, 2025, **14**, 733–752.
82. O. Ajayi, *Int. J. Soc. Res. Methodol.*, 2024, **12**, 573–584.
83. H. Yang, *et al.*, Epidemic informatics and control: a review from system informatics to epidemic response and risk management in public health, in *AI and Analytics for Public Health*, ed. H. Yang, R. Qiu and W. Chen, INFORMS-CSS 2020, Springer Proceedings in Business and Economics, Springer, Cham, 2022.
84. M. Bhanurangarao and R. Mahaveerakannan, *Migr. Lett.*, 2023, **20**, 1058–1072.
85. F. C. Udegbe, E. I. Nwankwo, G. T. Igwama and J. A. Olaboye, *Comput. Sci. IT Res. J.*, 2023, **4**, 525–545.
86. J. A. Olaboye, C. C. Maha, T. O. Kolawole and S. Abdul, *Int. Med. Sci. Res. J.*, 2024, **4**, 667–680.
87. I. Alrashdi and A. Alqazzaz, *Sustainable Mach. Intell. J.*, 2024, **7**(6), 1–28.
88. R. K. Debbadi and O. Boateng, *Int. J. Sci. Res. Arch.*, 2025, **14**, 778–796.
89. A. Joshi, R. Sparks, S. Karimi, S.-L. J. Yan, A. A. Chughtai, C. Paris and C. R. MacIntyre, *PloS one*, 2020, **15**, e0230322.
90. P. Gautret and R. Steffen, *Int. J. Infect. Dis.*, 2016, **47**, 46–52.
91. V. M. Mello, B. V. Lago, P. S. Sousa, F. C. Mello, C. B. Souza, L. C. Pinto, C. F. Ginuino, C. A. Fernandes, S. F. Aguiar and L. M. Villar, *Viruses*, 2019, **11**, 281.
92. J. G. Breman, *J. Infect. Dis.*, 2021, **224**, S379.
93. K. Nakajima, U. Desselberger and P. Palese, *Nature*, 1978, **274**, 334–339.
94. W.-N. Liang, T. Zhao, Z.-J. Liu, B.-Y. Guan, X. He, M. Liu, G. Liu, J. Wu, R. Huang and X. Xie, *Biomed. Environ. Sci.*, 2006, **19**, 445.
95. A. Schrauder, B. Schweiger, U. Buchholz, W. Haas, D. Sagebiel, A. Guignard and W. Hellenbrand, *Emerg. Infect. Dis.*, 2006, **12**, 1997.

96. M. A. Rolfes, B. Flannery, J. R. Chung, A. O'Halloran, S. Garg, E. A. Belongia, M. Gaglani, R. K. Zimmerman, M. L. Jackson and A. S. Monto, *Clin. Infect. Dis.*, 2019, **69**, 1845–1853.
97. C. Wang, Q. Li, X. Song and X. Dong, *Stat. Med.*, 2019, **38**, 3703–3718.
98. J. Whitehorn and S. Yacoub, *Clin. Med.*, 2019, **19**, 149–152.
99. M. Hoenigl, D. Seidel, A. Carvalho, S. M. Rudramurthy, A. Arastehfar, J.-P. Gangneux, N. Nasir, A. Bonifaz, J. Araiza and N. Klimko, *Lancet Microbe*, 2022, **3**, e543–e552.

Section VIII: Concluding Remarks

CHAPTER 19

Summary of Insights and Conclusions

YEŞEREN SAYLAN[a*], AHMET ÇAĞKAN İNKAYA[b] AND ADIL DENIZLI[a]

[a]Department of Chemistry, Hacettepe University, Ankara, Turkey;
[b]Department of Infectious Disease and Clinical Microbiology, Hacettepe University, Ankara, Turkey
*E-mail: yeseren@hacettepe.edu.tr

19.1 Summary of Key Findings and Insights

Viruses frequently emerge and circulate within human populations, often leading to disease outbreaks with varying levels of severity.[1] Over the course of human history, viral infections have imposed a heavy burden on public health systems and have triggered substantial humanitarian and economic consequences (Figure 19.1). When a novel virus is identified, it may lead to a small, localized outbreak—or, in more severe cases, escalate into a widespread epidemic or global pandemic, depending on how effectively it transmits among humans.[2] At a global level, these emergence events have repeatedly resulted in high rates of illness, loss of life, and significant financial impacts.[3]

The latest advances in virus detection using sensor technologies have yielded a wealth of insights that are reshaping diagnostic strategies across biomedical, environmental, and public health domains.[4] According to the

Detection Science Series No. 32
Advances in Virus Detection with Sensors
Edited by Adil Denizli
© The Royal Society of Chemistry 2026
Published by the Royal Society of Chemistry, www.rsc.org

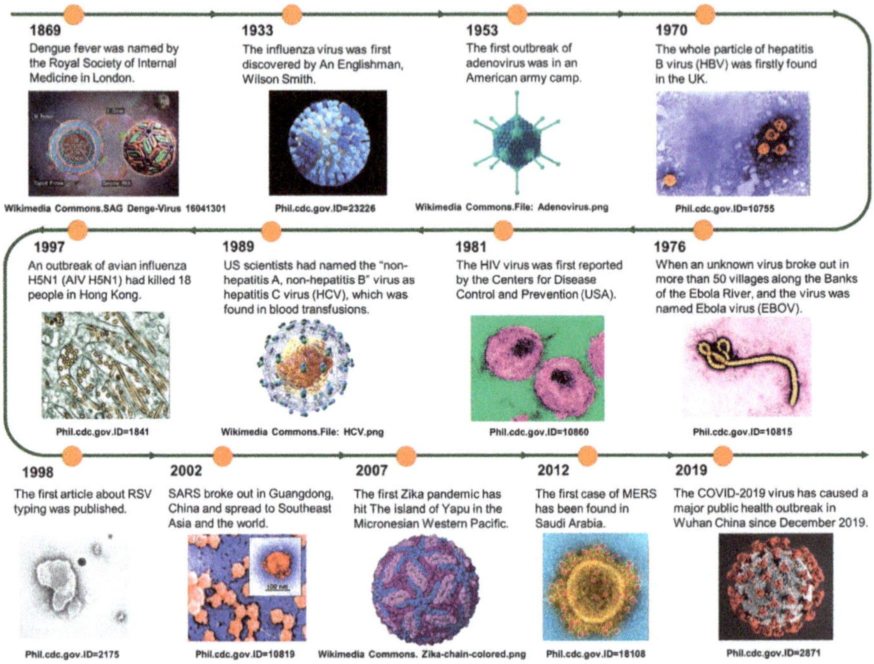

Figure 19.1 A historical overview of major outbreaks of notable infectious diseases.[1] Reproduced from ref. 1, https://doi.org/10.1002/advs.202105904, under the terms of the CC BBY 4.0 license, https://creativecommons.org/licenses/by/4.0/.

results of comprehensive analysis of current literature experimental developments in virus detection have revealed several critical findings and emerging trends.[5] For instance, the high sensitivity and specificity of functionalized materials, including gold nanoparticles, graphene, carbon nanotubes, and quantum dots, have been used to significantly enhance sensor sensitivity due to their unique electrical, optical, and surface properties.[6–8] Functionalization with specific recognition elements (*e.g.*, monoclonal antibodies, aptamers, and/or molecularly imprinted polymers) enables highly selective detection of target viruses.[9,10] In addition, multi-modal sensing platforms combining physical and chemical transduction show improved diagnostic performance compared with single-mode sensors.[11] The coupling of nanomaterials with biorecognition elements is a powerful approach for ultra-sensitive and also target-specific virus detection.[12–14] Several sensor-based platforms have demonstrated detection times within minutes, offering substantial improvements over traditional methods that require hours or days.[15] So, these platforms provide rapid detection capabilities and allow point-of-care (POC) applications.[16,17] Moreover, the portable and handheld sensor platforms, including smartphone-integrated platforms, enable testing in remote and resource-limited settings. Lab-on-a-chip (LOC) and microfluidic sensors streamline sample processing, reducing the need for external laboratory infrastructure.[18–21] The growth of fast, field-deployable sensors is crucial for early intervention and especially pandemic containment strategies.[22,23]

The sensor platforms offer versatility, such as using whole viruses, antigens, and nucleic acids in target recognition.[24] The whole virions [*e.g.*, severe acute respiratory system coronavirus 2 (SARS-CoV-2), dengue, human immunodeficiency virus (HIV)] are used with surface binding techniques.[25,26] Viral proteins and antigens are often utilized for early immune response monitoring. The viral RNA or DNA detected *via* hybridization-based sensors are combined with isothermal amplification (*e.g.*, loop mediated isothermal amplification and/or recombinase polymerase amplification) for enhanced virus detection.[27–29] The ability to target different viral components expands the application range of sensors across various clinical and surveillance needs.[30] In addition, the integration with artificial intelligence (AI), machine learning (ML), and data analysis generates large datasets and sensor outputs, especially in multi-analyte or multiplexed systems, which benefit from AI/ML-driven analysis.[31] Pattern recognition algorithms can enhance interpretation accuracy, differentiate closely related viruses, and reduce false positives.[32] AI integration enables predictive diagnostics, potentially identifying infection stages or allowing anticipation of outbreak dynamics. Smart sensor systems that integrate machine learning unlock powerful real-time diagnostic and epidemiological capabilities.[33,34]

With respect to the emergence of multiplexed and wearable virus detection systems, the results of research show increasing success in multiplexed detection, where a single platform identifies multiple viruses simultaneously—critical in co-infection scenarios or when symptoms are non-specific.[35] Wearable sensors, embedded in masks, patches, or clothing, are being developed for continuous monitoring of airborne viruses or biomarkers from sweat and saliva.[36] Despite technical success, few sensor technologies have achieved widespread clinical deployment, largely due to the limited clinical validation, real-world testing, challenges including mass production while retaining sensor performance, regulatory hurdles involving safety, sensitivity, reproducibility, and quality control.[37–40] Collectively, these findings underscore the transformative potential of sensor-based virus detection in reshaping diagnostic landscapes. The convergence of nanotechnology, biotechnology, data science, and engineering is driving a new generation of tools that are faster, smarter, and more accessible. Continued research and interdisciplinary collaboration are essential to refine these technologies, overcome translational barriers, and ensure their global effects on preparedness, response, and prevention of viral diseases.[41,42]

19.2 The Importance of Ongoing Research in Advancing Virus Detection with Sensors

Infectious diseases caused by viruses—such as influenza, HIV, Zika, severe acute respiratory syndrome (SARS), middle east respiratory syndrome (MERS), and most recently, SARS-CoV-2—pose persistent threats to public health worldwide.[43] Effective management of viral outbreaks hinges on rapid and reliable detection.[44] Early diagnosis not only enables prompt clinical intervention but also plays a crucial role in limiting transmission, especially during the asymptomatic or presymptomatic stages of infection.[45] Conventional virus detection techniques, including polymerase chain reaction (PCR), enzyme-linked immunosorbent assays (ELISA), and viral culture, are widely used and generally reliable.[46] However, they often require complex sample preparation, specialized laboratory settings, trained personnel, and significant processing time.[47,48] These limitations restrict their utility in low-resource settings or during large-scale screening efforts, such as in pandemics.

Ongoing research in sensor technology is revolutionizing the field of virus diagnostics.[49] Sensor-based detection systems offer a range of advantages including speed (many sensors can detect viral particles or biomarkers in minutes), sensitivity (advances in nanomaterials and molecular recognition elements have enabled detection at extremely low concentrations—sometimes at the single-virus level), portability (miniaturized and POC devices make it possible to conduct diagnostics in the field, airports, rural clinics, or even at home), and ease of use (many sensor platforms are being designed for non-experts, with automated or smartphone-integrated readouts).[50–53]

Several types of sensors are being actively developed and refined for virus detection.[54] For instance, biosensors can utilize biological recognition elements (*e.g.*, antibodies, nucleic acids, aptamers) to specifically bind viral targets. Biosensors can be electrochemical, optical, piezoelectric, or thermal, depending on the signal transduction method.[55] Nanomaterial-based sensors can incorporate materials like graphene, gold nanoparticles, or quantum dots to enhance sensitivity and signal amplification.[56] Lab-on-a-chip devices are miniaturized systems capable of performing complex assays on a single microfluidic platform, often integrating multiple sensing modalities.[57]

The convergence of sensor technologies with AI, ML, and internet-of-things (IoT) platforms is creating smart diagnostic systems capable of performing automated analysis, remote monitoring, and real-time reporting.[58] These integrations are particularly beneficial for outbreak surveillance and personalized healthcare. Despite promising progress, several challenges remain, such as standardization and validation. Many sensor systems are still at the research or prototype stage and require rigorous clinical validation.[59] In addition, scalability is still a challenge. Manufacturing high-performance sensors at scale without loss of sensitivity or specificity is a technical hurdle.[60] Also, stability and shelf-life are crucial for these systems. Ensuring sensor performance across varied environmental conditions and over extended periods is vital for real-world deployment.[61] Navigating the regulatory landscape for diagnostic devices is complex and time-consuming.[62]

Continued interdisciplinary collaboration among materials scientists, engineers, virologists, and clinicians is critical. Investment in research infrastructure and translational efforts is essential to bring these technologies from the lab to the field.[63] Ongoing research on sensor-based virus detection represents a pivotal shift toward faster, more accessible, and more accurate diagnostics.[64] These technologies hold promise not only for managing current infectious disease threats but also for preparing for future pandemics.[65] As innovation continues, sensor-based platforms are likely to become indispensable tools in global health surveillance and personalized medicine.[66]

19.3 Concluding Remarks

The detection of viruses with speed, accuracy, and scalability has become a cornerstone of modern public health, clinical diagnostics, and also biosurveillance. This chapter has presented an exploration of how sensor technologies are reshaping the way viruses are identified and monitored in medical settings. The urgency of detecting viral pathogens at early stages—often before symptoms arise—has never been clearer, as evidenced by the global disruption caused by the COVID-19 pandemic and the continued emergence of zoonotic diseases.

Sensor technologies have demonstrated immense promise in filling the gaps left by conventional diagnostic approaches. While PCR and ELISA remain gold standards in many clinical settings, their reliance on laboratory infrastructure and longer processing times limit their utility in decentralized or time-critical scenarios. In contrast, sensor-based diagnostics—particularly those built on electrochemical, optical, and mass-sensing platforms—offer the potential for real-time, on-site, and user-friendly detection, enabling faster medical decision-making and outbreak response. Innovations such as clustered regularly interspaced short palindromic repeats (CRISPR)-based diagnostics, quantum dot nanoprobes, and AI-integrated sensor systems represent a leap forward in both the sophistication and accessibility of virus detection tools. These advances are not merely incremental—they are indicative of a paradigm shift toward diagnostics that are portable, cost-effective, and intelligent. Furthermore, the growing integration of machine learning algorithms enables more nuanced interpretations of complex datasets, allowing for real-time analysis and pattern recognition that can be critical in managing community-wide transmission or monitoring emerging viral variants.

However, realizing the full potential of sensor-based virus detection requires confronting several critical challenges. Achieving high sensitivity and specificity, especially in complex biological matrices like saliva and/or blood, remains a scientific and engineering hurdle. Additionally, the standardization of sensor outputs, robust validation protocols, and ethical considerations—such as data privacy and equitable access—are paramount as these technologies transition from laboratories to clinics and communities. The case studies presented illustrate how sensor technologies have already affected disease surveillance, from monitoring swine fever in livestock to enabling rapid screening for avian influenza at ports of entry. These real-world applications underscore the value of rapid diagnostics in preventing economic losses, guiding public health interventions, and ultimately saving lives.

Looking ahead, the field stands at a crossroads of innovation and implementation. The coming years offer tremendous opportunity for deepening interdisciplinary collaborations between biotechnologists, materials scientists, data scientists, and clinicians; building open-source platforms and standardized protocols to accelerate validation and global dissemination; creating modular, upgradable sensor platforms capable of adapting to new viral threats through interchangeable recognition elements. Moreover, the integration of sensors with mobile health (mHealth) platforms and IoT networks will be key in building responsive, data-rich systems capable of not only detecting viruses but also tracking their spread and evolution.

In conclusion, sensor technologies for virus detection are at the forefront of a diagnostic revolution. As we strive toward greater preparedness for future pandemics, the insights, technologies, and methodologies explored in this chapter provide a strong foundation for innovation that is both scientifically rigorous and societally impactful. The commitment to ongoing research, ethical deployment, and global collaboration will ensure that sensor-based diagnostics become a pillar of resilient and responsive healthcare systems worldwide.

Acknowledgements

Yeşeren Saylan thanks the Young Scientist Awards Program (BAGEP) award from the Science Academy.

References

1. M. Xiao, F. Tian, X. Liu, Q. Zhou, J. Pan, Z. Luo and C. Yi, Virus detection: from state-of-the-art laboratories to smartphone-based point-of-care testing, *Adv. Sci.*, 2022, **9**(17), 2105904.
2. S. Roychoudhury, A. Das, P. Sengupta, S. Dutta, S. Roychoudhury, A. P. Choudhury and P. Slama, Viral pandemics of the last four decades: pathophysiology, health impacts and perspectives, *Int. J. Environ. Res. Publ. Health*, 2020, **17**(24), 9411.
3. R. E. Baker, A. S. Mahmud, I. F. Miller, M. Rajeev, F. Rasambainarivo, B. L. Rice and C. J. E. Metcalf, Infectious disease in an era of global change, *Nat. Rev. Microbiol.*, 2022, **20**(4), 193–205.
4. Y. Saylan, Ö. Erdem, S. Ünal and A. Denizli, An alternative medical diagnosis method: biosensors for virus detection, *Biosensors*, 2019, **9**(2), 65.
5. R. Eivazzadeh-Keihan, P. Pashazadeh-Panahi, T. Mahmoudi, K. K. Chenab, B. Baradaran, M. Hashemzaei and A. Maleki, Dengue virus: a review on advances in detection and trends–from conventional methods to novel biosensors, *Microchim. Acta*, 2019, **186**, 1–24.
6. S. Kumar, W. Ahlawat, R. Kumar and N. Dilbaghi, Graphene, carbon nanotubes, zinc oxide and gold as elite nanomaterials for fabrication of biosensors for healthcare, *Biosens. Bioelectron.*, 2015, **70**, 498–503.

7. M. Li, T. Chen, J. J. Gooding and J. Liu, Review of carbon and graphene quantum dots for sensing, *ACS Sen.*, 2019, **4**(7), 1732–1748.
8. F. R. Baptista, S. A. Belhout, S. Giordani and S. J. Quinn, Recent developments in carbon nanomaterial sensors, *Chem. Soc. Rev.*, 2015, **44**(13), 4433–4453.
9. O. Erdem, I. Eş, Y. Saylan and F. Inci, Unifying the efforts of medicine, chemistry, and engineering in biosensing technologies to tackle the challenges of the COVID-19 pandemic, *Anal. Chem.*, 2021, **94**(1), 3–25.
10. P. Singla, H. Kaur, S. Garg, N. Kaur, F. Canfarotta, R. K. Mahajan and M. Peeters, Early detection of SARS-CoV-2 with functionalized gold and molecularly imprinted polymeric nanoparticles: a mini review, *Mol. Syst. Des. Eng.*, 2023, **8**(11), 1337–1354.
11. Q. Han, H. Wang and J. Wang, Multi-mode/signal biosensors: Electrochemical integrated sensing techniques, *Adv. Funct. Mater.*, 2024, **34**(39), 2403122.
12. Y. Saylan, S. Akgönüllü, H. Yavuz, S. Ünal and A. Denizli, Molecularly imprinted polymer based sensors for medical applications, *Sensors*, 2019, **19**(6), 1279.
13. N. S. Shetty, V. Othayoth and A. U. Satyaprasad, Target-induced nanoparticle assemblies: a comprehensive review on strategies for nucleic acid functionalization, biosensing, and drug delivery applications, *Anal. Methods*, 2025, **17**, 3876–3901.
14. A. B. Ganganboina and E. Y. Park, Signal-amplified nanobiosensors for virus detection using advanced nanomaterials, *Trends Biosens. Res.*, 2024, 381–412.
15. A. S. Rao, M. Radanovic, Y. Liu, S. Hu, Y. Fang, K. Khoshelham and T. Ngo, Real-time monitoring of construction sites: Sensors, methods, and applications, *Automat. Construct.*, 2022, **136**, 104099.
16. S. M. Yang, S. Lv, W. Zhang and Y. Cui, Microfluidic point-of-care (POC) devices in early diagnosis: A review of opportunities and challenges, *Sensors*, 2022, **22**(4), 1620.
17. M. Zarei, Portable biosensing devices for point-of-care diagnostics: Recent developments and applications, *TrAC, Trends Anal. Chem.*, 2017, **91**, 26–41.
18. G. A. Akceoglu, Y. Saylan and F. Inci, A snapshot of microfluidics in point-of-care diagnostics: multifaceted integrity with materials and sensors, *Adv. Mater. Technol.*, 2021, **6**(7), 2100049.
19. Y. Saylan and A. Denizli, Molecularly imprinted polymer-based microfluidic systems for point-of-care applications, *Micromachines*, 2019, **10**(11), 766.
20. C. M. Ardila, Advancing healthcare through laboratory on a chip technology: Transforming microorganism identification and diagnostics, *World J. Clin. Cases*, 2025, **13**(3), 97737.
21. A. M. Mitrogiannopoulou, V. Tselepi and K. Ellinas, Polymeric and paper-based lab-on-a-chip devices in food safety: A review, *Micromachines*, 2023, **14**(5), 986.

22. G. Qiu, X. Zhang, A. J. deMello, M. Yao, J. Cao and J. Wang, On-site airborne pathogen detection for infection risk mitigation, *Chem. Soc. Rev.*, 2023, **52**(24), 8531–8579.
23. A. Suea-Ngam, L. Bezinge, B. Mateescu, P. D. Howes, A. J. deMello and D. A. Richards, Enzyme-assisted nucleic acid detection for infectious disease diagnostics: moving toward the point-of-care, *ACS Sens.*, 2020, **5**(9), 2701–2723.
24. R. Keçili, S. Büyüktiryaki and C. M. Hussain, Advancement in bioanalytical science through nanotechnology: past, present and future, *TrAC, Trends Anal. Chem.*, 2019, **110**, 259–276.
25. R. Bukasov, D. Dossym and O. Filchakova, Detection of RNA viruses from influenza and HIV to Ebola and SARS-CoV-2: a review, *Anal. Methods*, 2021, **13**(1), 34–55.
26. M. E. Abernathy, K. M. A. Dam, S. R. Esswein, C. A. Jette and P. J. Bjorkman, How antibodies recognize pathogenic viruses: structural correlates of antibody neutralization of HIV-1, SARS-CoV-2, and Zika, *Viruses*, 2021, **13**(10), 2106.
27. K. Heiss, J. Heidepriem, N. Fischer, L. K. Weber, C. Dahlke, T. Jaenisch and F. F. Loeffler, Rapid response to pandemic threats: immunogenic epitope detection of pandemic pathogens for diagnostics and vaccine development using peptide microarrays, *J. Proteome Res.*, 2020, **19**(11), 4339–4354.
28. L. T. N. Ngoc and Y. C. Lee, Current trends in RNA virus detection via nucleic acid isothermal amplification-based platforms, *Biosensors*, 2024, **14**(2), 97.
29. P. J. Asiello and A. J. Baeumner, Miniaturized isothermal nucleic acid amplification, a review, *Lab Chip*, 2011, **11**(8), 1420–1430.
30. M. Kheirollahpour, N. Shokoufi and M. Lotfi, The potential of optical technologies in early virus detection; prospects in addressing future viral outbreaks, *Crit. Rev. Anal. Chem.*, 2025, 1–29.
31. Z. Lu, K. Xu, K. Xiao, Q. Xu, L. Wang, P. Li and W. Huang, Biomolecule sensors based on organic electrochemical transistors, *npj Flexible Electron.*, 2025, **9**(1), 9.
32. S. Wang, B. Kang, J. Ma, X. Zeng, M. Xiao, J. Guo and B. Xu, A deep learning algorithm using CT images to screen for corona virus disease (COVID-19), *Eur. Radiol.*, 2021, **31**, 6096–6104.
33. A. P. Zhao, S. Li, Z. Cao, P. J. H. Hu, J. Wang, Y. Xiang and X. Lu, AI for science: predicting infectious diseases, *J. Safety Sci. Resilience*, 2024, **5**(2), 130–146.
34. A. V. L. N. Sujith, G. S. Sajja, V. Mahalakshmi, S. Nuhmani and B. Prasanalakshmi, Systematic review of smart health monitoring using deep learning and artificial intelligence, *Neurosci. Inf.*, 2022, **2**(3), 100028.
35. A. Qureshi and J. H. Niazi, Biosensors for detecting viral and bacterial infections using host biomarkers: a review, *Analyst*, 2020, **145**(24), 7825–7848.
36. J. Li, J. Yin, S. Ramakrishna and D. Ji, Smart mask as wearable for post-pandemic personal healthcare, *Biosensors*, 2023, **13**(2), 205.

37. B. Munos, P. C. Baker, B. M. Bot, M. Crouthamel, G. de Vries, I. Ferguson and P. Wang, Mobile health: the power of wearables, sensors, and apps to transform clinical trials, *Ann. N. Y. Acad. Sci.*, 2016, **1375**(1), 3–18.
38. X. Zhang, J. Chai, Y. Zhan, D. Cui, X. Wang and L. Gao, Design, fabrication, and application of large-area flexible pressure and strain sensor arrays: a review, *Micromachines*, 2025, **16**(3), 330.
39. Y. Luo, M. R. Abidian, J. H. Ahn, D. Akinwande, A. M. Andrews, M. Antonietti and X. Chen, Technology roadmap for flexible sensors, *ACS Nano*, 2023, **17**(6), 5211–5295.
40. N. Sharmile, R. R. Chowdhury and S. Desai, A comprehensive review of quality control and reliability research in micro–nano technology, *Technologies*, 2025, **13**(3), 94.
41. W. S. Bainbridge and M. C. Roco, Science and technology convergence: with emphasis for nanotechnology-inspired convergence, *J. Nanopart. Res.*, 2016, **18**(7), 211.
42. R. Prasad, A. Ghosh, V. Patel, B. Peng, B. B. Mendes, E. H. A. Win and J. Conde, Voices of nanomedicine: blueprint guidelines for collaboration in addressing global unmet medical needs, *ACS Nano*, 2025, **19**, 2979–2991.
43. S. De Gaetano, E. Ponzo, A. Midiri, G. Mancuso, D. Filippone, G. Infortuna and C. Biondo, Global trends and action items for the prevention and control of emerging and re-emerging infectious diseases, *Hygiene*, 2025, **5**(2), 18.
44. F. Sohrabi, S. Saeidifard, M. Ghasemi, T. Asadishad, S. M. Hamidi and S. M. Hosseini, Role of plasmonics in detection of deadliest viruses: A review, *Eur. Phys. J. Plus*, 2021, **136**(6), 675.
45. L. Ferretti, C. Wymant, M. Kendall, L. Zhao, A. Nurtay, L. Abeler-Dörner and C. Fraser, Quantifying SARS-CoV-2 transmission suggests epidemic control with digital contact tracing, *Science*, 2020, **368**(6491), eabb6936.
46. D. H. Reta, T. S. Tessema, A. S. Ashenef, A. F. Desta, W. L. Labisso, S. T. Gizaw and F. A. Reta, Molecular and immunological diagnostic techniques of medical viruses, *Int. J. Microbiol.*, 2020, **2020**(1), 8832728.
47. C. L. Charlton, E. Babady, C. C. Ginocchio, T. F. Hatchette, R. C. Jerris, Y. Li and S. J. Drews, Practical guidance for clinical microbiology laboratories: viruses causing acute respiratory tract infections, *Clin. Microbiol. Rev.*, 2018, **32**(1), 10–1128.
48. A. Cassedy, A. Parle-McDermott and R. O'Kennedy, Virus detection: a review of the current and emerging molecular and immunological methods, *Front. Mol. Biosci.*, 2021, **8**, 637559.
49. A. Kumar, S. Saini, A. Anvikar, N. Mishra and G. Misra, Evolving landscape of emerging virus diagnosis: challenges and innovations, *Mol. Biotechnol.*, 2025, 1–23.
50. T. Ozer, B. J. Geiss and C. S. Henry, Chemical and biological sensors for viral detection, *J. Electrochem. Soc.*, 2019, **167**(3), 037523.
51. A. Cassedy, A. Parle-McDermott and R. O'Kennedy, Virus detection: a review of the current and emerging molecular and immunological methods, *Front. Mol. Biosci.*, 2021, **8**, 637559.

52. H. Harpaldas, S. Arumugam, C. C. Rodriguez, B. A. Kumar, V. Shi and S. K. Sia, Point-of-care diagnostics: recent developments in a pandemic age, *Lab Chip*, 2021, **21**(23), 4517–4548.
53. D. V. Baker, J. Bernal-Escalante, C. Traaseth, Y. Wang, M. V. Tran, S. Keenan and W. R. Algar, Smartphones as a platform for molecular analysis: concepts, methods, devices and future potential, *Lab Chip*, 2025, **25**, 884–955.
54. F. Martinelli, R. Scalenghe, S. Davino, S. Panno, G. Scuderi, P. Ruisi and A. M. Dandekar, Advanced methods of plant disease detection: a review, *Agron. Sustainable Dev.*, 2015, **35**, 1–25.
55. Z. Gerdan, Y. Saylan and A. Denizli, Biosensing platforms for cardiac biomarker detection, *ACS omega*, 2024, **9**(9), 9946–9960.
56. Y. Saylan, N. Aliyeva, S. Eroglu and A. Denizli, Nanomaterial-based sensors for coumarin detection, *ACS omega*, 2024, **9**(28), 30015–30034.
57. D. S. Dkhar, R. Kumari, S. J. Malode, N. P. Shetti and P. Chandra, Integrated lab-on-a-chip devices: Fabrication methodologies, transduction system for sensing purposes, *J. Pharm. Biomed. Anal.*, 2023, **223**, 115120.
58. M. Alshamrani, IoT and artificial intelligence implementations for remote healthcare monitoring systems: a survey, *J. King Saud Univ.-Comput. Inf. Sci.*, 2022, **34**(8), 4687–4701.
59. J. C. Goldsack, A. Coravos, J. P. Bakker, B. Bent, A. V. Dowling, C. Fitzer-Attas and J. Dunn, Verification, analytical validation, and clinical validation (V3): the foundation of determining fit-for-purpose for Biometric Monitoring Technologies (BioMeTs), *npj Digital Med.*, 2020, **3**(1), 55.
60. X. Shi, H. Pu, L. L. Shi, T. C. He and J. Chen, Additive manufacturing technologies and translation of transistor-based point-of-care (POC) biosensors from laboratory to real-life applications, *Nanoscale*, 2025, **17**, 9804–9833.
61. P. Kumar, A. N. Skouloudis, M. Bell, M. Viana, M. C. Carotta, G. Biskos and L. Morawska, Real-time sensors for indoor air monitoring and challenges ahead in deploying them to urban buildings, *Sci. Total Environ.*, 2016, **560**, 150–159.
62. M. Heidari, S. Thangavel, P. Rahmanivahid, M. Khashehchi, A. Kumar, A. Kumar and A. Kumar, Navigating the landscape: Cutting-edge biomedical manufacturing techniques, in *Applications of biotribology in biomedical systems*, Springer Nature Switzerland, Cham, 2024, pp. 199–216.
63. R. G. L. da Silva, R. Chammas and H. M. D. Novaes, Rethinking approaches of science, technology, and innovation in healthcare during the COVID-19 pandemic: the challenge of translating knowledge infrastructures to public needs, *Health Res. Policy Syst.*, 2021, **19**, 1–9.
64. A. Kumar, S. Saini, A. Anvikar, N. Mishra and G. Misra, Evolving landscape of emerging virus diagnosis: challenges and innovations, *Mol. Biotechnol.*, 2025, 1–23.
65. D. E. Bloom, S. Black and R. Rappuoli, Emerging infectious diseases: A proactive approach, *Proc. Natl. Acad. Sci. U. S. A.*, 2017, **114**(16), 4055–4059.
66. S. Majumder and M. J. Deen, Smartphone sensors for health monitoring and diagnosis, *Sensors*, 2019, **19**(9), 2164.

Subject Index

www.ingramcontent.com/pod-product-compliance
Ingram Content Group UK Ltd.
Pitfield, Milton Keynes, MK11 3LW, UK
UKHW022101170926
8624IPUK00023B/14

9 781837 677764